AF664150

Soil Microflora

THE AUTHORS

Dr. Rajan Kumar Gupta (b. 1963) obtained his M.Sc. and Ph.D. degree from Banaras Hindu University and worked on Ecophysiology of Antarctic Cyanobacteria for his Ph.D. degree with Prof. A.K. Kashyap, HOD, Centre of Advanced Study in Botany, Banaras Hindu University, Varanasi. Since past twenty years he has been working on various aspects of Antarctic microflora.

Dr. Gupta was deputed by Govt. of India for his participation as Biological Scientist in Antarctica twice. He has participated in XIth and XIVth Indian Scientific Expeditions to Antarctica during 1991-92 and 1994-95. He has visited several countries like Mauritius, Japan, Nepal, Thailand, Belgium and South Africa for presentation of his work in the field of algal microflora. Dr. Gupta has worked on various aspects of cyanobacteria *i.e.* morphology, ecology and nitrogen fixation, biotechnological applications and published more then 40 technical papers in various National and overseas Journals and more then 18 chapters in various books. Dr. Gupta has published three Botany Practical Books one book on Paryavaran Adhyan (Environmental Studies) and two reference (research) books entitled *"Glimpses of Cyanobacteria"* and *"Advances in Applied Phycology"*. Two students have been awarded the D.Phil degree and many are working under his supervision for their D.Phil degree of HNB Garhwal University. He has worked on Use of Cyanobacteria as Biofertilizer in Antarctica as well as in Foot Hills of Garhwal Himalaya and is presently working on a project on Cyanobacteria of Paddy fields of Dehradun District of Himalaya. Dr. Gupta is member of a number of organizations in India and abroad. He is the Fellow of the Society for Environment and Ecoplanning and International Botanical Society and Chaired various sessions in the conferences in India and abroad. Presently Dr. Gupta is teaching Microbiology and Biotechnology in the Department of Botany, Govt. P.G. College, Rishikesh – 249 201 (Dehradun), Uttarakhand, India.

Dr. Mukesh Kumar (b. 1963) obtained his M.Sc and M. Phil. degrees in Botany, and M.Ed in Education from Meerut University, Meerut–250005 (now C.C.S. University, Meerut). He took his doctorate degree on Polyhouse Technology from H.N.B. Garhwal University, Srinagar – 246174 (Uttarakhand). Presently he is working as Reader at the Department of Botany, Sahu Jain Post-Graduate College, Najibabad – 246763 (Bijnor) U.P.

Dr. Kumar is well known for his researches in the fields of Polyhouse Technology and Cyanobacteria. He is a member/ fellow of several National and International research organizations. Working on various aspects of Cyanobacteria *i.e* distributional pattern, population dynamics and dominance of different genera, he has explored the biodiversity of Cyanophycean Flora of the Sub-Himalayan Belt of Garhwal and Kumaon regions of Uttarakhand state of India. He has accomplished a couple of Research Projects in the field of his specialization *i.e* Polyhouse Technology and Cyanobacterial Diversity sponsored by the University Grants Commission, New Delhi.

He has to his credit presented his original research findings on the platform of plant scientists in several conferences/seminars and workshops. He has published more than 21 technical papers to various journals and chapters in several reputed books. The great success of his previous book entitled *"Glimpses of Cyanobacteria"* has inspired him to work more on the architects of the earth–the cyanobacteria.

Four research scholars have already been awarded their Ph.D degrees and others are currently working under his supervision. Presently he is actively engaged in the studies on Ganga water pollution with particular reference to the quality of soil, water and phytoplankton growth in and around the stream along with the impact of tourism on various parameters.

Dr. Deepak Vyas (b. 1964) M.Sc., Ph.D. (B.H.U., Varanasi) is Senior Lecturer in Department of Botany, Dr. H.S. Gaur University, Sagar (M.P.). He has 10 years of teaching and 21 years of Research Experience. Nine students have obtained their Ph.D. degree under his guidance and six students are working for their Ph.D. Dr. Vyas has published more then 70 papers in Journals of National and International repute. He was recipient of International Award on Ozone Depletion Theory. Dr. Vyas has organized various National Seminar, Conferences and Workshops successfully. He is an expert in the Hydrogen Production by Microalgae and presently involve in fungal (VAM) research.

Soil Microflora

— *Editors* —

Rajan Kumar Gupta

Department of Botany
Pt. L.M.S. Government Post-Graduate College
(NAAC Accredited Grade "A" College)
(Affiliated to H.N.B. Garhwal University, Srinagar)
Rishikesh – 249 201, Dehradun, Uttarakhand, India

Mukesh Kumar

Department of Botany
Sahu Jain P.G. College
Najibabad – 246 769, U.P., India

Deepak Vyas

Department of Botany
Dr. H.S. Gaur University,
Sagar, M.P., India

2023

Daya Publishing House®

A Division of

Astral International Pvt. Ltd.

New Delhi - 110 002

First Published, 2009
Reprinted, 2023

ISBN: 978-81-7035-597-7 (Hardbound)

Published by : **Daya Publishing House®**
A Division of
Astral International Pvt. Ltd.
– ISO 9001:2015 Certified Company –
4736/23, Ansari Road, Darya Ganj
New Delhi-110 002
Ph. 011-43549197,
E-mail: info@astralint.com
Website: www.astralint.com

Printed at : **Replika Press Pvt. Ltd.**

— Dedicated to —

Late Prof. K.M. Vyas

Former Head, Department of Botany,
Dr. Hari Singh Gaur University, Sagar, M.P.

Professor G.S. Paliwal
D.Sc., Ph.D., FLS (London)

Ex Head
Department of Botany
HNB Garhwal University
SRINAGAR–246174, Pauri
(Uttarakhand)

Presently: Senior Consultant
Agricultural Finance Corporation Ltd.
Northern Regional Office
B-1/9, Community Centre,
Janak Puri, New Delhi

Off. & Mailing Address
D-37, Rana Pratap Road,
Adarsh Nagar
DELHI–110033 (India)

Foreword

The soil presents a unique environment which favours the growth of a rich and variable microbial flora in the different parts of the universe. A series of physico-chemical, biochemical and redox processes regulate the nature, frequency, properties and density of these forms. Thus, the study of all such microorganisms is significant not only in the context of agriculture but also in the sphere of health of both the animal and plant populations. Consequently, the discernible priority of the biologists of the day is identification, conservation, and management of the global microbial biodiversity in order to gain familiarity with additional forms, which can be of direct value to the mankind. This goal can only be achieved through consistent exploration, systematic recording and analysis of the various aspects of these biological entitles.

The second important aspect to which the soil microbes are related is the health and welfare of man himself and his pets. It is now common knowledge that there are a number of bacteria, fungi and algae which attach, inhabit and cause diseases in them. The pathologists have repeatedly argued that a broader outlook is called for in order to search the entire dynamics of the microbial ecology. The study of their life-cycle is a basic pre-requisite towards the proper understanding of the soil-related microorganisms in relation to their surroundings. As such the present day comprehension of the soil micro-flora requires complementary data from all the spheres of scientific pursuits with a bearing on the understanding of the conservation of the habitats. This is because the total environment of a microorganism in the soil is biologically complex and the ramificative interactions amongst them result in an array of events and products.

The present volume includes different aspects of soil micro-flora and is a qualified and sincere attempt towards improving our understanding of the ecology, diversity, seasonal behavior, role in crop production. nitrogen .fixation, and genetic engineering, spread over 30 Chapters, which have been authored by practicing scientists and experts in their respective fields of specialization.

I, therefore, have great pleasure in commending this volume containing holistic information on soil micro-flora to the attention of all interested in any way in this discipline with a bearing on Agriculture, Botany, Zoology and Forestry, The three members of the editorial team of the volume are known for their dedication and sincerity towards the execution of their researches and other academic pursuits. I am, therefore, assured of its quality and scientific standard and fervently hope that this work will provide incentive to several others for devoting themselves to this interesting field of study and analysis.

G.S. Paliwal

Preface

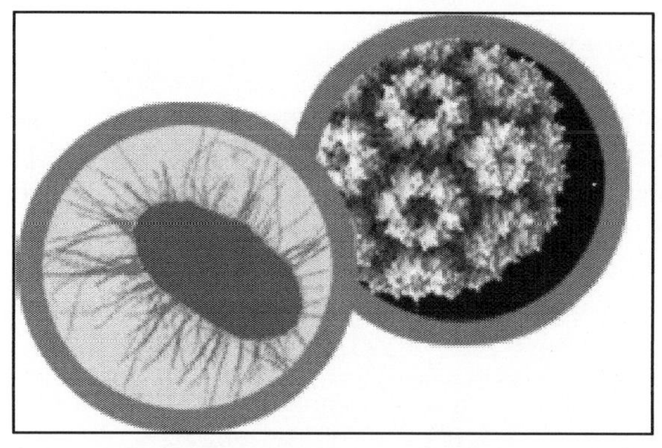

Microbes, the most beautiful and wonderful creatures might be tiny and hard to see, but they account for a large percentage of Earth's biodiversity. They have been living on the planet for 3.8 billion years compared to 200,000 for humans, and for most of the Earth's existence, they have been the only form of life around. In fact, all life on Earth today, including trees, fish and people, is thought to have evolved from the earliest microbes.

The term "microbe" describes bacteria, archaea, single-celled eukaryotic organisms such as amoebas, slime molds and parameciums, and even viruses by some broad definitions. Viruses are disputed because they are considered non-living and cannot replicate on their own, but the field of microbiology usually includes the study of viruses. Most microbes are unicellular, meaning one cell comprises each individual.

They are found almost everywhere on Earth, in soils, plants, geysers, ocean depths, frigid seas below Antarctic ice and in our bodies. Trillions of bacteria have been found in our guts. Some microbes, called extremophiles, are found in places where no other living organisms can survive–in boiling hot hydrothermal vents in the ocean and in rocks deep underground.

They can be helpful and/or harmful to other living things: bacteria such as *Streptococcus* and *E. coli* can infect and even kill humans, and algal blooms can be toxic to fish and deplete the oxygen of water, but other bacteria help us digest our food and replenish nutrients in soil, and some can help clean up oil spills.

Scientists are continually discovering new species, genuses, families and orders of microbes, with no end in sight. Because they have been around for so long, microbes evolve in more complicated ways than multicellular life–they can transfer genes between species and from one individual to another, something humans certainly can't do. They are so powerful that we can not underestmate the power of microbes.

Microbes are single-cell organisms so tiny that millions can fit into the eye of a needle. They are the oldest form of life on earth. Microbe fossils date back more than 3.5 billion years to a time when the Earth was covered with oceans that regularly reached the boiling point, hundreds of millions of years before dinosaurs roamed the earth. Without microbes, we couldn't eat or breathe. Without us, they'd probably be just fine.

Understanding microbes is vital to understand the past and the future of ourselves and our planet. Microbes <*my-crobes*> are everywhere. There are more of them on a person's hand than there are people on the entire planet! Microbes are in the air we breathe, the ground we walk on, the food we eat–they're even inside us! We couldn't digest food without them–animals couldn't, either. Without microbes, plants couldn't grow, garbage wouldn't decay and there would be a lot less oxygen to breathe. In fact, without these invisible companions, our planet wouldn't survive as we know it!

Our body is home to trillions of microbes. Run your tongue over your teeth–you're licking thousands of microbes that normally live on your teeth. Millions of them live on your tongue, too. A large part of "you" (that is, the mass of your body) is actually something else: bacteria, viruses and fungi.

Pick up a fistful of garden soil and you're holding hundreds if not thousands of different kinds of microbe in your hand. A single teaspoon of that soil contains over 1,000,000,000 bacteria, about 120,000 fungi and 25,000 algae. Microbes have been around for billions of years because they are able to adapt to the ever-changing environment. They can find a home anywhere and some of them live in places where we once thought NOTHING could survive. For example, scientists have discovered microbes living in the boiling waters of hot springs in Yellowstone National Park. These microbes "eat" hydrogen gas and sulfur and "breathe" hydrogen sulfide (a gas that smells like rotten eggs). Other heat-loving microbes live in volcanic cracks miles under the ocean surface where there is no light and the water is a brew of poisonous arsenic, sulfur and other nasty chemicals. Other microbes live in the permanently frozen ice of Antarctica. Microbes have been found living inside the stones that make up the walls of old cathedrals in Europe.

Some scientists even believe there is the possibility bacteria may have once lived on Mars. Algae, a large and diverse group of photosynthetic microorganisms, have been the subject of both basic and applied research over the years.

This group of microorganism has begun to emerge from the early descriptive stage into an experimental one, largely because of the current interest in their phototrophic metabolism, biological nitrogen fixation, environmental and ecological implications, source of innumerable products of commercial importance and prokaryotic genetic organization. Their peculiar features and potential applications in agriculture, aquaculture, bioremediation, bioenergy, human nutrition and pharmaceutical industries have attracted the attention of workers from diverse fields.

Different aspects of soil microbes and their potential biotechnological applications are being revealed by workers world over. India is richly endowed with the microbial flora, has long tradition of microbiological research. Microbes form a significant part of the subjects like Botany, Microbiology, Agriculture and Biotechnology being taught in the Indian Universities.

The research activities in Microbiology have expanded in several directions, during last two decades; there has been a natural explosion of information on various aspects of these organisms. The rapid advances and developments in microbiology prompted us to assemble the up-to-date information in the form of present book targeted to both post graduate students and researchers.

This volume consists of wide ranging 30 articles which encompasses topics which emphasises on microbial ecology, taxonomy, morphology, physiology, stress responses, bioremediation, heavy metal toxicity, bio-fuel production, biotechnology and molecular biology. Various information incorporated in the book by authors who are internationally acknowledged experts in the field of microbiology. The book with the intention of providing a sufficient depth of the subject to satisfy the needs at a level which will be comprehensive and interesting. We have tried to synthesise all the information which will be useful and hope that this book would be informative to the students, teachers, scientists and researchers in the field of basic and applied microbiology.

The authors wish to thank and give appreciations to all the scientists whose contributions have enriched this volume. We also express our deep sense of gratitude to our parents whose blessings have always prompted us to pursue academic activities deeply.

It is possible that in a work of this nature, some mistakes might have crept in text inadvertently and for these we own undiluted responsibility.

We are grateful to all authors for their contribution to this book. We are also thankful to Professor G. S. Paliwal, former Head Department of Botany, HNB Garhwal University, Srinagar (Uttarakhand) for his co-operation and valuable suggestions during the preparation of this volume. We thank him from the core of our heart.

We do hope that all those interested in the pursuit of soil science would find the volume useful and stimulating.

The editors profoundly thank M/s Daya Publishing House, New Delhi, which is known for its reputation in quality scientific publication for kindly accepting to publish this book on Soil Microflora and for their enthusiastic co-operation during the compilation of this volume.

The author (Rajan Kumar Gupta) is thankful to University Grants Commission, New Delhi for the financial support and wish to place on record his special thanks to his wife Mrs Alka and two little daughters Akriti and Ayushi for their cooperation in all his academic and scientific endeavors. The author (Mukesh Kumar) is also thankful to the University Grant Commission, New Delhi for providing the financial assistance in the form of a Major Research Project entitled, "Biodiversity of Cyanophyceae of the Sub-Himalayan Belt". His wife Dr. (Mrs.) Kiran Sharma, PGT Hindi Kendriya Vidyalaya, Moradabad and sons Ruchir and Nishaant also deserve for thanks from the core of his heart. Dr. Deepak Vyas thankfully acknowledge his wife Mrs. Nisha Vyas, two sons, Siddhart and Vedant for their enthusiastic cooperation during the compilation of this volume.

Finally we will always remain debtor to all our well wishers for their blessings without which this book would not have come to light.

Editors

Contents

List of Contributors

Abraham, G.
Centre for Conservation and Utilization of Cyanobacteria, Indian Agricultural Research Institute, New Delhi – 110 012

Adhikary, S.P.
P.G. Department of Botany and Biotechnology, Utkal University, Bhubaneswar – 751 004

Bahuguna, Ashutosh
Department of Biotechnology, Modern Institute of Technology, Dhalwala, Rishikesh – 249 201, Uttarakhand

Bhadauria, Seema
Microbiology Research Laboratory, Department of Botany, Raja Balwant Singh College, Agra, U.P.

Bilthare, Deepali
Lab of Microbial Technology and Plant Pathology, Department of Botany, Dr. H. S. Gour University, Sagar – 470 003, M.P.

Chaturvedi, U.K.
Department of Botany, Government Degree College, Budaun – 243 001, U.P.

Chauhan, Anuradha
Microbiology Research Laboratory, Department of Botany, Raja Balwant Singh College, Agra, U.P.

Choudhary, Kaushal Kishore
Department of Botany, Banaras Hindu University, Varanasi – 221 005

Dangwal, Koushalya
Department of Biotechnology, Modern Institute of Technology, Dhalwala, Rishikesh – 249 201, Uttarakhand

Das, Mihir Kumar
P.G. Department of Botany, G.M. College (Autonomous), Sambalpur – 768 004, Orissa

Dhar, Dolly Wattal
Centre for Conservation and Utilization of Cyanobacteria, Indian Agricultural Research Institute, New Delhi – 110 012

Ghosh, Paromita
G.B. Pant Institute of Himalayan Environment and Development, Garhwal Unit, Upper Bhaktiyana, P.O. Box. 92, Srinagar, Garhwal, Uttarakhand

Gupta, Rajan Kumar
Department of Botany, Pt. L.M.S. Govt. P.G. College, Rishikesh–249 201, Dehradun, Uttarakhand

Habib, Iqbal
Department of Botany, Government Degree College, Budaun – 243 001, U.P.

Jaiswal, Pranita
Department of Botany, University of Delhi, Delhi – 110 007

Khare, Anjali
Department of Botany, Advance Institute of Science and Technology, 179-Kalidas Road, Dehradun, Uttarakhand

Khare, Roshni
Lichenology laboratory, National Botanical Research Institute, CSIR, Rana Pratap Marg, Lucknow – 226 001

Kumar, Anita Suresh
A-12, Everest Flats, Waghawadi Road, Bhavnagar – 362 002, Gujarat

Kumar, Bijendra
Plant Pathology Section, College of Forestry and Hill Agriculture, GBPUA&T Hill Campus, Ranichauri – 249 199, Tehri Garhwal, Uttarakhand

Kumar, Mukesh
Department of Botany, Sahu Jain (P.G.) College, Najibabad, U.P.

Kumar, Narendra
School of Studies in Microbiology, Jiwaji University, Gwalior – 474 011, M.P.

Kumar, Pawan
School of Studies in Microbiology, Jiwaji University, Gwalior – 474 011, M.P.

Kumar, Promod
Department of Botany, Hindu College, Moradabad

Kumari, Preetesh
Microbiology Research Laboratory, Department of Botany, Raja Balwant Singh College, Agra, U.P.

Lily, Madhuri K.
Department of Biotechnology, Modern Institute of Technology, Dhalwala, Rishikesh – 249 201, Uttarakhand

Mishra, Arun Kumar
Laboratory of Microbial Genetics, Department of Botany, Banaras Hindu University, Varanasi – 221 005

Mohan, Dheeraj
Microbiology Research Laboratory, Department of Botany, Raja Balwant Singh College, Agra, U.P.

Nayaka, Sanjeeva
Lichenology laboratory, National Botanical Research Institute, CSIR, Rana Pratap Marg, Lucknow

Pandey, Usha
Department of Biotechnology, Faculty of Science and Technology, M.G. Kashi Vidyapith, Varanasi – 221 002

Paul, Bishwajeet
Division of Entomology, Indian Agricultural Research Institute, New Delhi – 110 012

Paul, Sangeeta
Division of Microbiology, Indian Agricultural Research Institute, New Delhi – 110 012

Rajhans, Ravi
Department of Biotechnology, Modern Institute of Technology, Dhalwala, Rishikesh, Uttarakhand

Richhariya, Pramod Kumar
Lab of Microbial Technology and Plant Pathology, Department of Botany, Dr. H. S. Gour University, Sagar – 470 003, M.P.

Saxena, Sudheer
Centre for Conservation and Utilization of Cyanobacteria, Indian Agricultural Research Institute, New Delhi – 110 012

Shah, Raghubir
Centre for Conservation and Utilization of Cyanobacteria, Indian Agricultural Research Institute, New Delhi – 110 012

Sharma, G.K.
P.G. Department of Botany, Hindu College, Moradabad

Singh, Anju
Laboratory of Microbial Genetics, Department of Botany, Banaras Hindu University, Varanasi – 221 005

Singh, Chatar
Department of Agroforestry, Institute of Agriculture Sciences, Bundelkhand University, Jhansi, U.P.

Singh, K.P.
Plant Pathology Section, College of Forestry and Hill Agriculture, GBPUA&T Hill Campus, Ranichauri – 249 199, Tehri Garhwal, Uttarakhand

Singh, Kaushal Pratap
Microbiology Research Laboratory, Department of Botany, Raja Balwant Singh College, Agra, U.P.

Singh, Satya Shila
Laboratory of Microbial Genetics, Department of Botany, Banaras Hindu University, Varanasi – 221 005

Singh, Shalini
Institute of Bioengineering and Biological Science, S-19/54, Varuna Bridge, Varanasi – 221 002, U.P.

Singh, Surendra
Centre of Advanced Study in Botany, Banaras Hindu University, Varanasi – 221 005, U.P.

Singh, Y.V.
Centre for Conservation and Utilization of Blue Green Algae, Indian Agricultural Research Institute, New Delhi

Singh, Yashveer
Department of Botany, Dr. S.P. Mukherjee Govt. Degree College, Bhadohi, U.P.

Srinivas, P.
Plant Pathology Section, College of Forestry and Hill Agriculture, GBPUA&T Hill Campus, Ranichauri – 249 199, Tehri Garhwal, Uttarakhand

Srivastava, Amrita
Laboratory of Microbial Genetics, Department of Botany, Banaras Hindu University, Varanasi – 221 005

Srivastava, Rachana
Institute of Bioengineering and Biological Science, S-19/54, Varuna Bridge, Varanasi – 221 002, U.P.

Tiwary, P.B.
Department of Botany, S.M. P.G. College, Chandausi, Moradabad, Uttar Pradesh

Tripathy, Pramila
P.G. Department of Botany and Biotechnology, Utkal University, Bhubaneswar – 751 004

Upreti, D.K.
Lichenology laboratory, National Botanical Research Institute, CSIR, Rana Pratap Marg, Lucknow – 226 001

Vala, Anjana K.
Department of Bioinformatics, Bhavnagar University, Bhavnagar – 364 002, Gujarat

Vyas, Deepak
Lab of Microbial Technology and Plant Pathology, Department of Botany, Dr. H. S. Gour University, Sagar – 470 003, M.P.

Yadav, Rajesh
Lab of Microbial Technology and Plant Pathology, Department of Botany, Dr. H.S. Gour University, Sagar – 470 003, M.P.

Yadav, Rekha
Microbiology Research Laboratory, Department of Botany, Raja Balwant Singh College, Agra, U. P.

Soil Microflora, 2009 *Pages* **1–5**
Editor: **Rajan Kumar Gupta, Mukesh Kumar & Deepak Vyas**
Published by: **DAYA PUBLISHING HOUSE, NEW DELHI**

Chapter 1

Soil Microflora: A General Aspect

Mukesh Kumar[1], Anjali Khare[2] and Rajan Kumar Gupta[3]
[1]Department of Botany, Sahu Jain (P.C.) College, Najibabad, U.P.
[2]Department of Botany, Advance Institute of Science and Technology, 179-Kalidas Road, Dehradun, Uttarakhand
[3]Department of Botany, Pt. L.M.S. Govt. P.G. College, Rishikesh – 249 201, Dehradun, Uttarakhand

ABSTRACT

Soil is a rich and varied biological laboratory, harbor a diverse population of living organisms including plants, animals and microorganisms. The soil microorganisms have been classified as microflora including bacteria, actinomycetes, fungi and algae, it is estimated that bacteria form about 90 per cent of the total population, actinomycetes about 9 per cent and fungi and algae together about 1 per cent .

Keywords: *Bacteria, Fungi, Microflora, Soil.*

Introduction

The region of Earth that supports microbial growth and brings about transformations is called biosphere. Soil is a natural entity, a biochemically weathered and synthesized product of nature. It is also a natural habitat for plants and large number of organisms. Pedology is the science concerned with soil, its classification and its description.

The physico-chemical properties of soil depend on the parent rock from which it has been formed by a weathering process. The chemical transformations or biotransformations taking place in the biosphere play a vital role in soil fertility. Soil is a complex mixture of various chemicals and contain organic, inorganic materials, oxides of iron aluminium and silicon with cations and anions. In the process of biotransformations microoganisms convert the soil to inorganic form.

Soil is a rich and varied biological laboratory; harbor a diverse population of living organisms including plants, animals and microorganisms. The study of living organisms of soil is called Soil Biology. A single gram of soil may contain from a mere hundred thousand to several billion bacteria. The quantity of living organisms is influenced by the physical, chemical and biological properties of soils. Although a number of plants and animals constitute the soil flora, we shall restrict ourselves to microflora present in the soil. So here we go!

Soil Microflora

Soil contains organic matter, which serves as a source of food for soil bacteria and fungi. But this raw organic matter in the soil is not directly used by the plants as food. It must be broken down into simpler products before it can be utilized and this work is done by different kinds of microorganisms in the soil. The microorganisms have been classified as:

Phytomicroflora

(*i*) Bacteria
(*ii*) Actinomycetes
(*iii*) Fungi
(*iv*) Algae

Zoomicroflora

(*i*) Protozoa
(*ii*) Nematodes

Of these, bacteria and fungi play key roles in maintaining a healthy soil. They act as decomposers that break down organic materials to produce detritus and other break down products.

In this section, we will take a simple approach to study soil microflora which includes bacteria, fungi, yeast, actinomycets and algae. A single gramme of soil can contain over 100 million bacteria, 1 million actinomycetes and 100000 fungi. The weight of the organisms would only account for 0.05 per cent of the weight of the soil. The exact proportions of each of these organisms will depend on soil conditions such as available moisture, aeration, organic matter levels and the type of plants present. Chemical conditions such as acidity and alkalinity will greatly affect organism populations. Fungi like acidic soils, actinomycetes prefer more alkaline conditions.

Although very small, microorganisms have an importance in soil which for outweighs their size. The activities of soil microorganisms are commonly influenced by (i) Their numbers in the soil, (ii) their weight per unit volume or area of soil-biomass and (iii) their metabolic activity. They decompose plant and animal residues; synthesize humus, cycle nutrients such as carbon and nitrogen. The chemical by products of microbial reactions bind together soil particles into stable aggregates that resist erosion. Microorganisms represent one of the largest reservoirs for essential soil nutrients.

Soil microflora plays a pivotal role in evaluation of soil conditions and in stimulating plant growth (Singh *et al.*, 1999). Microorganisms are beneficial in increasing the soil fertility and plant growth as they are involved in several biochemical transformation and mineralization activities in soil. Type of cultivation and crop management practices found to have greater influence on the activity of soil microflora (Mc Gill *et al.*, 1999). Continuous use of chemical fertilizers over a long period may cause imbalance in soil microflora and thereby indirectly affect biological properties of soil leading to soil degradation (Manickam and Venkataraman, 1972) The activity of soil microflora as comparitively more in surface than in subsurface horizons and decrease with depth due to decrease in organic matter (Rudramurthy and Gurumurthy, 2007).

Soil microflora play fundamental roles in many ecosystem processes including decomposition and nutrient cycling, and affect many important soil hydrological and chemical properties (Gallardo and Schlesinger, 1994; Hart *et al.*, 2005) Hence change in the soil microbial community may lead to changes in the structure and function of the overall ecosystem, and ultimately determine ecosystem sustainability (Bossio and Scow, 1995).

The soil organisms vary in number from a few per hectare to many millions per gramme of soil. The density of population is determined by food supply, moisture, temperature, physical condition and the reaction of the soil. In neutral soils, bacteria dominate over types of microscopic life on the other hand, fungi predominate in acidic and organic matter such soil. Algae abound on the soil in constantly moist or shady situation. Under favorable conditions, the bacteria multiply enormously. In sandy desert soils and under water-logged conditions they are very scarce.

As among the soil microflora, it is estimated that bacteria form about 90 per cent of the total population, actinomycetes about 9 per cent and fungi and algae together about 1 per cent. Let us now discuss soil microflora in brief:

Bacteria

Bacteria are single- celled organisms, and are the most numerous denizens of the soil, with populations ranging from 100 million to 3 billion in a gram. They reproduce so rapidly that one bacterium can produce 16 million more in just 24 hours. Most soil bacteria live in close proximity to plant roots and are often referred to as Rhizobacteria. Bacteria ıive in soil water, including the film of moisture surrounding soil particles. The majority of the beneficial soil- dwelling bacteria need oxygen and are known as Aerobic bacteria, while those that don not require air are referred to as anaerobic and tend to cause putrefaction of dead organic matter. Aerobic bacteria are most active in a soil that is moist and neutral soil pH, and where there is plenty of food available. The important roes that bacteria play are:

Nitrification

The soil microflora typically bacteria produce ammonia form organic compounds. Ammonia released in this manner is converted into nitrites by one group of organisms called *Nitrosomonas* and further converted into nitrates by another group of organisms called *Nitrobacter.* The process of conversion of nitrogen to nitrates is called nitrification. The nitrate forming bacteria are generally confined to the top 25 to 30 cm of the soil, where the content of organic matter is also more. These organisms are most active between 25 and 38^{0}C and under favorable conditions of tillage, aeration, neutral soil reaction, and moisture content at field capacity. They also control the nutrient cycling.

Nitrogen Fixation

Two other groups of bacteria intimately linked with the nitrogen problem of the soil take up free nitrogen from the air and convert it into nitrogenous compounds for the use of crop plants. This is termed as the process of nitrogen fixation. The process is carried out by free living nitrogen fixing bacteria in the soil or water such as Azatobacter, or by those which live in close symbiosis with leguminuous plants, such as *Rhizobia.* These bacteria form colonies in nodules they create on the roots of leguminous plants. The amount of nitrogen added to the soil varies from 50-150 kg per hectare.

Azatobacter and other non-symbiotic nitrogen- fixing bacteria work independently of any lost crop. Under optimum laboratory conditions, *Azatobacter* has been found to fix a considerable amount of nitrogen.

Denitrification

The conversion of nitrogen in the form of nitrate to gases such as nitrous oxide and dinitrogen by soil bacteria under anaerobic conditions is termed as the process of denitrification. The bacteria included are *Achromobacter* and *Pseudomonas*. A consequence of dentrification is that nitrogen is lost from the soil on the other hand, this process is a useful way to remove excess nitrate from waste water.

Other Potential Uses

Several genera of bacteria, especially the genus *Thiobacillus* can oxidize reduced sulphur compounds. *Thiobacillus thioxidans* can oxidize sulphur to sulphuric acid. Sulphur, therefore, can be used to decrease the pH of an alkaline soil.

Sulphur in the form of sulphate is used by anaerobic bacteria like the genus *Desulfo vibrio,* which convert it into hydrogen sulphide gas. Hydrogen sulphide gas reacts with metal ions and forms very insoluble metallic sulphides like pyrite (Fe_2S). In fact, it is possible that the pyrites associated with coal seams were deposited by the action of these bacteria years ago. The black colour of salt marsh soil and the rotted egg smell associated with them are a result of the activities of the sulphate reducing bacteria in these habitats. They attest to the occurrence of anaerobic conditions.

Actinomycetes

They are similar in size to bacteria but resemble moulds in their growth and physiology.They can grow in the deeper layers of the soil and under drier conditions and need less nitrogen. Actinomycetes are critical in the decomposition of organic mater and in humus formation, and their presence is responsible for the sweat "earthy" aroma which is associated with a good healthy soil.

Fungi

A gram of garden soil can contain around one million fungi, such as yeast and moulds. Fungi are chemoheterotrophic organisms *i.e.* they require a chemical source of energy as well as organic substrates to get carbon for growth and development.

Many fungi are parasitic. In terms of soil and humus creation, the most important fungi tend to be saprotrophic, that is, they live on dead or decaying organic matter, thus breaking it into simpler forms. They spread underground by sending their long threads known as mycelium throughout the soil, which may be found in the disintegrating organic matter on the surface of the soil or on the plant roots in the upper strata below the surface.

A fungus, *Armillaria bulbora*, discovered in the U.S. in the state of Michigan, could turn out to be earth's largest fungus growing among the roots of hard wood trees in a forest. The microscopic, branched filaments (called hyphae) of the fungus occupy a 14.8 ha (137 acre) area of land. Careful genetic analysis has shown the filaments constitute a single organism.

Algae

They are microscopic or larger plants containing chlorophyll. They are found in the top layer of constantly moist soils such as paddy fields. Some of them are capable of fixing atmospheric nitrogen especially blue green algae.

Conclusion

Soil microflora plays a pivotal role in evaluation of soil conditions and in stimulating plant growth. Type of cultivation and crop management practices found to have grater influence on the activity of soil microflora. Therefore, needed is an integrated approach that considers agriculture's potential impacts on soil biodiversity, one that maintains soil fertility productivity and crop protection by optimizing ecological synergies among biological components of the ecosystem and enhancing the biological efficiency of soil processes.

References

Bossio D and Scow K (1995). Impact of carbon and flooding on the metabolic diversity of microbial communities in soils. *Appl Environ Microbial,* 61: 4043–4050.

Gallardo A and Shlesinger W H (1994). Factors limiting microbial biomass in the mineral soil and forest floor of a warm-temperate forest. *Soil Biol Biochem,* 26: 1409–1415.

Hart S C, Delurca T H, Newman G S, Mackenzie D M and Boyle SI (2005). Post fire vegetative dynamics as drivers of microbial community structure and function in forest soil. *For. Ecol. Manage.*

Manickam T S and Venkataraman R (1972). Effect of continuous application of manures and fertilizers on some physical properties of soils II under irrigated conditions. *Madras Agric J,* 59: 508–512.

McGill W B, Cannon KR, Robertson J A and Cook F D (1980). Dynamics of soil microbial biomass and water stable organic carbon is Breton.L. after 50 years of cropping rotation. *Candian J Soil Sci,* 66: 1–19.

Rudramurthy H B and Gurumurthy B R (2007). Dynamics of soil microflora in different land use systems. *Karnataka J Agric Sci,* 20(1): 131–132.

Singh K, Borana J and Srivastava S (1999). Effect of thiram on root growth, root nodules and nitrogen fixation in *Glycine max* (1) merrit by *Brady Rhizobium japonicum. J Soil Boil and Ecol,* 19: 11–14.

Soil Microflora, 2009 *Pages* **6–20**
Editor: **Rajan Kumar Gupta, Mukesh Kumar & Deepak Vyas**
Published by: **DAYA PUBLISHING HOUSE, NEW DELHI**

Chapter 2

Distribution of Cyanobacteria in Coastal Sandy Soils and Alumina Mine Waste Soils of Orissa

S.P. Adhikary* and Pramila Tripathy
P.G. Department of Botany and Biotechnology, Utkal University, Bhubaneswar – 751 004

ABSTRACT

Cyanobacterial component in the soil crusts of two different habitats of Orissa state, in the east coast of India, *e.g.* one alumina mine waste burdened soils of Koraput district and the another, sandy soils of Ramchandi in the coast of Bay of Bengal, was investigated for their occurrence in the top soil as well as up to 30 cm below the soil cover. These soils showed wide variation in their nutrient content, pH and salinity levels. The crusts on the top layers of the soil of both the degraded habitats harboured principally cyanobacterial forms, though occasionally certain green algae and diatoms also occurred especially in the rainy season, hence were not the major component of the soil crusts of the localities through out the year. The sandy soils of Ramchandi harboured principally heterocystous forms, *e.g. Anabaena, Nostoc, Calothrix* and member of Stigonematales, and among the non heterocystous form, mainly a species of *Microcoleus* was the dominant organism in mine waste burdened soils. However, a species of *Nostoc, N. paludosum* and a species of *Westiellopsis* showed adaptation to both soil types showing wider acclimatization capability. Altogether 14 species of cynaobacteria were recorded in these two experimental sites. These were isolated into culture, morphometriaclly analysed, taxonomically enumerated, assigned with a strain number and deposited at the culture collection at Utkal University.

Keywords*: Soil crust, Mine burdened soil, Sandy soil, Cyanobacteria, Distribution pattern.*

* E-mail: adhikary2k@hotmail.com; Telefax: 0674 2587389, Telephone: 0674 2354054

Introduction

Orissa state is generally known as the land of mines. There are many mines like bauxite, coal, iron ore, alumina etc in the state. The Damonjodi alumina mine is one of those located near the district headquarter of Koraput. Due to mining activities in the area, the top layer vegetation was destroyed and the mine wastes are continuously dumped in the nearby areas. Increased soil erosion frequently occur following such disturbance. Result of this disturbance includes both loss of soil stability and loss of beneficial microorganisms, thus affect the soil productivity (Belnap and Eldridge, 2001). Recovery of microbial communities in such areas have been studied in different regions of the globe. Long term recovery studies indicated that recovery of the cyanobacterial community often took 5-30 years, and the moss and lichen components can take 40-100 years (Anderson *et al.*, 1982; Johansen *et al.*, 1984). Recovery of ecosystem function, such as nitrogen fixation activity may take longer than simple recovery of biomass (Belnap, 1996). In an experiment to study the cyanobacterial components along gradients of microtopography, desert samples were collected in the dead sea valley at Wadi bottoms (Israel) where the habitat has been periodically destroyed by the winter floods, and on progressively more elevated terraces where the habitat was more stable for a long time. At the lowest and thus geologically youngest terraces, only two cyanobacterial species *Microcoleus vaginatus* and *Schizothrix friessi* were found and on the elevated terraces cyanobacteria were more numerous, accompanied by lichens (Dor and Danin, 1996).

Cyanobacteria are an important component of many soils, including the surface crust that cover extensive areas in mine wastes. They have been reported from both on and below the surface. Depending upon the physical and chemical status of the soil, the biological components show diversity of occurrence pattern as blackish brown crusts on the upper millimeter of dry soils of barren lands in all regions of India showing tropical conditions (Tirkey and Adhikary, 2005). Cyanobacteria are usually restricted to the upper photic zone which extends up to 0.5 cm. Besides they also exist in the deeper horizons in dormant condition as spore or filament fragments (Chapman and Chapman, 1973; Roger and Reynaud, 1976). In the present work diversity of cyanobacteria in the soil crust of upper layer as well as from soils at various depths, *i.e.* from top soil to 1 cm, 5 cm, 10 cm, 20 cm, 30 cm depth was studied in an alumina mine waste area.

Studies of subsurface environment have received attention because they are important to human health, ecosystem functions, agriculture and environmental management (Hoyle and Arthur, 2000). Since microorganisms play essential role in subsurface geology, hydrology and ecology, knowledge about microbial community structure and composition is important to improve our conceptual and predictive understanding of sub-surface ecosystem process, function and management (Zhou *et al.*, 2004). Hence, cyanobacterial diversity in the mine wastes of Damonjodi was carried out.

Emerging from recent studies is the increased understanding that the ecosystem function of microbiotic crusts varies considerably by region and climate, and that assumptions true for one well-studied geographic region may not be true for other regions (Kubeckova *et al.*, 2003). Hence, in addition to Damonjodi mine waste soils, soil samples from sea shore area of Ramchandi coast was collected and analysed for occurrence of cyanobacteria at its different depths. Although the erodibility of soil with and without crusts has been quantified by several workers, very little work is available on the effects of microbial colonization in stabilization of sand dunes.

In India about 7 mh (million hectares) of available land are adversely affected by salt, leading to loss in crop production. Orissa state has an extensive coast line covering about 460 kms. To study the cyanobacterial diversity in the saline soils and sand dunes in this area, representative soils were collected from Ramchandi sea-shore located in the Puri district of Orissa state and analysed.

Materials and Methods

Surface soils containing the crusts whenever visible and also at different depths were collected at two different sites varying widely in their composition. Two sites were selected for the purpose. 1. In the coastal area of Ramachandi, 27 km from Puri and the other 2. in the Alumina mine-waste areas of Damonjodi, 35 km away from the district headquarters of Koraput district in Orissa state.The crusts and soils from different depths were collected from two different sites at Ramchandi, *e.g.* from the coast and ½ km away towards North from Ramchandi, and three sites at Damonjodi mine site, *e.g.*, at the mining area, ½ km away towards South from mine site and 5 km away towards South from mine site.

Surface vegetation at the collected sites were carefully removed followed by digging vertically upto 30 cm. Soil samples were collected from top-soil and at 1,5,10,20 and 30 cm depth from the surface. The crusts from soil surface and also the soils from different depths were collected using sterile scalpel and kept in capped specimen bottles assigning a collection number followed by date of collection and stored in a dark place. To analyse the composition of the crust and the soils, a pinch of each sample was inoculated to the agar plates prepared with the sterile BG ll medium with or without combined nitrogen (Rippka *et al.*, 1979) with 1.2 per cent (w/v) agar-agar. The cultures were incubated under fluorescent light at continuous light intensity of 7.5w/m^2 and 25±1°C temperature up to 30 days. The cyanobacterial forms appeared in culture were analysed.

The camera lucida diagram of each species was drawn. Microphotographs were taken in a Meiji ML-TH-05 trinocular research microscope using Fx-801 Nikon Camera. The organisms were identified following Desikachary (1959). The unialgal culture of each of the species was given a specific isolate number (UU-Utkal University) and maintained on agar slants in test tubes with 12 h L/12 hD cycle in the culture room at the Department of Botany, Utkal University.

Physico-chemical analysis of the soil samples collected from five different sites was carried out for measurement of soil pH. 1:2.5 proportions of the sample was prepared with distilled water and measured using a pH meter (Systronics model 335). Conductivity of the sample was measured using a conductivity bridge (621E, Electronics India). Soil organic carbon content was qualified following Walkley and Black as described by Kaushik (1987). Nitrate and phosphate content of the soil samples was measured following Kaushik (1987). The heavy metals *e.g.* Zn, Cu, Fe and Mn content of the soils was determined using Atomic Absorption spectrophotometry at the Central Institute for Freshwater Aquaculture, Kausalyaganga, Bhubaneswar.

Results

Ramchandi is located in the coastal Orissa in between Konark and Puri in the confluence of river Kushabhadra to Bay of Bengal. The area is sandy, and the sandy nature of the soil even extends up to more than a kilometre from this location. Surface soil samples were collected from two sites: one from the sand dunes, and another ½ km away from the shore and analysed. The soil was principally sandy (88-92 per cent) and contained very little silt (6-7 per cent) and clay (2-5 per cent), little amount of organic carbon (0.08-0.15 per cent), but with higher conductivity (0.33-0.73 m mho/cm). The pH was in the near acidic range. The N, P, Zn, Cu, Fe, Mn, however, were moderate as found in other soils. To the contrary the organic carbon, silt and clay of Damonjodi mine site as well as up to 5 kms of its surrounding area with dumped mine wastes showed the pH near alkaline range and the conductivity was lower (0.057-0.191 m mho/cm). The Zn, Cu, Fe, Mn as well as N, P values were higher in comparison to the soils of Ramchandi site (Table 2.1). On culturing all the soils from the sand dunes of Ramchandi and alumina mine red soil of Damonjodi, several organisms appeared in the culture of which few

Table 2.1: Physico-chemical Composition of Soils of the Study Sites at Ramchandi (Puri District) and Damonjodi (Koraput District) of Orissa State

Site of Collection	*pH*	*Conductivity (m mho/ cm)*	*Organic Carbon (%)*	*Sand (%) Average*	*Silt (%) Average*	*Clay (%) Average*	*N (ppm)*	*P (ppm)*	*Zn (ppm)*	*Cu (ppm)*	*Fe (ppm)*	*Mn (ppm)*
Ramachandi sand dune (S1)	6.7± 0.18	0.735± 0.05	0.15± 0.01	92	6	2	50± 6 2	8.5± 0.6	1.02± 0.04	1.04± 0.03	112.6± 6.2	88.4± 1.3
½ km. away from Ramachandi sea shore (S2)	6.2± 0.23	0.33± 0.03	0.08± 0.006	88	7	5	48± 4 5	9.2± 0.5	1.08± 0.02	1.17± 0.05	103.4± 8.1	58.4± 1.6
Damonjodi mine site (S3)	7.2± 0.31	0.111± 0.09	0.35± 0.04	70	20	10	62± 5 6	11.0± 0.8	1.18± 0.08	1.43± 0.08	120.4± 9.3	94.3± 2.0
½ km. away from Damonjodi mine site (S4)	7.5± 0.34	0.191± 0.08	0.39± 0.03	72	18	10	68± 6.1	11.5± 0.7	1.26± 0.1	1.27± 0.09	118.6± 8.4	98.6± 0.9
5 km. away from Damonjodi mine site (S5)	7.0± 0.26	0.057± 0.04	0.45± 0.05	63	22	15	82± 7.8	13.5± 0.9	1.29± 0.11	1.55± 0.11	120.8± 5.6	83.4± 0.8

Mean±S.D. values of 10 different samples is given.

species were specific to each site and few more occured in both the soils differing in physico-chemical characteristics. Altogether 14 species of cyanobacteria appeared in the culture from the soil of all the sites. These were identified following standard monographs and depicted in Table 2.2. These were *Lyngbya spirulinoides, Arthrospira nonconstricta, Plectonema puteale, Microcoleus vaginatus, Anabaena fertilissima, Nastoc paludosum, Nodularia spumigena, Calothirix crustacea, Calothrix contarenii, Scytonema leptobasis, Aulosira implexa, Westiellopsis prolifica, Westiellopsis* sp. *and Fischerella ambigua*. Of these, 4 species were lost during subculturing and the other 10 species were assigned with a specific strain number and maintained in the germplasm collection at the Department of Botany, Utkal University.

Table 2.2: Distribution of Cyanobacteria in Soils from Ramachandi Coast (Puri District) and Damonjodi Mine Site (Koraput District) of Orissa State

Organism	*Strain no.*	*Sand Dune Soil at Ramachandi*		*Aluminimum Mine Red Soil at Damonjodi*		*Generation Time (h)*
		Sand Dune	*Away from Sand Dune*	*Mine Area Soil*	*Away from Mine Area Soil*	
Lyngbya spirulinoides	Not in culture			++	+	200
Arthrospira nonconstricta	UU61242			+		190
*Plectonema puteale**	UU62246	++	++			272
*Microcoleus vaginatus**	UU61243			++	++	252
Anabaena fertilissima	Not in culture	++				
*Nastoc paludosum**	UU61240	++		++	++	147
Nodularia spumigena	Not in culture			++	++	
Calothirix crustacea	UU61238			++	+	260
Calothrix contarenii	UU62244	++	++			240
Scytonema leptobasis	Not in culture			++	++	
Aulosira implexa	UU61241			++	++	290
*Westiellopsis prolifica**	UU62247	+				156
Westiellopsis sp.	UU61239	+		+	+	230
Fischerella ambigua	UU62245	++				252

++: Dominant organism; +: Occurred in enriched soil culture upon prolonged incubation; *: Selected for further experiments.

Analysis of these results showed that, the sand dunes of Ramchandi harbour principally 2 species of cyanobacteria, *e.g. Plectonema puteale* and *Calothrix contarenii. Anaebaena fertilissima* as well as a species of *Westiellopsis, W. prolifica,* and *Fischerella ambigua* also occurred as the dominant component in the sand dune soils. To the contrary, many non- heterocystous filamentous forms *e.g. Lyngbya spirulinoides, Arthrospira nonconstricta, Microcoleus vaginatus*, in addition to the heterocystous, *Nodularia spumigena, Calothrix crustacea, Scytonema leptobasis* and *Aulosira implexa* occurred in the mine waste dumped soils at a radious of over 5 kms. Few species *e.g. Nostoc paludosum* and *Westiellopsis* sp. showed adaptation to both the types of soils, thus showing wide acclimatization capability. In general, the sand dune soils of Ramchandi harboured principally the heterocystous forms of which *Anabaena, Nostoc, Calothrix* and members of Stigonematales were dominant, where as in mine waste soils showing higher inorganic nutrition status encouraged growth of non-heterocystous filamentous forms. Further,

Microcoleus vaginatus which is a well known crust-forming species with cosmopolitan distribution was specific to the mine waste soils, and *Plectonema puteale* together with *Calothrix contarenii* were the dominant ones in the sandy soils and its nearby areas (Table 2.2).

Distribution of cyanobacteria at different depths from the surface of the coastal sandy soils as well as mine waste soils was analysed. Some of the cyanobacteria species occurred just on the top soil where as few more were detected in the culture of soils collected from different depths. *Nostoc paludosum* occurred up to 20 cm depth in the sandy soils of Ramchandi as well as up to 5 cm depth in the alumina waste soils. Similarly *Westiellopsis prolifica* occurred up to similar depth of the sandy soils and *Plectonema puteale* appeared on top soil of Ramchandi coast. In aluminium mine waste soils most of the species were confined within 5 cm from the top soil. *Arthrospira nonconstricta*, to the contrary, occurred from the soil collected at 5 cm depth and *Microcoleus vaginatus*, the cosmopolitan species occurred in abundance in the soil even at a depth up to 20 cms (Table 2.3). Almost all these organisms grew slowly in liquid culture. The doubling time in still culture, when grown in a flask of 100 ml capacity containing 50 ml of medium, varied from 156 h–290 h (Table 2.2). All these organisms isolated from soils and soil crust from different study sites were morphometrically analysed, the microphotograph was taken, camera lucida diagram for each species was drawn, and depicted in Plates 2.1 to 2.3. Systematic account of all these species is given below:

Table 2.3: Distribution of Cyanobacteria in Soils at Different Depths at Ramachandi Coast and Damonjodi Mine Sites

Organism	*Strain no.*	*Sand Dune Soil at Ramachandi*						*Aluminimum Mine Red Soil at Damonjodi*					
		Top Soil	*Depth*					*Top Soil*	*Depth*				
			1 cm	*5 cm*	*10 cm*	*20 cm*	*30 cm*		*1 cm*	*5 cm*	*10 cm*	*20 cm*	*30 cm*
Lyngbya spirulinoides	Not in culture							++	++				
Arthrospira nonconstricta	UU61242									+			
Plectonema puteale	UU62246	++											
Microcoleus vaginatus	UU61243							++	++	++	++	++	
Anabaena fertilissima	Not in culture	++	++	++									
Nastoc paludosum	UU61240	++	++	++	++	++		++	++	++			
Nodularia spumigena	Not in culture							++					
Calothirix crustacea	UU61238							++	++				
Calothrix contarenii	UU62244	++	++	++	++								
Scytonema leptobasis	Not in culture							++	++				
Aulosira implexa	UU61241							++	++				
Westiellopsis prolifica	UU62247	+	+	+	+	+							
Westiellopsis sp.	UU61239	+						+					
Fischerella ambigua	UU62245	+	+	+	+								

++: Dominant organism; +: Occurred in enriched soil culture upon prolonged incubation.

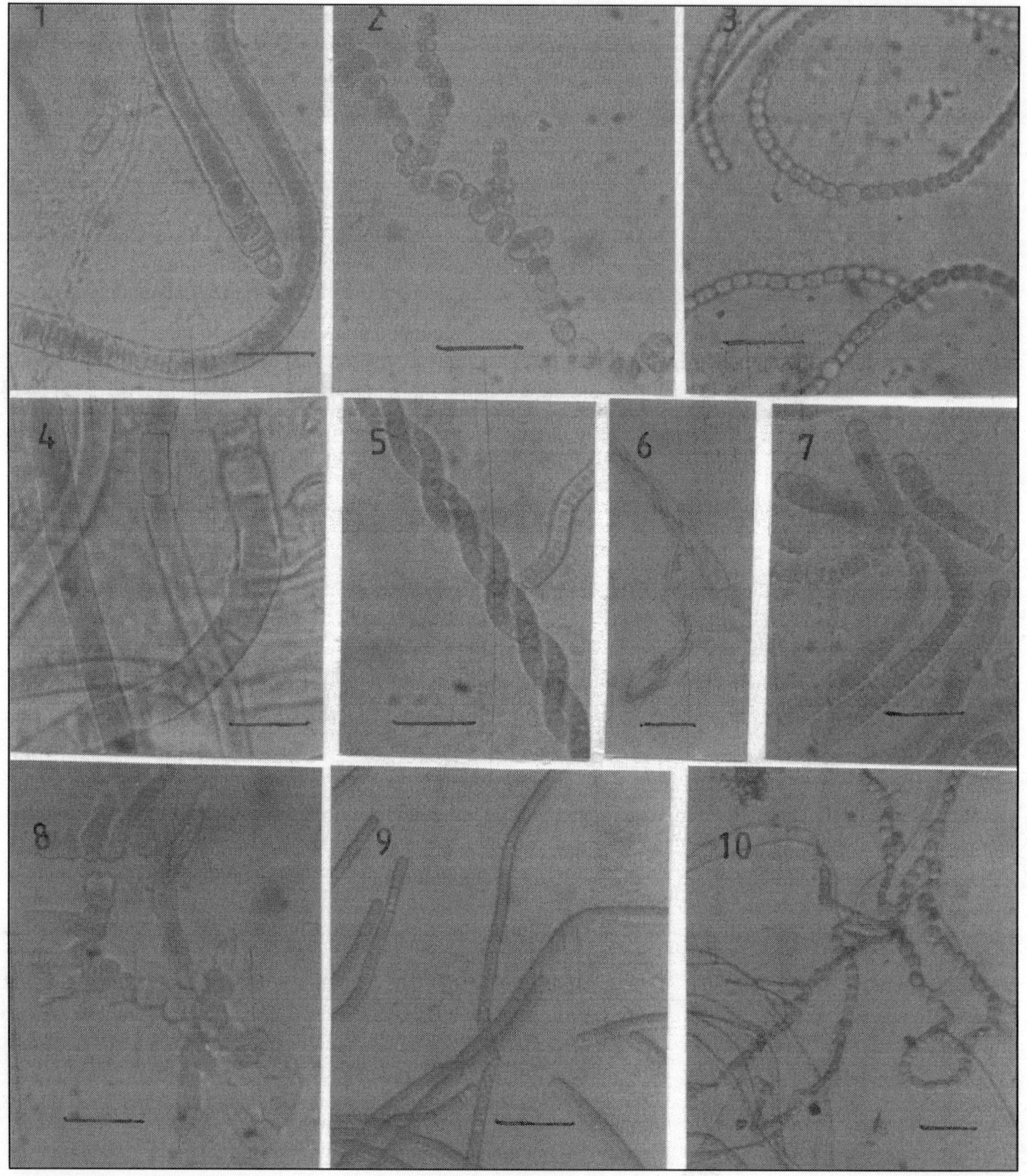

Plate 2.1

Figures 1-10. Light microscopic photograph of various organisms isolated from Damanjodi and Ramchandi soils and maintained in culture: 1. *Calothrix crustacea* **UU61238, 2.** *Westiellopsis* **sp. UU61239, 3.** *Nostoc paludosum* **UU61240, 4.** *Aulosira implexa* **UU61241, 5.** *Arthrospira nonconstricta* **UU61242, 6.** *Microcolecus vaginatus* **UU61243, 7.** *Calothrix contaranii* **UU62247, 8.** *Westiellopsis prolifica* **UU62247, 9.** *Plectonema puteale* **UU62246, 10.** *Fischerella ambigua* **UU62245.**

Scale bar: 1,2,3,4,5,7,8 and 9 = 20 μm; 6 and 10 = 30 μm.

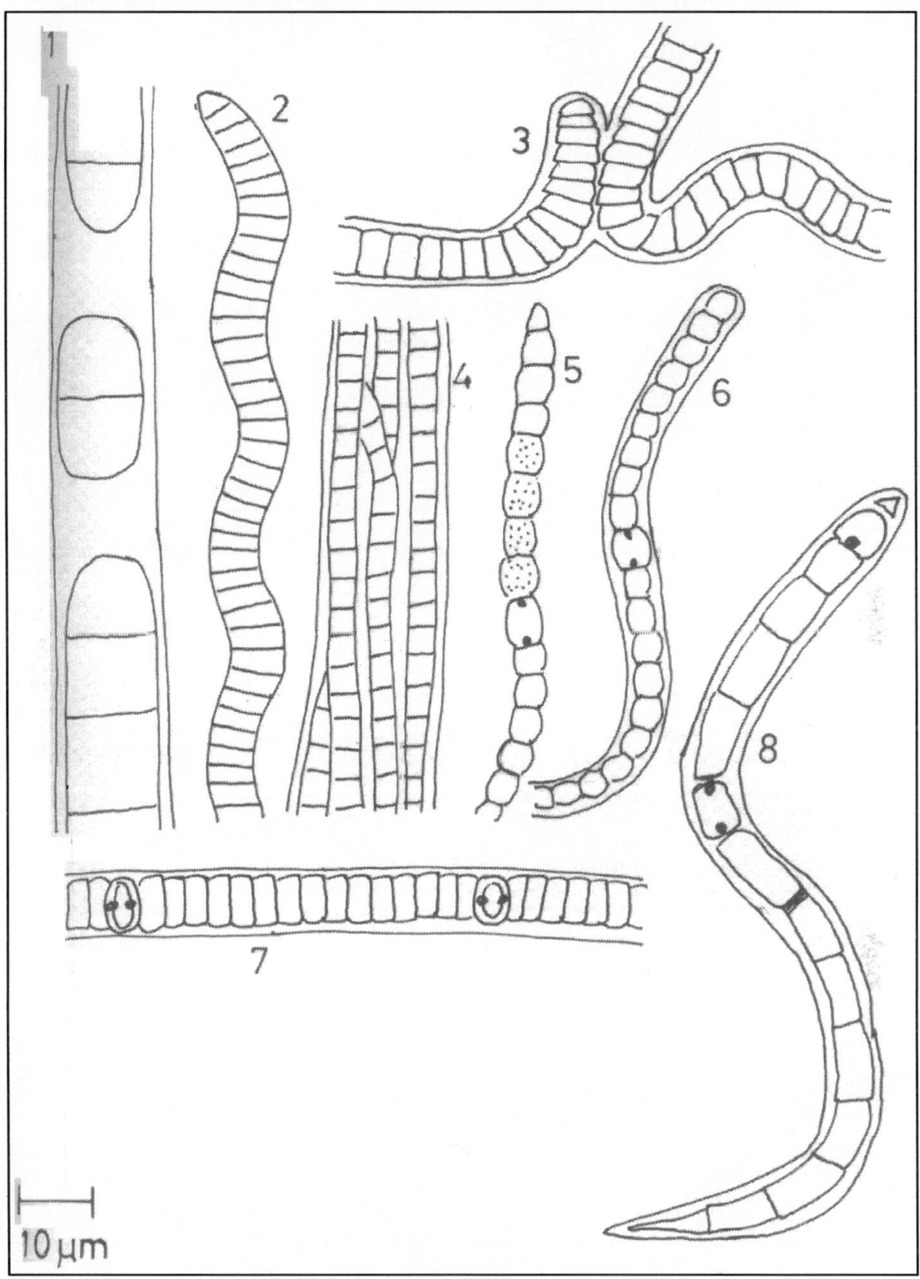

Plate 2.2

Figures 1-8. Camera lucida diagram of various cyanobacteria; 1. ***Lyngbya spirulinoides*** **2.** ***Anthrospira nonconstricta*** **UU61242, 3.** ***Plectonema puteale*** **UU62246, 4.** ***Microcoleus vaginatus*** **UU61243, 5.** ***Anabaena fertilissima,*** **6.** ***Nostoc paludosum*** **UU61240, 7.** ***Nodularia spumigena*** **8.** ***Calothrix crustacea*** **UU61238.**

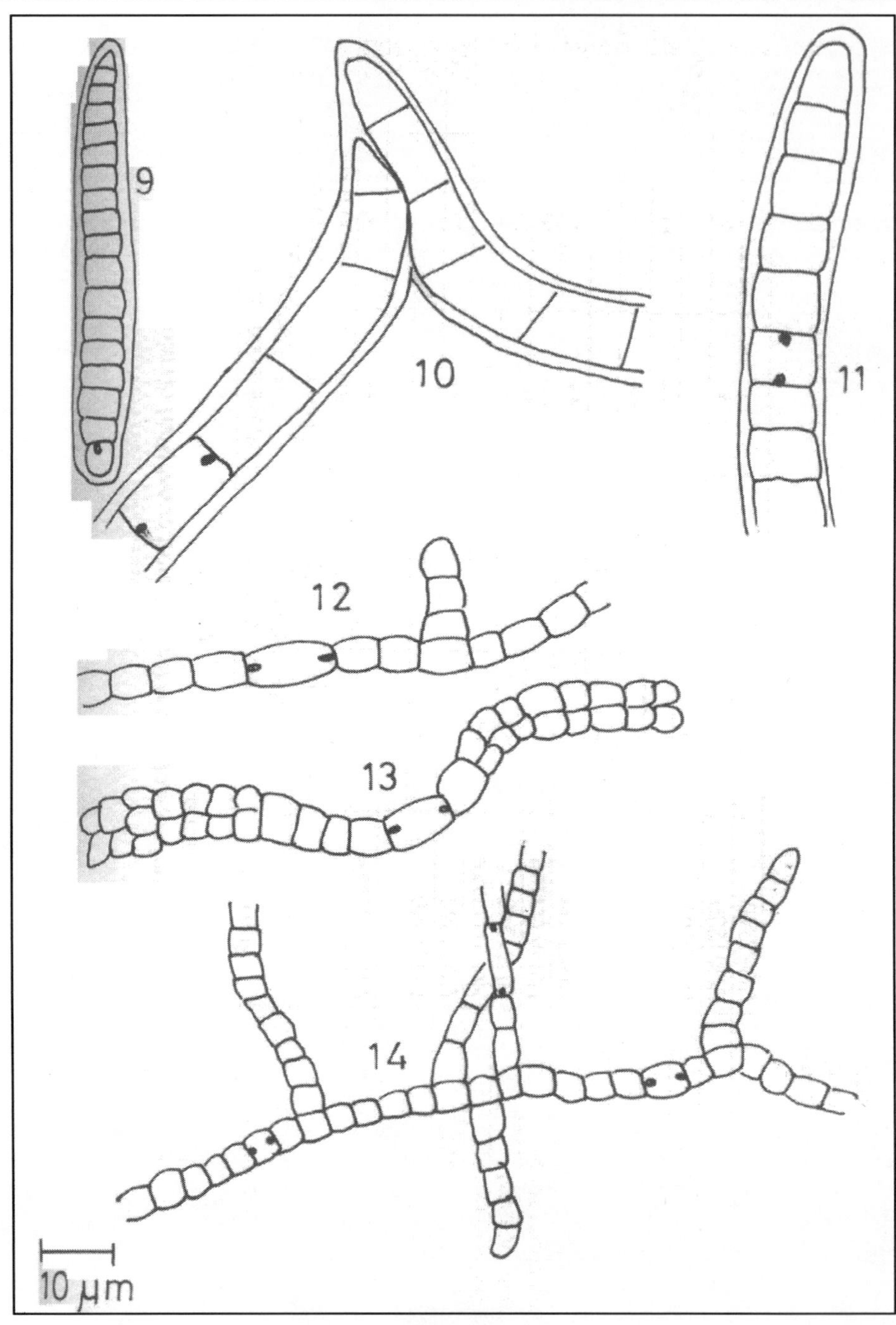

Plate 2.3

Figures 9-14. Camera lucida diagram of various cyanobacteria: 9. *Calothrix contarenii* UU62244, 10. *Scytonema leptobasis*, 11. *Aulosira implexa* UU61241, 12. *Westiellopsis prolifica* UU 62247, 13. *Westiellopsis* sp. UU61239, 14. *Fischerella ambigua* UU61245

Systematic Account

Systematic account of the cyanobacteria recorded in the crusts/soils from different depths collected from various sites of Ramchandi and Damonjodi is presented below. Mode of occurrence, sample number, date of collection and the assigned number of the culture collection of Utkal University (UU) to each species is also given.

***Lyngbya spirulinoides* Gomont**

Thallus yellowish brown, unbranched, cells slightly broader than long, breadth 15 mm, length 12.5 mm, heterocyst absent, sheath thin, 2mm, light yellow.

Occurred up to 1 cm depth in the soils of Damonjodi mine site, Koraput district. Date of collection 05.02.2004. (Plate 2.2, Figure 1)

***Arthrospira nonconstricta* Banerji**

Thallus bluish green, cells broader then long, breadth 4-6 mm, length 2-3.5 mm, heterocyst absent, sheath absent, regularly spirally coiled, end cells rounded.

Occurred at 5 cm depth under top soil of Damonjodi mine site, Koraput district. Date of collection 05.02.2004, strain number UU61242. (Plate 2.1, Figure 5; Plate 2.2, Figure 2).

***Plectonema puteale* (Kirchner) Hansgirg**

Thallus green, unbranched, cells broader then long, 2.5 mm to 5 mm broad, non-heterocystous, trichome variously bent, sheath thin, transparent, 0.5 mm wide around trichome.

Appeared in culture of top soil of Ramchandi area. Date of collection 29.02.2004, strain number UU62246 (Plate 2.1, Figure 9; Plate2.2, Figure 3).

***Microcoleus vaginatus* (Vaucher) Gomont**

Thallus green, unbranched, cells as long as broad, 5 mm, heterocyst absent, sheath colourless, number of trichomes many, enclosed in a sheath, coiled like a rope, occurred in the top soil as well as up to 20 cm depth in Damonjodi mine site, Koraput district.

Date of collection 05.02.2004, strain number UU61243 (Plate 2.1, Figure 6; Plate 2.2, Figure 4).

***Anabaena fertilissima* Rao**

Thallus green, unbranched, straight or bent, cells barrel shaped, breadth 4-7 mm, length 5-5.6 mm, heterocystous, heterocyst intercalary, spherical, 5-8.5 mm broad, spores in chain adjoining to the heterocyst, spherical, 6-9 mm broad, 5-8 mm long, sheath absent.

Occurred in the top soil as well as up to 5 cm depth of sandy soils in Ramchandi, Puri district, date of collection 29.02.2004 (Plate 2.2, Figure 5).

***Nostoc paludosum* Kützing ex Bornet et Flahault**

Thallus blue green, unbranched, coiled, cells as along as broad, 5 mm, heterocystous, heterocyst terminal and intercalary, as long as broad, 7.5 mm, sheath thin, transparent.

Occurred in the soils of Damonjodi mine site, Koraput district as well as in the soils of Ramchandi coast up to 5 cm depth in the former and up to 20 cm depth in the later samples. Date of collection 05.02.2004, strain number UU61240 (Plate 2.1, Figure 3; Plate 2.2, Figure 6).

Nodularia spumigena **Mertens ex Bornet et Flahault**

Thallus green, unbranched, cells short, discoid, breadth 7.5 mm, length 2.5 mm, heterocystous, hetorocyst broader than the vegetative cell, sheath thick, hyaline.

Occurred in the top soil of Damonjodi mine site, Koraput district.Date of collection 05.02.2004 (Plate 2.2, Figure 7).

Calothrix crustacea **Thuret**

Thallus blackish green, unbranched, cells longer than broad, length 12.5 mm, breath 5 mm, heterocystous, basal as well as intercalary, basal heterocyst oval, as long as broad, intercalary hetrocyst 5 mm broad, 10 mm long, heterocyst light yellow, sheath thick, 2 mm, light brown.

Occurred up to 1 cm depth in the mine site soil as well as 5 Km away from mine site of Damonjodi, Koraput district. Date of collection 05.02.2004, strain Number UU61238 (Plate 2.1, Figure 1; Plate 2.2, Figure 8).

Calothrix contarenii **(Zanard) Bornet et Flahault**

Thallus green, unbranched, cells broader than long, breath 7.5 mm, length 5 mm, basal heterocyst, rounded, 5 mm diameter, sheath firm, 2 mm thick, trichome slightly swollen at the base.

Occurred up to 10 cm depth in the soil of sea shore as well as ½ km away from the sand dunes at Ramchandi, Puri district. Date of collection 29.02.2004, strain number UU62244 (Plate 2.1, Figure 7; Plate 2.3, Figure 9).

Scytonema leptobasis **Ghose**

Thallus green, filamentous, heterocystous, showed false branching, sheath distinct, cells 12 mm long, 5 mm broad, heterocyst intercalary, light yellow, 10 mm long, 5 mm broad, sheath transparent, 2 mm thick.

Appeared in culture of soils up to 1 cm depth in Damonjodi mine area, Koraput district, the culture was subsequently lost. Date of collection 05.02.2004 (Plate 2.3, Figure 10).

Aulosira implexa **Bornet et Flahault**

Thallus green, unbranched, cells as long as broad, 10 mm, heterocystous, intercalary heterocyst, broader than long, breadth 10 mm, length some times longer, varies from 7.5 mm to 15 mm, sheath thin, yellowish.

Occurred in the top soil as well as at 1cm depth from top soil of Damonjodi mine site, Korput district. Date of collection 05.02.2004, strain number UU61241 (Plate 2.3, Figure 11).

Westiellopsis prolifica **Janet**

Thallus green, filamentous with true branching, filaments of two kinds, primary filament slightly thicker, more or less creeping, secondary filament thinner, erect, single row of cells, barrel shaped, 5-10 mm broad, 8-12 mm long, hetercystous, intercalary heterocyst, cylindrical, 5.5- 6 mm broad, 10-20 mm long, sheath thin.

Occurred in the top soil as well as up to 20 cm depth at the sea shore at of Ramchandi, Puri district. Date of collection 29.02.2004, strain number UU62247 (Plate 2.1, Figure 8; Plate 2.3, Figure 12).

Westiellopsis **sp.**

Thallus green, unbranched, cells of various size, breadth 4 mm–7.5 mm, length 4 mm to 7.5 mm, heterocystous, intercalary heterocyst, longer than broad, length 10 mm, breath 5 mm, yellowish, sheath absent.

Occurred in the top soil of Damonjodi, mine site, Koraput district as well as in the top soil at Ramchandi, Puri district. Date of collection 29.02.2004, strain number UU61239 (Plate 2.1, Figure 2; Plate 2.3, Figure 13).

***Fischerella ambigua* (Nägeli) Gomont**

Thallus creeping, branched, densely interwoven with thick broad and yellowish brown sheath, cells of the prostrate filament cylindrical, 4-5 mm broad, 5-7 mm long, cells of the erect filament cylindrical, at the apical zone cells quadrate, 2.3 mm broad, heterocyst intercalary, cylindrical, 5 mm broad, 7 mm long.

Occurred in the top soil as well as up to 10 cm depth of sandy soils at Ramchandi, Puri district. Date of collection 29.02.2004, strain number UU62245 (Plate 2.1, Figure 10; Plate 2.3, Figure 14).

Discussion

The arid and disturbed land of the world are characterized by incomplete plant cover or even its complete absence as in deserts, semi deserts and sand dunes. Between bushes and plants, the soil surface is often open and appears to be bare. But this appearance is deceptive. A detail analysis of such soils showed that the surface is densely populated by tiny organisms (Mult and Lange, 2004). Filamentous cyanobacteria reported to penetrate few millimeter of top soil is confirmed in the present work. These organisms are known to secrete slimy carbohydrate from their sheath with which they attach themselves to the soil particles on the ground, making them aggregated and bind the eroded soil.

Crust formation is extraordinarily important for the stability of soil surface because it protects against erosion by wind and water. There are reports that wind speeds 10 times higher than those as crust-less soils could not blown away soils with crusts faster. If the protective crust is removed, running water can also wash away materials and nutrients. Without this protection, many arid areas would transfer into dust bowls with constantly changing surface structure. Thus soil crust play very important role as pioneer in the recolonization of the disturbed soil by plant communities. Hence, in Damonjodi mine waste area where red mud is being constantly dumped in the barren land, and in Ramchandi dunes where sand seldom stabilises, the soil crust are of immense importance for their stabilisation.

Cyanobacteria are known to survive prolonged period of drought. Their successful revival has been reported after 35 years of storage (Trainer and Gladych, 1995). The ability of many forms to withstand adverse ecological condition, capacity to thrive well in hostile environment and respond to the onset of dry conditions by entering into a dormant resistant state has distinguished this group as pioneers of succession. This also represent a sizeable proportion of the microbiol phototrophic soil population, especially in arid soils where, because of their capability to withstand desiccation, high temperature, unsuitable pH and nutrient deficiency, they can surpass algal microflora (Tomaselli and Giovannetti, 1993). Of the prevailing cyanobacterial genera, principally the filamentous heterocystous forms utilize atmospheric nitrogen under a broad range of conditions. Thus the present interest is the practice of soil inoculation with N_2-fixing cyanobacteria for nitrogen enrichment of soil ecosystem and improve the fertility status of the soil (Venkataraman, 1961; Singh, 1978; Adhikary, 1998). Together with the other earlier reports (Belnap, 2001; Büdel, 2001) the present work showed that cyanobacteria are ubiquitous component of soil microflora. They are cosmopolitan in distribution as they require only little moisture and diffused light for their growth. However, being photo autotrophic in nature, they had restricted distribution along the soil profile. In general, in fallow fields they are generally found at about 2-3 cm depth. In light and sandy soils some unicellular algal forms like *Chlorella, Chlamydomonas, Chlorococcum* and Diatoms may be carried to the deeper layers by the percolating water. Such micro-stratification of algal and cyanobacterial population has also been reported earlier (Friedmann and Galun, 1974). Over a period of time, these forms have been seen resulting in the formation of soil crusts those are sufficiently thick and rich in organic matter to support growth of plants of next higher order of complexity. In cultivated soils, cyanobacteria have been found up to 20 cm depth because of turning of soil during ploughing. Occurrence of *Microcoleus vaginatus* in mine waste burden soils in greater abundance even up to 20 cm depth might be a reason of the continuous mixing of mine burden soils during dumping. Similarly in sandy soils of Ramchandi certain cyanobacteria also occur even at deeper layers up to 20 cm might be due to their migration along with water percolation as the soils are usually porous in the region.

The texture and chemical nature of soil also known to influence the cyanobacterial distribution. Since high moisture content and water holding capacity are the key factors to encourage cyanobacterial growth, and as sandy soils are poor in this regard, their incidence is poor in these soils in comparision to the mine waste soils of Damonjodi. Greater predominance of cyanobacteria was also observed when there was an increase in availability of nutrients and at alkaline pH which was also found in the present work.

A number of reports have shown that cyanobacterial growth greately influences the physical and chemical nature of soil. Nutrition independence for carbon in general, and for both carbon and nitrogen in the case of N_2-fixing cyanobacteria make these an important component of soil microflora. Cyanobacteria add substantial quantities of organic matter through primary production. Their growth not only narrows the C:N ratio making the soil more fertile but also increases the humus content of the soil. Further, they even enhance the absorption of soil organic matter, and can also reduce the effect of metals and other toxicants which are common components of mine waste areas.

Results of the present investigation showed that a diverse number of cyanobacteria mostly filamentous forms occurred on the top soil surface and few more can even penetrate up to few cm deep. Many of these organisms do posses thick sheath layers around their trichome and/or also produces copius mucilage which are carbohydrate in nature (Adhikary, 1998). The filamentous non-heterocystous forms like *Microcoleus vaginatus* and *Plectonema puteale* which occur in great abundance at Damonjodi mine waste soils and sandy soils of Ramchandi respectively form characteristic crusts, hence, are efficient soil binders that might be playing a very important role for stablization of soil. Further, the Stigonematales members like, *Westiellopsis* and *Fischerella* species occurred prominently in sandy soil with low nutrient availability might be playing an important role in N_2-fixation as these have been reported to be cosmopolitan in arid locations. Besides these, since many filamentous heterocystous cyanobacteria occurred at all these sampling sites suggest their role in N_2-fixation and improving the fertility status of the soils.

Acknowledgements

We thank the Ministry of Environment and Forests, Govt. of India for financial support through the AICOPTAX-algae project to carry out this investigation. One of us (PT) is grateful to the Govt. of Orissa, Education Dept. for grant of study leave for two years and to the Head of the P.G. Department of Botany, Utkal University for providing laboratory facilities.

References

Adhikary SP (1998). Polysaccharides from mucilaginous envelope layers of cyanobacteria and their ecological significance. *J Sci Ind Res,* 57: 454–466.

Anderson DC, Harper KJ and Rushforth SR (1982). Recovery of cryptogamic soil crusts from grazing on Utah winter ranges. *J Range Manag,* 35: 355–359.

Belnap J (1996). Soil surface disturbances in cold deserts: effects on nitrogenase activity in cyanobacterial lichen soil crusts. *Biol Fertil Soils,* 23: 362–367.

Belnap J (2001). Comparative structure of physical and biological soil crusts. In: *Biological Soil Crusts: Structure, Function and Management,* (Eds) Belnap J and Lange OL. Springer-Verlag, Berlin, pp 177–191

Belnap J and Eldridge DJ (2001). Disturbance recovery of biological soil crusts. In: *Biological Soil Crusts: Structure, Function and Management,* (Eds) Belnap J and Lange OL. Springer-Verlag, Berlin, pp 363–384.

Büdel B (2001). Comparative biography of soil crust biota. *Ecological Studs,* 150: 141–152.

Chapman VJ and Chapman DJ (1973). *Soil Algae and Symbiosis.* Macmillan, London, pp 381–387.

Desikachary TV (1959). *Cyanophyta.* ICAR Monograph on Algae, New Delhi, p 686.

Dor I and Danin A (1996). Cyanobacterial desert crusts in the dead sea valley. *Israel Algol Studs,* 83: 197–206.

Friedmann EI and Galun M (1974). Desert algae, lichens and fungi. In: *Desert Biology,* (Ed) Brown GW Jr. Academic Press, New York, pp 165–212.

Hoyle BI and Arthur EI (2000). Biotransformation of pesticides in saturated zone materials. *Hydrogeology J,* 8: 89–103.

Johansen JR, St. Clari LL, Webb BL and Nebeker GT (1984). Recovery patterns of cryptogamic soil crusts in desert range-lands following fire disturbance. *Biologist,* 87: 238–243.

Kaushik BD (1987). *Laboratory Methods for Blue-green Algae.* Associated Publishing Company, New Delhi, pp 29–38.

Kubeckova K, Johansen JR, Warren SD and Sparks R (2003). Development of immobilized cyanobacterial amendments for reclamation of microbiotic soil crusts. *Algol Studs,* 109: 341–362.

Mult HC and Lange OL (2004). When small organisms have a big effect. *German Research Life Sciences,* 1: 16–20.

Rippka R, Deruelles J, Waterbury JB, Herdman MA and Stainer RY (1979). Generic assignments, strain histories and properties of pure culture of cyanobacteria. *J Gen Microbiol,* 111: 1–61.

Roger PA and Reynaud PA (1976). Dynamics of the algal populations during a culture cycle in a Sanegal rice field. *Rev Ecol Biol Soil,* 13: 545–560.

Singh PK (1978). Nitrogen economy of rice soils in relation to nitrogen fixation by blue-green algae and *Azolla.* Report-Central Rice Research Institute, Cuttack.

Tirkey J and Adhikary SP (2005). Cyanobacteria in biological soil crusts of India. *Curr Sci,* 89: 515–521.

Tomaselli L and Giovannetti L (1993). Survival of diazotrophic cyanobacteria in soil. *World J Microbiol Biotech,* 9: 113–116.

Trainor FR and Gladych R (1995). Survival of algae in a desiccated soil: A 35 year study. *Phycologia*, 34: 191–192.

Venkataraman GS (1961). The role of blue-green algae in agriculture. *Sci Cult*, 27: 9–13.

Zhou J, Via B, Huang H, Palumbo AV and Tiedie MJ (2004). Microbiol diversity and heterogeneity in sandy subsurface soils. *Appl Environ Microbiol*, 70: 1723–1734.

Soil Microflora, 2009 *Pages* **21–29**
Editor: **Rajan Kumar Gupta, Mukesh Kumar & Deepak Vyas**
Published by: **DAYA PUBLISHING HOUSE, NEW DELHI**

Chapter 3

Significance of Soil Microorganisms in Sustainable Agriculture

Shalini Singh[1], Rachana Srivastava[1] and Y.V. Singh[2]

[1]Institute of Bioengineering and Biological Science, S-19/54, Varuna Bridge, Varanasi – 221 002, U.P.
[2]Centre for Conservation and Utilization of Blue Green Algae, Indian Agricultural Research Institute, New Delhi
E-mail: yvsingh63@yahoo.co.in yvsingh_algal@iari.res.in

ABSTRACT

Agriculture in a broad sense, is not an enterprise which leaves everything to nature without intervention. Rather it is a human activity in which the farmer attempts to integrate certain agro-ecological factors and production inputs for optimum crop and livestock production. Thus, it is reasonable to assume that farmers should be interested in ways and means of controlling beneficial soil microorganisms as an important component of the agricultural environment. Soil microorganisms are very important as almost every chemical transformation taking place in soil involves active contributions from soil microorganisms and they play an active role in soil fertility as a result of their involvement in the cycle of nutrients like carbon and nitrogen, which are required for plant growth. However, in most cases, soil microorganisms, beneficial or harmful, have often been controlled advantageously when crops in various agro-ecological zones are grown and cultivated in proper sequence (*i.e.*, crop rotations) and without the use of pesticides. This would explain why scientists have long been interested in the use of beneficial microorganisms as soil and plant inoculants to shift the microbiological equilibrium in a way that enhances soil quality and the yield and quality of crops.

Keywords: *Agricultural environment, Microbial inoculant, Microflora, microorganisms, Cyanobacteria.*

Introduction

The principal property of the soil fertility is determined by biological factors, mainly by microorganisms. The development of life in soil endows it with the property of fertility. "The notion of soil is inseparable from the notion of the development of living organisms in it". Soil is created by microorganisms. "Were this life dead or stopped, the former soil would become an object of geology" (Vining, 1990). Along with bacteria, a very large number of fungi, actinomycetes, algae, ultramicrobes, phages, protozoa, insects, worms and other living creatures inhabit the soil. Soil microorganisms are responsible for the decomposition of the organic matter entering the soil (*e.g.* plant litter) and therefore in the recycling of nutrients in soil. Certain soil microorganisms such as mycorrhizal fungi can also increase the availability of mineral nutrients (*e.g.* phosphorus) to plants. Other soil microorganisms can increase the amount of nutrients present in the soil. For instance, nitrogen-fixing bacteria can transform nitrogen gas present in the soil atmosphere into soluble nitrogenous compounds that plant roots can utilize for growth. Boussingault, showed that legume could obtain nitrogen from air when grown in the soil which was not heated. Fifty year later, a Dutch Scientist, Beijerinck, isolated bacteria from nodules of legume roots. These microorganisms, which improve the fertility status of the soil and contribute to plant growth, have been termed 'biofertilizers' and are receiving increased attention for use as microbial inoculants in agriculture. Similarly, other soil microorganisms have been found to produce compounds (such as vitamins and plant hormones) that can improve plant health and contribute to higher crop yield. These microorganisms (called 'phytostimulators') are currently studied for possible use as microbial inoculants to improve crop yield.

In contrast to these beneficial soil microorganisms, other soil microorganisms are pathogenic to plants and may cause considerable damage to crops. Large numbers of pathogenic microorganisms are routinely found in the soil and many of them can infect the plant through the roots and microorganisms present in the soil are antagonistic to these pathogens and can prevent the infection of crop plants. Antagonism against plant pathogens usually involves competition for nutrients and/or production of inhibitory compounds such as secondary metabolites (antimicrobial metabolites and antibiotics) and extracellular enzymes. Other soil microorganisms produce compounds that stimulate the natural defence mechanisms of the plant and improve its resistance to pathogens.

Classification of Soils Based on the Functions of Microorganisms

According to their indigenous microflora soils can be characterized into putrefactive, fermentative, synthetic and zymogenic reactions. Classification of soils based on the activities and functions of their predominant microorganisms is presented below:

Disease-Inducing Soils

In this type of soil, plant pathogenic microorganisms comprises 5 to 20 per cent of the total microflora if fresh organic matter with a high nitrogen content is applied to such a soil. Such soils tend to cause frequent infestations of disease organisms, and harmful insects. Thus, the application of fresh organic matter to these soils is often harmful to crops. Probably more than 90 per cent of the agricultural land devoted to crop production worldwide can be classified as having disease-inducing soil. Such soils generally have poor physical properties, and large amounts of energy are lost as "greenhouse" gases, particularly in the case of rice fields.

Disease-Suppressive Soils

These soils generally have excellent physical properties and readily, form water-stable aggregates and are well-aerated, and have a high permeability to both air and water. The microflora of disease-suppressive soils is usually dominated by antagonistic microorganisms that produce copious amounts of antibiotics. These include fungi of the genera *Penicillium*, *Trichoderma*, *Aspergillus*, and *Actinomycetes* of the genus Streptomyces. The antibiotics they produce can have biostatic and biocidal effects on soil-borne plant pathogens, including *Fusarium* which would have an incidence in these soils of less than 5 percent. Crops planted in these soils are rarely affected by diseases or insect pests. Crop yields in the disease-suppressive soils are often slightly lower than those in synthetic soils. Highly acceptable crop yields are obtained whenever a soil has a predominance of both disease-suppressive and synthetic microorganisms (Higa, 1991).

Zymogenic Soils

These soils are dominated by a microflora that can perform the breakdown of complex organic molecules into simple organic substances and inorganic materials. The organisms can be either obligate or facultative anaerobes. Such fermentation-producing microorganisms often comprise the microflora of various organic materials, *i.e.*, crop residues, animal manures, green manures and municipal wastes including composts. After these amendments are applied to the soil, their number: and fermentative activities can increase dramatically and overwhelm the indigenous soil microflora for an indefinite period. Based on the presence of microorganisms, the soil can be classified as a zymogenic soil which is generally characterized by (*a*) pleasant, fermentative odors, (*b*) favorable soil physical properties, (*c*) large amounts of inorganic nutrients, amino acids, carbohydrates, vitamins and other bioactive substances which can directly or indirectly enhance the growth, yield and quality of crops, (*d*) low occupancy of *Fusarium* fungi which is usually less than 5 percent, and e) low production of greenhouse gases (Higa, 1994).

Synthetic Soils

These soils contain significant populations of microorganisms which are able to fix atmospheric nitrogen and carbon dioxide into complex molecules such as amino acids, proteins and carbohydrates and such microorganisms include photosynthetic bacteria which perform incomplete photosynthesis anaerobically, certain Phycomycetes (fungi that resemble algae), and both green algae and blue-green algae which function aerobically. All of these are photosynthetic organisms that fix atmospheric nitrogen. These soils have a low *Fusarium* occupancy and they are often of the disease-suppressive type. The production of gases from fields where synthetic soils are present is minimal, even for flooded rice (Parr *et al.*, 1994)

Application of Beneficial and Effective Microorganisms

The misuse and excessive use of chemical fertilizers and pesticides have often adversely affected the environment and created many (*a*) food safety and quality and (*b*) human and animal health problems. Consequently, there has been a growing interest in nature farming and organic agriculture by consumers and environmentalists as possible alternatives to chemical-based, conventional agriculture.

Agricultural systems which conform to the principles of natural ecosystems are now receiving a great deal of attention in both developed and developing countries. New concepts such as alternative agriculture, sustainable agriculture, soil quality, integrated pest management, integrated nutrient management and even beneficial microorganisms are being explored by the agricultural research establishment (Reganold *et al.*, 1990; Parr *et al.*, 1992). Although these concepts and associated methodologies hold considerable promise, they also have limitations. For example, the main limitation in using microbial inoculants is the problem of reproducibility and lack of consistent results.

Unfortunately certain microbial cultures have been promoted by their suppliers as being effective for controlling a wide range of soil-borne plant diseases and some suppliers have suggested that their particular microbial inoculant is a kin to a pesticide that would suppress the general soil microbial population while increasing the population of a specific beneficial microorganism. One might speculate that if all of the microbial cultures and inoculants that are available as marketed products were used some degree of success might be achieved because of the increased diversity of the soil microflora and stability that is associated with mixed cultures. While this, of course, is a hypothetical example, the fact remains that there is a greater likelihood of controlling the soil microflora by introducing mixed, compatible cultures rather than single pure cultures (Higa, 1991).

Even so, the use of mixed cultures in this approach has been criticized because it is difficult to demonstrate conclusively which microorganisms are responsible for the observed effects, how the introduced microorganisms interact with the indigenous species, and how these new associations affect the soil/plant environment. Thus, the use of mixed cultures of beneficial microorganisms as soil inoculants to enhance the growth, health, yield, and quality of crops has not gained widespread acceptance by the agricultural research establishment because conclusive scientific proof is often lacking.

The use of mixed cultures of beneficial microorganisms as soil inoculants is based on the principles of natural ecosystems which are sustained by their constituents; that is, by the quality and quantity of their inhabitants and specific ecological parameters, *i.e.*, the greater the diversity and number of the inhabitants, the higher the order of their interaction and the more stable the ecosystem. The mixed culture approach is simply an effort to apply these principles to natural systems such as agricultural soils, and to shift the microbiological equilibrium in favor of increased plant growth, production and protection (Parr *et al.*, 1994).

It is important to recognize that soils can vary tremendously as to their types and numbers of microorganisms. These can be both beneficial and harmful to plants and often the predominance of either one depends on the cultural and management practices that are applied. It should also be emphasized that most fertile and productive soils have a high content of organic matter and, generally, have large, populations of highly diverse microorganisms (*i.e.*, both species and genetic diversity). Such soils will also usually have a wide ratio of beneficial to harmful microorganisms (Higa and Wididana, 1991).

Types and Functions of Beneficial Microorganisms in Agriculture

Microorganisms we find that more life is living below the surface than above in most environments. Soil are put into six groups; earthworms, nematodes, arthropods, bacteria, fungi and protozoa with each having their own function. They do play an important role in plant and water. The cycle of nutrients growing through our environment are mainly driven by these microorganisms. They maintain: Decomposition, Mineralization, Nitrogen cycling, Storage and release of nutrients, Carbon cycling and take the pollutants out of the water before it releases underground or surface water.

Actinomycetes

Actinomycetes can be found in surface soil, as well as in lower horizons to some depth. They are numerous and are second only to the bacteria in abundance (Alexander, 1961; Baecker *et al.*, 1983). Actinomycetes are counted in hundreds of thousands and millions in one gram, algae–in thousands and tens of thousands. The total mass of these organisms in the upper surface layer of the soil may amount to two-three tons in one hectare.This microorganisms are responsible for distinctive scent of freshly exposed, moist soil, preferred neutral to alkaline soil; high oxygen requirement; prevelant in dry region and also release carbon, nitrogen and ammonia during the decomposition of organic matter. They help for humus, associate with non-leguminous plants like bitter brush to fix nitrogen and make it available to other plants in the area.

The actinomycetes show characteristics of both bacteria and fungi. Growth of actinomycetes is hyphal and they may form a mycelium characteristic of the fungi (Alexander, 1961). In addition to hyphal growth, actinomycetes are capable of producing an aerial mycelium and conidia characteristic of fungal species. When they are grown in pure liquid culture the turbidity usually associated with bacterial growth is absent, and in many species the growth rate is not exponential, as it is in bacteria (Alexander, 1961). However, unlike fungi, actinomycetes are prokaryotic and in this respect are more closely related to the bacteria. The controversy over the taxonomic placement of this group has been similar to that of the cyanobacteria (formerly the blue-green algae). Actinomycetes are rather difficult to identify to the species level and most of the interest in them over the years has been in obtaining antibiotic compounds. Thus, the search for novel species, and the mass screening of species for secondary metabolites has been a major focus of pharmaceutical companies worldwide. Goodfellow and Cross (1974) have suggested that actinomycetes seem to form a distinctive part of forest liter, playing a small but vital role in nutrient cycling. Baecker and King (1978) showed that 20 different types of actinomycetes were able to break down cellulose when grown on a cellulose containing medium. As actinomycetes show hyphal growth, they are probably well adapted to penetrating insoluble substrates such as wood.

Bacteria

Microorganisms found all over the world and even found down the earth as far as one mile. It thrives under moist condition but found near plant roots as an important food source. The ability to fix molecular nitrogen is a widespread characteristic of prokaryotic cells, being established among various groups of bacteria including some archaea (Belay *et al*, 1984). The distribution of BNF among archaea and eubacteria indicates that nitrogen fixation is an ancient innovation, which developed early in the evolution of microbial life on earth. Within the eubacteria, nitrogen fixation has been described for members of the proteobacteria, cyanobacteria, actinobacteria, spirochaetes, clostridiales, purple-sulfur (Chromatiales) and green-sulfur (Chlorobiales) bacteria.

Nitrogen-fixing microorganisms are highly diverse, ranging from "free-living" autotrophic bacteria of the genus Azotobacter to symbiotic, heterotrophic bacteria of the genus *Rhizobium*, and blue-green algae (now mainly classified as blue-green bacteria), all of which function aerobically. Photosynthetic microorganisms fix atmospheric carbon dioxide in a manner similar to that of green plants (Dalton and Kramer, 2006). They are also highly diverse, ranging from blue-green algae to green algae that perform complete photosynthesis aerobically to photosynthetic bacteria which perform incomplete photosynthesis anaerobically. The photosynthetic bacteria, which perform incomplete photosynthesis anaerobically, are highly desirable, beneficial soil microorganisms because they are able to detoxify soils by transforming reduced, putrefactive substances such as hydrogen sulfide into useful substrates. This helps to ensure efficient utilization of organic matter and to improve soil fertility. Photosynthesis involves the photo-catalyzed splitting of water which yields molecular oxygen as a by-product. Thus, these microorganisms help to provide a vital source of oxygen to plant roots. Reduced compounds such as methane and hydrogen sulfide are often produced when organic materials are decomposed under anaerobic conditions. These compounds are toxic and can greatly suppress the activities of nitrogen-fixing microorganisms.

Earthworm

Earthworms are the region for huge industries today. People are raising and selling earthworm all over the world, especially whole sale earthworms. The work of the earths in the soil is what earthworms it. Producing casting that the earth equivalent to the compost as an organic fertilizer and soil supplements. Earthworms shape soil quality by consuming plant and organic matter and converting it to humus. Earthworm "castings," the product of their digestion, are richer and less acidic than the surrounding soil. Earthworms excrete calcium carbonate which lowers soil acidity. Earthworm castings also contain more calcium, nitrogen, phosphate, and potassium than other soil. Earthworms also contribute to soil quality by burrowing, which helps to loosen and aerate the soil.

Although the abundance of earthworms may be affected by relatively few turf pesticides, earthworm distribution and behavior may be altered to a greater degree. Litter and surface soils treated with certain pesticides have a repellent effect on earthworms, and this reduces the breakdown and incorporation of organic matter into the subsurface horizons. Benomyl and carbendazim are particularly lethal to earthworms and also exhibit this repellent effect, which results in the avoidance of feeding in treated soils. Consequences include reduction in the amount of available nutrients in the root zone, decreased porosity and aeration of the soil, decreased water-holding capacity, and poor drainage (Crump, 1969).

Fungi

Basically two types of fungi-mycorrhizal and normal. Fungi thrived in well drained, neutral to acidic aerated soil. Normal fungi help, decompose the organic matter in litter and soil, but play less of an overall role. Because fungi are everywhere, agricultural practices have profound effects on their growth, distribution and survival. The impact may be positive or negative. The amendment of soil with organic material will enhance the activity of decomposer fungi in the soil. Some of these fungi may also be antagonistic to fungal pathogens of plants and lead to suppression of disease. In any event, the community structure of fungi will be changed through this enrichment process. However, excessive deep plowing may separate the organic material from the fungal decomposers which occur in the top few centimeters of the soil. This tillage will lead to a general decline in fungal biomass. The addition of pesticides to the soil will also affect the survival of selected fungi (Harley and Smith, 1983).

Mycorrhizal fungi help develop healthy root system by growing on plant root. The fungus is actually a network of filaments that grow in and around the plant root cells, forming a mass that extents considerably beyond the plants root systems. These essentially extents the plant's reach to water and nutrients, allowing it to utilize more of the soil resources. Mycorrhizae are mutualistc associations between plant roots and fungi. These beneficial symbioses are ubiquitous in nature and almost all plant species have some form of mycorrhizal association with fungi. Herbaceous and tree species, both deciduous and coniferous, are receptive to infection by mycorrhizal fungi (Jones, 1993). The endomycorrhizal fungi generally associated with the roots of agricultural crops are in the Class Zygomycetes to which the common black bread mold belongs. However, these fungi are obligate symbionts and cannot be cultivated outside the living roots of plants. Their colonization is internal to the root and cannot be seen without staining and microscopy. The common genera are Glomus and Gigaspora producing large, distinctive azygospores that can be wet sieved from the soil. These spore germinate in the presence on plant roots and infect the outer cortical cells. However, the cell is not killed and although the plant cell wall is penetrated the cell membrane is not disrupted. The endomycorrhizal fungus produces a highly branched hyphal structure called an arbuscule within the plant cell by invaginating its cell membrane. This infection creates an absorptive structure with a very high surface area of transfer for nutrients between the plant and the fungus (Metting, 1993)

Nematodes

Four of every five multicellular animals on the planet are nematodes. Many species are well known as important and devastating parasites of humans, domestic animals and plants. However, most species are not pests; they occupy any niche that provides an available source of organic carbon in marine, freshwater and terrestrial environments. There may be 50 different species of nematodes in a handful of soil and millions of individuals can occupy 1 m^2. Of the nematodes in soil that do not feed on higher plants, some feed on fungi or bacteria; others are carnivores or omnivores.

Soil nematode communities may also provide useful indicators of soil condition. Nematodes vary in sensitivity to pollutants and environmental disturbance. Nematodes respond rapidly to disturbance and enrichment of their environment; increased microbial activity in soil leads to changes in the proportion of opportunistic bacterial feeders in a community. Over time the enrichment opportunists are followed by more general opportunists which include fungal feeders and different genera of bacterial feeders (Bongers and Ferris, 1999).

Bacterial-feeding nematodes have a higher carbon:nitrogen (C:N) ratio (±5.9) than their substrate (±4.1), so that in consuming bacteria they take in more N than necessary for their body structure. The C:N ratio of fungal-feeding nematodes is closer to that of their food source. However, for nematodes of both feeding habits, a considerable proportion of the C consumed is used in respiration (perhaps 40 per cent of the food intake (Ingham *et al.*, 1985). The N associated with respired C that is in excess of structural needs is also excreted. The excreted N is available in the soil solution for uptake by plants and by microbes. Because microbivorous nematodes exhibit a wide range of metabolic rates and behavioral attributes, the contribution of individual species to nitrogen cycling and soil fertility may vary considerably.

Arthopods

Arthropods that live in the soil perform a number of critical soil functions. Soil is made mostly of the feces of arthropods. It includes bugs-like insect (ants), crustacean (sow-bugs) arachids (spiders), and myriapods (centipeds). Their functions are widely:

1. To stir and churn the soil.
2. To stir organic matter so as to assist other microorganism in the decomposition.
3. To help to distribute beneficial microorganisms in the soil.
4. Through consumption, digestion and excretion of organic matter arthropods help improve soil, structure and change nutrients into forms available to plants.
5. Help regulate the population of other microorganisms.

A number of beneficial arthropods show at least some level of adaptation to agroecosystems. These include predatory Hymenoptera (ants and wasps), Coleoptera (carabid, coccinellid, and staphylinid beetles), Heteroptera (pirate, assassin, and ambush bugs), Neuroptera (lacewings), Diptera (syrphid and chamaemyiid flies) as well as mites and spiders. The major advantage of natural enemies is suppression of phytophagous insect pests at little or no cost and minimal harm to humans or the environment. A disadvantage, from a producer's perspective, is that they are better suited for long-term suppression of pest populations than as a reactive control strategy when outbreaks occur. The majority of soil-dwelling arthropod species probably feeds principally upon the fungi growing on the surface of fecal pellets (Anderson, 1975). As they graze on the hyphae they often physically abrade the pellets themselves, exposing additional nutrients to microbial attack. Each time an annellid worm or arthropod ingests solid food it secretes copious carbohydrate lubricants (mucus). These carbohydrates do act as energy sources for intestinal microbes, but in most instances the mucus ultimately surrounds and embeds the faecal material as well (Lee, 1985). When the content of the faecal material is largely inorganic, as it is for most deep-dwelling earthworms, this input creates a branching "drilosphere" throughout the soil of additional resources and subsequent microbial growth (Barois *et al.*, 1993).

Conclusion

Microorganisms can help to improve and maintain the soil chemical and physical properties. The proper and regular addition of organic amendments are often an important part of any strategy to exercise such control. Previous efforts to significantly change the indigenous microflora of a soil by introducing single cultures of extrinsic microorganisms have largely been unsuccessful. Even when a beneficial microorganism is isolated from a soil, cultured in the laboratory, and reinoculated into the same soil at a very high population, it is immediately subject to competitive and antagonistic effects from the indigenous soil microflora and its numbers soon decline. Thus, the probability of shifting the "microbiological equilibrium" of a soil and controlling it to favour the growth, yield and health of crops is much greater if mixed cultures of beneficial and effective microorganisms are introduced that are physiologically and ecologically compatible with one another. When these mixed cultures become established their individual beneficial effects are often magnified in a synergistic manner.

References

Alexander M (1961). *Introduction to Soil Microbiology*. John Wiley and Sons, Inc, USA, 472 pp.

Anderson JM (1975). Succession, diversity and trophic relationships of some soil animals in decomposing leaf litter. *Journal of Animal Ecology*, 44: 475–495.

Baecker, AAW, Dyker, RMP and King, B (1983). The role of actinomycetes in the biodeterioration of wood. In: *Biodeterioration: Papers Presented at the 5th International Biodeterioration Symposium Aberdeen* (Eds) TA Oxley and SB Barry. John Wiley and Sons, Chichester, 64–74 pp.

Baecker, AAW and King, B (1978). Decay of wood by actinomycetes. In: *Biodeterioration: The Proceedings of the Fourth International Biodeterioration Symposium*, (Eds) TA Oxley, G Becker and D Allsopp. Pitman Publishing, Ltd, Berlin, London, 53–57 pp.

Barois I, Villemin G, Lavelle P and Toutain F (1993). Transformation of the soil structure through *Pontoscolex corethrurus* (Oligochaeta) intestinal tract. *Geoderma*, 56: 57–66.

Belay N, Sparling R, Daniels L (1984). Dinitrogen fixation by a thermophilic methanogenic bacterium. *Nature*, 312: 286–288.

Bongers T and Ferris H (1999). Nematode community structure as a bioindicator in environmental monitoring. *Trends in Evolution and Ecology*, 2: 18–23.

Crump D. R. (1969). Earthworms: A profitable investment. *NZ J Agric*, 119(2): 84–85

Dalton DA and Kramer S (2006). Nitrogen-fixing bacteria in non-legumes. In: *Plant-Associated Bacteria.* Springer, Netherlands, 105–130 pp.

Goodfellow M and Cross T (1974). Actinomycetes. In: *Biology of Plant Litter Decomposition*, Vol 2 (Eds) CH Dickinson and GJF Pugh. Academic Press, New York, 269–302 pp.

Harley JL and Smith SE (1983). *Mycorrhizal Symbiosis*. Academic Press, London.

Higa T (1991). Effective microorganisms: A biotechnology for mankind. In: *Proceedings of the First International Conference on Kyusei Nature Farming*, (Eds) JF Parr, SB Hornick and CE Whitman. U.S. Department of Agriculture, Washington DC, USA, 8–14 pp.

Higa T and Wididana GN (1991). The concept and theories of effective microorganisms. In: *Proceedings of the First International Conference on Kyusei Nature Farming*, (Eds) JF Parr, SB Hornick and CE Whitman. US Department of Agriculture, Washington DC, USA, 118–124 pp.

Higa T.(1994). Effective microorganisms: A new dimension for nature farming. In: *Proceedings of the Second International Conference on Kyusei Nature Farming*, (Eds) JF Parr, SB Hornick and ME Simpson. US Department of Agriculture, Washington DC, USA, 20–22 pp.

Ingham RE, Trofymow JA, Ingham ER and Coleman DC (1985). Interactions of bacteria, fungi and their nematode grazers: Effects on nutrient cycling and plant growth. *Ecological Monographs*, 55: 119–140.

Jones DG (1993). *Exploitation of Microorganisms*. Chapman and Hall, London.

Lee KE (1985). *Earthworms: Their Ecology and Relationships with Soils and Land Use*. Academic Press, Sydney, Australia, 411 pp.

Metting FB (1993). *Soil Microbial Ecology: Applications in Agricultural and Environmental Management*. Marcel Dekker Inc, New York.

Parr JF, Papendick RI, Hornick SB and Meyer RE (1992). Soil quality: Attributes and relationship to alternative and sustainable agriculture. *Amer J Alternative Agric*, 7: 5–11.

Parr JF, Hornick SB and Kaufman DD (1994). Use of microbial Inoculants and organic fertilizers in agricultural production. In: *Proceedings of the International Seminar on the Use of Microbial and Organic Fertilizers in Agricultural Production*. Food and Fertilizer Technology Center, Taipei, Taiwan.

Reganold JP, Papendick RI and Parr JF (1990). Sustainable agriculture. *Scientific American*, 262(6): 112–120.

Vining L (1990). Functions of secondary metabolites. *Ann Rev Microbiol*, 44: 395–427.

Soil Microflora, 2009 *Pages* ***30–39***
Editor: **Rajan Kumar Gupta, Mukesh Kumar & Deepak Vyas**
Published by: **DAYA PUBLISHING HOUSE, NEW DELHI**

Chapter 4

Impact of Herbicides on Cyanobacterial Flora of Rice Fields

Mihir Kumar Das
P.G. Department of Botany, G.M. College (Autonomous), Sambalpur – 768 004, Orissa
E-mail: mihir_gmc@rediff.com

ABSTRACT

The extensive use of herbicides has become imperative for achieving higher production, but at the same time, indiscriminate use of these herbicides is known to affect the cyanobacterial population in the rice fields. Cyanobacteria are known to contribute to the maintenance of soil fertility and are consequently considered to be an important component of soil biota. These organisms are more sensitive to herbicides than other groups of microorganisms. It is therefore, essential that regular monitoring be undertaken to evaluate the influence of herbicides on cyanobacteria. In the present review, range of herbicide-induced biological responses, induction of biochemical changes in the organisms, the relative tolerance of cyanobacteria to herbicides etc. are highlighted.

Keywords: *Cyanobacteria, Herbicides, Rice field, Toxicity.*

Introduction

The global energy crisis and dwindling mineral oil reserves have widened the gap between supply and demand to nitrogenous fertilizers. An introduction of fertilizer responsive high yielding crop varieties has increased the demand of this important crop nutrient. This has resulted in further burden on small and marginal farmers, especially in developing countries. This has become necessary to look for alternative sources to meet at least a part of nitrogen requirement for crop production. In

India, rice is cultivated on about 40 million hectares of area, which constitutes about 37-40 per cent of total area under cereals. Though, rice cultivation is an age-old practice in our country, the average production is only about 1.7 t/ha. This is because more than 85 per cent of the total area of rice is owned by small and marginal farmers. These farmers cannot afford to use various inputs needed to harvest maximum yield of rice. They do not get full returns/unit nitrogenous fertilizers in their fields because of high nitrogen losses in the ecosystem.

Cyanobacteria

The past few decades have widened remarkable advancement in harnessing some of the potentially useful microorganisms to build up the fertility of the soil to increase the crop yield. In recent year cyanobacteria (blue-green-algae), a group of soil microorganisms have been shown to be agriculturally important, particularly in topical rice field soils. This is because of capacity of some of the algae to synthesize organic substances and also to fix atmospheric nitrogen. Submerged conditions of a rice field produce congenial habitat for cyanobacteria where they form most different system providing biologically fixed nitrogen to the crop. The importance of cyanobacteria was recognized by De (1939) who reported that these microorganisms are responsible for natural fertility of tropical rice field soil. Since then series of reports have been appeared emphasizing their role in nitrogen cycle in general and rice field on particular. The propagation of cyanobacteria will not only enrich the nitrogen status of the soil by their fixation process, but also provide organic matter and biologically potent substances for plant growth. These algae form a living constituent of the soil microflora and continue their activity year after year. Besides they release oxygen for the paddy roots and increase soil phosphate. Some of them prevent the loss of soil ammonia and leaching out of nitrates by converting them into organic nitrogen. Cyanobacteria also produce surface humus after death and exert a solvent action on certain minerals–maintaining a reserve supply of elements in a semi-available form for higher plants either by secretion or upon death and decomposition. They have also been tremendously important in shaping the course of evolution and ecological change throughout earth's history. The oxygen atmosphere that we depend on was generated by numerous cyanobacteria during the Archaean and Proterozoic Eras. Many factors, namely light, temperature desiccation, rewetting, pathogens, antagonisms, grazers, p^H and nutritional status of the soil and water and toxicants determine the abundance of cyanobacteria in any ecosystem (Padhy, 1985). Pesticides are used throughout the globe on a large scale to combat mainly insects, fungi that destroy partially or fully crop plants and their products. Herbicides, nematocides, rodenticides etc. are also in use in different occasion in present day civilization.

Pesticide Pollution

Though pesticides were the miracle drugs for the plants during 1950s, doubts were raised concerning the hazardous effects of these chemicals on the environment from the early 1960s. Richael Carson was one of the first to become aware of the danger of pesticide pollution and in her famous book entitled "Silent Spring", she produced evidences of toxic effects of pesticides upon wild life and potential hazards to man (Misra and Mani, 1994). Despite increasing research efforts towards crop improvement by methods obviating the use of pesticides (Day,1987) agriculture remains heavily dependent on these agrochemicals (Gadkari, 1988). The soil being the ultimate sink on the rice fields harbors many beneficial Cyanobacteria, which are potentially susceptible to pesticides. Cyanobacteria are known to contribute to the maintenance of soil fertility (Singh, 1961; Venkatarman, 1972). It is therefore, imperative that regular monitoring be undertaken to evaluate the influence of pesticides on these organisms (Das and Adhikary, 2006).

Many works have been carried out in various laboratories of the world on the responses of pesticides/herbicides to cyanobacteria through *in vitro* determination and/or under *in vivo* condition. Most of the works relate to the determination of lethal doses of pesticides on different species/strain of cyanobacteria. The more recent studies have been more biochemical and physiological in nature than the earlier ones, departing from examination of pesticide-induced qualitative and quantitative fluctuations in soil microbial populations and concentrating more on factors such as the impairment of nitrogen-fixation (Kapoor and Sharma, 1980, Kaushik and Venkataraman, 1983). The reviews relating to the toxicity of insecticides (Das and Adhikary, 2006), heavy metals (Das, 2007), fungicides (Das, 2008), have been discussed earlier. However, the present review considers the impact of herbicides on cyanobacteria and factors affecting nitrogen fixation, pesticide metabolization, cyanobacteria biosensor for detection of herbicides, introduction of herbicide-resistance in cyanobacteria etc.

Herbicides

Chemicals, used for killing or inhibiting the growth of unwanted plants, are called herbicides. Selective herbicides kill specific targets while leaving the desired crops relatively unharmed. Some of these act by interfering with the growth of the weed and are often based on plant hormones. Herbicides used to clear waste ground are non-selective and kill all plant materials with which they come into contact. Some plants produce natural herbicides, such as the genus Juglans (walnuts). They are applied in total vegetation control programmes for maintenance of highways and rail roads. Smaller quantities are used in forestry, pasture systems and management of areas set aside as wildlife habitat. Herbicides are widely used in agriculture and in landscape turf management. In the U.S., they account for about 70 per cent of all agricultural pesticides used (Kellogg *et al.,* 2000). The first widely used herbicide was 2, 4–dichlorophenoxyacetic acid (2, 4-D). It is easy and inexpensive to manufacture and kills many broadleaf plants while leaving grasses largely unaffected. The low cost of 2, 4-D has led to continued usage today and it remains one of the most commonly used herbicides in the world. The 1950s saw the introduction of the triazine family of herbicides, which includes atrazine. It has become the greatest concern regarding ground water contamination. Atrazine does not break down readily (within a few weeks) after being applied to soils of above neutral p^H. Glyphosate frequently sold under the brand name Roundup was introduced in 1974 for non-selective weed control. Many modern chemical herbicides for agriculture are specifically formulated to decompose within a short period after application. However, during their stay in the crop fields, they affect a variety of soil inhabiting microflora, especially nitrogen-fixing cyanobacteria, thereby lowering the productivity of the soil. Many workers have studied the response of cyanobacteria to herbicides (Fitzgerald *et al.,* 1952; Venkataraman and Rajyalakshmi, 1971; Das, 1977; Wright *et al.,* 1977; Cullimore and Mc Cann, 1977; Allen *et al.,* 1983; Stratton, 1984; Singh and Datta, 2005) and many others.

Herbicides and Cyanobacteria

Many of the studies conducted with herbicides have focused on cyanobacteria, in particular those of paddy field soils. Sharma (1986), in a review of the pesticides effects on cyanobacteria in the paddy fields in India, refers to the need for understanding the effects of pesticides in general on cyanobacteria and draws attention to the enormous increase in the use of pesticides in modern agriculture. Growth of several non-nitrogen fixing cyanobacteria is inhibited by 2, 4-D, which is an aryloxyalkanoic acid herbicide. This has auxin-activity and is used mainly as a post-emergence chemical to control large weeds with many crops (Padhy, 1985). Many cyanobacteria were reported to tolerate high concentrations of 2, 4-D; the stimulation of growth at low concentrations was also noted (Mc Cann and Cullimore, 1979; Mishra and Tiwari, 1986). Fitzgerald *et al.,* 1952 found that the growth of bloom-forming *Microsystis aeruginosa* could not be controlled by 2, 4-D at 250 ppm. Growth of *Anacystis nidulans* was inhibited at levels between 90 and 100 ppm, but the level of 50 ppm supported its growth (Voight and Lynch, 1974). It tolerated 200 kg/ha 2, 4-D under field conditions (Venkataraman and Rajyalakshmi, 1971). Das and Singh, 1977 reported that a concentration of 10 µg per ml of 2, 4-D showed stimulation of growth and nitrogen fixation of bloom forming blue-green alga *Anabaenopsis raciborskii* and they could tolerate up to 800 µg per ml in liquid culture media with and without nitrate nitrogen and up to 9 µg ml on to agar plates. *Microcystis flos-aquae* tolerated up to 1200 ppm 2, 4-D and 1,500 ppm was reported to be completely lethal. The LD_{50} for *M. flos-aquae* was 400 ppm (Das, 1977). *M.aeruginosa*, collected from the temple tanks of Puri and Bhubaneswar, India could tolerate and grew even at 500 µg ml^{-1} of 2, 4-D, while 1000 µg ml^{-1} concentration was found to be sub-lethal and further increase of the herbicide dose was lethal to the organism (Swain *et al.*, 1994).

Hamdi *et al.*, 1970 treated cultures of *Tolypothrix tenuis* with 0.045, 0.45 and 4.5 ppm of 2, 4-D either at the time of culture inoculation or 10 days after inoculation. At both times there was an inhibition on the dry weight and nitrogen content and a stimulation of chlorophyll-a content. Under rice-field conditions, *T.tenuis* tolerated 200 kg/hectare of 2, 4-D (Venkataraman and Rajyalakshmi, 1971). This observation was further supported by Mishra and Tiwari (1986) who found that *Mastigocladus laminosus* and *T.tenuis* tolerated 5000 ppm 2, 4-D; the former organism actually exhibiting stimulation at this level (Khalil *et al.*, 1980). Under nitrogen fixing condition, *Nostoc linckia* was eliminated completely at 2000ppm 2, 4-D (Tiwari *et al.*, 1981). Mishra *et al.*, (1989) observed that 2, 4-D at 100,500 and 1000 ppm stimulated L methionine-DL-sulfoximine (MSX) induced photo-production of ammonia by *N.linckia*. Lethal levels of 2, 4-D to *Anabaena doliolum*, *Nostoc calcicola*, *N. linckia* and *Nostoc* sp. were found to vary from 1500-2000 ppm, but toxicity was alleviated by high p^H levels and the presence of certain carbon and nitrogen sources on the culture (Mishra and Pandey, 1989).

Many of the herbicides used in the studies are commonly used paddy field chemicals known to inhibit photosynthesis. The growth and cellular nitrogen level of *Anabaena doliolum* were reduced by concentrations of the uracils, isocil and terbacil at 0.01 per cent (100 ppm) (Kapoor and Sharma, 1980). In the same study, the benzoic acid herbicide, nitrofen was slightly less toxic than the uracils tested to *Anabaena doliolum*. A slight-reduction in growth and cellular nitrogen was observed at 100 ppm nitrofen. Monuron, however, was completely growth-inhibitory to *Nostoc muscorum* at 0.34 mM (Vaishampayan,1984a). Mutagenic activity was also observed in *Nostoc muscorum* using paraquat (Vaishampayan, 1984b). The effect of the anilide herbicide propanil on the same bacterium was observed by Singh and Tiwari (1988a) to depend on the size of the inoculum of the organism. When an inoculum of 2.0 x10^7 cells mL^{-1} was used, propanil at 5 ppm was lethal to *Nostoc muscorum*. This concentration was inhibitory but not lethal, when larger inocula (2.5 and 3.0 x 10^7 cells mL^{-1}) were used. They also observed that sub-lethal levels of propanil were found to cause an inhibition of photosynthetic oxygen evolution and a stimulation of respiratory oxygen uptake in both organisms (Singh and Tiwari, 1988b). This herbicide also inhibited the nitrogenase, nitrate reductase and glutamine synthetase activities.

The influence of photosynthesis inhibiting herbicides Goltix and Sencor on growth and nitrogenase activity of *Anabaena cylindrica* and *Nostoc muscorum* was studied(Gadkari, 1987). *A. cylindrica* was entirely inhibited even in the presence of low field concentrations (Sencor 10 ppm, Goltix 50 ppm). In contrast, Goltix (50 ppm and 100 ppm) and Sencor (10 ppm, 20 ppm, 50 ppm and 100 ppm) did not exert an inhibitory influence on growth and nitrogen activity of *N.muscorum*. Among the herbicides with activities not involving the disruption of photosynthesis butachlor and fluchloralin (cell division inhibitors) were examined for their effect on *Nostoc muscorum* and *Gloeocapsa* sp. (Singh and Tiwari 1988a, 1988b). The butachlor was used in the form of Machete (R) a commercial preparation containing 0.5 gmL^{-1} active ingredient and fluchloralin was in the form of Basalin (R) a commercial preparation with 0.48 gmL^{-1} active ingredient. Machete(R) at 6 ppm and Basalin (R) at 25 ppm were lethal to *Nostoc muscorum* when the inoculation size was 2.0 x 10^7 cells mL$^{-1.}$ Larger inocula were more tolerant of these levels. *Gloeocapsa* sp. (1.0 x 10^6 cells mL^{-1}) survived 2000 and 300 ppm Machete (R) and Basin (R) respectively (Singh and Tiwari, 1988a). Chinnaswamy and Patel (1983) observed the inhibition of growth of *Anabaena flos-aquae* by 10 and 25 ppm Basalin (R) but not by 1 ppm. Khalil *et al.* (1980) reported a high tolerance of *Mastigocladus laminosus* and *Tolypothrix tenuis* to Basalin (R). These cyanobacteria could grow at 100 and 50 ppm respectively. If we consider the influence of butachlor and fluchloralin on metabolic processes in *Nostoc muscorum* and *Gloeocapsa* sp., sub-lethal levels of both herbicides had little effect on photosynthetic oxygen evolution in either of the test cyanobacteria, but did cause a stimulation of respiratory oxygen uptake (Singh and Tiwari, 1988b).

Kashyap and Pandey (1982) observed that although 0.05 ppm of butachlor was stimulatory to *Anabaena doliolum*, 20 ppm were lethal to the same organism. A sub-lethal concentration of 5 ppm caused a decline in protein and phycobilin levels and an inhibition of heterocyst differentiation and nitrogen fixation. The stimulation observed at very low levels of butachlor was suggested to have been a result of the increased availability of nutrients brought about by herbicide–induced alterations in membrane permeability. Roychoudhury and Kaushik (1986), when examined the response of halotolerant strain of *Tolypothrix ceylonica* and *Scytonema circinnatum* to herbicides, Butachlor (Machete) and Stam F-34 found that growth and chlorophyll synthesis of these two cyanobacteria were markedly reduced at p^H 7 as compared to p^H 9, both in absence or presence of herbicides. In the absence of herbicides, *Scytonema* registered over *Tolypothrix* 33, 26 and over 15 percent increase in growth, chlorophyll and nitrogen fixation respectively. Butachlor and Stam F-34 inhibited the growth and chlorophyll synthesis equally, in both the strains of algae, however, nitrogen fixation was severely affected by stam F-34 as compared to machete. Up to 10 ppm of both the herbicides, growth and chlorophyll synthesis of algae were not affected, however, beyond 10 ppm, a marked inhibition was recorded.

Mishra and Pandey (1989), however, noted that *Anabaena doliolum, Nostoc calcicola, N. linckia* and *Nostoc* sp. were unable to tolerate butachlor at from 6–8 ppm. Butachlor and alachlor were noticed by Vaishampayan (1985) to have mutagenic activity in *Nostoc muscorum*. This organism could not servive at 80 ppm butachlor and produced only a few viable colonies on solid medium at 150 ppm alachlor (Vaishampayan, 1985). These two herbicides were reported by the author to cause rupture of the cell wall of *Nostoc muscorum*, followed by lysis, rather than eliciting any apparent physiological disorder.

Kolte and Goyal (1992), when applied butachlor (Machete), benthiocarb (Saturn), pandimethalin (Stomp) and oxadiazon (Romester) at half, full and double rate of recommended concentrations against the axenic cultures of *Anabaena khannae, Calothrix morchica, Nostoc calcicola* and *Tolypothrix limbata* observed differential effect on growth and nitrogen fixation. Out of these four weedicides, pandimethalin proved to be the only one which adversely affected the growth at 0.5 and 1.0 ppm concentrations. *Anabaena khannae* remained unaffected at the super-normal level of pandimethalin, but the growth of the other three algae was drastically reduced at this concentration. However, in *Nostoc calcicola*, this weedicide at sub-normal level appreciably increased the growth. Butachlor and oxadiazon were completely harmless to *Calothrix marchica* and *Tolypothrix limbata* even at super-normal concentrations. Growth and nitrogen fixation did not seem to be inter-linked phenomena in all the cases with respect to the effect of weedicides. In *Anabaena khannae*, pandimethalin significantly decreased nitrogen fixation but the growth remained unaffected even at super-normal level. Butachlor had no effect on growth of

Calothrix marchica and *Tolypothrix limbata*, but nitrogen fixation increased with the increasing concentration of the chemical. Only *Nostoc calcicola* showed a direct correlation with the growth and nitrogen fixation in the presence of all the weedicides (Kolte and Goyal, 1992). This study showed that these four weedicides do not interfere with the algal activity at their field application doses; rather some of them accelerated growth and nitrogen fixation even at super-normal levels.

Singh and Datta (2005) studied the effect of graded concentrations of four common rice field herbicides (Arozin, Butachlor Alachlor, 2, 4-D) on diazotrophic growth, macromolecular contents, heterocyst frequency and tolerance potentials of Ca-alginate immobilized cyanobacterial isolates *Nostoc punctiformae*, *N. calcicola*, *Anabaena variabilis*, *Gloeocapsa* sp., *Aphanocapsa* sp. and laboratory strain *N. muscorum* ISU and compared with free living cultures. Cyanobacterial isolates showed progressive inhibition of growth with increasing dosage of herbicides in both free and immobilized status. There were significant differences in the relative toxicity of four herbicides. Arozin proved to be more growth toxic in comparison to alachlor, butachlor and 2, 4-D. Growth performance of the immobilized cyanobacterial isolates under herbicide stress showed a similar diazotrophic growth pattern to free cells with no difference in lethal and sub-lethal doses. However, at lethal concentrations of herbicides, the immobilized cells exhibited prolonged survivability of 14-16 days as compared to their free-living counterparts (8-12 days). The decline in growth, macromolecular contents and heterocyst frequency was found to be similar on both the states in graded dosages of herbicides. The organism, *A. variabilis* showed maximum natural tolerance towards all the four herbicides tested. Evidently immobilization by ca-alginate seems to provide protection to the diazotrophic cyanobacterial inoculants to a certain extent against the growth-toxic action of herbicides.

Pandey and Kashyap (2007) reported that the growth of *Anacystis nidulans*, *Nostoc muscorum* and *Anabaena doliolum* was completely inhibited at 2.5, 5.0 and 20 μg/ml of rice field herbicide machete respectively, while a slight stimulation of growth was observed at lower concentrations. Stimulation of cyanobacterial growth in the presence of low concentrations of machete was associated with an increase in the cellular level of phycobilins and RNA, while there was little impact on the levels of chlorophyll-a and DNA. Photosynthetic pigments were degraded at lethal concentrations. The toxicity of the herbicide towards *N.muscorum* and *A. doliolum* could be reversed by supplementing the growth medium with nitrate, nitrite or ammonia. This did not apply for *A. nidulans*. It is suggested that machete inhibited nitrogen- fixation on the former two strains while availability of nutrients was affected in the latter strain. In either case, death of the organisms was most likely due to nitrogen starvation.

Vaishampayan (1984) attempted to introduce herbicide-resistance in *Nostoc muscorum*. The herbicide monuron [3-(4-chlorophenyl)-1, 1-dimethylurea] inhibits growth and heterocyst formation in the nitrogen fixing *N.muscorum*.These inhibitory effects are glucose-reversible in both N_2-free and NO_3 media. A spontaneous mutant of this cyanobacterium, resistant to a dose of monuron normally applied to the rice-field of North India, has been isolated and preliminarily characterized as having no requirement for an organic carbon source for its normal physiology under monuron treated conditions.

Allen *et al.* (1983) studied the effect of photosystem II herbicides diuron and atrazine on the photosynthetic membranes of the cyanobacteria *Aphanocapsa* 6308 and compared with that of higher plant, *Spinacia oleracea*. The inhibition of photosystem II electron transport by these herbicides was investigated by measuring the photo-reduction of the dye 2, 6-dichlorophenol–indophenol spectrophotometrically using isolated membranes. The concentration of herbicide that caused 50 per cent inhibition of electron transport (I_{50} value) in *Aphanocapsa* membranes for diuron was 6.8 x 10^{-9} molar and I_{50} value for atrazine was 8.8 x 10^{-8} molar. ^{14}C-labelled diuron and atrazine were used to investigate herbicide binding with calculated binding constants (K) being 8.2x10^{-8} molar for atrazine and 1.7 x 10^{-7} molar for diuron. Experiments involving the photo- affinity label (^{14}C) azidoatrazine and autoradiography of polyacrylamide gel indicated that the herbicide atrazine binds to a 32-kilodalton protein an *Aphanocapsa* 6308 cell extract.

About the one-half of the herbicides presently used in agriculture act by inhibiting the light reactions in photosynthesis mostly by targeting the photosystem II(PSII) complex (Draber *et al.*, 1991)·A series of algal and cyanobacterial PS II based-whole-cell(Brewster *et al.*, 1995;Campanella *et al.*, 2000) and tissue biosensors have therefore been developed for detection of a class of herbicides which inhibit photosynthetic electron transport. In these systems, herbicides are detected by testing inhibition of the Hill reaction (Campanella *et al.*, 2000), inhibition of 2,6-dichlorophenol indophenol photo-reduction or change in chlorophyll fluorescence, which can be correlated with the pollutant concentration. Alternatively, the effect of herbicides was measured directly through inhibition of maximum growth rate (Abdol–Hamid, 1996; Shao *et al.*, 2002)

Shao *et al.* (2002) tried to generate a cyanobacterial biosensor that could be used to detect herbicides and other environmental pollutants. A representative freshwater cyanobacterium, *Synechocystis* sp. of strain PCC 6803, was chromosomally marked with the luciferase gene luc (from the firefly *Photinus pyralis*) to create a novel bioluminescent cyanobacterial strain. Successful expression of the luc gene during growth of *Synechocystis* sp. strain PCC 6803 cultures was characterized by measuring optical density and bioluminescence. Bioluminescence was optimized with regard to uptake of the luciferase substrate, luciferin and the physiology of the cyanobacterium. Bioassays demonstrated that a novel luminescent cyanobacterial biosensor has been developed which responded to a range of compounds including different herbicide types and other toxin. This biosensor is expected to provide new opportunities for the rapid screening of environmental samples or for the investigation of potential damage. It has numerous advantages in ecotoxicity testing.

Now-a-days, attention has been paid to the potential interaction of soil microorganisms with herbicides and their degradation products. These studies are important since, as pointed out by Stratton (1983,1985) that pesticides are rarely found alone in the environment, but rather with a mixture of xenobiotics including other pesticides, miscellaneous degradation products, pesticide carrier solvents, heavy metals and fertilizers. Interaction may also occur between pesticides and soil chemicals such as inorganic plant nutrients. Stratton (1984) selected the herbicide, atrazine and four of its degradation products and tested their effect on *Anabaena cylindrica, A. inequalis* and *A. variabilis.* It revealed that any mixture containing the parent compound exerted an antagonistic and synergistic effect on photosynthesis and growth respectively. Stratton and Corke (1982) and Stratton (1984) stressed the need for further studies of this nature, but pointed out the difficulty of interpreting the data because of continuous nature of degradation of these toxic chemicals in the environment. Toxicity of the heavy metals on combinations with pesticides were also investigated in wild and tolerant species of *Synechococcus cedrorum* (Bisen *et al.*, 1983). They reported that the inhibition of uptake of different nitrogenous nutrients was mainly because of metal or pesticide conduced depletion of ATP and reductants. A synergistic inhibitory pattern of interaction was noted in the growth, nutrient uptake pigment composition and photosynthetic rate of *Synechococcus cedrorum* when Zn^{2+}, DCMU and 2, 4-D were used in combination with Hg^{2+} in the growth medium (Bisen *et al.*, 1983).

Conclusion

In general, herbicide toxicity is regarded only as the lethal effects of the chemical upon a particular organism. The direct effects concern changes in growth rate and changes on specific metabolic rates *i.e.* photosynthesis and respiration. Secondly, herbicides have toxicity which is selective for certain groups of microorganisms (Padhy, 1985). The effects of this kind may alter population equilibrium in an indirect manner by changing the competitive efficiency of one or other group. Thirdly, herbicide may promote either directly or indirectly, the growth of one or more types of organism. The type favored, could be either beneficial or undesirable in paddy field ecosystem (Das and Adhikary, 2006). Finally, toxic action of herbicides causing death of certain groups of organism would result in the release of nutrients by the decay of these organisms, thus inducing a corresponding increase in the number of pesticide/herbicide tolerant forms.

Acknowledgement

The author thanks Dr. S.P. Adhikary, Professor of Botany and Biotechnology, Utkal University, Bhubaneswar for moral support and to the Principal, G.M. College (Autonomous), Sambalpur for providing laboratory facilities to carry out the UGC funded project during which all these literatures were collected. UGC is also acknowledged for financial assistance to carry out the projects.

References

Abdel-Hamid MI (1996). Development and application of a simple procedure for toxicity testing using immobilized algae. *Water Sci Technol*, 33: 129–138.

Allen MM, Turnburke AC, Lagace EA and Steinback KE (1983). Effects of Photosystem II herbicides on the photosynthetic membranes of the cyanobacterium *Aphanocapsa* 6308. *Plant Physiol.*, 71: 388–392.

Bisen PS, Shukla HD and Gupta A (1983). Cyanobacterial research at Barkatullah University. In: *Proc Natl Sem on "Cyanobacterial Research-Indian Science"*, Tiruchirapalli, pp. 5–11.

Brewster JD, Lightfield AR and Bermel PL (1995). Storage and immobilization of photosystem II reaction centers used in an assay for herbicides. *Anal Chem*, 67: 1296–1299.

Campanella LF, Cubadda F, Sammartino MP and Saoncella A (2000). An algal biosensor for the monitoring of water toxicity in estuarine environments. *Water Res*, 35: 69–76.

Chinnaswamy R and Patel RJ (1983). Effect of pesticide, mixtures on the blue-green alga *Anabaena flos-aquae*. *Microb Letter*, 24: 141–143.

Cullimore DR and McCann AE (1977). Influence of four herbicides on the algal flora of a prairie soil. *Plant and Soil*, 46: 455.

Das B (1977). Effect of herbicides and pesticides on the freshwater blue-green algae. Utkal Univ, Vani Vihar (*Ph.D. Dissertation*).

Das MK (2007). Heavy metal toxicity to rice field cyanobacteria. In: *Advances in Applied Phycology*, (Eds) Gupta RK and Pandey VD. Daya Publishing House, New Delhi, p. 186–195.

Das MK (2008). Toxicity of different fungicides to the rice-field cyanobacteria. In: *Environmental Biotechnology and Biodiversity Conservation*, (Ed.) Das MK. Daya Publishing House, New Delhi, p. 101–114.

Das MK and Adhikary SP (2006). Toxicity of different insecticides to rice-field cyanobacteria. In: *Glimpses of Cyanobacteria*, (Eds) Gupta RK, Kumar M and Paliwal GS. Daya Publishing House, New Delhi, p. 122–134.

Das B and Singh PK (1977). The effect of 2,4- dichlorophenoxy acid on growth and nitrogen fixation of blue-green alga *Anabaenopsis raciborskii*. *Arch Environ Contam Toxicol*, 5: 445.

Day PR (1987). Crop improvement: Constraints and challenges. In: *British Crop Protection Conference on Weeds*. BCPC Pub Surrey, UK, p. 3–12.

De PK (1939). The role of Blue-green algae in nitrogen fixation in rice-fields. *Proc Roy Soc*, London, 127: 121–134.

Draber WT, Tietjen JF, Kluth JF and Trebst A (1991). Herbicides in photosynthesis research. *Angew Chem*, 3: 1621–1633.

Fitzgerald GP, Gerloff GC and Skoog F (1952). Studies on chemicals with selective toxicity to blue-green algae. *Sewage Ind Wastes*, 24: 888.

Gadkari D (1987). Influence of the photosynthesis-inhibiting herbicides Goltix and Sencor on growth and nitrogenase activity of *Anabaena cylindrica* and *Nostoc muscorum*. *Biol Fertil Soils*, 3: 171–177.

Gadkari D (1988). Assessment of the effects of photosynthesis-inhibiting herbicides Diuron, DCMU, Metamitron and Metribuzin on growth and nitrogenase activity of *Nostoc muscorum* and a new cyanobacterial isolate strain G4. *Biol Fertil Soils*, 6: 50–54.

Hamdi YA, El-Nawawy AS and Tewfix MS (1970). Effect of herbicides on growth and nitrogen fixation of alga *Tolypothrix tenuis*. *Acta Microbiol*, 2: 53–56.

Kapoor K and Sharma VK (1980). Effect of certain herbicides on survival, growth and nitrogen fixation of blue-green alga *Anabaena doliolum* Bharadwaja. *Z Allg Mikrobiol*, 20: 465–469.

Kashyap AK and Pandey KD (1982). Inhibitory effects of rice field herbicides Machete on *Anabaena doliolum* Bharadwaja and protection by nitrogen sources. *Z Pflanzen Physiol*, 107: 339–345.

Kaushik BD and Venkataraman GS (1983). Response of cyanobacterial nitrogen fixation to insecticides. *Curr Sci*, 52: 321–323.

Kellogg RL,Nehring R, Grube A, Goss DW and Plotkin S (2000). *Environmental Indicators of Pesticides Leaching and Runoff from Farm Fields*. United States Department of Agriculture Natural Resources Conservation Service.

Khalil K, Chaporkar CB and Gangawane LV (1980). Tolerance of blue-green algae to herbicides. In: *Proc Natl Work on Algal Systems*. Ind Soc Biotech IIT, New Delhi, p. 36–39.

Kolte SO and Goyal SK (1992). On the effect of herbicides on growth and nitrogen fixation by cyanobacteria. *Acta Bot India*, 20: 225–229.

Mc Cann AE and Cullimore DR (1979). Influence of pesticides on the soil algal flora. *Residue Reviews*, 72: 1–31.

Mishra AK and Pandey AB (1989). Toxicity of three herbicides to some nitrogen-fixing cyanobacteria. *Ecotox Environ Saf*, 17: 236–246.

Mishra AK, Pandey AB and Kumar HD (1989). Effects of the three pesticides on MSX-induced ammonia photo–production by the cyanobacteria *Nostoc linckia*. *Ecotox Environ Saf*, 18: 145–148.

Misra SC and Mani D (1994). *Agricultural Pollution*, Vol 2. Ashish Publ. House, New Delhi, pp. 11.

Padhy RN (1985). Cyanobacteria and Pesticides. *Residue Reviews,* 95: 1–44.

Pandey KD and Kashyap AK (2007). Differential sensitivity of three cyanobacteria to the rice field herbicides Machete.*J Basic Microb,* 26(7): 421–428.

Roy choudhary P And Kaushik BD (1986). Response of cyanobacterial growth and nitrogen fixation to herbicides. *Phykos,* 25: 36–43.

Shao CY, Howe CJ, Porter AJR and Glover LA (2002). Novel cyanobacterial biosensor for detection of herbicides. *Appl Environ Microbial,* 68(10): 5026–5033.

Sharma VK (1986). A review of recent work on pesticide studies on the nitrogen-fixing algae. *J Environ Biol,* 7: 171–175.

Singh LJ and Tiwari DN (1988a). Some important parameters in the evaluation of herbicide toxicity in diazotrophic cyanobacteria. *J Appl Bacteriol,* 64: 365–376.

Singh LJ and Tiwari DN (1988b). Effects of selected rice field herbicides on photosynthesis, respiration and nitrogen assimilating enzyme systems of paddy soil diazotrophic cyanobacteria. *Pestic Biochem Physiol,* 31: 120–128.

Singh RN (1961). Role of blue-green algae in nitrogen economy of Indian agriculture. *ICAR,* New Delhi.

Singh S and Datta P (2005). Growth and survival potentials of immobilized diazotrophic cyanobacterial isolates exposed to common rice-field herbicides. *World J Microbiol and Biotech,* 21(4): 441–446.

Stratton GW (1984). Effect of the herbicide atrazine and its degradation products, alone and in combination on phototrophic microorganisms. *Arch Environ Contam Toxicol,* 13: 35–42.

Stratton GW and Corke CT (1982). Toxicity of the insecticides permethrin and some degradation products towards algae and the cyanobacteria. *Environ Poll,* 29: 71–80.

Swain N, Rath B and Adhikary S.P (1994). Growth response of the cyanabactrium *Microcystis aeruginosa* to herbicides and pesticides. *J Basic Microbiol,* 34: 197–204.

Tiwari DN, Pandey AK and Mishra AK (1981). Action of 2,4-dichlorophenoxyactetic acid and rifampicin on heterocyst differentiation in the blue-green alga *Nostoc muscorum. J. Biosci,* 3: 33.

Vaishampayan A (1984a). Biological effects of a herbicide on a nitrogen-fixing cyanobacterium (blue-green alga): An attempt for introducing herbicide resistance. *New Phytol,* 96: 7–11.

Vaishampayan A (1984b). Powerful mutagenecity of a bipyridylium herbicide in a nitrogen fixing blue–green alga *Nostoc muscorum. Mutation Res,* 18: 39–46.

Vaishampayan A (1985). Mutagenic activity of alachlor, butachlor and carbaryl to a nitrogen–fixing cyanobacterium *Nostoc muscorum. J Agric Sci,* 104: 571–576.

Venkataraman GS (1972). *Algal Biofertilizer and Rice Cultivation.* Today and Tomorrow's Printers and Publishers, New Delhi.

Venkataraman GS and Rajyalakhmi B (1972). Relative tolerance of nitrogen-fixing blue-green algae to pesticides. *Ind J Agric Sci,* 42: 119–121.

Voight RA and Lynch DL (1974). Effect of 2, 4-D and DMSO on prokaryotic and eukaryotic cells. *Bull Environ Contam Toxicol,* 12: 400.

Wright SJL, Stainthorpe AF and Downs JD (1977). Interactions of the herbicides propanil and metabolite 3, 4-dichloro-aniline, with blue-green algae. *Acta Phytopathol Acad Sci Hung,* 12: 51–60.

Soil Microflora, 2009 *Pages* **40–52**
Editor: **Rajan Kumar Gupta, Mukesh Kumar & Deepak Vyas**
Published by: **DAYA PUBLISHING HOUSE, NEW DELHI**

Chapter 5

Mycorrhizae: Benefits and Practical Applications in Forest Management

K.P. Singh*, P. Srinivas and Bijendra Kumar
Plant Pathology Section, College of Forestry and Hill Agriculture, GBPUA&T Hill Campus, Ranichauri – 249 199, Tehri Garhwal, Uttarakhand

In Indian Himalayas, the Hill and Mountain ecosystem extends over twelve states from Jammu and Kashmir in the northwest to Nagaland in the northeast and is about 2800 km long and 222 to 300 km wide. It covers a total area of about 591 thousand km^2 inhabiting nearly 51 million people accounting for 18 percent and 6 percent of area and population of the country, respectively. The region is endowed with diverse climatic conditions, which permit the successful plantation of all kinds of forest trees. The soils of mid and high hills are acidic, deficient in P and N and low in bacterial and actinomycetes population. These soils are light textured, poor in exchangeable bases, high in sequioxides with extremely high phosphorus fixing capability. Phosphorus being one of the major nutrients in addition to nitrogen, always becomes a limiting nutrient in acid soils due to formation of unavailable complexes with Fe and Al. The availability of P in these soils can be enhanced by various chemical and biological means. Among the biological means, mycorrhiza has been reported for mobilization of P to forest trees and nurseries plants along with other nutrients like S, K, Ca, and Zn. Several fungi form beneficial associations with forest tree species which invade the feeder root tissues and form modified roots called mycorrhizae (fungus-roots), which greatly increases efficiency of nutrient and water uptake. Most plants require mycorrhizae for normal growth and development in natural soils. Earlier it was thought that roots absorb soil nutrients and water primarily through root hairs on the feeder roots, however, recent evidence indicates that most absorption occurs through feeder roots that are infected by beneficial fungi. The mycorrhizal fungus parasitizes the cortical tissues of the young roots, but the

* E-mail: kps60@rediffmail.com; Fax: 01376-252150 , Telephone: 01376- 252138

presence or absence of specialized beneficial fungi will dramatically affect seedling form and size and tree development. The formation of mycorrhizae aids water and mineral absorption for the tree, and the fungus in turn receives needed organic compounds from its association with the tree. They are usually emphasized for their beneficial effect on root function and barriers to infection by other destructive soil borne root pathogens in the nursery and in the field. Mycorrhizae are now known to be ubiquitous in the roots of trees and also in the roots of almost all higher plants.

The term mycorrhizae is derived from two Greek words, mycos meaning fungus and rhizome meaning root and is applied to structures resulting from the association of the mycelium of certain fungi with the small roots of a higher plant. Mycorrhizae is a beneficial association between a fungus and a plant root and function as a mutualistic, symbiotic biotophy between a fungus and a higher plant.

Mycorrhizae were first described by Theodore Harting on coniferous trees, but he did not investigate their function. Frank described the relation of mycorrhizae to the growth of trees and to the growth of fungi in the forests and coined the term mycorrhiza in 1885. Several pioneer workers *i.e.* Melin in Sweden, Bjorkman in Poland, Harley in Great Britain, and Hatch and Doak in the United States, did their research on mycorrhizae of forest trees. The practical applications of these fungus-root relationships is to maximize yields in forest and agricultural systems. After infection of mycorrhizal fungi in conifers, the roots become swollen, branch dichotomously (Pinaceae), and may become colored due to the presence of fungal tissue. Angiosperm mycorrhizae do not often exhibit the increase in volume and degree of branching as in conifer mycorrhizae. Mycorrhizae enhance plant growth through increased nutrient uptake, stress tolerance and disease resistance. VAM fungi can alter root exudation pattern, enhance chitinolytic activity and alter photosynthetic/respiratory deficiencies (Dehne, 1982).

Type of Mycorrhizae

Mycorrhizae are classified into three main types based on the physical relationship of the fungus and the root cells. These are as follows:

Ectomycorrhizas (Ectotrophic Mycorrhiza)

Ectomycorrhizae form a structure known as the Hartig net between the cells, as well as a fungus mantle or cover on the surface of feeder roots. The ectomycorrhizal feeder roots develop a swollen appearance, and in pines they normally have a forking habit. Ectomycorrhizae are characterized by the presence of tightly interwoven hyphae covering the surface of infected feeder roots, called the fungus mantle. The fungi also grow between the cortex cells of their hoot roots, forming the Hartig net. This type of mycorrhizal association is commonly found in some forest trees belonging to the families *Fragaceae, Butalaceae, Salicaceae* etc. The common genera of the plant species include Papulus, Salix, Cedrix, Pinus, spruce, fir, beech, eucalyptus, alder, oak, and hickory. Both the mantle and Hartig net (named for Robert Hartig, who first described the association) provide protection for the roots from pathogenic root fungi. Ectomycorrhizae on pines are often forked or branched (bifurcate), but may occur unbranched (monopodial) or repeatedly branched (coralloid) as well (Singh *et al.*, 2007; Wilcox, 1983).

Fungi which are now known to form ectomycorrhizae with higher plants are in three classes within the Eumycota (Basidiomycotina, Ascomycotina and Zygomycotina). The largest number by far are Basidiomycotina, few genera of Ascomycotina are known but only one genus and a few species have been identified in the zygomycotina. The entire families of Basidiomycotina have evolved as mycorrhizal symbionts of higher plant families (Miller *et al.*, 1992). Ascomycotina tentatively implicates

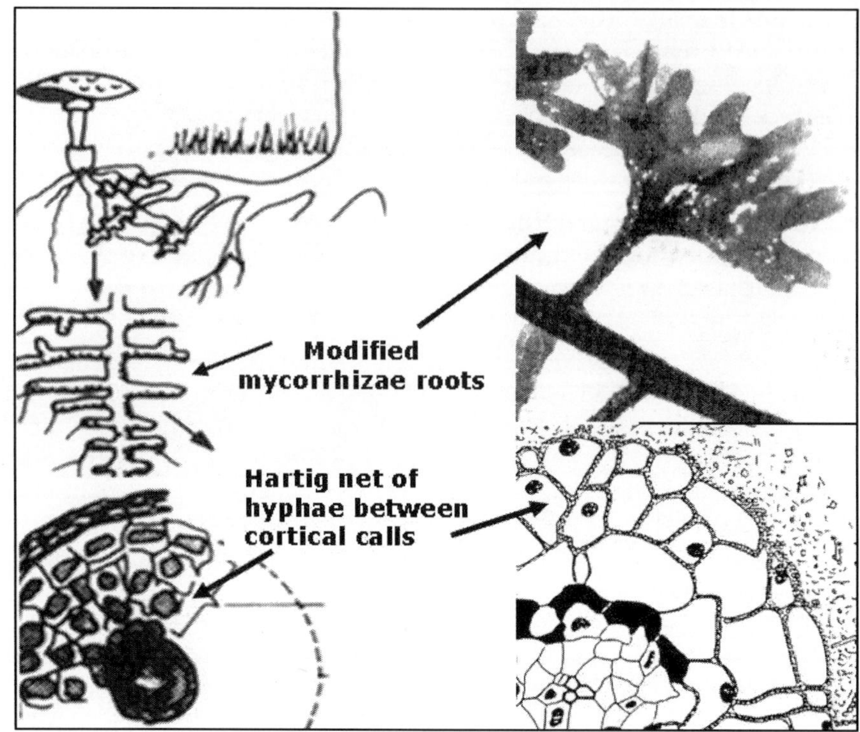

Figure 5.1: Ectotrophic Mycorrhiza

more species as mycorrhizal associates. In the Zygomycotina, the Endogonaceae in the sense of Gerdemann and Trappe contain only two ectomycorrhizal species, both in the genus Endogone. Most ectomycorrhizal association in trees is caused by basidiomycetes in the mushroom group (Agaricales) or the puffball group (Gasteromycetes) and occasionally Deuteromycetes, Ascomycetes and Phycomycetes. Spores of ectomycorrhizal fungi are produced aboveground and are wind disseminated. Ectomycorrhizal are usually short-lived, lasting from a few months to a maximum of 3 years. Conifer and some hardwood species that form mycorrhizae are of this type. Ectomycorrhizae appear white, brown, yellow, or black, depending on the color of the fungus growing on the root.

Endomycorrihzae (Endotrophic Mycorrhizae)

Endomycorrhizae roots externally appear similar to nonmycorrhizal roots in shape and color, but the fungus invades the cortical cells of the feeder roots. The hyphae within the cells (intracellular) grow in the cortical cells of the host root for a period of time and then disintegrate or are digesting by the host. Endomycorrhizae do not form a mantle or Hartig net but instead form absorbing hyphae (haustoria), called arbuscules, or by forming large swollen food storing hyphal swellings, called vesicles, in the living cortical cells of the roots. A loose weft of mycelium may occur around the outside of the root. Most endomycorrhizae contain both vesicles and arbuscules and are, therefore, called vesicular-arbuscular (VA) mycorrhizae. The colonization of endomycorrhizae usually does not cause any obvious changes in the root although their role in root absorption is thought to be similar to ectomycorrhizae. Vesicular-arbuscular endomycorrhiza occurs everywhere in all terrestrial habitats

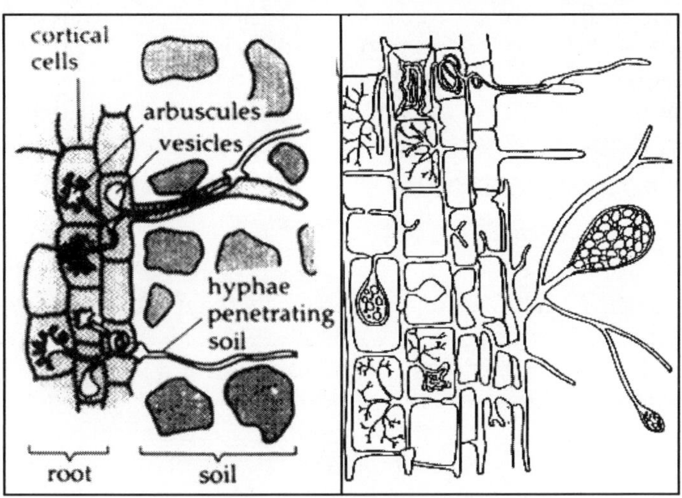

Figure 5.2: Endotrophic Mycorrhizaa

and most common in monocots and dicots plants. It also occurs in many forest plants including Texus, Podocarpus, Cupresus and Araucaria etc.

Vesicular-arbuscular mycorrhizas are formed by members of all phyla of land plants. The fungual symbionts appear to be restricted to a few genera of the Phycomycetous family Endogonaceae. In contrast, the host plants are very diverse, not only in taxonomic position but also in life form and geographical distribution. Herbaceous plants, shrubs and trees of temperate and tropical habitats may all from vesicular-arbuscular mycorrhizas and there is little evidence for specificity between particular fungi and host plants. Only a few families and genera of plants do not generally form vesicular-arbuscular mycorrhizas. Endomycorrhizae are characterized by the obligate relationship of orchids with the fungus *Rhizoctonia* spp. and a nonobligate but very normal relationship between grasses, legumes, some hardwood trees, and many other plants with *Endogone* spp. These are produced on most cultivated plants and on some forest trees mostly by zygomycetes, primarily of the genus *Glomus*, but also other fungi, such as *Acaulospora*. There are clear indications that mycorrhiza formation induces alteration in host cells which make them more resistant to soil and root-borne pathogens.

Vesicular-arbuscular mycorrhizal fungi were members of the Endogonales, rather than chytrids (Peyronel 1923, 1924). The family was monographed in 1974 with segregation of the genus Endogone into seven genera. The Endogonales (Zygomycotina) as presently conceived consists of a single family, the Endogonaceae. The genera are separated on the basis of spore formation. The type genus, *Endogone*, forms zygospores but is not known to produce VA mycorrhizae. *Acaulospora, Entrophospora, Gigaspora, Glaziella, Glomus* and *Sclerocystis* have not been demonstrated to form zygospores and contain either proven or presumed VAM species. The spores of Acaulospora, Entrophospora, and Gigaspora have been termed azygospores, *i.e.* parthenogenic zygospores, because they rather resemble the zygospores of *Endogone* spp., but no sexual origin has been observed (Ames and Schneider, 1979). The spores of *Glaziella, Glomus* and *Sclerocystis* are regarded as chlamydospores.

Ectendomycorrizae (Ectendotrophic Mycorrhiza)

The ectendomycorrhizae are an intermediate group which form haustoria and a Hartig net and may sometimes form a mantle. They are caused by fungi of unknown identity that grow into and also around the cortical cells of the root. Fungus shown to form ectendomycorrhizae is species of Ascomycotina. Ectendomycorrhizae are found on a few tree species but little is known about this group of mycorrhizae. It is generally found in well fertilized nurseries. Both conifers and broadleaf plants have this type of mycorrhizae. They are found sometimes in the root system of beech, lodge pole pine and pondersa pine. This sort of mycorrhizal association is considered to be transitional between ectotrophic and endotrophic forms, where infection is typically ectotrophic (intercellular) alongwith endotrophic penetration of hyphae.

Benefits of Mycorrhizae

The major benefits to the plant of having mycorrhizae, regardless of the type, are (a) enhanced nutrient and water absorption, (b) increased drought/stress tolerance, (c) improved transplantability, (d) favoring growth of beneficial microbiota in the rhizosphere and (e) reduced susceptibility to root diseases. The physiological changes due to mycorrhizae formation can include increased rate of photosynthesis, altered nutritional state, phytohormone balance, chemical constituents, and improved membrane permeability. However the net effects of the formation of mycorrhizae may vary depending on the fungus, host plant, soil, and climatic conditions (Linderman and Hoefnagels, 1992).

Enhanced Nutrient and Water Absorption

Mycorrhizae apparently improve plant growth by increasing the absorbing surface of the root system. Mycorrhizal root system benefits their respective hosts by increasing the capacity of the roots to absorb nutrients from the soil. Mycorrhizae can greatly increase the absorptive surface area of roots. In turn, the increase in roots increases the moisture and mineral element absorbing surface and provides the plant with a better capacity to survive and grow. Measurements indicated that in some instances total root surface was increased thirty times more than an uninfected root. *Cenococcum graniforme* produces black ectomycorrhizae, whose abundance can easily be measured. In a spruce stand this fungus produced 41 billion sclerotia in the top 1 cm. In the upper 10 cm, there were 148-209 kg/ha dry weight of sclerotia (Tainter and Baker, 1996). Adjoud *et al.* (1996) tested three AM fungi (*Glomus intraradices*, *G. mosseae* and *G. caledonium*) on 11 *Eucalyptus* species and found positive effects on growth in 21 per cent of the plant-fungus combinations. Chen *et al.* (2000) also found positive growth effects in *E. urophylla* with three AM fungi (*Glomus invermaium*, *Acaulospora laevis* and *Scutellospora calospora*).

In addition, the fungus-root association is a more efficient mineral element absorbing organ can selectively absorb nutrients, providing further benefits to the host tree. More nitrogen and phosphorus are absorbed from the soil and accumulated in plants with mycorrhizae, and by making feeder roots more resistant to infection by certain soil fungi such as *Phytophthora*, *Pythium*, and *Fusarium*. The fungus capacity for extraction of elements from the soil organic matter is assumed to be part of the increased efficiency. The association helps in increased availability and absorption of plant nutrients like phosphorus, potassium, calcium magnesium sulphur, iron, manganese, zinc and copper etc. The mycorrhizal fungi are known to increase growth of eucalyptus through a process of improved nutrient acquisition, (especially P and N), although the effect varies with the species of *Eucalyptus* studied and the fungi inoculated (Adjoud *et al.*, 1996; Lu *et al.*, 1998).

Respiration of mycorrhizal roots influences the rhizosphere or mycorrhizosphere. High respiration shows active ion uptake and sporocarps contain greater concentrations of calcium, potassium, nitrogen, sodium, phosphorus, and zinc than do pine needles. The fungus parasitizes the plant for most of its carbohydrate and vitamin needs. The saprophytic fungi cannot penetrate the physical and chemical defense barriers of living roots. The parasitic fungi penetrate and cause a series of disruptive reactions leading toward the death of the host. The mycorrhizal fungi, like other obligate parasites, induce a limited host chemical defense reaction which is not totally disruptive to the host and is tolerated by the fungi. Mycorrhizae also play a role in corbon cycling. Labeled carbon moves readily from host to mycobiont, usually in the form of sugars. It can also be transferred to other adjacent green plants and achlorophyllous plant via the mycobionts.

Increased Drought/Stress Tolerance

It is well documented that the plants with mycorrhiza are more tolerant to stress such as soil salinity, alkalinity, acidity and drought conditions. The mycorrhizal plants having access and reach to larger soil volume by extended root growth and increased absorptive area, exhibit better growth than the nonmycorrhizal ones especially in the arid and semi arid regions where low moisture and high temperature are very critical for survival and growth of the plants (Mishra and Mishra, 2004). Mycorhizal plants are also more tolerant to toxic heavy metals than the non-mycorrhizal plants (Smith and Read, 1997).

The structures of mycorrhizae are also important in the nutrient cycling process because they can extend farther and faster than can roots. The mycorrhizal fungal mycelium that grows into the soil effectively extends the root system and larger root tips. The outgrowing hyphae produce extracellular auxins, vitamins, cytokinins, enzymes, and so on, and influence root tissue and ion uptake. Mycorrhizae have an important influence on water relations in that mycorrhizal seedlings resist drought.

Improved Transplantability

Though the exact role of the mycorrhizal association in improved growth of the plants is not yet fully understood, it is for certain that if the nursery plants are transplanted in the soils which are loaded with mycorrhizal species, far more growth is observed. Inoculation of young trees (mainly *Quercus* spp) with the valuable late stage fungus *Tuber melanosporum* is an established management method with a good success rate (Hall *et al.*, 1994).This may be due to prompt protection from pathogens and better nutrition that may be provided by the immediate mycorrhizal association. The capacity of undisturbed forest soils to exclude or suppress soil borne pathogen like *Fusarium*, compared to nursery soils, is well documented (Figure 5.3). Even when plants with known pathogens in or on their tissues are transplanted into the wild, the pathogens disappear within a relatively short period. Smith (1967) planted pine seedlings with high levels of pathogenic *Fusarium oxysporum* on or in their roots out into forest sites. Subsequent isolations from those seedlings documented the pathogen's disappearance within 1.5 years.

In many countries, the failure of coniferous plantations has been ascribed to the absence of proper mycorrhizal fungi. Though, soil fumigation, practiced in most forest nurseries, effectively controls weeds, insects, and root pathogens, but may also reduce populations of antagonists, including mycorrhizal fungi and rhizobacteria or fungi. Both antagonistic bacteria and mycorrhizal fungi may need to be re-introduced at time of seeding to effectively block pathogen infection which occurs during the first weeks after conifer seed germination. It is possible that reestablishment of antagonistic microbes can be enhanced by providing part of the natural forest environment, such as forest needle litter (Schisler and Linderman, 1989; Linderman, 1989) could encourage the functioning of antagonists. The introduction of humus from forest stands or pure cultures of mycorrhizal fungi has also enabled trees to grow in these areas. The succession of mycorrhizas did not appear to compromise growth

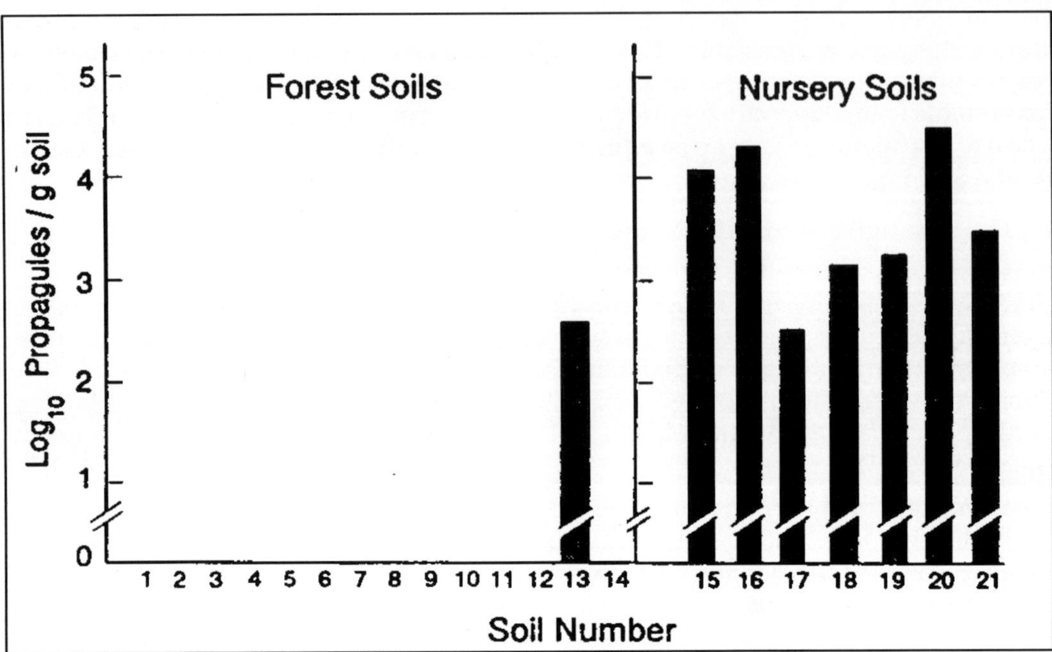

Figure 5.3: Relative Populations of Fusarium Isolated from Conifer Forest Soils vs Conifer Nursery Soils (from Linderman and Hoefnagels, 1992)

effects of the host. Moreover, the greatest growth responses were seen in plants colonized by both types of mycorrhiza. Verma. and Jamaluddin (1995) studied the effect of AM on growth, biomass and per cent root infection of teak seedlings after 7 months of inoculation and found that mixed inoculation with AM had more positive effect on growth parameters like height, collar diameter, biomass, and per cent root infection as compared to control seedlings. Therefore, in theory, an ideal strategy for outplanting would be to inoculate the seedlings with both types of mycorrhiza (Brundrett, 2000).

Favoring Growth of Beneficial Microbiota in the Rhizosphere

The changes in membrane permeability influences the quality and quantity of root exudation, which in turn affects the quality and quantity of microorganisms proliferating in the rhizosphere soil, now more appropriately termed the "mycorrhizosphere" (Linderman, 1988a). It also helps survival and proliferation of beneficial microorganisms like phosphorus solublizers, organic matter decomposers and nitrogen fixers etc. Sporocarps are usually eaten within a few weeks, thus providing a concentrated nutrient source for decomposers and consumers.

Reduced Susceptibility to Root Diseases

Presence of mycorrhizal fungi imparts a degree of protection or resistance against certain root pathogen. Infective propagules (hyphae or spores) of feeder root pathogens (*Phytophthora, Pythium, Rhizoctonia,* and *Fusarium*) are chemically stimulated by feeder roots and pathogenically infect these tissues by ramifying into meristematic, primary cortex, and occasionally vascular tissues, causing limited or extensive necrosis. If a pathogen infects and destroys this feeder root prior to infection by an ectomycorrhizal fungus, an ectomycorrhiza cannot be formed. However, if the ectomycorrhizal fungus infects and transforms the feeder roots into an ectomycorrhiza prior to pathogenic infection, the tissues of this transformed root are no longer vulnerable to attack by the pathogen. It has been reported that the mycorrhiza reduced the infection by root invading fungi. It is also reported that *Glomus mosseae* induces higher chitinase and arginine accumulation which caused development of resistance in mycorrhizal plants (Mosse, 1973). Zak (1964) and Marx (1972, 1975) postulated different mechanisms whereby ectomycorrhizae may suppress root pathogens by providing mechanical barrier, the fungal mantle, to penetration of primary cortical cells by the pathogens, production of antibiotics, inducing production of antifungal substances in the cortical cells and inhibit infection and spread of pathogens, ectomycorrhizal fungi can utilize various chemicals in the root and at the root surface, thereby reducing the amount of nutrients available to pathogens, and favouring proliferation of antagonistic microflora in the mycorrhizosphere soil. The disease suppression may be because of any of the above stated mechanisms or combination of one or more of these mechanisms however enhanced nutritional status of the host, physical fortification of host roots and production of antibiotics in rhizosphere, merit maximum attention.

More then hundred species of ectomycorrhizal fungi are reported to produce antifungal, antibacterial, or antiviral compounds. Inhibitors present in ectomycorrhizae may be antibiotics produced directly by the fungal symbiont (Marx and Davey, 1969) or antibiotics produced by the host as a result of stimulation by fungus infection (Krupa *et al.*, 1973). Ectomycorrhizal development causes an increase in quantities of volatile terpenes produced on roots (Krupa and Fries, 1971). Mycorrhizal fungi such as *Lectarious delicious* and *Boletus* sp. antagonize *Rhizoctonia solani, Lectarious camphorales, Lectarious* and *Cortinarious* sp. have been found to produce antibiotics known as 'chloromycorrhiza' and 'Mycorrhizin A' which are antifungal to the phytopathogens like *Rhizoctonia solani, Pythium debarynum* and *Fusarium oxysporum* etc. (Mishra and Mishra, 2004). The evidence for a role of mycorrhizal fungi in protecting plants against disease concerns the sheathing mycorrhizas of pine trees and the pathogen *Phytophthora cinnamomi* (Mars, 1991; Marx and Davey, 1969). Several other researchers have also highlighted the potential of mycorrhizal system as a tool for biocontrol of soil borne pathogens (Krishna and Bagyaraj, 1983; Natarajan and Govindaswamy, 1990; Reddy and Sreevani, 2003, Jalali, 2001).

Several workers reported the interaction of VA mycorrhizal fungi with other microorganisms present in soil. These fungi interact with soil inhabiting microorganisms such as other fungi, bacteria, actinomycetes, insects, and nematodes. VA mycorrhizae have neither an external mechanical barrier, such as a fungus mantle, nor do they produce any apparent antibiotics. Instead, most studies indicate that changes within the root tissue influence disease development more than alterations in the rhizosphere (Schenck, 1991). He had suggested the protective mechanisms in mycorrhizal plants are cell wall thickenings, changes in amino acids and reducing sugars, increased chitinase activity, and a general change in plant physiology.

A variety of responses have been described which can broadly be summarized as under.

1. VAM infection in general protects plants from soil borne fungi
2. Higher nutrient concentration in mycorrhizal plants makes them more susceptible to foliar pathogens.
3. No definite relationship appears to exist between bacterial infection and mycorrhization.
4. Pre-mycorrhizal infection of transplanting crops protects the plant from nematode infection.

In addition to phosphate, VAM enhance uptake of Ca^{2+}, Cu^{2+}, So_4^{2-} and Zn^{2+} (Smith and Gianinazzi-Pearson 1988). Host susceptibility to infection and tolerance to disease is influenced by the nutritional status of the host and fertility level of the soil (Cook and Baker 1982). For example, nematode damaged plants frequently show deficiency of boron, nitrogen, iron, magnesium and zinc (Good, 1968). In the absence of VAM, phosphate can combine with minor elements to create deficiencies which would predispose plants to root-knot nematodes (Smith *et al.*, 1986). Thus, plants infected with VAM fungi are likely to be more resistant to subsequent attack by pathogenic fungi and nematodes. VAM has also been known to protect plants against infection potential of *Meloidogyne incognita* on black henbane (*Hyoscyamus niger*) (Pandey *et al.*, 1999) and other plant parasitic nematodes (Sikora, 1979; Hussey and Roncadori, 1982). These fungi exert a selective pressure on microbial population in the mycorrhizosphere, some of this can result in specific effect on root pathogens.

Proposed Model for Relationship Between Mycorrhizae-Soil Borne Diseases

Based on our current understanding of the role of mycorrhizae in forest/plant disease, a model is proposed that takes into account soil and environmental factors, plant vigour and intensity of mycorrhizal infection.

1. Low Disease Severity
   (*a*) Mycorrhizal strain effective and present in soil at optimum dose level
   (*b*) High susceptibility of host towards mycorrhizae
   (*c*) Potential dose of pathogen is moderate
   (*d*) Environmental and soil factors favour mycorrhizae development
2. Moderate Disease Severity
   (*a*) Moderate susceptibility of host towards mycorrhizae
   (*b*) Pathogen population density low/high
   (*c*) Enviromental conditions favourable
   (*d*) Soil conditions favorable
   (*e*) Mycorrhizae strain effective resulting in low/high root infection

   In spite of favorable soil and environmental factors, the above situation will lead to slight loss in yield because mycorrhizae proliferation is not extensive and normal/moderate host vigor.
3. High Disease Severity
   (*a*) Moderate/high susceptibility of the host towards mycorrhizae.
   (*b*) Soil conditions unfavourable leading to low mycorrhizae infection
   (*c*) Environmental factors favourable.
   (*d*) Pathogen population high

These conditions would lead to low host vigour and result in high loss in yield.

This proposed model is an over-simplification of the complex interaction but should provide a working base to make a realistic assessment of the role of mycorrhizae in especially soil borne plant diseases.

Conclusion

Plants with mycorrhizae have been invariably found to be much larger and much more vigorous that nonmycorrhizal plants. The symbiosis between the host plant and the mycorrhizal fungus is generally viewed as providing equal benefits to both partners. Generally mycorrhizae do not cause disease, but absence of mycorrhizae in certain fields result in plant stunting and poor growth, which can be avoided if the appropriate fungi are added to the plants. The benefits to plants like increased mineral and water availability resulting from nutrient exchange with the fungus, and increased root surface area, resistance to root pathogens through mantle formation and reduced carbohydrate levels in the plant root and increased populations of nonpathogenic microorganisms in the rhizosphere of the plant, resulting from mycorrhizal activities, are received with out any significant investment. The benefits of mycorrhizal system can be fully reaped if they are well preserved and their proliferation is encouraged in both nursery and forest soils. However it should also be noted that mycorrhizal fungi should not be seen as the sole provider of all the nutritional and plant protective benefits, but it should be integrated in a wider forest and nursery management strategy. Conservation of mycorrhizal fungi in forest and nursery soil should also be given priority. Factors detrimental to mycorrhizal growth have to be documented and managed accordingly. Pollution (especially ammonia pollution) may reduce the diversity of the ectomycorrhizal community (Kowalski *et al.* 1989, Schaffers and Termorschuizen 1989; Peter Ayer and Egli, 2001). Liming has also been shown to have a detrimental effect on the fruiting of ectomycorrhizal fungi (Agerer, 1989).

Many more extensive studies on different influences of mycorrhizal fungi on plants, their mode of action, their mass production and delivery system may prove more productive in harnessing better benefits from these useful fungi, in future.

References

Adjoud D, Plenchette C, Halli-Hargas R. and Lapeyrie F (1996). Response of 11 eucalyptus species to inoculation with three arbuscular mycorrhizal fungi. *Mycorrhiza*, 6: 129–135.

Agerer R (1989). Impacts of artificial acid rain and liming on fruitbody production of ectomycorrhizal fungi. *Agriculture Ecosystems and Environment*, 28: 3–8.

Ames RN and Schneider RW (1979). *Entrophospora*: A new genus in the Endogonaceae. *Mycotaxon*, 8: 347–352

Brundrett MC (2000). What is the value of ectomycorrhizal inoculation for plantation-grown Eucalypts? In: *Mycorrhizal Fungi Biodiversity and Applications of Inoculation Technology*, (Eds) MQ Gong, D Xu, C Zhong, YL Chen, B Dell and MC Brundrett. China Forestry Publishing House, Beijing, pp 151–160.

Chen YL, Brundrett MC and Dell B (2000). Effects of ectomycorrhizas and vesicular-arbuscular mycorrhizas, alone or in competition, on root colonization and growth of *Eucalyptus globulus* and *E urophylla*. *New Phytologist*, 146: 545–556.

Cook RJ and Baker KF (1982). *The Nature and Practice of Biological Control of Plant Pathogens*. The American Phytopathological Society, St Paul, MN, USA.

Dehne HW (1982). Interaction between vasicular-arbuscular mycorrhizal fungi and plant pathogens. *Phytopathology*, 72: 1114–1119.

Good JM (1968). Relation of plant parasitic nematodes to soil management practices. In: *Tropical Nematology*, (Eds) Smart GC and Perry VG. University of Florida, Gainsville, pp 113–138.

Hall I, Brown G and Byars J (1994). *The Black Truffle*. New Zealand Institute for Crop and Food Research, Christchurch, New Zealand.

Harley JL and Smith SE (1983). *Mycorrhizal Symbiosis*. Academic Press, London, New York, 483 pp.

Hussey RS and Roncadori RW (1982). Vesicular-arbuscular mycorrhizae may limit nematode activity and improve plant growth. *Plant Disease*, 66: 9–14.

Jalali BL (2001). Mycorrhiza and plant health-need for paradigm shift. *Indian Phytopath*, 54: 3–11.

Kowalski S, Wojewoda W, Bartnik C and Ripsk A (1989). Mycorrhizal species composition and infection patterns in forest plantations exposed to different levels of industrial pollutants. *Agriculture, Ecosystems and Environment*, 28: 249–255.

Krishna KR and Bagyaraj DJ (1983). Interaction beteen *Glomus fasciculatum* and *Sclerotium rolfsii* in peanut. *Canadian J Bot*, 67: 2349–257.

Krupa S and Fries N (1971). Studies on ectomycorrhizae of pine. I. Production of volatile organic compounds. *Can J Bot*, 49: 1425–1431.

Krupa S, Andersson J and Marx DH (1973). Studies on ectomycorrhizae of pine. IV. Volatile organic compounds in mycorrhizal and nonmycorrhizal root systems of *Pinus echinata* Mill. *Eur J For Path*, 3: 194–200.

Linderman RG and Hoefnagels M (1992). Controlling root pathogens with mycorrhizal fungi and neneficial bacteria. In: *Paper Presented at the Western Forest Nursery Associations Meeting*. Stanford Sierra Camp, Fallen Leaf Lake, CA, September 14–18, 4 pp.

Linderman RG, (1988a). Mycorrhizal interactions with the rhizosphere microflora: The mycorrhizosphere effect. *Phytopathology*, 78: 366–371.

Linderman RG (1989). Organic amendments and soilborne diseases. *Can J Plant Pathol*, 11: 180–183.

Lu X, Malajczuk N and Dell B (1998). Mycorrhiza formation and growth of *Eucalyptus globulus* seedlings inoculated with spores of various ectomycorrhizal fungi. *Mycorrhiza*, 8: 81–96.

Manion PD (1981). *Tree Disease Concepts*. Prentice Hall, Inc, Englewood Chiffs, New Jersey, 399 pp.

Mars DH (1991). Mycorrhizae in interactions with other microorganisms. B. Ectomycorrhizae. In: *Methods and Principles of Mycorrhizal Research*, (Eds) Schenck, NC. APS Press, St Paul, Minnesota, USA, pp 225–228.

Marx DH (1972). Ectomycorrhizae as biological deterrents to pathogenic root infections. *Annu Rev Phytopathol*, 10: 429–454.

Marx DH and Davey CB (1969a). The influence of ectotrophic mycorrhizal fungi on the resistance of pine roots to pathogenic infections. III. Resistance of aseptically formed mycorrhizae to infection by *Phytophthora cinnamomi*. *Phytopathology*, 59: 549–558.

Marx DH and Davey CB (1969b). The influence of ectotrophic mycorrhizal fungi on the resistance of pine roots to pathogenic infections. IV. Resistance of naturally occurring mycorrhizae to infection by *Phytophthora cinnamomi*. *Phytopathology*, 59: 559–565.

Miller JH, Barber B, Thompson ML, McNabb KL, Bishop LM and Taylor JWJ (1992). Pest and pesticide management on southern forest. *USDA Forest Service*, Asheville, NC, SRS MB R8–MB 60.

Mishra BB and Mishra SN (2004). Mycorrhiza and its significance in sustainable forest development. *Orissa Review*, p 52–55.

Mosse B (1973). Advances in the study of vesicular-arbuscular mycorrhiza *Annu Rev Phytopathol*, 11: 171–196.

Natarajan K and Govindaswamy V (1990). Antagonism of ectomycorrhizal fungi in some common root pathogens. In: *Current trends in Mycorrhizal Research,* (Eds) Jalali, BL and H Chand. Haryana Agri Univ Press, Hisar, pp 98–99.

Pandey R, Gupta ML, Singh HB and Kumar S (1999). The influence of vesicular-arbuscular mycorrhizal fungi alone or in combination with Meloidogyne incognita on *Hyoscyamus niger* L. *Bioresource Technology,* 69: 275–278.

Peter M, Ayer F and Egli S (2001). Nitrogen addition in a Norway spruce stand altered macromycete sporocarp composition measured by PCR–RFLP analysis of the ribosomal ITS–region. *New Phytologist,* 149: 311–326.

Peyronel B (1923). Fructificatio de l'endophyte a arbuscules et a vesicules des mycorrhizes endotrophes. *Bull Soc Mycol, France,* 39: 1–8.

Peyronel B (1924). Specie de Endogone protructrici di micorize endotrofiche. *Boll Mens R Staz Patol Veg Roma,* 5: 73–75.

Reddy BN and Sreevani A (2003). Biocontrol of dampinf off of tomato by using arbuscular mycorrhizal fungi. *J Mycol Pathol,* 3: 492 (Abstract).

Schaffers AP and Termorschuizen AJ (1989). A field survey on the relations between air pollution, stand vitality and the occurrence of fruitbodies of mycorrhizal fungi in plots of *Pinus sylvestris. Agriculture, Ecosystems and Environment,* 28: 449–454.

Schenck NC (1991). *Methods and Principles of Mycorrhizal Research.* APS Press, St Paul, Minnesota, USA, 243 pp.

Schisler DA and Linderman RG (1984). Evidence for the involvement of the soil microbiota in the exclusion of *Fusarium* from coniferous forest soils. *Can J Microbiol,* 30: 142–150.

Schisler DA and Linderman RG (1989). The influence of humic-rich organic amendments to coniferous nursery soils on Douglas-fir growth, damping-off, and associated microorganisms. *Soil Biol Biochem,* 21: 403–408.

Sikora RA (1979). Predisposition to Meloidegyne infection by the endotrophic mycorrhizal fungus Glomus mosseae. In: *Root-knot Nematodes (Meloidogyne Species) Systematics, Biology and Control,* (Eds) Lamberti, F and CE Taylor. AP, New York, pp 399–404.

Singh KP, Kumar J and Srinivas P (2007). Laboratory manual of forest pathology. *CFHA, GBPUAT,* Ranichauri, pp 45–46.

Smith GS, Roncadori RW and Hussey RS (1986). Interaction of endomycorrhizal fungi, superphosphate and Meloidogyne incognita on cotton in microplot and field studies. *J Nematology,* 18: 208–216.

Smith RS, Jr (1967). Decline of *Fusarium oxysporum* in roots of *Pinus lambertiana* seedlings transplanted into forest soils. *Phytopathology,* 57: 1265.

Smith SE and Gianinazzi-Pearson V (1988). Physiological interactions between symbionts in vesicular–arbuscular mycorrhizal plants. *Annual Rev Pl Physiol and Molecular Biology,* 39: 221–244.

Smith SE and Read DJ (1997). *Mycorrhizal Symbiosis.* Academic Press, San Diego, CA 605 pp.

Tainter FH and Baker FA (1996). *Principles of Forest Pathology.* John Wiley and Sons, INC, New York, 805 pp.

Verma RK and Jamaluddin (1995). Association and activity of arbuscular mycorrhiza of teak (*Tectona grandis*) in Central India. *Indian Forester*, 121(6): 536–537.

Wilcox HE (1983). Fungal parasitism of woody plant roots from mycorrhizal relationships to plant disease. *Annu Rev Phytopathol*, 21: 221–242.

Zak B (1964). Role of mycorrhizae in root disease. *Annu Rev Phytopathol*, 2: 377–392.

Soil Microflora, 2009 *Pages* **53–63**
Editor: **Rajan Kumar Gupta, Mukesh Kumar & Deepak Vyas**
Published by: **DAYA PUBLISHING HOUSE, NEW DELHI**

Chapter 6

Role of Earthworms in Soil Biology and Crop Production

***Y.V. Singh*[1], *Shalini Singh*[2] *and Rachana Srivastava*[2]**
[1]*Centre for Conservation and Utilization of Blue Green Algae, Indian Agricultural Research Institute, New Delhi*
E-mail: yvsingh63@yahoo.co.in yvsingh_algal@iari.res.in
[2]*Institute of Bioengineering and Biological Science, S-19/54, Varuna Bridge, Varanasi – 221 002, U.P.*

ABSTRACT

Earthworms are found in a wide range of habitats throughout the world, having adapted to many different soil types as well as to lakes and streams. Earthworms-often called night crawlers, garden worms, red worms or simply worms are a valuable resource to many people. They provide bait for fishing, a source of protein for food, and most importantly, they play a unique and important role in conditioning the soil. With the advent of chemical pest control, however, earthworms have become non-target recipients of many pesticides. Some of the most effective pesticides are broad spectrum in action, and they may inadvertently harm earthworms and other beneficial soil organisms. Harmful substances ingested by earthworms also may be concentrated up the food chain. Earthworms belong to the phylum Annelida and the class Oligochaeta, which consists of over 7000 species. Although one acre of soil may hold up to eight million earthworms, most people pay little attention to these productive and beneficial animals. They mostly go unnoticed from day to day, unless a heavy rain forces them to the surface of the soil, an angler needs some bait, or their casts disrupt a game of golf.

Keywords: *Earthworm, Earthworm casts, Soil aeration, Soil fertility enhancement, Vermiculture.*

Introduction

While nature works slowly in the production of topsoil, often over centuries, man, through poor agricultural practices, may deplete this valuable resource within an individual's lifetime. In the absence of a rich population of soil animals, 500 to 1000 years may be required to create an inch of topsoil. However, under favourable conditions, earthworms, lowly creatures to many people, can speed up this process to only five years. As agriculture, and ultimately civilization, depend on the maintenance of a fertile topsoil (Hyams, 1952), it is in our best interest to encourage earthworms in their soil building activities. The earthworm is truly an amazing little creature and the best friend the farmer and gardener ever had.

Much of the work regarding earthworm effects on other organisms has focussed on the functional significance of microbial-earthworm interactions, and little is known on the effects of earthworms on microfloral and faunal diversity. Earthworms can effect soil microfloral and fauna populations directly and indirectly by three main mechanism: (1) comminution, burrowing and casting; (2) grazing; (3) dispersal. These activities change the soils physico-chemical and biological status and may cause drastic shifts in the density, diversity, structure and activity of microbial and faunal communities within the drilosphere. Certain organisms and species may be enhanced, reduced or not be effected at all depending on their ability to adapt to the particular conditions of different earthworm drilospheres. A large host of factors (including $CaCo_3$, enzymes, mucus and antimicrobial substances) influence the ability of preferentially or randomly ingested organisms to survive (or not) passage through the earthworm gut, and their resultant capacity to recover or proliferate (or not) in earthworm casts.

Long before the invention of agricultural implements, earthworms ploughed the soil, mixing, tilling and building topsoil as they burrowed through the earth. Their importance has been clearly recognized for nearly 200 years, and even in the Fourth Century B.C., Aristotle, it is said, aptly referred to earthworms as "the intestines of the earth" though he may well have been referring to their appearance rather than to their function. But what do we know about these animals? The following will help us to understand earthworms and how we may be able to benefit from their activities.

Taxonomy

Earthworms are scientifically classified as animals belonging to the order Oligochaeta, class Chaetopoda, phylum Annelida. In this phylum there are about 1,800 species of earthworms grouped into five families and distributed all over the world (Lee, 1985). The most common worms in North America, Europe, and Western Asia belong to the family Lumbricidae, which has about 220 species. Only a few types are of interest to the commercial earthworm grower, and of these only two are raised on a large-scale commercial basis. Some of the more common species used for bait are the following:

Nightcrawlers

This earthworm is common to the northern states and may be picked from fields and lawns at night for commercial fish-bait sale. Although very popular with fishermen, they are not commonly raised on a commercial basis because they reproduce slowly and require special production and control procedures.

Field Worms

Also known as garden worms. These make excellent fish bait and are often preferred by those who want a small number of worms for their own use. They are not prolific breeders, so are not recommended for commercial enterprises.

Manure Worms

Also known as bandlings, red wigglers, or angleworms because of their squirming reactions when handled. These are particularly adaptable to commercial production and are one of the two types most commonly grown by successful worm farmers.

Red Worms

These are basically another type of manure worm, differing mainly in size and colour from their larger and darker cousins. They are also very adaptable to commercial production, and together with manure worms constitute about 80 to 90 percent of commercially-produced worms.

Manure worms and red worms can adapt to living in many different environments. They will eat almost any organic matter at some stage of decomposition, as well as many other types of materials which contain organic substances that can be ingested.

These worms may be found in manure piles or in soils containing large quantities of organic matter, but the new grower should purchase breeding stock from a reputable grower or distributor. Breeder worms may be purchased in lots as small as 1,000 worms. One 8-foot by 3-foot by 1-foot deep bin, however, may contain 100,000 worms or more.

Abundance, Distribution and Biology of Earthworms

The majority of temperate and many tropical soils support significant earthworm populations. A square yard of cropland in the United States can contain from 50-300 earthworms, or even larger populations in highly organic soils. A similar area of grassland or temperate woodlands will have from 100-500 earthworms (Edwards and Lofty, 1972). Based on their total biomass, earthworms are the predominant group of soil invertebrates in most soils. Earthworms are found all over the world. Australia, the Sahara Desert, Greenland and China are among only a few countries that have their own distinct indigenous species. Although several species live in various horizons (layers) of the soil or in the surface layer, others can be found in rotting logs, in the axils of tree branches (the upper angle between the branch and the trunk, sometimes up to 10 m above ground) or along the moist soil surrounding bodies of water (lakes, rivers, springs, ponds).

The family of earthworms that is most important in enhancing agricultural soil is Lumbricidae, which includes the genuses *Lumbricus, Aporrectodea*, and several others. Lumbricids originated in Europe and have been transported by human activities to many parts of the world. The United States has only one or two known native species of lumbricids. Others were brought to this country by settlers (probably in potted plants from Europe), and were distributed down the waterways. Generally, lumbricids are much more common in the north and east than in the drier south and west of the United States. They tend to be more abundant in loam and clay loam and even in silty soil, than in sandy soil and heavy clay. Populations also build up in irrigated soil. Earthworm populations tend to increase with soil organic matter levels and decrease with soil disturbances, such as tillage and potentially harmful chemicals (Crump, 1969).

Despite this wide variety of habitats, there are still certain environmental conditions which must be maintained for an earthworm to survive. For example, all earthworms need *an adequate food supply* to be close at hand. Earthworms generally remain close to their food supply. Since earthworms breath through their skin (they have no lungs), it is important that their environment is *moist* to allow for respiration. Earthworms release internal fluids (like perspiration) which traps the dissolved oxygen. Too much moisture (heavy rainfall) however takes the place of the valuable oxygen dissolved in the soil (also needed for survival), which may cause the earthworms to crawl to the soil surface. Here at the soil surface, earthworms will be exposed to ultra-violet radiation (sunlight) which is lethal to earthworms in a short period of time (Reynolds, 1973).

Earthworms are light sensitive and prefer moist slightly warm soil to grow and reproduce. Since earthworms live and travel around in the soil they form burrows as they move. Some species make deep vertical burrows. These earthworms are anecic species. Other species burrow continuously to form a network of channels-some vertical and some horizontal in the rhizosphere and are called endogeic species. Some earthworm species are not strong burrowers and live in the uppermost layer of soil in the litter layer. These earthworms are called epigeic species and they form shallow vertical burrows where they temporarily escape from drought, heat and disturbances. Though small, earthworms are fighters. They have developed certain survival strategies which help them cope with nasty environmental conditions. When the weather gets cold and the soil starts to freeze, earthworms move deeper down and overwinter in a state called aestivation. To aestivate, the earthworm generates a natural antifreeze and then curls up in a little knot. Earthworms also aestivate when conditions become dry or hot (Reynolds, 1977)

Earthworms breathe through their skin and must be in an environment that has at least 40 per cent moisture (at least as damp as a wrung out sponge). If their skin dries out, they cannot breathe and will die. Earthworms prefer a near-neutral soil pH. Instead of teeth, earthworms have a gizzard like a chicken that grinds the soil and organic matter that they consume. Their main intent is to eat the soil microorganisms that live in and on the soil and organic matter. Worm excrement is commonly called worm casts or castings. These soil clusters are glued together when excreted by the earthworm and are quite resistant to erosive forces. Their castings contain many more microorganisms than food sources because their intestines inoculate the casts with microorganisms. Earthworms become sexually mature when the familiar band (the clitellum) appears around their body, closer to their mouth. Each worm with a clitellum is capable of mating with other worms and producing cocoons that contain baby worms. Cocoons are lemon shaped and slightly smaller than a pencil eraser. Anecic worms are capable of burrowing to depths of 6 feet, often dragging surface litter (organic matter) into their burrows (Satchell, 1967).

Benefits of Earthworm

Builders of Soil

Earthworms benefit the soil in many ways, primarily due to the physical and chemical effects of their casts and burrows. Earthworm casts, consisting of waste excreted after feeding, are composed mostly of soil mixed with digested plant residues. Casts modify soil structure by breaking larger structural units (plates and blocks) into finer, spherical granules. An exception to this has been reported from some Canadian clay soils in which, during wet weather, worms can convert a soil structure to a massive paste. As plant material and soil passes through an earthworm's digestive system, its gizzard breaks down the particles into smaller fragments. These fragments, once excreted, are further decomposed by other worms and microorganisms. Earthworm casts can contribute up to 50 percent of the soil aggregates in some soils (Tashiro, 1987).

Cast production is most abundant in moist spring and fall seasons when earthworms inhabit surface layers of the soil. During this time, 20 casts per square foot of soil surface are not uncommon, and as much as 40 pounds of casts per 1000 square feet per year have been recorded. Under conditions of extreme temperatures or moisture stress during summer and winter, earthworms migrate downward into subsoil horizons, and enter a resting state called aestivation. In irrigated areas, such as golf course greens, fairways, and tees, this behavior may be altered and earthworms may not descend during the summer months. Thus, their activity may be regarded as a problem requiring management (Ware, 1978).

Many species of earthworms deposit their casts beneath the soil surface within their burrows, where casts contribute to pedogenesis. Species that excavate permanent, vertical burrows, however, deposit their casts on the soil surface, where they play a greater role in soil profile development. In addition to benefitting soil structure, casts also provide nitrogen in a useable form for other organisms that decompose organic matter on the soil surface. This interaction stimulates an accelerated decomposition rate, which helps reduce thatch buildup (Zoran, 1986).

The burrowing of earthworms improves the physical structure of the soil, creating channels through which plant roots may more easily penetrate the soil. In addition to increasing soil porosity and aeration, this activity also improves soil drainage and water penetration while eliminating hardpan conditions. Earthworms may also enhance soil structure through the formation of aggregates. Secretions in earthworm intestines cement soil particles together into aggregates which aid in erosion control. Man, through agricultural practices, such as cultivation, may temporarily improves soil structure, but the earthworm has longer-term effects in maintaining soil tilth (Darwin, 1881)

Field experiments conducted at IARI, New Delhi on organic farming during 2003-06 have revealed that vermicompost is a very effective organic amendments for organic farming. Its combined application with other bio-inoculants like Blue Green Algae (BGA), *Azolla* and Farm Yard Manure (FYM) could increase the grain yield in rice in the range of 65 to 102 per cent and 100 to 112 per cent, compared to absolute control (Table 6.1). Application of all these 4 organic amendments altogether increased the rice grain yield in the range of 114 to 116.8 per cent over total control.The rice grain yield (4.05 t/ha) obtained under combined application of four organic amendments was at par with the yield recorded under recommended dose of chemical fertilizer application (4.38 t/ha). Similar trend was recorded in grain yield of wheat but yield of wheat was lower as compared to its optimum yield level. Interestingly, there was no serious incidence of any insect pest or disease in organically grown rice as well as wheat crop. Microbial population (Actinomycetes, Bacteria, Fungi and BGA) increased due to the application of organic amendments in comparison to total control and recommended fertilizer application which accordingly resulted in a notable enhancement in dehydrogenase and phosphatase enzyme activities. Microbial population of Actinomycetes, Bacteria, Fungi and BGA in a composite soil sample before starting of experimentation in was 2003 was 74,203,14 and 3, respectively. Rice grain analysis for Iron and Manganese contents showed a significant increase in the treatments having 2 or more organic amendments added altogether over control (Table 6.1). Zinc, copper and Manganese content in grain also increased due to organic treatments but the increment was not significant over control. Soil organic carbon and available phosphorus contents were also found to be significantly increased due to organic farming over control as well as chemical fertilizer application.

Soil Fertility Enhancement

Earthworms are also important to nutrient availability of the soil. As they feed, they deposit digested organic matter and associated minerals along their burrows in the form of casts, a rich source of nutrients is placed in close proximity to the plant roots that grow through the burrows.

Comparative analyses of casts and surrounding soil have shown that casts contain five times more nitrogen, seven times more phosphorus, 11 times more potassium, three times more exchangeable magnesium, and one-and-one-half times more calcium. One explanation for this dramatic increase is that earthworms liberate nutrients from particles of both organic and mineral matter that would otherwise remain unavailable to plants. *Lumbricus terrestris*, the common nightcrawler, excretes pinhead-sized calcareous concretions that may raise the pH of the soil. Another factor is soil microbial activity within the casts, which promotes rapid transformation of soluble nitrogen into microbial proteins, thereby reducing the leaching of available nitrogen (Gaddie and Douglas, 1975).

Table 6.1: Effect of Different Organic Treatments on Rice Grain Yield, Concentration of Iron (Fe), Zinc (Zn), Copper (Cu) and Manganese (Mn) in Rice Grain and Microbial Population and Enzymatic Activity in Soil at Mid Crop Stage of Rice (mean of 3 years)

TrNo	*Treatments*	*Rice Grain Yield (t/ha)*	*Concentration in Rice Grain (ppm)*				*Microbial Population (CFU/gm of Soil) and Enzymatic Activity**				
			Fe	*Zn*	*Cu*	*Mn*	*1**	*2**	*3**	*4**	*5**
1	*Azolla*(A)*	2.54	35.1	32	12	34	332	369	31	59	131
2	BGA (B)	2.46	34.8	31	12	33	341	356	63	74	124
3	FYM (F)	2.24	35.2	32	12	34	261	322	51	61	110
4	Vermicompost (V)	2.66	35.3	32	13	35	276	365	43	48	108
5	A+B	3.25	37.2	34	12	35	287	380	32	23	121
6	A+F	3.42	36.2	33	13	36	279	364	33	42	134
7	A+V	3.85	36.9	33	14	34	195	321	32	35	113
8	B+F	3.26	36.1	33	13	35	267	386	34	55	113
9	B+V	3.50	37.1	34	14	36	243	364	37	68	127
10	F+V	3.58	37.4	34	13	35	267	368	34	57	112
11	A+B+F	3.66	38.9	35	15	37	256	376	41	78	122
12	A+F+V	3.70	37.6	35	16	39	380	402	65	98	124
13	B+F+V	3.82	38.3	35	16	38	376	378	75	86	132.52
14	A+B+F+V	4.05	39.8	36	17	40	301	334	61	87	125
15	$N_{80}P_{30}K_{30}$	4.38	33.1	32	13	36	164	332	69	23	101
16	$N_0P_0K_0$	1.84	32.4	31	12	32	160	312	29	12	101
	C.D (at 5%)	0.48	3.4	3	4	4					

1*: Actinomycetes x 10^3; 2*: Bacteria x 10^3; 3*: Fungi x 10^3 4*: BGA x10^3; 5*: Dehydrogenase enzyme activity.

Rate of application/ha: *Azolla* 1.0 t; BGA 2 kg; FYM 5.0 t; Vermicompost 5.0 t.

In soils populated by earthworms, accelerated decomposition of organic matter and an increase in available nitrogen results in greater numbers of nitrogen-fixing bacteria. Phosphorus availability also increases, due to earthworms' ingestion of phosphate rock particles and the consequent movement down burrows of phosphorus-containing casts. Furthermore, an abundance of earthworms means an abundance of decomposed organic matter–decomposition is limited only by the amount of material available, not by earthworms' capacity to ingest plant material (Ghabbour, 1966).

Aeration and Drainage

Earthworm burrows, too, exert both physical and chemical effects on soil. Burrows are of two types. Temporary burrows are made by earthworms moving from one feeding site to another. Permanent burrows are homes to individual worms, are usually more extensive, and are open to the surface, allowing the resident earthworm to select the most favorable microenvironment for feeding. Permanent burrows are fastidiously rebored by earthworms removing casts, organic matter and soil that have washed in.

As they burrow, earthworms excavate networks of passageways throughout the soil, which improves the soil's porosity. Up to two-thirds of all pore space in some soils are estimated to be the result of earthworm burrows, which can increase a soil's moisture-holding capacity–in some cases by as much as 400 percent. Because of the large diameter and low surface-tension of most burrows, they also serve as drainage systems during irrigation and heavy rainfall. This may account for better mixing of soluble nutrients throughout the soil profile (Hopp, 1973).

Earthworms also act as effective agents of soil aeration. As they penetrate the topsoil and proceed downward into the subsoil, they may increase the soil-to-air ratio by eight to thirty percent.

Earthworm Attrition

With so many benefits to the soil accrued from the activity of earthworms, why are they given so little consideration when pesticides are selected, pesticides that ultimately bring them harm? Pesticide registration guidelines initially gave little consideration to the potential impact of pesticides on non-target species. This has changed dramatically in recent years, and the Environmental Protection Agency now gives considerable attention to the impact of pesticides on earthworms and other non-target species during the registration process. Use patterns that negatively impact non-target species are unlikely to obtain registration; in fact, at present there are no pesticides registered by the EPA specifically for earthworm control.

Additionally, the applicator is often unaware of the detrimental effects that various pesticides have on earthworms. To be sure, the acute effects of various pesticides on earthworm distribution and abundance have been the topic of very little research in this country. Even less is known about pesticides' chronic effects on earthworms.

To meet the demands of greater use, more sophisticated means of pest control–and more advanced chemicals–are needed to maintain tees, fairways and greens under heavy use. Finally, early chemicals with broad-spectrum pesticidal activity and long-term residual effects, such as chlordane, resulted in the chronic reduction of earthworm activity. A single treatment could hold earthworm numbers in check for multiple seasons, depending on soil type and climatic conditions. By comparison, pesticides in use today are generally less toxic to earthworms; consequently, earthworm activity is more noticeable (Kevan, 1968).

Pesticides and Earthworms

Toxicity to earthworms varies widely among types of pesticides classified by use–insecticides and related compounds, fungicides, herbicides, fumigants, and vermicides. Two groups of pesticides are extremely toxic to earthworms and most other soil organisms–fumigants, such as chloropicrin, dichloropropane, and methyl bromide, and vermicides (designed intentionally to kill worms), such as ammonium sulphate, lead arsenate, and mercuric chloride.

Herbicides, at the other extreme, pose relatively little threat of earthworm toxicity. Their modes of action are directed toward plant regulation, and physiological processes of plants differ significantly from those of animals. This leaves fungicides and insecticides responsible for the most extensive pesticide impact on earthworms.

Insects, like earthworms, may be beneficial inhabitants of the soil in that they decompose organic matter; they may also act as predators or parasites to harmful insects. However, they can also be serious pests and must be maintained below damaging levels. Root- and shoot-feeding insects, which pose the greatest threat to golf course turf, are presently managed with organophosphate and carbamate insecticides to reduce their populations to non-injurious levels. However, a determination of non-injurious population densities is purely arbitrary.

Insecticide Toxicity

Earthworms are generally susceptible to carbamate compounds, which will significantly reduce their populations. Carbaryl, a carbamate pesticide often used for insect control, acts as a cholinesterase inhibitor, thereby producing long-lasting immobility and rigidity. Bendiocarb (Turcam) and propoxure (Baygon) are two other carbamate insecticides that cause paralysis in earthworms at normal dose rates. Carbofuran, another carbamate, is also very toxic to earthworms. Moreover, a sublethal response, characterized by weight loss, delayed clitellum development, and absence of cocoon production, has also been observed at recommended rates of carbofuran application.

Organophosphates are the most widely used class of turf insecticides. They have been successful in controlling white grubs, mole crickets, chinch bugs, and sod webworms, to name a few. Of the organophosphates, ethoprop is the most toxic to earthworms. In contrast, chlorpyrifos, isofenphos, and trichlorfon are considered non-toxic to earthworms when applied at normal dose rates.

Understanding how particular classes of biocides act upon target species may yield insights as to their effects on other living organisms. Organophosphates, as well as carbamates, mimic the structure of the acetylcholine molecule, an important component in the transmission of nerve impulses across synaptic gaps in many animals. Cholinesterase, an important enzyme in the nervous system, is responsible for the destruction of acetylcholine once a nerve impulse has crossed the synapse, thus preparing the synapse for another impulse. The presence of organophosphates or carbamates results in the phosphorylation of cholinesterase, thereby suppressing the destruction of acetylcholine. This results in a continuous firing of nerve impulses across the synapse, which is manifested as tetany. Because the axillary neuromuscular junctions of insects and other lower animals do not contain acetylcholine or cholinesterase, organophosphate and carbamate insecticides act instead on the central nervous system. The result is hyperexcitability, tremors convulsions, paralysis, and eventually death. Experimental evidence shows that long-term disruptions of the nervous system, such as excision of the brain, indicates that respiration in earthworms is not dependent on muscular contraction as in insects. Rather, it is the circulation of blood by rhythmic peristaltic muscle contractions that is affected. Thus, organophosphate and carbamate insecticides are believed to cause death by anoxia, not as a function of respiration but as a function of reduced blood circulation.

Fungicide Toxicity

Of the numerous fungicides registered for use on turf, only those in the benzimidazole class have demonstrated any remarkable toxicity to earthworms. This class includes benomyl, thiabendazole, thiophonate-methyl, and carbendazim, which is a metabolite of benomyl, and thiophonate-methyl. These compounds are used as broad-spectrum protectants. Their mode of action is primarily systemic; the ester metabolites of these compounds interfere with DNA synthesis by disrupting microtubule formation, which results in delayed mitosis. In addition to the acute toxicity of the benzimidazoles, other, sublethal effects, have been noted in treated worms, including reduced feeding, retarded growth rates, reduced cocoon production, and reduced nerve conduction velocity.

Although the abundance of earthworms may be affected by relatively few turf pesticides, earthworm distribution and behavior may be altered to a greater degree. Litter and surface soils treated with certain pesticides have a repellent effect on earthworms, and this reduces the breakdown and incorporation of organic matter into the subsurface horizons. Benomyl and carbendazim are particularly lethal to earthworms and also exhibit this repellent effect, which results in the avoidance of feeding in treated soils. Consequences include reduction in the amount of available nutrients in the root zone, decreased porosity and aeration of the soil, decreased water-holding capacity, and poor drainage.

Managing Earthworms

Earthworms, though often regarded as an annoyance by golfers and golf course superintendents, also provide several benefits to turf, as we have just seen. Reduction in the number of earthworms, whether intentional or not, can have a detrimental effect on both the physical and the chemical properties of the soil. Therefore, to maintain good soil structure capable of sustaining optimum plant growth, it would appear that attempts should be made to reduce the application of biocides known to adversely affect earthworm populations.

Clearly, the earthworm and its presence on the golf course raises many more questions than there are answers. Earthworms are generally thought to be beneficial; however, as with any other species, populations which are too high or out of place may warrant control actions. Currently there is insufficient data to determine at what levels earthworms become pests. In addition, a scientifically-based benefit: pest ratio has yet to be determined. Alternative management options need to be devised and the feasibility of such options evaluated. Chemical compounds can be developed specifically for earthworm control, but they may have a greater adverse effect on non-target organisms than pesticides registered for insect or pathogen control. All of these issues should be addressed and research carried out to answer the many questions that have arisen over the understanding of earthworm ecology (Rodale, 1961).

How do Agricultural Activities Affect Earthworms?

As earthworms are a measure of soil fertility, so are they indicators of soil management practices. Consequently, the use of earthworms to our benefit depends not only upon a knowledge of their activities but also upon an awareness of how our own activities, in particular agricultural practices, may influence their distribution.

Earthworm population may be increased or decreased by the following agricultural practices:

1. Cultivation
2. Cropping
3. Fertilizers
4. Pesticides

Earthworms are generally more numerous in grasslands than in arable land. Evidence indicates, however, that earthworm populations do not decline from mechanical damage during tillage operations, but rather from a reduction in the organic matter content of the soil. Repeated row cropping will reduce the number of earthworms, while the inclusion of grass or field crops in a rotation and intercropping will counter this effect.

Limestone generally increases earthworm populations and, in poor soils, nitrogen fertilizers may also benefit these indirectly. Most other mineral fertilizers have little effect on earthworm numbers, while organic matter such as manure, crop residues, or mulches favour earthworm multiplication by providing them with a source of food.

Many (though not all) of the insecticides, herbicides, and fungicides that are used to control agricultural pests are toxic to earthworms and may conflict with the natural biological control of pests. For example, earthworms play an important role in the control of apple scab, caused by the fungus *Venturia inequalis*, which overwinters on fallen leaves and twigs. Apple scab may be culturally controlled by burning these disease-carrying materials in the fall, or it may be chemically prevented through the use of copper sulphate, which is also toxic to earthworms. A less expensive, but equally effective means of controlling apple scab, however, is the introduction of earthworms, preferably Lumbricus terrestris, into orchard soils. These animals take the fallen leaves and twigs into their burrows where the vegetation eventually decomposes and ceases to be a source of disease. (One researcher found that earthworms may remove up to 90 per cent of leaf-fall in orchards (Raw, 1962).

Other pesticides that are lethal to earthworms include arsenic and copper compounds, chloropicrin, metham sodium, methyl bromide, D-D, chlordan, heptaclor, phorate and carbamate insecticides (Edwards and Lofty, 1972). Although other compounds may be less toxic to earthworms, these chemicals are concentrated in their bodies and may be lethal to birds and mammals when they are eaten.

How may Earthworms be Used to Increase Soil Fertility?

Earthworms have been successfully introduced into areas where they are absent and have been found to increase the yield of crops. The long-term benefits of encouraging earthworms can be translated into dollars. Researchers have estimated that for every dollar invested in earthworms on New Zealand sheep farms, the farmer can expect a return of $3.34 and an increase in carrying capacity of 2.5 stock units/hectare or an increase in productivity of 25-30 per cent.

When considering the use of earthworms to improve soil fertility it is important to remember that these animals thrive only under certain conditions. Most are unable to survive in sandy, dry, acid soils and all need organic matter for food. In addition, not all earthworm species are suitable for land reclamation. Species that are the easiest to cultivate, *i.e.,* those grown on compost or manure piles, are usually not suitable for inoculation of arable lands.

What are Other Uses of Earthworms?

Earthworms are familiar to the fisherman and poultry producer as bait or animal feed, but few North Americans realize that earthworms are regarded as a source of dietary protein (Waldon, 1978). Earthworms have also been used for medicinal purposes since ancient times in the treatment of illnesses such as bladder stones, jaundice, rheumatism, fever and impotency. Their efficiency, however, requires proper scientific investigation!

Vermiculture, the art of breeding and raising earthworms, is a billion dollar enterprise, supplying eager fishermen, zoos, fish hatcheries, poultry producers, and biology classrooms. The production of earthworms requires large amounts of organic matter with which to feed them. Consequently, vermiculture could be easily integrated with industries such as canneries, breweries, slaughterhouses, and papermills where large quantities of organic waste are produced. Rabbit breeders have found that earthworms placed under hutches are very effective in controlling odours from animal droppings and provide extra income if the earthworms are sold. Similarly, many of our urban wastes could be recycled through earthworms, solving many of our current problems in respect of solid waste disposal and water pollution.

Conclusions

Whether we are backyard gardeners or fully fledged farmers, it must be remembered that earthworms are not the antidote to infertile soils and poor management. If soils are to be improved through the use of earthworms, we must provide them with sufficient food and moisture. Only then may we profit from their activities as ploughmen and builders of the soil

References

Crump DR (1969). Earthworms: A profitable investment. *NZ J Agric,* 119(2): 84–85.

Darwin C (1881). *Darwin on Humus and the Earthworms: The Formation of Vegetable Mould through the Action of Worms with Observations on their Habits*. Faber and Faber, London, 153 pp.

Edwards CA and Lofty JR (1972). *Biology of Earthworms*. Chapman and Hall, Ltd. (Available from John Wiley, 605 Third Ave, New York, NY 10022.

Gaddie RE and Douglas DE (1975). Earthworms for ecology and profit. In: *Earthworms and the Ecology,* Bookworm Publ, Ontario 180 pp.

Ghabbour SI (1966). Earthworms in agriculture: A modern evaluation. *Rev Ecol Biol Soc,* 3(2): 259–271.

Hopp H (1973). *What Everyone who Gardens should Know about Earthworms*. Garden Way Publ, Charlotte, Vermont, 39 pp.

Hyams E (1952). *Soil and Civilization*. Thames and Hudson, London 312 pp.

Kevan DK (1968). *Soil Animals,* 2nd edn. Witherby, London, 244 pp.

Lee KE (1985). *Earthworms: Their Ecology and Relationships with Soils and Land Use.* Academic Press. Sydney.

Raw F (1962). Studies of earthworm populations in orchards. I. Leaf burial in apple orchards. *Ann App Biol,* 50: 389–404.

Reynolds JW (1973). Earthworms (Annelida:Oligochaeta) ecology and systematics. In: *Proc 1st Soil Microcommunities Conference,* (Ed) DL Dindal. US Atomic Energy Commission , 95–120 pp.

Reynolds JW (1977). *The Earthworms (Lumbricidae and Sparganophilidae) of Ontario.* Royal Ontario Museum, Toronto 141 pp.

Rodale R (1961). *The Challenge of Earthworm Research.* Soil and Health Foundation, Emmaus, Pennsylvania 102 pp.

Satchell JE (1967). Lumbricidae. In: *Soil Biology,* (Eds) A Burges and F Raw. Academic Press, New York, 259–322 pp.

Tashiro H (1987). *Turfgrass Insects of the United States and Canada.* Cornell University Press, Ithaca.

Waldon B (1978). Vermiculture is good for you. *Harrowsmith,* 3(1): 47–50.

Ware GW (1978). *Pesticides: Theory and Application.* W.H. Freeman and Company, San Francisco.

Zoran MJ (1986). Teratogenic effects of the fungicide benomyl on posterior segmental regeneration in the earthworm *Eisenia fetida. Pesticide Science,* 17: 641–652.

Soil Microflora, 2009 *Pages* **64–75**
Editor: **Rajan Kumar Gupta, Mukesh Kumar & Deepak Vyas**
Published by: **DAYA PUBLISHING HOUSE, NEW DELHI**

Chapter 7

Diversity of Soil Lichens in India

Roshni Khare[1], D.K. Upreti[1]*, Sanjeeva Nayaka[1] and R.K. Gupta[2]

[1]*Lichenology laboratory, National Botanical Research Institute, CSIR, Rana Pratap Marg, Lucknow – 226 001*

[2]*Pt. L. M. S. Government Post-Graduate College, Rishikesh – 249201*

ABSTRACT

India has a widespread occurrence of terricolous lichen communities represented by 51 and 14 macro and micro lichen genera respectively. Some of the species have only restricted occurrence on soil while some species share both soil and mosses. Out of 22 terricolous families in India Cladoniaceae, Collemataceae, Peltigeraceae, Parmeliaceae, Stereocaulaceae, Physciaceae and Lobariaceae are dominant. The genera *Cladonia, Collema, Peltigera, Leptogium, Lobaria* and *Stereocaulon* are the dominant soil lichens. Terricolous lichens grow on diverse habitat at different altitudinal gradients. The alpine regions of the country exhibit maximum diversity of soil lichens. Ecologically soil lichens are very important as they are primary colonizers and stabilizers of the soil.

Introduction

Lichens are the most fascinating and widely distributed organisms on this planet. Lichen is a consortium of a fungal and algal component resulting in a stable thallus. Their specific structure and unique physiology enable them to colonize on a number of substrates in varied climatic conditions. On the basis of size of thallus the lichens can be classified into macrolichens and microlichens. Lichens which are bigger in size and shape can be easily recognized as leaf like (foliose) and thread

* E-mail: upretidk@rediffmail.com

like (fruticose) structures are respectively called macrolichens while the group of lichens, which form a crust over the substratum and are quite smaller in size falls in the microlichens category.

The successful combination of alga and fungus resulted out in more than 20,000 lichen species in the world. India has a rich diversity of lichens represented by more than 2000 species (Awasthi, 2000), which is about 10 per cent of the total known lichen species from the world. A wide range of habitat differentiation found in lichens. Mostly lichen genera preferred specific type of substratum but many of them grow on more than one type of substratum. A certain period of time is needed for the establishment of thallus on a substratum and therefore, it is essential that the substratum remains undisturbed for that period. The form of lichen vegetation depends largely upon shape, structure, water relations and the chemistry of the substratum. On the basis of substratum lichens classified in to saxicolous (rocks and stones inhabiting), corticolous (growing on tree bark), terricolous (soil inhabiting), ramicolous (growing on twigs), lignicolous (growing on wood), muscicolous (growing over moss tuffs and cushion), omnicolous (inhabiting various substrates) and on man made substrate.

Terricolosus/Soil Lichens

The term Terricolous is rather ambiguous as it includes lichens of mineral soil, those of humid organic soil and those which live on terricolous bryophytes. Several species which are mainly found on rocks and are epiphytic can occasionally becomes terricolous. In India out of about 70 macrolichens genera 51 exhibit their growth on soil while only 14 microlichens genera out of 160 genera are soil inhabiting.

Species of macro lichen genera *Acsoscyphus, Allocetraria, Anaptychia, Baeomyces, Bryoria, Bulbothrix, Canomaculina, Catapyrenium, Cetraria, Cladia, Cladonia, Collema, Dibaeys, Evernia, Everniastrum, Flavocetraria,Flavocetrariella, Flavoparmelia, Fuscopannaria, Heppia, Heterodermia, Hypogymnia, Hypotrachyna, Icmadophila, Lecanora, Leprocaulon, Leptogium, Lethariella, Melanelia, Melanilixia, Nephroma, Oropogon, Parmeliella, Parmelinella, Parmotrema, Peltigera, Phaeophyscia, Physcia, Physconia, Pseudocyphellaria, Peltula, Psora, Ramalina, Siphula, Solorina, Stereocaulon, Sticta, Teloschistes, Thamnolia and Xanthoparmelia* colonize on soil. Some species of *Caloplaca, Candeleriella, Diplotomma, Diploschistes, Endocarpon, Lecidella, Lepraria, Leproloma, Mycobilimbia, Ochrolechia, Peccania, Pertusaria, Rinodina, Rusavakia* and *Toninia* are terricolous microlichens in India (Table 7.1)

In India soil lichens belongs to many families such as Alectoriaceae, Bacidiaceae, Baeomycetaceae, Candelariaceae, Cladoniaceae, Collemataceae, Heppiaceae, Icmadophilaceae, Lecanoraceae, Lichinaceae, Lobariaceae, Nephromataceae, Pannariaceae, Peltigeraceae, Pertusariaceae, Parmeliaceae, Physciaceae, Porpidiaceae, Siphulaceae, Stereocaulaceae, Teloschistaceae and Verrucariaceae. The dominant terricolous lichen families are Cladoniaceae, Collemataceae, Peltigeraceae, Parmeliaceae, Stereocaulaceae, Physciaceae and Lobariaceae.

The terricolous lichens growing on diverse habitat at different altitudinal gradients. Some lichens colonize the bare undisturbed soil specially along road side in subtemperate and temperate regions such as crustose lichen species of *Baeomyces, Diploschistes, Endocarpon, Mycobilimbia* and some Lecidioid lichens. Species of lichen genera such as *Psora decipiens, Diploschistes scruposus, Endocarpon pusillum* and species of *Cladonia* and *Peccania* colonize on calcareous soil. Species of *Cladonia, Peltigera* and *Lobaria* grows on siliceous soil. Several species of *Cladonia* grows on decaying stumps of trees over soil. *Diploschistes muscorum, Pertusaria bryontha* and *Rinodina conradi* are amongst those species which are more or less terricolous but also grows spreading over dead or living terrestrial mosses. Certain species of *Cladonia, Collema, Cetraria, Leptogium, Heterodermia, Lecanora, Lobaria, Peltigera, Thamnolia* and Parmeliaceae are also found on soil in temperate and alpine regions.

Table 7.1: List of Soil Lichen Species in India

Species Name	Habitat
Acroscyphus sphaerophoroides Lév.	Mainly terricolous
Allocetraria ambigua (Bab.) Kurok. and M. J. Lai	Mainly terricolous
Allocetraria stracheyi (Bab.) Kurok. and M.J. Lai	Mainly terricolous
Anaptychia psuedoroemeri D. D. Awasthi and S. R. Singh	Mainly terricolous
Baeomyces pachypus Nyl.	Mainly terricolous
Baeomyces sorediifer Nyl.	Mainly terricolous
Bryoria bicolor (Ehrh.) Brodo and D. Hawksworth	Mainly terricolous
Bryoria implexa (Hoffm.) Brodo and D. Hawksworth	Mainly terricolous
Bryoria smithii (Du Rietz) Brodo and D. Hawksworth	Rarely terricolous
Bryoria tenuis (Dahl) Brodo and D. Hawksworth	Mainly terricolous
Bulbothrix meizospora (Nyl.) Hale	Rarely terricolous
Caloplaca heterospora Poelt and Hinteregger	Mainly terricolous
Caloplaca maura Poelt and Hinteregger	Sometimes terricolous
Caloplaca citrina (Hoffm.) Thr. Fr.	Mainly terricolous
Candeleriella vitellina (Ehrh.) Müll. Arg.	Mainly terricolous
Canomaculina subtinctoria (Zahlbr.) Elix	Mainly terricolous
Catapyrenium cinereum (Pers.) Körb.	Mainly terricolous
Cetraria aculeata (Schreb.) Fr.	Mainly terricolous
Cetraria islandica (L.) Ach. subsp. *islandica*	Mainly terricolous
Cetraria muricata (Ach.) Eckfeldt	Mainly terricolous
Cladia aggregata (Sw.) Nyl.	Mainly terricolous
Cladonia awasthiana Ahti and Upreti	Mainly terricolous
Cladonia cartilaginea Müll. Arg.	Mainly terricolous
Cladonia cariosa (Ach.) Spreng.	Mainly terricolous
Cladonia chlorophaea (Flöerke ex Sommerf.) Spreng.	Mainly terricolous
Cladonia coccifera (L.) Willd	Mainly terricolous
Cladonia coniocraea (Flörke) Spreng.	Mainly terricolous
Cladonia corniculata Ahti and Kashiwadani	Mainly terricolous
Cladonia corymbescens Nyl. ex. Height.	Mainly terricolous
Cladonia fenestralis Nuno	Mainly terricolous
Cladonia fimbriata (L.) Fr.	Mainly terricolous
Cladonia furcata (Huds.) Schrad.	Mainly terricolous
Cladonia macroptera Räsänen	Mainly terricolous
Cladonia mongolika Ahti	Mainly terricolous
Cladonia ochrochlora Flöerke	Mainly terricolous
Cladonia pocillum (Ach.) Grognol	Mainly terricolous

Contd...

Table 7.1–Contd...

Species Name	Habitat
Cladonia pyxidata (L.) Hoffm.	Mainly terricolous
Cladonia ramulosa (With.) Laundon	Mainly terricolous
Cladonia rangiferina (L.) Wigg.	Mainly terricolous
Cladonia singhii Ahti and Dixit	Mainly terricolous
Cladonia scabriuscula (Del. in Duby) Nyl.	Mainly terricolous
Cladonia squamosa Hoffm.	Mainly terricolous
Collema coccophorum Tuck.	Mainly terricolous
Collema limosum (Ach.) Ach.	Mainly terricolous
Collema rugosum Kremp.	Mainly terricolous
Collema subflaccidum Degel.	Mainly terricolous
Collema tenax (Sw.) Ach. *tenax*	Mainly terricolous
Collema tenax var. *vulgare* (Schaer.) Degel.	Mainly terricolous
Dibaeys baeomyces (L. F.) Rambold and Hertel	Mainly terricolous
Dibaeys pulogensis (Vain.) Kalb and Gierl	Mainly terricolous
Diplotomma geophilum (Florke and Sommerr.) Szat. ex. D. D. Awasthi and S. Singh	Mainly terricolous
Diploschistes scruposus (Schreb.) Norman	Mainly terricolous
Diploschistres muscorum (Scop.) R. Sant.	More or less terricolous
Diploschistes scruposus (Schreber) Norman	More or less terricolous
Evernia mesomorpha Nyl.	Rarely terricolous
Everniastrum cirrhatum (Fr.) Hale	Mainly terricolous
Everniastrum nepalense (Taylor) Hale	Mainly terricolous
Everniastrum vexans (Zahlbr.) Hale	Rarely terricolous
Flavocetraria cucullata (Bell.) Kärnefelt and Thell	Mainly terricolous
Flavocetraria nivelis (L.) Kärnefelt and Thell	Mainly terricolous
Flavocetrariella leucostigma (Lév.) D. D. Awasthi comb. nov.	Mainly terricolous
Flavocetrariella melaloma (Nyl.) D. D. Awasthi comb. nov.	Terricolous with mosses
Flavoparmelia caperata L. Hale	Mainly terricolous
Fuscopannaria saltuensis P. M. Jørg.	Mainly terricolous
Heppia lutosa (Ach.) Nyl.	Mainly terricolous
Heterodermia boryi (Fée) Kr. P. Singh and S. R. Singh	Mainly terricolous
Heterodermia diademata (Taylor) D. D. Awasthi	Mainly terricolous
Heterodermia firmula (Nyl.) Trevis.	Mainly terricolous
Heterodermia japonica (Sato) Swinsc. and Krog	Mainly terricolous
Heterodermia leucomelos (L.) Poelt	Mainly terricolous

Contd...

Table 7.1–Contd...

Species Name	Habitat
Heterodermia obscurata (Nyl.) Trevis.	Mainly terricolous
Heterodermia pseudospeciosa (Kurok.) W. Culb.	Mainly terricolous
Hypogymnia hypotrypha (Nyl.) Rass.	Mainly terricolous
Hypogymnia fragillima (Hillm.) Rass.	Mainly terricolous
Hypogymnia physodes (L.) Nyl.	Rarely terricolous
Hypotrachyna costaricensis (Nyl.) Hale	Mainly terricolous
Hypotrachyna koyaensis (Asahina) Hale	Mainly terricolous
Icmadophila ericetorum (L.) Zahlbr.	Mainly terricolous
Lecanora chondroderma Zahlbr.	Mainly terricolous
Lecanora himalayae Poelt	Mainly terricolous
Lempholemma chalazanum (Ach.) B. de Lesd.	Mainly terricolous
Leprocaulon arbuscula (Nyl.) Nyl.	Mainly terricolous
Leprocaulon pseudoarbuscula (Asahina) I. M. Lamb and Ward	Mainly terricolous
Leptogium burnetiae Dodge var *burnetiae*	Rarely terricolous
Leptogium burnetiae Dodge var. *hirsutum*(Sierk) P. M. Jørg.	Rarely terricolous
Leptogium corniculatum (Hoffm.) Minks	Mainly terricolous
Leptogium cyanescens (Rabenh.) Körb.	Mainly terricolous
Leptogium denticulatum Nyl.	Mainly terricolous
Leptogium moluccanum (Pers.) Vain.	Mainly terricolous
Leptogium platynum (Tuck.) Herre	Mainly terricolous
Leptogium trichophorum Müll. Arg.	Mainly terricolous
Lethariella cladonioides (Nyl.) Krog.	Mainly terricolous
Lobaria isidiosa (Müll. Arg.) Vain.	Mainly terricolous
Lobaria kurokawae Yoshim.	Rarely terricolous
Lobaria pseudopulmonaria Gyeln.	Mainly terricolous
Lobaria retigera (Bory) Trev.	Mainly terricolous
Melanelia hepatizon (Ach.) Thell	Mainly terricolous
Mycobilimbia hunana (Zahlbr.) Awasthi	Mainly terricolous
Nephroma expallidum (Nyl.) Nyl.	Mainly terricolous
Nephroma helveticum Ach. var. *helveticum*	Mainly terricolous
Nephroma isiodosum (Nyl.) Gyeln.	Mainly terricolous
Nephroma parile (Ach.) Ach.	Mainly terricolous
Oropogon formosanus Asahina	Mainly terricolous
Parmeliella tryptophylla (Ach.) Müll. Arg.	Mainly terricolous
Parmelinella wallichiana (Taylor) Elix and Hale	Rarely terricolous

Contd...

Table 7.1–Contd...

Species Name	*Habitat*
Parmotrema crinitum (Ach.) Choisy	Rarely terricolous
Parmotrema grayanum (Hue) Hale	Rarely terricolous
Parmotrema mellissi (C. W. Dodge) Hale	Mainly terricolous
Parmotrema nilgherrense (Nyl.) Hale	Mainly terricolous
Parmotrema pseudonilgherrense (Asahina) Hale	Rarely terricolous
Parmotrema reticulatum (Taylor) Choisy	Mainly terricolous
Peltigera canina (L.) Willd.	Mainly terricolous
Peltigera collina (Ach.) Schrad.	Mainly terricolous
Peltigera didactyla (With.) J. R. Laundon	Mainly terricolous
Peltigera dolichorhiza (Nyl.) Nyl.	Mainly terricolous
Peltigera dolichospora (Lu) Vitik.	Mainly terricolous
Peltigera horizontalis (Hudson) Baumg.	Mainly terricolous
Peltigera leucophlebia (Nyl.) Gyeln.	Mainly terricolous
Peltigera macra Vainio	Mainly terricolous
Peltigera malacea (Ach.) Funck	Mainly terricolous
Peltigera membranacea (Ach.) Nyl.	Mainly terricolous
Peltigera polydactylon (Neck.) Hoffm. var. *polydactylon*	Mainly terricolous
Peltigera polydactylon (Neck.) Hoffm. var. *polydactyloni* Gyeln.	Mainly terricolous
Peltigera praetextata (Flörke) Zopf.	Mainly terricolous
Peltigera rufescens (Weiss) Humb.	Mainly terricolous
Peltigera venosa (L.) Hoffm.	Mainly terricolous
Pertusaria bryontha (Ach.) Nyl.	More or less terricolous
Phaeophyscia constipata (Norrl. and Nyl.) Moberg	Mainly terricolous
Phaeophyscia hispidula (Ach.) Moberg var. *hispidula*	Mainly terricolous
Physcia adscendens (Fr.) Oliv.	Rarely terricolous
Physcia albinea (Ach.) Malbr.	Terricolous over rocks
Physconia muscigena (Ach.) Poelt	Mainly terricolous
Pseudocyphellaria crocata (L.) Vain.	Sometimes terricolous
Psora decipiens (Hedwig) Hoffm.	Mainly terricolous
Psora himalayana (Bab.) Timdal	Mainly terricolous
Ramalina intermedia (Delise ex Nyl.)Nyl.	Rarely terricolous
Ramalina roesleri (Hochst) Hue	Sometimes terricolous
Ramalina cfr *taitensis* Nyl.	Mainly terricolous
Rinodina conradi Körber	More or less terricolous
Rusavakia sorediata (Vain.) S. Kondratyuk and Kärnefelt	Rarely terricolous
Siphula ceratites (Wahlemb) Fr. var. *ceratites*	Mainlmy terricolous

Contd...

Table 7.1–Contd...

Species Name	*Habitat*
Solorina sinensis Hochst.	On soil over boulders, terricolous
Stereocaulon alpinum Laurer	Mainly terricolous
Stereocaulon austroindicum I. M. Lamb	Mainly terricolous
Stereocaulon coniophyllum I. M. Lamb	Rarely terricolous
Stereocaulon foliolosum Nyl. Var. *foliolosum*	Rarely terricolous
Stereocaulon glareosum (Sav.) H. Magn.	Mainly terricolous
Stereocaulon himalayense D. D. Awasthi and I. M. Lamb	Mainly terricolous
Stereocaulon myriocarpum Th. Fr.	Mainly terricolous
Stereocaulon paradoxum I. M. Lamb	Mainly terricolous
Stereocaulon piluliferum Th. Fr.	Rarely terricolous
Stereocaulon sasaki Zahlbr. var. *sasaki*	Mainly terricolous
Stereocaulon sasaki Zahlbr. var. *tomentosoides I. M. Lamb*	Mainly terricolous
Sticta cyphellulata (Müll. Arg.) Hue	Rarely terricolous
Sticta limbata (Sm.) Ach.	Rarely terricolous
Sticta nylanderiana Zahlbr.	Rarely terricolous
Sticta orbicularis (R. Br.) Hue	Mainly terricolous
Sticta weigelii (Ach.) Vain. var weigelii	Sometimes terricolous
Teloschistes flavicans (Sw.) Norm.	Rarely terricolous
Thamnolia vermicularis (Sw.) Schaer. var. *vermicularis*	Mainly terricolous
Thamnolia vermicularis (Sw.) Schaer. var. subuliformis (Ehrh.) Schaer.	Mainly terricolous
Toninia aromatica (Sm.) A. Massal	Terricolous in open habitats
Xanthoparmelia bellatula (Kurok and Filson) Elix Johntson and Armstrong	Mainly terricolous
Xanthoparmelia terricola Hale, Nash and Elix	Mainly terricolous

Collection of Soil Lichens

The collection of soil lichens is slightly different from the collection of other groups of lichens. When the soil dries up or when little pressure is applied on it lichens growing on soil crumble into small pieces. So crustose lichen growing on soil should be taken up along with a 1-2 cm thick layer of soil in size of about 5 x 5 cm and then tightly packed within absorbent paper so that the whole material remains in an intact condition. Another method is that soil lumps with the crustose lichen are placed in a thin layer of aqueous glue, which gradually seeps into the soil from below and on drying does not allow the soil crumbling, and retains the lichen in its position. Special attention is essential for crustose lichens which are growing on dry, gravelly soil as crumbling starts during collection. In these cases water can be sprayed to delay crumbling. For collection of foliose and fruticose forms, the crumbling of the soil does not affect much of lichen thallus except the natural appearance so they should collect carefully.

All the terricolous lichen species occur on more or less calciferous soils especially clayey and sandy soil, either quite naked or in association with mosses. Due to having weak in competition

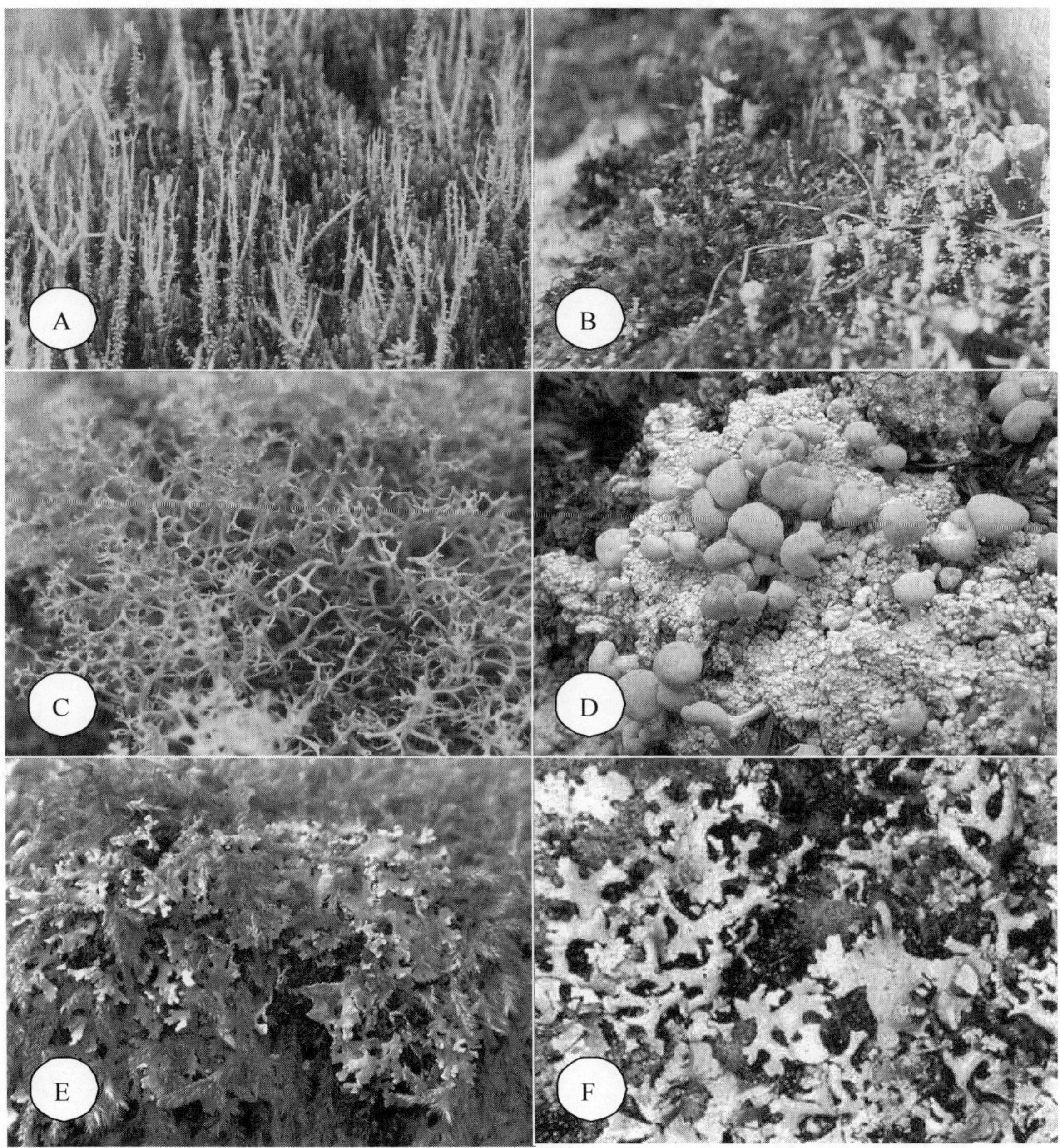

Plate 7.1

(A) *Cladonia scabriuscula;* **(B)** *Cladonia coccifera;* **(C)** *Cladia aggregata;* **(D)** *Dibaeis baeomyces;* **(E)** *Heterodermia obscurata;* **(F)** *Hypogymnia physodes*

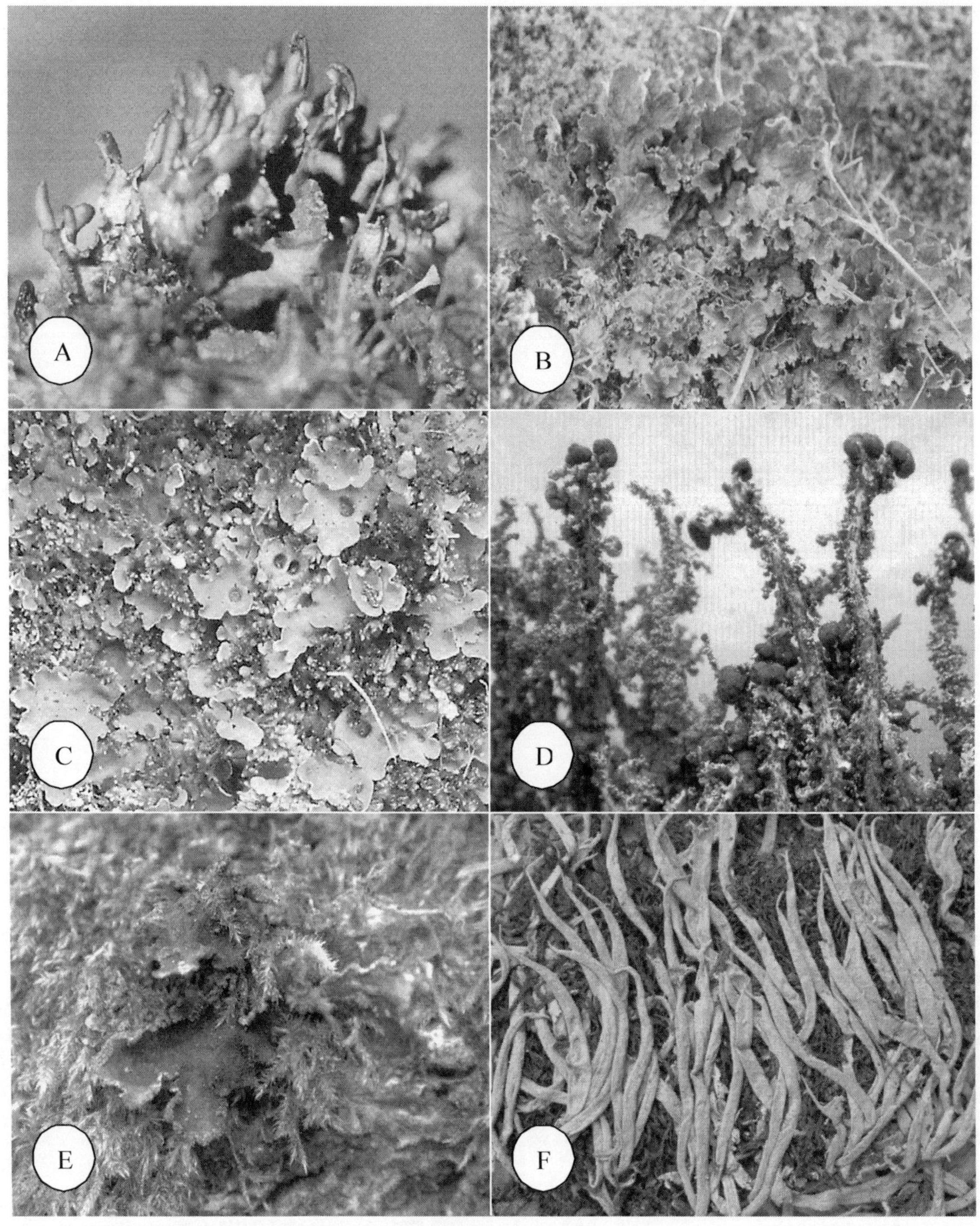

Plate 7.2

(A) *Peltigera canina;* (B) *Peltigera rufescens;* (C) *Solorina* sps.*;* (D) *Stereocaulon foliolosum;* (E) *Sticta limbata;* (F) *Thamnolia vermicularis*

lichens have to grow on naked soil or soil with a sparse vegetation of small plants. Both biotic and abiotic factors are responsible for the growth of lichens in each environment. Stable substrates are necessary prerequisite for the establishment and growth of lichens (Rogers, 1974) while they are generally absent from landscapes dominated by unstable soil or clay (Eldridge, 1996).

Types of Terricolous Lichens

Based on the type of soil the terricolous lichens can be categorized as basic, coastal and acidic soil lichens. These soil lichens may be sporadic within vascular plant dominated communities. Pebbles which are not constant or in movable stage rarely colonized by lichens. Only immobile pebbles are colonized by them. Species of brown *Cetraria* are common on pebbles in alpine regions of India.

In basic soils for successful colonization the soil lichens requires open and sufficiently stable habitats. In these habitats vascular plant flora is either reduced in height by grazing or by the dryness of the soil. Some lichens generally treated as terricolous are characteristic of bryophyte rich communities and are themselves predominantly bryophilous rather than strictly terricolous. *Caloplaca citrina, Collema tenax, Collema crispum* and *Lecidioid* species are the common lichens on basic soil in India. The amount of calcium carbonate in the soil determines the abundance and species diversity of lichen dominated communities in calcareous sites. *Cladonia pocillum* and *Collema tenax* occur on soils with 100 mg ca/100g and a pH usually above 6.2 while *Lecidea* species require a pH of 8.2 to colonize.

Very few lichen communities have ability to colonize coastal soil. Mostly lichens of these conditions require open habitats with bare but generally more or less stable soil. At coastal sites soil lichens are most frequent on wind wept, cliff tops, eroded banks or on thin layer of soil associated with rock outcrops. *Cladonia* and *Peltigera* are the most common genera in coastal soil.

The soils adjacent to rotting trees, logs and stumps to provide large amount of humus to the soil for colonization of lichen species. In acidic soils having areas the age of the community and frequency of burning are of considerable importance in relation not only to the diversity of the vascular plant flora but also to that of cryptogams including lichens. Open communities show a sequence of lichen development after burning. Common lichen species are *Cladonia arbuscula, Cladonia implexa, Icmadophila ericetorum* and other species.

Distribution of Soil Lichens in India

Soil lichen communities are best developed in alpine and arctic conditions and other habitats where overgrowing trees are lacking. The foothills of the Himalayas below an altitude of 600 m usually have fewer lichens in moist places. The terricolous lichen vegetation changes as the altitude increases. Evergreen moist forests have more luxuriant growth of lichen as compared to dry deciduous forests. In the moist open places *Collema* spp., *Cladonia* spp., *Diploschistes scruposus* and *Leptogium cyanescens* are the common soil lichens found in tropical areas.

In temperate regions of India the corticulous lichens dominates over saxicolous and terricolous. The ground flora under coniferous forest at lower temperate areas remains mostly dry thus favour scanty to poor growth of soil lichens. The evergreen temperate forest and coniferous forest of upper temperate regions provide a moist shady environment suitable for species of *Lobaria, Peltigera, Stereocaulon* and *Cladonia* to colonize on soil or among mosses. Common crustose soil lichens genera are *Caloplaca, Diploschistes, Diplotomma* and *Pertusaria.*

Most of the alpine regions in the Himalayas exhibit dominant growth of the terricolous communities of lichens. Fruticose species *Cladonia rangiferina* and *Cladia aggregata* grow luxuriantly in moist slopes in alpine regions. The cold desert in the Himalayas also exhibit good growth of terricolous lichens. Out of 81 species of lichens recorded from the cold desert of Lahul and Spiti area of Himachal Pradesh, 18 were soil inhabiting (Shrivastava *et al.*, 2006). *Cetraria* spp., *Hypogymnia hypotrypa*, *Lethariella cladonioides* and *Thamnolia vermicularis* are the most common terricolous lichen species in Eastern Himalayan region.

Ecology of Soil Lichens

Soil lichens were most strongly associated with total content of annual rainfall, soil pH, calcium carbonate levels while to a lesser content their distribution was associated with plant cover, soil texture and organic carbon levels of soil. They are highly vulnerable to environmental changes caused by growth or increase in density of grasses, shrubs or trees and accumulation of humus and litter decomposition. Soil colonization starts with the growth of crustose and squamulose lichens consisting of pioneer lichen community. According to Linnaeus "Crustaceous lichens are the first foundation of vegetation". Soil crust lichens bind the top soil and prevent erosion so play an important role in ecology of tropical, temperate and alpine regions. The stage of soil colonization leading to increase runoff and decrease erosion is followed by higher plants colonization and beginning of true soil formation. Lichens occupy the empty spaces left by higher plants and generally do not share same habitats. It was observed that lichens were most common on soils with a fine texture which either crusted or hard setting while rare on self mulching sands and gravels.

Lichens are important agents for soil stabilization. The most stabilizing group is of squamulose species of *Catapyrenium, Heppia, Peltula, Psora* and *Diploschistes*. They grow appressed to the soil and effectively stabilize it. The different factors which determine the colonization of lichens on soil are drought, nutrient status and various environmental factors. Ecological and biogeographical studies of soil lichens revealed that the major environmental factors influencing their distribution are rainfall amount and distribution and other soil factors such as calcium carbonate content and sodium content. Most of the soil lichens have ability to fix the atmospheric nitrogen, through their own phycobiont (if a blue green algae) or because of accessory phycobiont housed in cephalodia. Such lichens include species of *Collema, Leptogium, Peltigera, Solorina* and *Stereocaulon*.

In succession on soil, the lichens exhibit considerable variations at different altitudinal zones. In alpine regions, the exposed glacial moraine may quickly covered by the crustose lecidioid lichens, followed by Cetrarioid and Parmelioid lichen species. Soil lichens on soil usually form an intimate part of already higher plant associations. (Schartz and Martin, 1960) considered that lichens were among the primary biochemical weathering agents involved in the conversion of bare rock into soil. Lichens are believed to increase soil microtopography and therfore enhance the capacity of the soil to pond and retain water (West, 1990). Terricolous lichens present much greater problems of phytosociological delimitation than community characteristics of other substrates. The areas where lichens form the primary soil communities will be of great importance to ecologists in future. The soil lichen crust contribute significantly to the fertility of desert soils and also helps in the growth of vascular plants. The cyanophycean terricolous lichens enrich the nitrogen contents of soil. The soil lichens in desert areas provide shelter to a number of insects.

The destruction of soil surface lichens by trampling of grazing animals have far reaching effects as the surface soil containing nutrients is easily blown away by wind or air. Due to the change in the soil texture the microflora of the soil changed radically which will in turn effect the germination of seeds and the growth of many phanerogames. In arid areas the soil lichens play an important role in soil stability, hydrology and fertility. The lichens are resistant to extremes of heat and drought thus enables to occupy habitats unavailable to colonize by other group of plants. Mostly the crustose squamulose and sometimes the foliose lichens covers the soil surface in desert region and stabilize the soil. The desert soils which are more prone to wind and water erosion; the lichen cover retard such erosion to a marked extent. The rhizines present on the lower side of the lichen thallus binds the friable soil particles prevent the rain drop impact on to the soil.

Ecologically lichens are probably most important as primary colonizers and stabilizers. They are the only plants which can grow on a bare rock as an initiator of colonization and often remain the sole colonizers with no real successors. On soil crust in semi arid areas, the lichens togather with other cryptrogams play a major role in holding the soil in place, preventing the soil erosion with stabilizing the surface and building up humus to form the soil more fertile. Their very important function has been to be little appreciated in the past.

Acknowledgements

We are thankful to Dr. R. Tuli, Director, National Botanical Research Institute, Lucknow (CSIR) for providing laboratory facilities to work. The authors are also grateful to the Ministry of Environment and Forest, New Delhi for financial support as most of the soil lichens from alpine regions of India were collected during the course of this project.

References

Awasthi DD (2007). *A Compendium of the Macrolichens from India, Nepal and Sri Lanka*. Bishen Singh and Mahendra Pal Singh, Dehradun, pp 1–580.

Awasthi DD (1991). A key to the microlichens of India, Nepal and Sri Lanka. *Biblthca Lichenol*, 40: 1–336.

Eldridge DJ (1996). Distribution and floristics of lichens in arid and semi arid New South Wales, Austria. *Australian Journal of Botany*, p 44.

Melbourne and Ludwig JA (1994). Small scale patch heterogenety in semi arid landscapes. *Pacific Conservation Biology*, 1: 201–208.

Mukherji KG, Chamola BP, Upreti DK and Upadhyay RK (Eds) *Biology of Lichens*. Aravali Books International, New Delhi, pp 3–48.

Rogers RW (1971). Distribution of the lichen chondropsis in relation to its heat and drought resistance. *New Phytologist*, 70: 1069–1077.

Rogers RW (1972). Soil surface lichens in arid and semi arid South Eastern Australia III. The relation between distribution and environment. *Australian Journal of Botany*, 20: 301–316.

Rogers RW (1974). Lichens from the TGB. Osborn vegetation reserve at Koonamore in arid South Australia. *Transaction of the Royal Society of South Australia*, 98: 113–124.

Rogers RW (1977). Lichens of hot arid and semi-arid lands. In: *Lichen Ecology*, (Ed) MRD Seaward. Academic Press, London, pp 211–252.

Tongway DJ (1994). *Rangeland Soil Condition Assessment Manual*.

West NE (1990). Structure and function of microphyte soil crusts in wild land ecosystems of arid and semi arid regions. *Advances in Ecological Research*, 20: 179–223.

Soil Microflora, 2009 *Pages* **76–85**
Editor: **Rajan Kumar Gupta, Mukesh Kumar & Deepak Vyas**
Published by: **DAYA PUBLISHING HOUSE, NEW DELHI**

Chapter 8

Bacterial and Fungal Diversity in Alkali Affected Soil

Kaushal Pratap Singh*, Dheeraj Mohan and Seema Bhadauria
Microbiology Research Laboratory, Department of Botany,
Raja Balwant Singh College, Agra, U.P.

ABSTRACT

Soil microflora is affected by various edaphic factors, which include soil type, soil moisture and soil reactions; seasonal changes like temperature, rainfall, light and the type of ground vegetation. In the present study, on the occurrence and distribution of bacterial and fungal microflora in 'Usar' soils of different pH values revealed that the soils with low alkalinity contained large number of micro fungi, bacteria and their number decreased with progressive increase in soil pH. The diversity of fungal and bacterial species was affected by seasonal variation. The maximum bacterial count was found in rainy season at all the sites while minimum was in summer season. The maximum fungal colonies were isolated during March and August. During summer fungal population were found to be minimum in numbers. Species belonging to *Ascomycetes*, *Aspergillus flavus*, *Aspergillus nidulans* were found almost in every season of the year.

Keywords: *Mainpuri district, microbial diversity, alkali-affected soil, seasonal variation, nitrifiers.*

Introduction

India have 2.8 million hectare of alkali lands, which are affected by high alkali conditions (Abrol and Bhumbla, 1971). These soils are deficient in organic matter, nitrogen, zinc and soluble calcium

* E-mail: kaushalmpi1978@yahoo.com; Phone: +91-9412660984

(Sharma and Bhargava, 1978). Such soil contains high amounts of sodium carbonate and sodium, leads to high pH and adversely affects both physical and nutritional properties of the soil and thus makes these soils very inhospitable for plant growth. The reason for high ESP values is that excess sodium carbonate results in the precipitation of soluble and most of the exchangeable calcium resulting in high values of exchangeable sodium percentage.

High salt content is mainly in the surface layers 0 to 30 cm soil depth. Sometimes the EC of the saturation extract is very high (around 45 mmhos/cm). These soils have poor physical condition *i.e.* highly dispersed having very low infiltration rate and hydraulic conductivity. In Northern India these soils are known by name of Usar, Kallar, Kshan and in Southern India by the name of Chopan, Kshar and Choudu.

The physical and chemical properties of a soil determine the nature of the environment in which microorganisms are found. These environmental characteristics in turn affect the composition of microbial diversity both qualitatively and quantitatively. The soil reaction plays a vital role in the growth of microorganisms. The total number of bacteria, actinomycetes and fungi are usually the highest in productive normal soil followed by saline and alkali soil. (Neelkantan and Pahwa, 1977). Soil remains in a state of dynamic equilibrium with its environment and the microbial diversity of a particular soil is governed by various edaphic factors.

Alkalinity tolerant microbial strains are to be searched out for better reclamation of the salt affected soil. Microorganisms play a very vital role in mobilization of different nutrients from organic and inorganic non-available forms to available forms and vice-versa. To sustain the fertility status of soil, maintenance of appropriate status of microflora is very essential. The knowledge about the pattern of microbial diversity and their various activities may be of considerable significance in helping the exploitation of microbial activities in building up the fertility of these soils.

Material and Methods

Microbiological Analysis of the Soil

Total Bacterial Count

Soil samples were analysed for their bacterial count by serial dilution method (Johnson *et al.*, 1959). One gram of fresh air dried and sieved soil was transferred to 100 ml sterilized conical flask having 99 ml sterilized distilled water to get the dilution of 1:100. After shaking transferred 1 ml soil in the original dilution to the tube having 9 ml sterilized distilled water possessing tube. This was diluted serially to obtain dilutions *viz.* 0.0001, 0.00001, 0.000001, 0.0000001. This last dilution was used for enumeration by transferring its one ml in sterilized petriplates with sterilized pipette and added about 15-20 ml of melted nutrient agar medium (having near about 40°C temperature) by opening the plates slightly. Mixed the medium thoroughly with the samples in petriplates. When the medium solidified, inverted the petriplates and kept for incubation at 35±1°C for 48 hours. Four replicates for each sample were considered and the observations were taken by counting the colonies with the help of colony counter and calculated as follows:

$$\text{Total bacterial count/g} = \frac{\text{Total bacterial colonies in all plates}}{\text{Number of replicates}} \times \text{Dilution used}$$

Total Fungal Count

The serial dilution method was used for enumeration of fungal population. The serial dilutions of 1 g homogenized soil in 99 ml sterilized distilled water was prepared to get 0.01 dilution of soil which was further diluted serially to obtain the dilution ratio 1:10000. One ml of this dilution was placed in sterilized petriplates and 15-20 ml of Czapek's dox agar, Rose Mortin and Modified Czapek's Rose-Bengal agar medium was poured, mixed thoroughly. For each sample 30 replicates were prepared and all petriplates were incubated at 28±1°C for 7 days and observed till 14 days (Smith, 1969). Chloramphenicol was added to the medium to prevent bacterial growth. The total count was calculated as follows:

$$\text{Total colonies/g} = \frac{\text{Total number of fungal colonies}}{\text{Number of replicates}} \times \text{Dilution used}$$

Isolation, Purification and Identification of Specific Microbes

Bacteria

Isolation and Purification

Isolation of specific bacterial species was made by serial dilution method (Johnson *et al.*, 1959) as described in total bacterial count. The colonies appeared on the plates were repeatedly streaked 2-3 times for purification on nutrient agar plates (Smith, 1969). The purified cultures were maintained on nutrient agar slant.

Identification

The identification of bacterial isolates was done by Gram's staining, morphology of cells, colonial characters and physiological and biochemical tests. For cell morphology, colony size, shape, chromogenesis, elevation, margin, opacity, motility etc. were considered. Biochemical tests comprised of IMVIC (Inodole, Methyl red, Vogesproskaur and Citrate reduction tests).

Fungi

Isolation and Purification

Fungi were isolated by serial dilution method given previously in case of bacterial isolation (Johnson *et al.*, 1959) and purified by single spore technique. The pure cultures were maintained on Czapek's dox agar slants and stored at 4±1 °C.

Identification

The fungal isolates were identified depending on their colonial characteristics, types of spores, their arrangements and other important characteristics given in literature (Gilman, 1967; Smith, 1969; Ainsworth *et al.*, 1973; Ellis, 1976) confirmed through the courtesy of Director of Common Wealth Mycological Institute, Kew, Surrey, England.

Results

Soil provides a natural environment for the biodegradation of waste materials through complex physical, chemical and microbiological processes. Biodegradation in influenced by the microorganisms that are present in the soil. Some microorganisms are directly related to plant growth. These are better known as biofertilizer in Agro-commercial sector have become vitally important simply because these are the source to fix nitrogen available in the atmosphere and to raise solubility of soil nutrients (phosphorus).

The results obtained on the microbial status of the study sites are given in Table 8.1.

Table 8.1: Bacterial Analysis of the Soil of Experimental Sites in Different Seasons

Sites	Total Bacterial Count x 10^5			Analysis of Variance					CD for Significant Treatments			
	Winter	Summer	Rain	Source	DF	SS	MS	F	Treat	S.M. (M)	t-val at 5%	CD at 5%
S_1 (Dannahar)	16.00	15.20	14.00	S	3	9.178	3.0593	115.444	S	0.04203	1.96	0.116506
S_2 (Ishwarpur)	18.90	16.20	13.00	C	2	181.5	90.773	3425.4	C	0.0364	1.96	0.100897
S_3 (Nouner)	17.00	16.00	12.80	S x C	6	22.23	3.7043	139.784	S x C	0.0728	1.96	0.201794
S_4 (Ujhaia)	18.00	16.10	13.20	Error	48	1.272	0.0265		C.V. =	0.01049		
Sites	**Nitrifiers x 10^3**			**Analysis of Variance**					**CD for Significant Treatments**			
	Winter	Summer	Rain	Source	DF	SS	MS	F	Treat	S.M. (M)	t-val at 5%	CD at 5%
S_1 (Dannahar)	3.80	2.60	2.00	S	3	3.542	1.1806	49.1921	S	0.04	1.96	0.110874
S_2 (Ishwarpur)	4.10	3.20	2.70	C	2	30.66	15.329	638.715	C	0.03464	1.96	0.09602
S_3 (Nouner)	4.00	3.70	2.30	S x C	6	2.534	0.4223	17.5949	S x C	0.06928	1.96	0.19204
S_4 (Ujhaia)	3.90	2.60	2.20	Error	48	1.152	0.024		C.V. =	0.05065		
Sites	**Azotobacter x 10^1**			**Analysis of Variance**					**CD for Significant Treatments**			
	Winter	Summer	Rain	Source	DF	SS	MS	F	Treat	S.M. (M)	t-val at 5%	CD at 5%
S_1 (Dannahar)	0.00	0.00	0.00	S	3	15.26	5.0867	337.238	S	0.03171	1.96	0.087897
S_2 (Ishwarpur)	1.30	1.10	1.00	C	2	0.382	0.1912	12.674	C	0.02746	1.96	0.076121
S_3 (Nouner)	1.40	1.20	1.00	S x C	6	0.271	0.0452	2.99488	S x C	0.05492	1.96	0.152242
S_4 (Ujhaia)	1.20	1.10	1.00	Error	48	0.724	0.0151		C.V. =	0.14063		

The soil samples were analysed for their total bacterial and fungal counts. The bacterial count was determined by serial dilution method on nutrient agar medium and on yeast mannitol agar medium for *Rhizobium*. The bacterial colonies appeared were purified by streak method and maintained on nutrient agar slants and on yeast extract mannitol agar medium by keeping them in refrigerator at 4±1 °C. The isolates were identified by the methods prescribed by Jacobs and Gerstein (1960) and Bryan and Bryan (1961). The fungal count was also determined by serial dilution method. Fungi were purified by single spore method and maintained on Czapek's dox agar medium slants at 4±1 °C. The isolates were identified by comparing the characteristics given in literature (Smith, 1969; Subramanian, 1971; Barnett and Hunter, 1972; Domsch and Gams, 1972; Ainsworth *et al.*, 1973; Ellis, 1976) and confirmed through the courtesy of Director of Common Wealth Mycological Institute, Kew, Surrey, England. The results obtained on microbiological status of the soil samples of all sites are given in Table 8.2.

Table 8.2: Microbiological Analysis of Alkaline Soils of Mainpuri District in Winter

Sl.No.	Fungi	Total Population	% Frequency	Analysis of Variance (Population)		
		Per gram of soil		Source	Treatment	Error
1.	*Aspergillus flavus*	92	72	DF	15	64
2.	*A. terreus*	75	70	SS	129385.888	184
3.	*A. candidus*	82	62	MS	8625.726	2.875
4.	*A. fumigatus*	70	55	F	3000.252	
5.	*A. niger*	65	60	S. E. (M) =		0.7583
6.	*A. clavatus*	62	49	C D AT 1% level		2.762
7.	*A. humicola*	80	35	C D AT 5% level		210.2
8.	*Alternaria fasciculatus*	51	42	C. V. =		2.08%
9.	*A. tenuis*	60	60	*Analysis of Variance (Frequency)*		
10.	*Cladosporium epiphyllum*	48	62	*Source*	*Treatment*	*Error*
11.	*C. longicolum*	42	61	DF	15	64
12.	*C. herbarum*	48	62	SS	18044.688	160
13.	*Fusarium candatum*	52	70	MS	1202.979	2.5
14.	*Mucor racemosus*	175	96	F	481.192	
15.	*Rhizopus arrhizus*	150	82	S. E. (M) =		0.7071
16.	*Syncephalastrum racemosum*	155	85	C D AT 1% level		2.576
	Total Population	1307		C D AT 5% level		190
				C. V. =		2.47%

Bacterial Count

The count of all sites varied with the changes in seasons. At site S_1 the bacterial count varied from 14.0 x 10^5 to 16.0 x 10^5. The count was found maximum in winter. In case of nitrifiers the count was also higher in winter *i.e.* 3.8 x 10^3 and *Azotobacter* was nil in all the seasons. At site S^2 the bacterial count was higher in winter (18.9 x 10^5) and minimum was in rainy season (13.0 x 10^5). The nitrifiers count

Figure 8.1: Close View of White Layer of Salt and Cracks on Surface Soil Due to Percolation of Water and Loss of Moisture in Alkali Soil

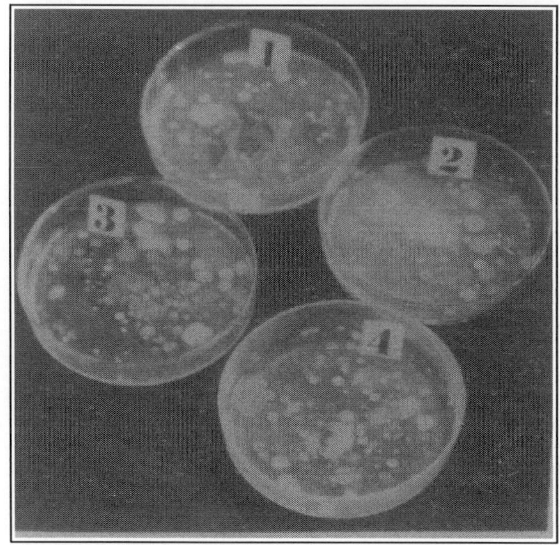

Figure 8.2: Bacterial Colony on Nutrient Media Isolated from Alkali Soil

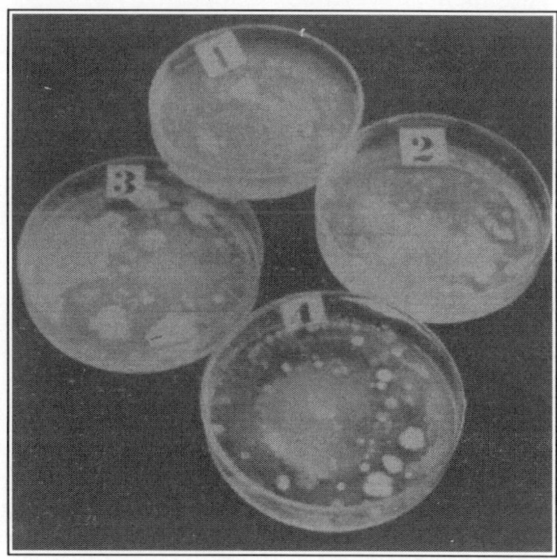

Figure 8.3: Fungal Colony on Nutrient Media Isolated from Alkali Soil

was higher in winter *i.e.* 4.1×10^{3} and it was minimum in rain *i.e.* 2.7×10^{3}. At S_3 same results were shown as S_2 and S_1. Bacterial population was found maximum in winter. At S_3 the maximum bacterial population was 17.0×10^{5} in winter and minimum in rainy season *i.e.* 12.8×10^{3}. Maximum population of nitrifiers was found 4.0×10^{3} and minimum was 2.1×10^{3}. *Azotobacter* showed highest population 1.4×10^{1} in winter and minimum was 1.0×10^{1}. Likewise at S_4 maximum total bacteria was found18.0 $\times 10^{5}$ and minimum was 13.2×10^{3} in rainy season. Maximum nitrifier's population was found 3.9×10^{3} in winter and minimum was 2.0×10^{3} in rainy season and *Azotobacter* population at Ujhaia (S_4) was highest in winter season 1.2×10^{1} and minimum in rainy season *i.e.* 10×10^{1} per gram of soil.

Fungal Count

The qualitative analysis of the soil samples in total showed 26 fungal forms in soil sample. Groupwise categorization revealed that fungal forms belonging to *Phycomycetes, Ascomycetes* and *Deuteromycetes* were present as fungal population.

Fungal flora *Aspergillus flavus, A. candidus, A. terreus, A. fumigatus, A. nidulans, Absidia* sps., *Cladosporium epiphyllum, C. herbarum, C. lignicolum, Fusarium candatum, F. chlamydosporum, Penicillium candidum, Mucor racemosus, M. humalis* and *Rhizopus arrhizae* were isolated in rainy season. Among them *Fusarium candatum* was to be found in maximum number *i.e.* 195/g and *Penicillium candidum* was to be found in highest frequency.

Table 8.3: Microbiological Analysis of Alkaline Soils of Mainpuri District in Summer

Sl.No.	Fungi	Total Population	% Frequency	Analysis of Variance (Population)		
		Per gram of soil		Source	Treatment	Error
1.	*Aspergillus flavus*	65	60	DF	12	52
2.	*A. candidus*	40	44	SS	8598.462	130
3.	*A. fumigatus*	50	48	MS	716.538	2.5
4.	*A. terreus*	26	40	F	286.615	
5.	*Alternaria fasciculatus*	30	41	S. E. (M) =		0.7071
6.	*A. tenuis*	40	33	C D AT 1% level		2.576
7.	*Cladosporium herbarum*	26	20	C D AT 5% level		1.96
8.	*C. epiphyllum*	55	64	C. V. =		3.78%
9.	*Fusarium chlamydosporum*	40	33	*Analysis of Variance (Frequency)*		
10.	*Penicillium candidatum*	51	45	*Source*	*Treatment*	*Error*
11.	*Penicillium notatum*	50	40	DF	12	52
12.	*Penicillium humicola*	30	52	SS	9504.738	137.2
13.	*Chaetomium indicum*	41	63	MS	792.062	2.638
	Total Population	544		F	300.198	
				S. E. (M) =		0.7264
				C D AT 1% level		2.646
				C D AT 5% level		2.014
				C. V. =		3.61%

Fungal flora *Aspergillus flavus, A. terreus, A. candidus, A. fumigatus, A. niger, A. clavatus, A. humicola, Alternaria fasciculatus, A. tenius, Cladosporium epiphyllum, C. logicolum, C. herbarum, Fusarium candatum, Mucor racemosus, Rhizopus arrhizus* and *Syncephalastrum racemosus* were isolated in winter season. Among them *Mucor racemosus* was to be found in maximum number *i.e.* 175/g and highest frequency *i.e.* 96 per cent. *Cladosporium epiphyllum* was to be found in minimum population *i.e.* 42/g.

In case of summer season fungal flora *Aspergillus flavus, A. candidum, A. fumigatus, A. terreus, Alternaria fasciculatus, A. tenius, Cladosporium herbarum, C. cladosporiodes, C. epiphyllum, Penicillium candidatum, P. notatum, P. humicola, Chaetomium indicum* were isolated. Out of them *Aspergillus flavus* was to be found in maximum population 65/g. While *Cladosporium chlamydosporum* was to be found in highest frequency 64 per cent. *Aspergillus terreus* and *Cladosporium herbarum* were to be found in least population *i.e.* 26/g and 27/g respectively and showed least frequency *i.e.* 40 per cent and 20 per cent respectively (Tables 8.2–8.4).

Table 8.4: Microbiological Analysis of Alkaline Soils of Mainpuri District in Rain

Sl.No.	Fungi	Total Population	% Frequency	Analysis of Variance (Population)		
		Per gram of soil		Source	Treatment	Error
1.	*Aspergillus flavus*	65	80	DF	15	64
2.	*A. candidus*	51	72	SS	108719.588	167.6
3.	*A. terreus*	95	80	MS	7247.973	2.619
4.	*A. fumigatus*	94	72	F	2767.722	
5.	*A. nidulans*	40	60	S. E. (M) =		0.7237
6.	*Absidia corymbifera*	76	76	C D AT 1% level		2.636
7.	*Absidia sps.*	65	64	C D AT 5% level		2.006
8.	*Cladosporium epiphyllum*	95	84	C. V. =		1.73%
9.	*C. herbarum*	70	76	Analysis of Variance (Frequency)		
10.	*C. longicolum*	100	64	Source	Treatment	Error
11.	*Fusarium candatum*	195	74	DF	15	64
12.	*F. chlamydosporum*	85	70	SS	4996.5	159.2
13.	*Penicillium candidum*	155	88	MS	333.117	2.488
14.	*Mucor racemosus*	105	84	F	133.916	
15.	*Rhizopus hiemalis*	108	84	S. E. (M) =		0.7053
16.	*Rhizopus arrhizus*	100	81	C D AT 1% level		2.57
	Total Population	1499		C D AT 5% level		1.955
				C. V. =		2.09%

Discussion

Soil microflora is affected by various edaphic factors, which include soil type, soil moisture and soil reactions; seasonal changes like temperature, rainfall, light and the type of ground vegetation. Studies on the occurrence and distribution of bacterial and fungal microflora in 'Usar' soils of different pH values revealed that the soils with low alkalinity contained large number of micro fungi, bacteria and their number decreased with progressive increase in soil pH.

The pH value for optimum growth of microorganisms varied from organism to organism. *Actinomycetes* required alkaline pH for optimum growth (Waksman, 1952). However, there are certain bacteria, which can tolerate acidity or alkalinity such as the bacterial count of all sites varied with the changes in seasons. The maximum bacterial count was found in rainy season at all the sites while minimum was in summer season. Nitrifying bacterial population was found in all the sites. While, *Azotobacter* was absent in site S-1 in all the seasons.

In case of fungal microflora, *Aspergillus flavus* was found to be most dominant species in all the seasons. Maximum fungal colonies were isolated during March and August. During rainy season the rapid leaching of salts with consequent fall in pH, high moisture content favours sufficient fungal growth. During summer these soils are extremely dry with a very high temperature and marked increase in alkalinity often showing deposition in alkali in some areas in the form of 'Reh'. During summer fungal population were found to be minimum in numbers. The total number of fungus was affected by soil phosphorus while is composition altered (Kaufman, 1963) showed a considerable increase in fungal population with the decrease in the pH. Species belonging to Phycomycetes represented by *Absidia corymbifera*, *Absidia* sp., *Mucor hiemalis*, *Mucor racemosus*, and *Rhizopus arrhizus* showed highest frequency percentage of occurrence during and after the rains. Species belonging to *Ascomycetes*, *Aspergillus flavus*, *Aspergillus nidulans* were found almost in every season of the year.

References

Abrol IP and Bhumbla DR (1971). Saline and alkaline soils in India: their occurrence and management. Paper presented at FAO/UNDP Semin, 'Soil Fertility Research' Feb. 15–20, New Delhi, *FAO World Soil Resources Rep* 41: 42–51.

Ainsworth GC, Sparrow FK and Sussman AS (1973). *The Fungi: An Advanced Treatise,* Vol IVA and B. Academic Press, New York, London, p 621.

Barnett HL and Hunter BB (1972). *Illustrated Genera of Imperfect Fungi*, 3rd edn. Burgess Publ Co, Minneapolis, USA, p 241.

Bryan AH and Bryan CG (1961). *Bacteriology: Priciples and Practices*. Burnes and Noble Inc, New York, p 425.

Domsch KH and Gams W (1972). *Fungi in Agriculture*. Longman Group Ltd, p 290.

Ellis MB (1976). *More Dematiaceous Hyphomycetes*. CMI, Kew, Surrey, England, p 507.

Gilman JC (1967). *A Manual of Soil Fungi*, Revised 2nd edn. Oxford and IBH Publishing Company, Calcutta, New Delhi, Bombay, p 450.

Jacobs MB and Gerstein MJ (1990). *Handbook of Microbiology*. Princeton, New Jersey, New York, p 322.

Johnson FL, Curl AE, Bond HJ and Fribourg AH (1959). *Method for Studying Soil Microflora-Plant Disease Relationships*. Burgess Publ Co, USA.

Kaufman DD (1963). The effect of mineral fertilization and soil reaction and C:N ratio on soil fungi. *Dis Abst*, 24: 459–464.

Neelkantan S and Pahwa PR (1977). Note on the distribution of microorganisms in normal saline and alkali soils. *Curr Agric*, 1: 77–78.

Sharma SK and Bhargava OP (1978). Soil survey of the operational research project area for the reclamation of alkali soils, Rep. No. 8, CSSRI, pp 72.

Smith G (1969). *An Introduction to Industrial Mycology*, 5th edn. Edward Arnold, London.

Subramanian CV (1971). *Hyphomycetes*. ICAR, New Delhi, p 930.

Waksman SA (1952). *Soil Microbiology*. John Wiley and Sons, Inc, New York, pp 355.

Soil Microflora, 2009 *Pages* ***86–101***
Editor: **Rajan Kumar Gupta, Mukesh Kumar & Deepak Vyas**
Published by: **DAYA PUBLISHING HOUSE, NEW DELHI**

Chapter 9

Frankia-Actinorhizal Symbiosis: An Overview

Amrita Srivastava, Anju Singh, Satya Shila Singh and Arun Kumar Mishra*
Laboratory of Microbial Genetics, Department of Botany, Banaras Hindu University, Varanasi – 221 005

ABSTRACT

Frankia, a member of actinomycetes, is a remarkable non-legume nitrogen-fixing symbiont contributing a voluminous 15 to 25 per cent of the global biologically fixed nitrogen. Its host range varies within 24 genera belonging to 8 distinct families of plants known as actinorhizas and occupies different habitat. On the basis of morphological and molecular data, they have been grouped into 3 distinct phylogenetic clusters, depending on the hosts that they infect. Its morphology includes characteristic root nodules that are quite similar to lateral roots. Nitrogen-fixation process is regulated at physiological, biochemical as well as genetic level equally by actinorhizal and microbial symbiont. Genomes of three *Frankia* strains have been sequenced and further studies on the evolution of plant-microbe symbiosis and the physiological and genetic interactions that take place in nitrogen fixing symbioses are being conducted. The ability of this microbe to survive under varying stress conditions such as salinity, extremely low temperature, heavy metal pollution etc. highlights its probable utility as a biofertilizer.

Keywords: *Frankia, Actinorhizal, Nodules, Nitrogen-fixation, Biofertilizer.*

Introduction

Frankia draws interest because of multiplicity of its mode of existence in nature. There are descriptions of its occurrence in three closely associated but functionally diverse niches. It is found as

* Corresponding Author: E-mail: akmishraau@rediffmail.com/akmishraau@hotmail.com;
Telephone: 0542-3296111/09335474142

a symbiont (in the root nodules of actinorhizal plants) or as a free-living associate of the non-host plants (in rhizosphere) or as a saprophyte (in soil). Its cells can also accumulate in river and lake sediments (Huss-Danell *et al.*, 1997). Apart from its normal life-sustaining functions, it performs a dynamic function of nitrogen fixation in association with certain higher plants. During this process it converts free dinitrogen of the atmosphere into ammonia, which can be assimilated by plants. Unlike other nitrogen fixing symbionts whose distribution is restricted to one or few particular families of plants, *Frankia* can colonize plants of about 24 genera belonging to 8 distinct families. The group of plants inhabited by *Frankia* is termed as 'actinorhizal' plants (Torrey and Tjepkema, 1979). 'Actinorhizal plants' is a self-explanatory term where 'actino' refers to *Frankia* symbionts and 'rhiza' refers to host roots. *Frankia* in association with actinorhizal plants contributes 15 to 25 per cent of global terrestrial biological nitrogen fixation every year (Dixon and Wheeler, 1986).

Frankia is a gram-positive, aerobic, filamentous microorganism with a high G+C content which ranges from 66 to 75 mol per cent. Sporangiophores can be terminal or intercalary in position but they always occur in multilocular sporangia. Production of colored pigments (like red, yellow, orange, pink, brown, green, and black), siderophore and antibiotics has been reported in some members of the *Frankia* species (Lechevalier and Lechevalier, 1990; Haansuu *et al.* 2001; Sarma *et al.*, 2006; Singh *et al.*, 2008)

Phylogeny

Being Gram positive filamentous nitrogen fixing bacterium, growing by branching and tip extension, *Frankia* closely resembles the antibiotic-producing *Streptomyces* sp. Various tools have been applied for construction of *Frankia* phylogeny, important ones being comparative sequence analysis of the 16S rRNA gene, the genes for nitrogen fixation (*nif* genes) and other genes (Benson and Clawson, 2000), DNA-DNA homology studies (An *et al.*, 1985; Dobritsa and Stupar, 1989; Fernandez *et al.*, 1989; Normand *et al.*, 1996; Benson and Clawson, 2000) and rDNA sequencing. *Frankia* isolates have been accordingly grouped in three clusters: Cluster 1 includes the strains which infect members of higher Hamamelidae, belonging to order Fagales including Betulaceae, Myricaceae and Casuarinaceae. Cluster 2 comprises of the strains which infect members of Coriariaceae, Datiscaceae, Rosaceae and Rhamnaceae (*Ceanothus* sp.). Cluster 3 includes the strains infecting members of Myricaceae, Rhamnaceae, Elaeagnaceae and Casuarinaceae (*Gymnostoma* sp.). In addition to this, its members are also isolated from the nodules of the Betulaceae, Rosaceae and other members of Casuarinaceae and Rhamnaceae (*Ceanothus* sp.) as poorly effective or non-effective strains.

Frankia strains were likely to be related to members of the Dermatophilaceae within the Actinomycetales as per their sporangium morphogenesis (Lechevalier and Lechevalier, 1979). It has been placed closer to *Anabaena* and the gram-negative proteobacteria than to the low G+C gram-positive microbes on the basis of *nifH* genes and amino acid sequence studies (Normand and Bousquet, 1989) representing a recent lateral gene transfer from an ancestral gram-negative organism. *glnA* sequence of *Frankia* glutamine synthetase gene shows proximity of the sequence with *Streptomyces*, then to *Anabaena* and then to gram-negative microbes (Kumada *et al.*, 1993).

Structural Details of *Frankia*

These actinobacteria grow in the form of dense filaments showing presence of hyphae, sporangia and vesicles in free-living mode whereas additional structures called nodules are present in symbiotic mode. They have a type III cell wall containing mesodiaminopimelic acid, alanine, glutamic acid, muramic acid, and glucosamine (Lechevalier and Lechevalier, 1990). In addition, they also contain amphiphilc compounds called hopanoids, specially in vesicles, which stabilize their cell membranes.

Hyphae

Hyphae are branched, septate, ranging in width from 0.5 to 1.5 μm. They contain numerous small granules which might be those of glycogen (Benson and Eveleigh, 1979; Lancelle *et al.*, 1985). Cell walls in *Frankia* have been observed to be composed of two layers of electron-dense material, a base layer and an outer layer (Baker *et al.*, 1980). Cross-walls originate from the base layer. Membranous layers are sometimes visualized outside the outer cell wall. Treatment of *Frankia* hyphae with lysozyme have been reported to produce protoplasts, which provided indirect evidence that, a peptidoglycan layer is present in hyphae (Tisa and Ensign, 1987). Inside the hyphal cells, numerous rosette shaped granules, presumed to be glycogen have been observed. In addition, lipid droplets have also been observed (Schwintzer and Tjepkema, 1991). Abeysekera *et al.* (1989) had reported the fine structure of *Frankia* hyphae by performing freeze fracture electron microscopic studies on *Alnus incana* sub sp. *rugosa* root nodules. It was interpreted that *Frankia* hyphae appear cylindrical in oblique fracture planes and the hyphal plasma membrane is covered by a thin zone of lipid laminae. The lipid laminae in turn were encapsulated by a capsule, which contained microfibrils with an average width of 9 nm within an amorphous region. The microfibrils resembled the nodule cortical cell wall microfibrils which were presumably cellulosic in nature and were arranged in orderly layers (Abeysekera *et al.*, 1989). Hyphae bear terminal or intercalary sporangia (Newcomb *et al.*, 1979) which become multilocular due to internal segmentation.

Sporangium

Sporangia in *Frankia* develop as terminal or intercalary structures. These are produced readily in culture by all *Frankia* strains isolated so far (Benson and Silvester, 1993). Many often segmentation occurs within a large multilocular sporangium containing innumerable spores (Horriere *et al.*, 1983; Benson and Silvester, 1993). Sporogenesis is mostly enterothallic which is analogous to spore formation in fungi. In this type of sporogenesis, old and loosely packed spores are distal to younger, more densely packed, developing spores. The intersporal matrix develops from the inner layer of the cell wall. The younger spores are somewhat irregular shaped compared to older spores which are spherical in shape and measure 1-5 μm in diameter. These spores are non-motile compared to spores in multilocular sporangia of the genus *Geodermatophilus* which are motile and bear tufts of flagella (Benson and Silvester, 1993). Developing spores inside the sporangia show electron-translucent nucleoid regions with dispersed fibrils and numerous lipid droplets. The matured spores have evenly dispersed cytoplasm but lack tubules, which are prominent features in *Frankia* hyphae. Structural morphology of spores in a few *Frankia* strains isolated from *Alnus incana* sub sp. rugosa (Horriere, 1984), strain ArI3 from *Alnus* rubra, strain HFPCcI3 from *Casuarina cunninghamiana*, strain EuI1 from *Elaeagnus umbellata* and strain HFPCpI1 from *Comptonia peregrina* have shown the existence of sporangia that develop due to the hyphal thickening and septa formation from inner layer of a double layered sporangial cell wall (Schwintzer and Tjepkema, 1991).

Vesicles

Frankia vesicles have been considered as the most definitive structures that characterize the genus. Although functionally analogous in many ways to heterocysts of cyanobacteria, the vesicle is a unique developmental structure designed for physiological compartmentalization and has never been described in any other prokaryotic group (Benson and Silvester, 1993). Vesicle formation is induced under nitrogen starved condition (Benson and Silvester, 1993). Vesicles usually develop as terminal swellings on hyphae or on short side branches and initially develop as provesicles (Schwintzer and Tjepkema, 1991). Provesicles are spherical, dark in appearance under phase contrast microscope, generally 1.5-2.0 µm in diameter with dense cytoplasm which does not exhibit nitrogenase activity. Sometimes these provesicles show internal compartmentalization. Provesicles rapidly develop into mature vesicles which are 2.0-4.0 µm in diameter and are characterized by two important structural elements, *i.e.* the vesicle envelope and internal septation, both of which appear to be necessary for nitrogenase activity in cultured *Frankia* cells although the necessity of these under natural conditions have been nullified (Benson and Silvester, 1993). The structure of vesicle envelope was first analyzed in detail by Torrey and Callaham (1982), who observed that enveloped vesicles showed a marked birefringence under plane-polarized light. Their interpretations imply the presence of highly structured lipid enveloped laminated layer which could be identified through freeze fracture techniques and electron microscopy. The multilaminated lipid envelope is composed of several monolayers, each approximately 4 µm in thickness (Torrey and Callaham, 1982). Although vesicle formation is induced only during nitrogen limitations, there is however exceptions whereby a few *Frankia* strains *i.e.* CcI.17 and CpI.2 isolated from *Casuarina cunninghamiana* and *Comptonia peregrina* have shown to induce vesicle formation even in presence of nitrogenous source like ammonia (NH_4^+) (Gauthier *et al.*, 1984; Meesters *et al.*, 1985). Abeysekera *et al.* (1989) had reported an average number of lipid laminae in *Frankia* vesicles which differed altogether when the strains were grown at different oxygen tensions. On the other hand fine structure of symbiotic vesicles from root nodules of *Alnus incana* sub sp. rugosa had depicted an uneven distribution of lipid laminae with occasional fusion and presence of internal membrane septa under freeze fractured electron microscopy (Abeysekera *et al.*, 1989). The symbiotic vesicles were observed to attach themselves to the subtending hyphae by the stalks, which contained a very thick layer of lipid laminae between *Frankia* plasma membrane and capsule (Abeysekera *et al.*, 1989). The function of the lipid laminae had been described as the sole diffusion barrier to the ingress of oxygen and therefore it was presumed that one could expect more lipid laminae to occur on the envelope of *Frankia* vesicles exposed to high partial pressure of oxygen (pO_2) (Tjepkema *et al.*, 1980; Abeysekera *et al.*, 1989). The lipid nature of *Frankia* vesicles has also been confirmed by the use of Nile red fluorescent dye (Schwintzer and Tjepkema, 1991). The relative lipid content of *Frankia* vesicles and hyphae was analyzed by comparing the lipids of isolated vesicles to those of hyphal cultures and the analysis showed that vesicles have a higher proportion of neutral lipids and glycolipids but lower levels of polar lipids than vegetative cells. Analysis of the neutral lipid fraction revealed the presence of abundant unidentified compounds which were suggested to be the long-chain fatty acids or the alcohols (Benson and Silvester, 1993). The fatty acid composition of the lipid envelops in *Frankia* have been studied by employing techniques such as Phospholipid-fatty-acid (PLFA) and Fatty-acid methyl-ester (FAME) analyses whereby it has been evidenced that *Frankia* cells contain a high concentration of pentacyclic hopanoid lipid bacteriohopanetetrol (Berry *et al.*, 1989). Further work on isolated vesicle envelops from *F. alni* HFPCpI1 indicated that they consists mainly of bacteriohopanetetrol and bacteriohopanetetrol phenylacetate monoester in approximately equal amounts (Berry *et al.*, 1989). Bacteriohopanes are common in many bacterial groups and have been proposed to play a direct role in membrane stability and fluidity. Their presence in the vesicle envelops of *Frankia* has been reported to maintain a number of interesting implications including oxygen protection (Benson and Silvester, 1993). Vesicles in *Frankia* appear to have a limited life although this has not been fully investigated by using continuous or semi-continuous culture techniques (Benson and Silvester, 1993). During senescence, internal septa become irregular, nitrogenase activity is lost and turn into ghosts which are recognizable in culture as deformed, thin-walled, non-functioning cells (Burggraaf and Shipton, 1982). Although the main physiological significance of *Frankia* vesicles is nitrogenase activity, however a proportion of vesicles have been observed to germinate and produce hyphae in culture (Shantaram *et al.*, 1987).

Nodule

Symbiotic association between *Frankia* and actinorhizal plants is marked by specific structures known as nodules. Nodule is actually an assemblage of hyphae and vesicles covered with plant cell wall-like material. The bundles of hyphae and vesicles present in an infected plant are called "vesicle clusters" or "hyphal clusters". The capsule is considered rich in pectin, cellulose and hemicellulose (xylans) (Lalonde and Knowles, 1975). Functionally, actinorhizal nodules are analogous to legume nodules but there is a basic structural difference between them. Legume nodules have a cortical layer of tightly packed cells within which there is a central infected zone covered by an endodermis, hence the vascular tissue is peripheral. In contrast to this, actinorhizal nodules have a central stele that has infected tissue adjacent to it, or that lies within a cylinder of infected cortical cells. Within the nodule, the hyphae can be differentiated into two–the hyphae that penetrate cell walls are termed as "colonizing hyphae" and the ones that branch in host cells are termed as "proliferating hyphae" (Newcomb and Wood, 1987). Thus, the basic difference between non nitrogen fixing free living and nitrogen fixing symbiotic strains is that of vesicles and nodules are the centre of nitrogen fixing activity.

Symbiosis and Nitrogen Fixation

The importance of *Frankia* strains lies in this particular function. It is possible that some molecular mechanisms could be common to actinorhizal, rhizobial and mycorrhizal symbiosis (Gianinazzi-Pearson *et al.*, 1991; Balaji *et al.*, 1994). There are two modes of entry into the host system–either by root hair infection (intracellular) or by intercellular penetration.

Root hair infection has been studied in *Alnus, Casuarina, Comptonia, Myrica* (Torrey, 1976; Callaham *et al.*, 1979; Berry and Torrey, 1983). The first step in root hair infection is branching and curling of root hair. Once the root hairs get distorted, *Frankia* enters through it and causes hydrolysis of the cell wall (Diouf *et al.*, 2003). This is followed by cell division in the hypodermis and cortex leading to the formation of a prenodule (Becking, 1975). The *Frankia* hyphae penetrate through the prenodule tissue into the inner cortex of the root. The prenodule is the site of early nitrogen fixation and represents a primitive symbiotic organ (Laplaze *et al.*, 2000) and the nodule formation does not involve prenodule cells. There is a marked similarity between nodule and lateral root with reference to the development as well as the internal anatomy. As in the case of lateral root, primary nodule lobe primordia arise in the pericycle, endodermis or cortex at the same time as prenodule development (Benson and Silvester, 1993). The nodules are colonized by vegetative hyphae that differentiate into diazo-vesicles. The entire process is accompanied by three important cellular events *i.e.* nuclear migration in the cell centre, cytoplasmic movement and appearance of phragmoplast-like structures.

Infection by intercellular penetration has been studied in *Elaeagnus, Shepherdia* and *Ceanothus* (Miller *et al.*, 1985; Racette *et al.*, 1989; Liu and Berry, 1991). In this case, there is a direct penetration of root epidermis cells and cortex. This mode of infection generally proceeds in the absence of the root hair. In case the root hairs are present, they remain uninfected as in the case of *Ceanothus*. For the sake of entry, hyphae utilize the intercellular space in between adjacent epidermal cells and move through the middle lamella. Moving through the cortex, it enters the cell and becomes intracellular. No prenodule formation has been reported except *Ceanothus* where mitotic divisions occur but they are not associated with infection or enlargement of the cell (Liu and Berry, 1991). In either case this initial entry is followed by encapsulation of hyphae during invasion and vesicles in the nodule (Lalonde *et al.*, 1975, 1977). The capsule is generally similar to the cell wall in composition except in *Ceanothus*, where the presence of galacturonans distinguishes it from the cell wall (Liu and Berry, 1991).

While entering through the root hair, the hyphae get encapsulated in continuation with the host cell wall and are eventually covered by host plasmalemma. The encapsulating layer consists of plant cell wall-like material. During intercellular infection, the hyphae are covered in a pectic capsule (Liu and Berry, 1991) extending through the intercellular space where the infective hyphae occur in massive amounts. The capsule consists of pectin in *Alnus* (Lalonde, 1975), large amount of galacturonans in *Ceanothus* (Liu and Berry, 1991) and also cellulose and hemicellulose. One of the differences between legume and actinorhizal symbiosis is the kind of nodule formed. On the basis of differential development and morphology, legume nodules are termed "determinate" and actinorhizal ones are termed "indeterminate".

In case of indeterminate root nodules, the site of initiation of nodule formation is inner cortex of the root which is several cell layered apart from the root hair containing an infection thread. It is the inner cortex that gives rise to nodule primordium through cell divisions. Meanwhile, there is elongation of infection thread and after leaving the root hair cell it enters the vacuolated root cortical cells. Crossing the cytoplasmic bridge present in outer cortical cells (Bakhuzien, 1988; Hirsch, 1992), branches of infection thread enter the nodule primordium cells, as a result of which the process of cell division gets replaced by differentiation (Newcomb *et al.*, 1979; Hirsch, 1992). Nodule meristem is formed due to the mitotic divisions of the cells lying next to the nodule primordium. Apart from the nodule cortex (Bond, 1948) and the basal tissues of the nodule, other nodular tissues develop from nodule meristem.

Structurally, indeterminate nodule can be divided into four zones- Zone I is the nodule meristem which gives rise to cells of the central tissue. Zone II consists of 'invasion zone' through which the infection threads proceed and it is also the site of expression of early nodulin genes. It also consists of a 'pre-fixing zone', also known as 'early symbiotic zone'. The 'interzone region' lying between Zone II and III contains numerous amyloplasts (Hirsch, 1992). Zone III can further be differentiated into 'nitrogen fixing zone' which consists of maximum number of symbionts and the 'inefficient zone' where there is limited nitrogenase activity and its cells are probably involved in assimilation of the fixed nitrogen and exchange of carbon (Wall, 2000). Following this zone is the Zone IV *i.e.* 'senescent zone', lying towards the side of vascular bundle closest to the point of attachment. It is the zone of degeneration of plant as well as symbiont cells.

Differential screening of nodule cDNA libraries with root and nodule cDNA has been used to isolate several nodule specific or nodule enhanced plant genes in various actinorhizal plants including *Alnus, Datisca, Eleagnus* and *Casuarina* (Powlowski and Bisseling, 1996; Franche *et al.*, 1998; Wall, 2000; Obertello *et al.*, 2003).

Parallel with the infection of roots, several actinorhizal nodulin genes get activated (Mullin and Dobritsa, 1996). There are two groups of nodulin genes, those involved in plant infection or in nodule organogenesis are termed 'early nodulin genes' whereas those involved in regulating all sorts of functions of the nodule are termed 'late nodulin genes' (Obertello *et al.*, 2003). Hemoglobin genes, one of the late nodulin genes, have been characterized in actinorhizal nodules (Flemming *et al.*, 1987). Its presence is first of all noticed in young infected cells lying next to the apical meristem. Maximum amount of hemoglobin (truncated) is detected in cells which are completely filled with the symbiont (Gherbi *et al.*, 1997; Obertello *et al.*, 2003). While conducting its major function of transporting oxygen to microbial cells, it also aids in protection of oxygen sensitive nitrogenase enzyme (*See review Sarma et al.*, 2005)

Regulation of Nitrogen Fixation

Actinorhizal host plants and *Frankia* are the two counterparts of nitrogen fixation. The process of nitrogen fixation by this actinomycete is very closely integrated with photosynthetic activity of the host. Such dependence is because of the high amount of energy required in nitrogen fixation, the ultimate source of which is energy assimilated by plants. There is possibility of either equal roles of both or dominance of any one of these which might be regulating the process single-handedly. Obviously, the genotype of the host as well as the symbiont is the chief factor governing the extent of their role and involvement in the phenomenon.

It was found that when a combination of high and low nitrogen fixing *Frankia* strains were selected and tested on three *Casuarina* host clones, the greatest number of nodules were always developed on a particular host and it did not depend on the strain used (Akkermans, 1971; Sougoufara *et al.*, 1992; Verghese and Mishra, 2002). Likewise, there was a counterpart of above host, the number of nodules was independent of the strain used and which always had the minimum nodular presence (Sougoufara *et al.*, 1992; Verghese and Mishra, 2002). A marked difference is generally seen in the nitrogen fixing capabilities of the same strain either grown in cultures or in vicinity of host. Moreover, when a particular strain infects more than one host, the morphological appearance of the nodule varies. Probably the differences in nitrogen-fixing capabilities and nodular morphology are governed by certain factors provided by the host. The symbiont has a direct dependence on the physiological state of the host because it is the host that provides *Frankia* with required environment and conditions. Steps of nitrogen fixation varying from initial recognition and selection of strain, suppression of its defense genes so as to allow the infection to begin, nodule development, protection of nitrogenase from oxidizing environment, supply of ATP and carbon compounds for fulfilling the microbial needs are solely under the host command (Verghese and Mishra, 2002).

On the other hand, *Frankia* being the bearer of nitrogen fixing apparatus also plays a leading physiological role (Murry *et al.*, 1984). Many strains are infective in nature but not all of them possess nodule forming capacity. Thus, there is a difference between infective and effective strains, regulating mechanism of which must lie in control of the symbiont.

As a matter of fact, it is the coordination between the actinorhizal host and the *Frankia* symbiont that governs the occurrence as well as the extent of nitrogen fixation. Basically, nitrogen necessity of the host is catered by the symbiotic associate and the mechanism runs in accordance with it. Conversely, it is the *Frankia* which makes provision for the actual process and a considerable share of the mechanism governance lies in its control. Apart from the contribution of these two, soil condition, pH, soluble phosphorus, calcium levels, available nitrogen, conductivity of water, climate, light availability, canopy cover etc. also affect the process (Quesada and Valiente, 1996; Han and New, 1998; Quesada *et al.*, 1997 and 1998; Olson *et al.*, 1998).

Genome Characteristics

Frankia genetics has been a fascinating field because of its economic and ecological importance. In this field, work has been rather slow because of technical drawbacks. Till date, genome sequences for three *Frankia* strains have been obtained and they show the effect of evolutionary forces in shaping the microbial genomes (Normand *et al.*, 2007). Three *Frankia* strains: the *Casuarina* infective CcI3 by National Science Foundation, the *Elaeagnus* infective EAN1pec (EAN) by the Department of Energy/

Joint Genome Institute and the *Alnus* infective *F. alni* strain ACN14a (ACN) by the Genoscope were sequenced in 2003. The genomes are circular and the (G+C) content of DNA is 66 to 75 mol per cent (An *et al.*, 1989). The genome size for CcI3, ACN and EAN has been shown to be 5,433,628, 7,497,934 and 9,035,218 nucleotides respectively with a marked size difference of 3.6 Mb (Normand *et al.*, 2007). The most remarkable difference is in the number of transposes, with CcI3 and EAN having almost 10 times as many transposes as ACN. Plasmids have also been reported with the proportion of strains carrying plasmids varying from 10 per cent or less (Normand *et al.*, 1983, Simonet *et al.*, 1984) to 62 per cent in a study of 16 isolates from *Myrica pensylvanica* (Bloom *et al.*, 1989). There has been no generalization regarding the function of these extrachromosomal elements.

Various strains tend to loose certain genes and gain others by lateral transfer due to niche specialization. The three genomes show high levels of conservation at the origin of replication and the genes tend to be more species specific at the replication terminus where there are more repeats and insertion sequences (Normand *et al.*, 2007).

Genes related to symbiosis include those encoding nitrogenase (*nif*), uptake hydrogenase (*hup*), squalene hopane cyclase (*shc*) involved in bacteriohopane biosynthesis and *nod* genes.

Work has been conducted to search the rhizobial *nod*-homologous genes in the *Frankia* genome: *nod*A homologues were absent, *nod*C homologues were found at 24-43 per cent levels of identity, two to three *nod*B homologues were present at 32-48 per cent levels of identity and *nod*D and *nod*Q homologues were present at low levels of identity (Normand *et al.*, 2007).

Studies of *Klebsiella pneumoniae* give description of at least 20 *nif* genes involved in nitrogen-fixation and many of these genes have homologues in other nitrogen-fixing organisms (Dean and Jacobson, 1992). *Frankia* is a high-mol per cent G+C gram-positive genus and the availability of its *nif* gene sequence makes it valuable in phylogenetic studies of gene organization. *Nif* gene expression is intimately tied to the developmental cycle of *Frankia* vesicles (Benson and Silvester, 1993). There is one cluster of *nif* genes per genome and is present at the same position, closer to the origin of replication and away from all other genes related to symbiosis. *nifH* codes for the Fe protein and *nifD* and *nifK* genes code for MoFe protein of enzyme nitrogenase. In some *Frankia* strains *nifHDK* are clustered on the chromosome (Mullin and An, 1990). Products of *nifB*, *nifW* and *nifZ* are all involved in FeMo-cofactor biosynthesis (Dean and Jacobson, 1992). *nifV* codes for a homocitrate essential for the FeMo-Cofactor synthesis and lies upstream of *nifH* in CcI3 and ACN, but away from it in EAN (Normand *et al.*, 2007). Eight ORF namely *nifX*, *orf3*, *orf1*, *nifW*, *nifZ*, *nifB*, *orf2* and *nifU* have been identified and characterized in *Frankia alni* CpI1 starin (Harriott *et al.*, 1995).

The energy efficiency for nitrogen fixation is increased through the oxidation of dihydrogen (Mattson and Sellstedt, 2002). Dihydrogen is a byproduct of nitrogenase activity and its oxidation yields electrons which are utilized for ATP production. Genes have been discovered in the sequenced *Frankia* genomes for uptake hydrogenase in the form of two distantly related hydrogenase clusters (Normand *et al.*, 2007).

Nitrogenase is a highly oxygen sensitive enzyme. The diazotrophs have evolved several mechanisms to prevent the nitrogenase from vicinity of molecular oxygen. One such mode is presence of squalene hopane cyclase in vesicle cell envelope of *Frankia* (Berry *et al.*, 1993). There are two copies of the gene in ACN (Alloisio *et al.*, 2005) and EAN but only one in CcI3 (Normand *et al.*, 2007).

Some other inferences have been drawn *i.e.* plasmid transfer and integration into the genome has occurred in the past- at least 8 integrated plasmids have been identified (Normand *et al.*, 2007). There are a number of DNA methylase genes, the function of which is to encode a protein that could recognize and restrict methylated DNA (Waite-Rees *et al.*, 1991), limiting the possibility of transforming DNA. Bacteriophage integration into the genome has also occurred in the past (Normand *et al.*, 2007). Presence of two phylogenetically diverse catalases and two distantly related hydrogenase clusters in the three sequenced genomes give clear indications of lateral gene transfer and specialization for particular niches (Leul *et al.*, 2007).

Ecological Significance

Several patches of lands on earth are considered worthless as far as plantation is concerned. These lands are considered barren and devoid of the normal flora. Such lands include high altitude cold lands, densely and continuously cultivated lands or salty sea shores etc. At these places, the soil is devoid of assimilable form of nitrogen which greatly hampers the plant growth and productivity. Sometimes it becomes deficient in some or many of the important nutrients due to continuous cultivation on the same land for consecutive periods. Thus, further growth of any sort of plantation becomes very difficult. For long periods of time, nitrogenous fertilizers and supplements have been used to maintain the nutrient levels and fertility of the soil. But there have been several side effects of excessive use of such supplements. Generally salinity of soil increases to such an extent under these circumstances that water uptake gets hampered and most forms of plantations fail to give positive result on lands highly fertilized for long periods.

Being a nitrogen fixing organism, *Frankia* plays a major ecological role as symbionts of plants which are useful in reclaiming and conditioning soil. The woody actinorhizal plants may contribute as much as 15 to 25 per cent of the total biological nitrogen fixed globally in terrestrial ecosystems each year (Dixon and Wheeler, 1986). *Frankia* constitute the major nitrogen-fixing symbioses in temperate forests (*Alnus*), dry chaparral and matorral (*Ceanothus, Trevoa, Talguenea, Chamaebatia, Cercocarpus, Purshia*), coastal dunes (*Casuarina, Myrica, Hippophae*), alpine communities (*Alnus*) and in colder regions (*Alnus, Dryas*). They assist in root exploration, water retention, mineral uptake and resource sharing. It also shows a tripartite relationship where it remains associated with plants as well as mycorrhiza. Thus, *Frankia*, in association with actinorhizal plants, has the potentiality to integrate into schemes for addressing issues of pyrodenitrification (Crutzen and Andreae, 1990) and reforestration (Diem and Dommergues, 1990).

Due to present rate of development, a novel problem of metal pollution has cropped up. Excessive accumulation of heavy metals is toxic to most plants. When heavy metal ions present in the environment at an elevated level are excessively absorbed by roots and translocated to shoot, they lead to the impaired metabolism and reduced growth (Foy *et al.*, 1978; Bingham *et al.*, 1986). Excessive metal concentrations in contaminated soils result in decreased soil microbial activity and soil fertility (McGrath *et al.*, 1995).

One of the best remedies for the removal of heavy metal contamination is through phytoremediation. Latter may occur via phytoextraction, rhizofilteration, phytostabilization and phytovolatilization. Among these, rhizofilteration is the use of plant roots to absorb, concentrate or precipitate metals from effluents (Jing *et al.*, 2007). Microbial population present in association with roots proves to be a great help in the above process. Microbial populations such are known to affect mobility of heavy metals and availability to the plant through release of chelating agents, acidification, phosphate solubilization, and redox changes (Smith and Read, 1997; Abou-Shanab *et al.*, 2003). Toxicity is also because of iron deficiency in various plant species grown in heavy metal contaminated soil. Although iron is present in abundance, it is unavailable to plants because of its ferric state under normal aerobic conditions. The high demand of iron from the unavailable source is made available by several microbes through secretion of high affinity low molecular mass iron chelating agents termed siderophore (Neilands, 1981; Meyer *et al.*, 1987; Arahou *et al.*, 1998; Dave and Dube, 2000). Microbial iron-siderophore complexes can be taken up by the plants and thereby serve as an iron source for plants (Reid *et al.*, 1986; Bar-Ness *et al.*, 1991; Wang *et al.*, 1993). Siderophore production has been reported in many members of rhizobiaceae (Guerinot, 1994), several nitrogen fixing microorganisms such as cyanobacteria (Boyer *et al.*, 1987), *Azotobacter* (Page and Huyer, 1984), *Azospirillum* (Bachhawat and Gosh, 1987) and *Frankia* (Aronson and Boyer, 1992 and 1994; Singh *et al.*, 2008).

Frankia is present at high altitudes approximating 1400 to 1500 mts asl which are generally prone to landslides. For example, Lachen and Zema regions of north eastern Himalayan range (India) experiences regular landslides where *Frankia* is found in association with *Hippopheae salicifolia* D. Don. Proliferation of root system gets greatly enhanced during symbiosis. These roots bind the minute soil particles together serving as a medium of holding and preventing soils from sliding past.

Apart from its existence in close association with actinorhizal plants, several techniques have confirmed the widespread presence of *Frankia* strains in many soils with and without host plants (Huss-Danell and Frej, 1986; Smolander, 1990). It is significant that, in addition to hyphae, both sporangia and vesicles are formed in the absence of added carbohydrate, suggesting that *Frankia* cells can both proliferate (sporulate) and presumably fix nitrogen, as well as grow outside root nodules in soil (Smolander and Sarsa, 1990). This characteristic is highly beneficial for soil health. In the presence of heavily polluted or nutrient deficient soil, presence of *Frankia* replenishes soil to some extent. Thus, it finds a rich application as biofertilizer.

Future Prospects

Phylogenetic studies with confirmed data can reveal clear picture of evolution of *Frankia* and actinorhizal plants, as well as some insight into the origin of nitrogen fixation process. Recent phylogenetic data show that all nodule-forming plants belong to a single clade indicating a common origin of the symbiotic nitrogen fixation in the angiosperm (Soltis *et al.*, 1995; Obertello *et al.*, 2003). Most of the work has been focused on *Rhizobium*-legume symbiosis. Once the entire process of actinorhizal nitrogen fixation and its pathway gets elucidated, it would be likely to induce infection of nitrogen fixing organisms in other non-legume plants. Further, the better understanding of the origin and location of molecular signals and their long and short range exchange may aid the enhancement in the nitrogen fixing capacity of other artificially induced systems. The capacity of *Frankia* to fix nitrogen in free living condition makes it an appropriate biofertilizer. Deducing and realizing the potential of *Frankia* in reclamation of unusable lands is perhaps the most potential field of *Frankia* study.

Acknowledgements

The Head, Department of Botany, BHU, Varanasi is gratefully acknowledged for providing laboratory facilities.

References

Abou-Shanab RA, Angle JS, Delorme TA, Chaney RL, van Berkum P, Moawad H, Ghanem K and Ghozlan HA (2003). Rhizobacterial effects on nickel extraction from soil and uptake by *Alyssum murale*. *New Phytol*, 158 (1): 219–224.

Abeysekera RM, Newcomb W, Silvester WB and Torrey JG (1989). A freeze-fracture electron microscopic study of *Frankia* in root nodules of *Alnus incana* grown at three oxygen tensions. *Can J Microbiology,* 36: 97–108.

Akkermans ADL (1971). Nitrogen fixation and nodulation of *Alnus* and *Hippophaë* under natural conditions. *PhD Thesis*, Leiden State University, The Netherlands.

Alloisio N, Marechal J, Heuvel B, Normand P and Berry AM (2005). Characterization of a gene locus containing squalene-hopene cyclase (*shc*). in *Frankia alni* ACN14a, and an *shc* homolog in *Acidothermus cellulolyticus*. *Symbiosis*, 39: 83–90.

An C S, Riggsby WS and Mullin BC (1985). Relationships of *Frankia* isolates based on deoxyribonucleic acid homology studies. *Int J Syst Bacteriol,* 35: 140–146.

An C S, Wills JW, Riggsby WS and Mullin BC (1985). Deoxyribonucleic acid base composition of 12 *Frankia* isolates. *Can J Bot,* 61: 2859–2862.

Arahou M, Diem HG and Sasson A (1998). Influence of iron depletion on growth and production of catechol siderophore by different *Frankia* strains. *W J Microbiol Biotechnol,* 14: 31–36.

Aronson DB and Boyer GL (1992). *Frankia* produces a hydroxamate siedrophore under iron limitation. *J Plant Nutr,* 15: 2193–2201.

Aronson DB and Boyer GL (1994). Growth and siderophore formation in six iron-limited strains of *Frankia*. *Soil Biol Biochem,* 26: 561–567.

Bachhawat AK and Gosh S (1987). Iron transportin *Azospirillum brasilense*; role of siderophore spirilobactin. *J Gen Microbiol,* 133: 1759–1765.

Bakhuzien R (1988). The plant cytoskeleton in the *Rhizobium*-legume symbiosis. *PhD Thesis*, University of Leiden, The Netherlands.

Baker D, Newcomb W and Torrey JG (1980). Characterization of an ineffective actinorhizal microsymbiont, *Frankia* sp. EuI1 (Actinomycetales). *Can J Microb,* 26: 1072–1089.

Balaji B, Ba AM, Larue TA, Tepfer D and Piché Y (1994). *Pisum sativum* mutants insensitive to nodulation are also insensitive to invasion *in vitro* by the mycorrhizal fungus, *Gigaspora margarita*. *Plant Sci,* 102: 195–203.

Bar-Ness E, Chen Y, Hadar Y, Marchner H and Romheld V (1991). Siderophores of *Pseudomonas putida* as an iron source for dicot and monocot plants. *Plant Soil,* 130 (1–2): 231–241.

Becking J H (1975). Root nodules in nonlegumes. In: *Development and Function of Roots*, (Eds) Torrey JG and Clarkson DT. Academic Press Ltd, London, p 508–566.

Benson DR, and Clawson ML (2000). Evolution of the actinorhizal plant symbioses. In: *Prokaryotic Nitrogen Fixation: A Model System for Analysis of Biological Process*, (Ed) Triplett EW. Horizon Scientific Press, Wymondham, UK, p. 207–224.

Benson DR and Silvester WB (1993). Biology of *Frankia* strains, actinomycete symbionts of actinorhizal plants. *Microbiol Rev,* 57 (2): 293–319.

Berry A, Harriott O, Moreau R, Osman S, Benson D and Jones A (1993). Hopanoid lipids compose the *Frankia* vesicle envelope, presumptive barrier of oxygen diffusion to nitrogenase. *Proc Natl Acad Sci, USA,* 90: 6091–6094.

Berry AM, Kahn RKS and Booth MC (1989). Identification of indole compounds secreted by *Frankia* HFPArI3 in defined culture medium. *Plant and Soil,* 118: 205–209.

Berry AM, and Torrey JG (1983). Root hair deformation in the infection process of *Alnus rubra*. *Can J Microbiol*, 61: 2863–2976.

Bingham FT, Pareyea FJ and Jarrell WM (1986). Metal toxicity to agricultural crops. *Metal Ions Biol Syst*, 20: 119–156.

Bloom RA, Mullin BC and Tate RL (1989). DNA restriction patterns and DNA–DNA solution hybridization studies of *Frankia* isolates from *Myrica pensylvanica* (bayberry). *Appl Environ Microbiol*, 55: 2155–2160.

Bond L (1948). Origin and developmental morphology of root nodules of *Pisum sativum*. *Bot Gaz*, 109: 411–434.

Boyer GL, Gillam A and Trick CG (1987). Iron chelation and uptake. In: *Cyanobacteria*, (Eds) Fay P and Van Baalen C. Elsevier, London, p. 415–436.

Burggraaf AJP and Shipton WA (1982). Estimates of *Frankia* growth under various pH and temperature regimes. *Plant and Soil*, 69: 135–147.

Callaham D, Newcomb W, Torrey JG and Peterson RL (1979). Root hair infection in actinomycete–induced root nodule initiation in *Casuarina*, *Myrica*, and *Comptonia*. *Bot Gaz*, 140(Suppl.): S1–S9.

Crutzen PJ, and Andreae MO (1990). Biomass burning in the tropics: Impact on atmospheric chemistry and biogeochemical cycles. *Sci*, 250: 1669–1678.

Dave BP and Dube HC (2000). Regulation of siderophore production by iron Fe (III) in certain fungi and fluorescent Pseudomonads. *Ind J Exp Biol*, 38: 297–299.

Dean DR and Jacobson MR (1992). Biochemical genetics of nitrogenase. In: *Biological Nitrogen Fixation*, (Eds) Stacey G, Burris RH and Evans HJ. Chapman and Hall, New York, p. 763–834.

Diem HG and Dommergues YR (1990). Current and potential uses and management of Casuarinaceae in the tropics and subtropics. In: *The Biology of Frankia and Actinorhizal Plants*, (Eds) Schwintzer CR and Tjepkema JD. Academic Press, Inc San Diego, Calif, p. 317–342.

Diouf D, Diop TA and Ndoye I (2003). Actinorhizal, mycorhizal and rhizobial symbioses: How much do we know? *Afr J Biotechnol*, 2(1): 1–7

Dixon ROD and Wheeler CT (1986). *Nitrogen Fixation in Plants*. Chapman and Hall, Blackie, New York, p. 101–103.

Dobritsa SV and Stupar OS (1989). Genetic heterogeneity among *Frankia* isolates from root nodules of individual actinorhizal plants. *FEMS Microbiol Lett*, 58: 287–292.

Fernandez MP, Meugnier H, Grimont PAD and Bardin R (1989). Deoxyribonucleic acid relatedness among members of the genus *Frankia*. *Int J Syst Bacteriol*, 39: 424–429.

Fleming AI, Wittenberg JB, Wittenber BA, Dudman WF, Appleby CA (1987). The purification, characterization and ligand-binding kinetics of hemoglobins from root nodules of the non-leguminous *Casuarina glauca-Frankia* symbiosis. *Biochem Biophys Acta*, 911: 209–220.

Foy CD, Chaney RL and White MC (1978). The physiology of metal toxicity in plants. *Ann Rev Plant Physiol*, 29(1): 511–566.

Franche C, Laplaze L, Duhoux E, Bogusz D (1998). Actinorhizal symbioses: Recent advances in plant molecular and genetic transformation studies. *Crit Rev Plant Sci*, 17: 1–28.

Gauthier D, Frioni L, Diem H G, Dommergues Y (1984). The Colletia spinosissima-*Frankia* symbiosis. *Acta Oecol Oecol Plant*, 5: 231–239.

Gianinazzi–Pearson V, Gianinazzi S, Guillemin JP, Trouvelot A, Duc G (1991). Genetic and cellular analysis of resistance to vesicular (VA). mycorrhizal fungi in pea mutants. In: *Advances in Molecular Genetics of Plant–Microbe Interaction,* Vol. 1, (Eds) Hennecke H and Verma DPS. Kluwer Academic Publishers, Dordrecht, p. 336–342.

Gherbi H, Duhoux E, Franche C, Pawlowski K, Nassar A, Berry A, Bogusz D (1997). Cloning of a full-length symbiotic hemoglobin cDNA and *in situ* localization of the corresponding mRNA in *Casuarina glauca* root nodule. *Physiol Plant,* 99: 608–616.

Guerinot ML (1994). Microbial iron transport. *Ann Rev Microbiol,* 48: 743–772.

Haansuu JP, Klika KD, Söderholm PP, Ovcharenko VV, Pihlaja K, Haahtela KK and Vuorela PM (2001). Isolation and biological activity of frankiamide. *J Ind Microbiol Biotechnol,* 27: 62–66.

Han SO and New PB (1998). Variation in nitrogen fixing ability among natural isolates of *Azospirillum. Microb Ecol,* 36: 193–201.

Harriott OT, Hosted TJ and Benson DR (1995). Sequences of *nifX*, *nifW*, *nifZ*, *nifB* and two ORF in the *Frankia* nitrogen fixation gene cluster. *Gene,* 161: 63–67.

Hirsch AM (1992). Developmental biology of legume nodulation. *New Phytol,* 122: 211–237.

Horriere F, Lechevalier MP and Lechavalier HA (1983). In vitro morphogenesis and ultrastructure of a *Frankia* sp. ArI3 (Actinomycetales). from *Alnus rubra* and a morphologically similar isolate (AirI2). from *Alnus incana* subsp. rugosa. *Can J Bot,* 61: 2843–2854.

Horrière F (1984). *In vitro* physiological approach to classification of *Frankia* isolates of *'the Alnus group'*, based on urease, protease, and ß-glucosidase activities. *Plant Soil,* 78: 7–13.

Huss-Danell K and Frej A (1986). Distribution of *Frankia* strains in forest and afforestation sites in Northern Sweden. *Plant Soil,* 90: 407–418.

Huss-Danell K, Uliassi D, Renberg (1997). River and lake sediments as source of infective *Frankia* (*Alnus*). *Plant Soil,* 197: 35–39.

Jing Yan-de, He Zhen-li and Yang Xiao-e (2007). Role of soil rhizobacteria in phytoremediation of heavy metal contaminated soils. *J Zhejiang Univ Sci,* B8(3): 192–207.

Kumada Y, Benson DR, Hillemann D, Hosted TJ, Rochefort DA, Thompson C, Wohlleben W and Tateno Y (1993). Evolution of the glutamine synthetase gene: one of the oldest genes ever studied. *Proc Natl Acad Sci, USA,* 90: 3009–3013

Lalonde M (1977). The infection process of the *Alnus* root nodule symbiosis. In: *Recent Developments in Nitrogen Fixation,* (Eds) Newton WE, Postgate JR and Rodriguez-Barrueco C. Academic Press, London, p. 569–589.

Lalonde M and Calvert HE (1979). Production of *Frankia hyphae* and spores as an infective inoculant for *Alnus* species. In: *Symbiotic Nitrogen Fixation in the Management of Temperate Forests,* (Eds) Gordon JC, Wheeler CT and Perry DA. Forest Research Laboratory, Oregon State University, Corvallis, p. 95–110.

Lalonde M and Knowles R (1975). Ultrastructure of the *Alnus crispa* var. mollis Fern. root nodule endophyte. *Can J Microbiol,* 21: 1058–1080.

Lancelle SA, Torrey JG, Hepler PK and Callaham DA (1985). Ultrastructure of freeze-substituted *Frankia* strain HFPCcI3, the actinomycete isolated from root nodules of *Casuarina cunninghamiana. Protoplasma,* 127: 64–72.

Laplaze L, Duhoux E, Franche C, Frutz T, Svistoonoff S, Bisseling T, Bogusz D and Pawlowski K (2000). *Casuarina glauca* prenodule cells display the same differentiation as the corresponding nodule cells. *Mol Plant Microbe Interact*, 13: 107–12.

Lechevalier MP and Lechevalier HA (1979). The taxonomic position of the actinomycetic endophytes. In: *Symbiotic Nitrogen Fixation in the Management of Temperate Forests*, (Eds) JC Gordon, CT Wheeler, and DA Perry. Forest Research Laboratory, Oregon State University, Corvallis, p. 111–122.

Lechevalier MP and Lechevalier HA (1990). Systematics, isolation, and culture of Frankia. In: *The Biology of Frankia and Actinorhizal Plants*, (Eds) Schwintzer CR and Tjepkema JD. Academic Press, Inc, New York, p. 35–60.

Leul M, Normand P and Sellstedt A (2007). The organization, regulation and phylogeny of uptake hydrogenase genes in *Frankia*. *Physiol Plant*, 130: 464–470.

Liu Q and Berry AM (1991). The infection process and nodule initiation in the *Frankia-Ceanothus* root nodule symbiosis: A structural and histochemical study. *Protoplasma*, 163: 82–92.

Liu Q and Berry AM (1991). Localization and characterization of pectic polysaccharides in roots and root nodules of *Ceanothus* spp during intercellular infection by *Frankia*. *Protoplasma*, 163: 93–101.

Mattsson U and Sellstedt A (2002). Nickel affects activity more than expression of hydrogenase protein in *Frankia*. *Curr Microbiol*, 44: 88–93.

McGrath SP, Chaudri AM and Giller KE (1995). Long-term effects of metals in sewage sludge on soils, microorganisms and plants. *J Ind Microbiol*, 14 (2): 94–104.

Meyer J, Halle F, Hohnadel D, Lemanceau P and Ratefiarivel H (1987). Siderophore of *Pseudomonas*, Biological properties. In: *Iron Transport in Microbes, Plants and Animals* (Eds) Winkelmann G, Vander helm D and Neiland JB. VCH Verlagsgesellschaft, Weinheim, p. 189–205.

Miller IM and Baker DD (1985). The initiation, development and structure of root nodules in *Elaeagnus angustifolia* L. (Elaeagnaceae). *Protoplasma*, 128: 107–119.

Mullin BC and An CS (1990). The molecular genetics of *Frankia*. In: *The Biology of Frankia and Actinorhizal Plants*, (Eds) Schwintzer CR and Tjepkema JD. Academic Press, Inc, New York, p. 195–214.

Mullin BC and Dobritsa SV (1996). Molecular analysis of actinorhizal symbiotic systems: Progress to date. *Plant Soil*, 186: 9–20.

Murry MA, Fontaine MS, Tjepkema JD (1984). Oxygen protection of nitrogenase in *Frankia* sp. HFPArI3. *Arch Microbiol*, 139 (2–3): 162–166.

Newcomb W, Callaham D, Torrey JG and Peterson RL (1979). Morphogenesis and fine structure of the actinomycetous endophyte of nitrogen-fixing root nodules of *Comptonia peregrina*. *Bot Gaz*, 140(Suppl.): S22–S34.

Newcomb W, Sippel D and Peterson RL (1979). The early morphogenesis of *Glycine max* and *Pisum sativum* root nodules. *Can J Bot*, 57: 2603–2616.

Newcomb W and Wood SM (1987). Morphogenesis and fine structure of *Frankia* (Actinomycetales): The microsymbiont of nitrogen-fixing actinorhizal root nodules. *Int Rev Cytol*, 109: 1–88.

Normand P and Bousquet J (1989). Phylogeny of nitrogenase sequences in *Frankia* and other nitrogen-fixing microorganisms. *J Mol Evol*, 29: 436–447.

Normand P, Lapierre P, Tisa L, Gogarten JP, Alloisio N, Bagnarol E, Bassi CA, Berry AM, Bickhard DM, Choisne N, Couloux A, Cournoyer B, Cruveiller S, Daubin V, Demange N, Francino MP,

Goltsman E, Huang Y, Kopp O, Labarre L, Lapidus A, Lavire C, Marechal J, Martinez M, Mastronunzio JE, Mullin BC, Niemann J, Pujic P, Rawnsley T, Rouy Z, Schenowitz C, Sellstedt A, Tavares F, Tomkins JP, Vallenet D, Valverde C, Wall L, Wang Y, Medigue C, Benson DR (2007). Genome characteristics of facultatively symbiotic Frankia sp. strains reflect host range and host plant biogeography. *Genome Research,* 17: 7–15

Normand P, Orso S, Cournoyer B, Jeannin P, Chapelon C, Dawson J, Evtushenko L and Misra AK (1996). Molecular phylogeny of the genus *Frankia* and related genera and emendation of family Frankiaceae. *Int J Syst Bacteriol,* 46: 1–9.

Normand P, Queiroux C, Tisa LS, Benson DR, Rouy Z, Cruveiller S and Medigue C (2007). Exploring the genomes of *Frankia. Physiologia Plantarum,* 130: 331–343.

Normand P, Simonet P, Butour JL, Rosenberg C, Moiroud A and Lalonde M (1983). Plasmids in *Frankia* sp. *J Bacteriol,* 155: 32–35.

Obertello M, Oureye SY, Laplaze L, Santi C, Svistoonoff S, Auguy F, Bogusz D and Franche C (2003). Actinorhizal nitrogen fixing nodules: infection process, molecular biology and genomics. *Afr J Biol,* 2(12): 528–538.

Olson JB, Steppe TF, Litaker RW and Pearl HW (1998). N_2-fixing microbial consortia associated with the ice cover of lake Bonney, Antarctica. *Microb Ecol,* 36: 231–238.

Page WJ and Huyer M (1984). Depression of the *Azotobacter vinelandii* siderophore system, using iron–containing minerals to limit iron repletion. *J Bacteriol,* 158: 496–502.

Pawlowski K, Bisseling T (1996). Rhizobial and actinorhizal symbioses: What are the shared features? *Plant Cell,* 6: 1899–1913.

Quesada A, Leganes F and Valiente EF (1997). Environmental factors controlling N_2 fixation in Mediterranean rice fields. *Microb Ecol,* 34: 39–48.

Quesada A, Nieva M, Leganes F, Ucha A, Martin M, Prosperi C and Valiente EF (1998). Acclimatization of Cyanobacterial communities in rice fields and response of nitrogenase activity to light regime. *Microb Ecol,* 35: 147–155.

Quesada A and Valiente EF, (1996). Relationship between abundance of nitrogen–fixing cyanobacteria and environmental features of spanish rice fields. *Microb Ecol,* 32: 59–71.

Racette S andTorrey JG (1989). The isolation, culture and infectivity of a *Frankia* strain from *Gymnostoma papuanum* (Casuarinaceae). *Plant Soil,* 118: 165–170.

Reid CP, Szaniszlo PJ and Crowley DE (1986). Siderophore Involvement in Plant Iron Nutrition. In: *Iron Siderophores and Plant Diseases,* (Ed) TR Swinburne. Plenum Press, New York, p. 29–42.

Sarma HK, Sharma BK, Singh SS, Tiwari SC and Mishra AK (2006). Polymorphic distribution and phenotypic diversity of *Frankia* strains in nodule of *Hippöpahe salicifolia* D.Don. *Curr Sci,* 90: 1516–1520.

Sarma HK, Sharma BK, Tiwari SC and Mishra AK (2005). Truncated hemoglobins: A single structural motif with versatile functions in bacteria, plants and unicellular eukaryotes. *Symbiosis,* 39: 151–158.

Schwintzer CR and Tjepkema JD (1991). Actinorhizal plants: *Frankia-* symbioses. In: *Biology and Biochemistry of Nitrogen Fixation,* (Eds) Dilworth MJ and Glenn AR. Elsevier Science Publishing, Inc, New York, p 350–372.

Shantaram S, DuTeau NM, Prakash RK and Atherly AG (1987). Comparative cytology of effective and ineffective root nodules of North American cultivars of *Glycine max* L. formed by two *Rhizobium fredii* strains. *Cytobios,* 50: 181–190.

Silvester WB and Harris SL (1989). Nodule structure and nitrogenase activity of *Coriaria arborea* in response to varying pO2. *Plant and Soil,* 118: 97–109.

Simonet P, Capellano A, Navarro E, Bardin R and Moiroud A (1984). An improved method for lysis of *Frankia* with achromopeptidase allows detection of new plasmids. *Can J Microbiol,* 30: 1292–1295.

Singh A, Mishra AK, Singh SS, Sarma HK and Shukla E (2008). Influence of Iron and chelator on siderophore production in *Frankia* strains nodulating *HippÖphae salicifolia* D. Don. *J Basic Microbiol,* 48: 104–111.

Smith SE and Read DJ (1997). *Mycorrhizal Symbiosis.* Academic Press Inc, San Diego.

Smolander A (1990). *Frankia* populations in soils under different tree species with special emphasis on soils under *Betula pendula. Plant Soil,* 121: 1–10.

Smolander A and Sarsa M (1990). *Frankia* strains in soil under *Betula pendula*: behavior in soil and in pure culture. *Plant Soil,* 122: 129–136.

Soltis DE, Soltis PS, Morgan DR, Swensen SM, Mullin BC, Dowd JM and Martin PG (1995). Chloroplast gene sequence data suggest a single origin of the predisposition for symbiotic nitrogen fixation in angiosperms. *Proc Natl Acad Sci, USA,* 92: 2647–2651.

Sougoufara B, Maggia L, Duhoux E and Dommergues YR (1992). Nodulation and nitrogen fixation in nine *Casuarina* clone-*Frankia* strain combinations. *Acta Oecol,* 13: 497–503.

Tisa LS and Ensign JC (1987). Isolation and nitrogenase activity of vesicles from *Frankia* sp. strain EAN1pec. *J Bacteriol,* 169 (11): 5054–5059.

Tjepkema JD, Ormerod W and Torrey JG (1980). Vesicle formation and acetylene reduction (nitrogenase activity). in *Frankia* sp. CpI1 in culture. *Can J Microbiol,* 27: 815–823.

Torrey JG (1976). Initiation and development of root nodules of *Casuarina* (Casuarinaceae). *Amer J Bot,* 63: 335–344.

Torrey JG and Callaham D (1982). Structural features of the vesicle of *Frankia* sp. CpI1 in culture. *Can J Bot,* 28: 749–757.

Torrey JG and Tjepkema JD (1979). Symbiotic nitrogen fixation in actinomycetes-nodulated plants. *Bot Gaz,* 140(Supplement): i–ii.

Verghese S and Mishra AK (2002). *Frankia*-actinorhizal symbiosis with special reference to host-microsymbiont relationship. *Curr Sci,* 83 (4): 404–408.

Waite-Rees PA, Keating CJ, Moran LS, Slatko BE, Hornstra LJ and Benner JS (1991). Characterization and expression of the *Escherichia coli* Mrr restriction system. *J Bacteriol,* 173: 5207–5219.

Wall LG (2000). The actinorhizal symbiosis. *J Plant Growth Regul,* 19: 167–182.

Wang Y, Brown HN, Crowley DE and Szaniszlo PJ (1993). Evidence for direct utilization of a siderophore, ferroxamine B, in axenically grown cucumber. *Plant Cell Environ,* 16(5): 579–585.

Soil Microflora, 2009 *Pages* **102–107**
Editor: **Rajan Kumar Gupta, Mukesh Kumar & Deepak Vyas**
Published by: **DAYA PUBLISHING HOUSE, NEW DELHI**

Chapter 10

Cyanobacterial Surfactants Enhance the Fertility and Stability of Tropical Soils

Usha Pandey
Department of Biotechnology, Faculty of Science and Technology,
M.G. Kashi Vidyapith, Varanasi – 221 002
E-mail: Usha_pandey28@yahoo.co.in

ABSTRACT

Increasing consciousness about future sustainable agriculture has lead scientists all over the world to search for more eco-friendly natural alternatives which can be used to maintain the fertility stability of agricultural lands. The present study was an effort to investigate the efficacy of a natural surfactant of cyanobacterial origin in enhancing water availability and regulating adsorption/retention of the nutrients in the root environment of agricultural crops. The chosen determinants were two branched diazotrophic cyanobacteria (*Hapalosiphon welwitschii* and *Westiellopsis prolifica*) and two prime nutrients (N and P) required for maintaining agricultural productivity. Observations indicated that the surfactants derived from these cyanobacteria significantly reduced the surface tension with substantially improved water retention and hydraulic conductivity in the root environment. These surfactants also enhanced mobilization and availability of N and P in the soil solution. These observations form the base line data for the use of cyanobacterial surfactants towards sustainable agriculture.

Introduction

Recent global attention has shown serious concern towards increasing demand for food and causes of the declining agricultural sustainability. The rising global food demand by burgeoning human population coupled with the diminishing agricultural soil fertility stability has led agricultural sustainability to emerge as one of the most important socio-economic-environmental issues of global concern. Furthermore, the growing consciousness for safe and nutritionally balanced food and environment–induced dietary contamination have forced scientists to search alternative ecofriendly–fertilization to secure agriculture sustainability and safe food availability. In this respect, cyanobacteria have emerged as an important input for biofertilizer industries. Commonly referred to as blue green algae, cyanobacteria constitute the most well adapted ubiquitous organisms in the biosphere. The dual capacity of fixing atmospheric carbon and nitrogen signify many cyanobacteria to be the attractive source of biofertilizer. Their natural population not only adds nitrogen to the soil but helps reducing soil erosion, decreasing soil compaction, adding moisture, organic matter and liberating growth regulators (Goyal, 1993; Boussiba, 1991).

The mucilage of cyanobacterial origin constitutes a crucial component of the rhizosphere, contributing to various fundamental plant–soil interactions, microbial dynamics and nutrient–turnover (Whitton, 2000). Cyanobacterial mucilage affects soil physical properties in the rhizosphere making the soil environment more conducive for soil microbes to proliferate and function. According to Passioura (1988) the soil contains sizeable quantity of surface active materials of biological origin. In many cases cyanobacteria contribute to surfactants in a major way. These surface active materials modify the surface tension of soil solution, which, in turn, increase the ability of roots to take up water, especially under the condition of soil water deficit.

The lipid surfactants in the rhizosphere help root water retention, increase hydraulic conductivity, regulate chemical adsorption and microbial processes. Such surfactants also help regulating matric potential, one of the most important components of the water potential in soil/plant system and thus help controlling the amount and distribution of water in the soil pore space and mediating many other soil processes, directly and indirectly (Young and Ritz, 2000). In addition to the effects on physical properties, surfactants may also modify the chemical properties of the rhizosphere. Rhizosphere mucilage has been found to affect the rates of nitrogen immobilization and mineralization (Mary *et al.*, 1993).

Although tropical soils are very rich in these diversified group of microorganism (Pandey, 2002), no proper attention has been given towards exploiting these properties of cyanobacteria in agriculture. Since this aspect bears immense opportunity in regulating soil fertility stability, an effort was made to investigate the use of cyanobacterial culture filtrates as surfactants and its effects on phosphate and nitrate adsorption/retention in tropical soils.

Materials and Methods

The cyanobacterial species, *Hapalosiphon welwitschii* and *Westiellopsis prolifica* considered in this study are branched soil inhabiting diazotrophs. These cyanobacteria were raised to unialgal populations in Fogg's medium at pH 7.5. Stock cultures were maintained in the medium under day/light fluorescent tube with 14: 10 h light/dark cycles at a temperature of 25±2°C. After 20 days of growth, cyanobacterial filaments were taken on moist filter papers in petri dishes and mucilage was collected from the filter paper using a drawn glass Pasteur pipette in a laminar flow cabinet. The mucilages thus collected were centrifuged at 12,000 rpm for 30 min and filtered to remove all the insoluble cyanobacterial filaments using 0.2 μ m nylon syringe filter.

The harvested mucilages were freeze dried immediately. Cyanobacterial phospholipids were eluted from total lipid in whole cells as per the method of Rouser *et al.* (1976). Commercially available phosphatidyl choline surfactant known as lecithin was used as a standard. Different concentrations (0.1 to 1.0 mg ml^{-1}) of lecithin was used for measurement of surface tension. For experimental purpose, 1 to 3.5 mg fresh weight of cyanobacterial cells was boiled and filtered in 1 litre distilled water.

Surface tension was measured by the capillary rise method using small precision base capillaries (radius = 0.315 mm) thoroughly cleaned in sodium hydroxide solution. Capillary rise was measured at 25°C using the equilibrium position of the recedimg meniscus (Nelkorn and Ogborn, 1978). Sandy loamy soils collected from tropical agricultural fields were used throughout the experiments. The dull brown light textured sandy loams are formed from parent materials rich in lime sand stone and are alkaline in nature. Soil pH and water holding capacity were determined according to the methods described in Pandey and Sharma (2002). Soil samples were autoclaved before experimentation. Each set of experiments containing 100 g of autoclaved soil and 50 ml of cyanobacterial filtrate, lecithin and water separately were treated for 24 hours. Phosphate and nitrate adsorption/desorption was measured using the method described by Rowell (1994). Soil samples were treated with 500 µg/ml P and N for 24 h and soil suspensions were prepared and filtered through Whatman No. 41 filter papers. The P and N contents were determined following phosphomolybdate and diphenylamine method respectively.

Results and Discussion

Cyanobacteria constitute one of the most diversified groups of organisms. Probably the diversified and varied composition of lipids helps this group of organisms to survive various environmental extremes. The major lipids of cyanobacteria are glycolopids and phospholipids. The proportion of phospholipids is relatively smaller than those of the glycolipids (Nicholas, 1973). Although relatively lower in content, the phospolipid plays important role characterizing morphological and physiological diversity of cyanobacteria (Piorreck and Pohl, 1984). Stable membrane formation is a sensitive indicator of the presence of phospholipids as a surfactant (Ballard *et al.*, 1986). In the present study, 15 and 21 per cent phospholipids were observed in the cell cultures of *Hapalosiphon* and *Westiellopsis* after 20 days of growth. However, it declined in both cultures after this phase. Variation in surface tension was studied in cyanobacterial filtrates and in lecithin and both species proved their efficacy in reducing surface tension. Surface tension was reduced to 48 m Nm^{-1} in the presence of highest concentration of lecithin considered in this study. Similar trends appeared in case of *Westiellopsis prolifica* feltrate. Slightly high surface tension (52 m Nm^{-1}) was recorded in the filtrates of *Hapalosiphon welwitschii* (Figure 10.2).

The water holding capacity and pH was improved in the presence of surfactants (Table 10.1). The presence of surfactant could substantially improve the water retention and hydraulic conductivity of rhizosphere as well as the chemical adsorption and microbial processes. Water potential of soil regulates the amount and distribution of water in pore spaces and regulate many other soil processes (Campbell, 1985). Addition of cyanobacterial filtrates caused significant improvement in water holding capacity. In soils, supplemented with *W. prolifica* filtrates, the soil pH reduced from 7.8 to 7.22 and water holding capacity enhanced from 42.2 to 57.6 per cent. Rhizosphere surfactants, depending upon the quality of soil, help regulating availability of nitrogen by increasing the concentration of N in soil solution. Agrochemical studies using tracer techniques have shown that after application of nitrogen fertilizers, plants grown in the fields take up only 30 to 40 percent nitrogen. Slow mineralization of the immobilized nitrogen and its partial uptake by crops affect the yield. Nitrogen fertilizers are applied to almost all form crops even in legumes, because at the beginning of growth, when nodule bacteria are still underdeveloped, legumes need a source of available nitrogen in the root environment. Cereal crops are highly sensitive to nitrogen availability which promotes vegetative growth, development of reproductive organs and increases the tillering energy, grain yield and protein content of the grains.

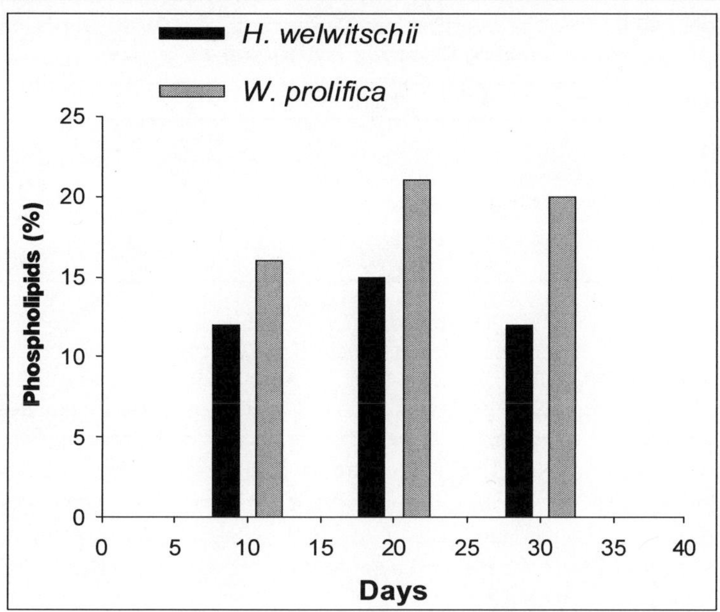

Figure 10.1: Phospholipid Concentration of Two Cyanobacteria Considered in this Study

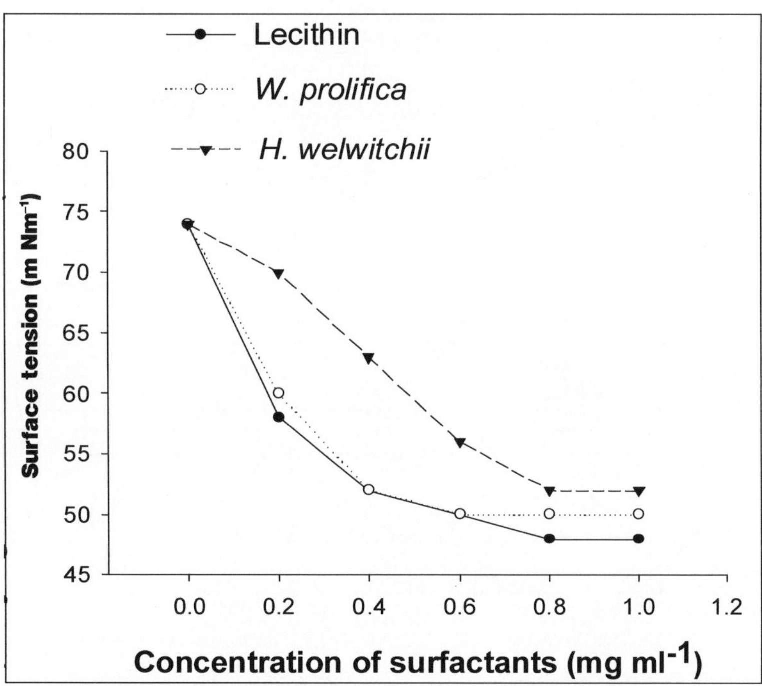

Figure 10.2. Changes in Surface Tension due to Lecithin and Cyanobacterial Surfactants in Aqueous Solution

Table 10.1: Effect of Lecithin and Cyanobacterial Surfactants on Water Holding Capacity (WHC) and pH of Soil

Treatment	*WHC (%)*	*Soil pH*
Control	42.2±3.50	7.80±0.16
Soil +*Hapalosiphon* filtrate	53.5±4.12	7.35±0.11
Soil + *Westiellopsis* filtrate	57.6±4.18	7.22±0.09
Soil + Lecithin	58.5±4.25	7.40±0.11

Values are mean (n = 5)±1SE.

Cyanobacterial filtrates were found to be effective in mobilizing N in available form in the soil solution (Table 10.2). In the soil treated with filtrates of *W. prolifica,* N in the soil solution was found to be 158 µg/ml. The values were 140 µg/ml in soil solution treated with filtrates of *Hapalosiphon welwitschii* and 152 µg/ml in soil solution treated with lecithin, a commercially established surfactant. These observations indicate the role of cyanobacteria to be used as surfactants in promoting crop growth apart from their massive contribution as biofertilizer. Similar observations were made with respect to P availability in soil solution. For instance, 160 µg/ml P was recorded in the soil treated with filtrates of *W. prolifica.* However, for soil treated with filtrates of *H. welwitschii,* this effect was comparatively lower (P in soil solution remained 130 µg/ml). Phosphorus plays a crucial role in the growth of plants and the effects of phosphorus on vital activities of plants are manifold. Adequate phosphorus availability not only raises the crop yields considerability, but also improves the quality of edible produce. The grain yield in cereal crops increases and the grains become richer in starch and proteins. Fruits and tap roots accumulate more carbohydrates. In fibre crops, it help develop long, strong and fine fibres. Phosphates in the soil are often regulated by organic matter, moisture content and temperature. The presence of microbial surfactants in the soil helps develop gradients of matric potential and hydraulic conductivity which may extend from the plant rhizosphere into the bulk soil. By regulating soil water relationships, surfactants may enable plants to withstand periods of water scarcity in the root environment. The increase in phosphorus/nitrogen desorption may be linked to the soil properties and disaggregation by surfactants in rhizosphere.

Table 10.2: Effects of Cyanobacterial Surfactants and Lecithin on P and N Adsorption/Desorption Values

Trearment	*Soil N Adsorption ($\mu g\ g^{-1}$)*	*Solution N Concentration ($\mu g\ ml^{-1}$)*	*Soil P Adsorption ($\mu g\ g^{-1}$)*	*Solution P Concentration ($\mu g\ ml^{-1}$)*
Control	390.0±26.0	112.0±9.6	438.1±36.5	63.0±5.1
Soil +*Hapalosiphon* filtrate	360.2±21.4	142.0±11.0	376.5±20.6	130.4±7.8
Soil + *Westiellopsis* filtrate	388.5±30.0	158.2±13.2	388.4±30.2	160.2±11.5
Soil + Lecithin	350.2±27.2	152.5±11.6	322.0±25.2	175.0±13.2

Values are mean (n = 5)±1SE.

Such surfactants may also help increasing the availability of other nutrients in the region of root uptake. It is interesting to note that the phospholipids are such powerful surfactants that, even at these low concentrations considered in the present study, significant reduction in surface tension were observed. In the soil where the growing root tips are subjected to abrasion and mechanical impediments, presence of mucilage protects these growing tips and function as surfactant to promote growth (Groleau *et al.*, 1999). The present study provides evidence that the cyanobacterial exudates can be very useful soil surfactant promoting crop growth by decreasing soil surface tension and mobilizing nutrients to soil solution in the rhizosphere. It is important to maintain *in situ* cyanobacterial diversity to improve soil fertility through biologically fixed atmospheric nitrogen as well as through added surfactants. This dual capacity of certain cyanobacteria makes them to be a novel resource of biofertilizer industries.

References

Ballard RE, Jones J, Reed D and Inchley A (1986). He (I) photoelectron studies of lipid layers. *Chemical Physics Letters*, 135: 119–122.

Boussiba S (1991). Nitrogen-fixing cyanobacteria: Potential uses. *Plant and Soil*, 137: 177–180.

Campbell GS (1985). Soil physics with basic transport models for soil–plant systems. Amsterdam, The Netherlands, Elsevier

Fogg GE (1951). Studies on nitrogen fixation by blue green algae II: Nitrogen fixation by *Mastigocladus laminosus*. *J Expt Bot*, 2: 117.

Goyal SK (1993). Algal biofertilizers for vital soil and free nitrogen. *Proc. Indian Natn Sci Acad*, 859: 295–302.

Groleau RV, Plantureux S and Guckert A (1999). Influence of plant morphology on root exudation of maize subjected to mechanical impedence in hydroponic conditions. *Plant and Soil*, 201: 231–239.

Marry B, Fresneau C, Morel JL, Mariotti A (1993). C&N cycling during decomposition of root mucilage, roots and glucose in soil. *Soil Biol and Biochem*, 25: 1005–1014.

Nelkorn M and Ogborn JM (1978). *Advanced Level Practical Physics*, 4th edn. UK Heinemann Educational Books, London.

Nicholas B W (1973). Lipid composition and metabolism. In: *The Biology of Blue Green Algae*, (Eds) Carr NG and Whitton BA. Blackwell Scientific Publishers, Oxford Press, p. 144–161.

Pandey J and Sharma MS (2002). *Environmental Science: Practical and Field Manual*. Yash Publishing House, Bikaner.

Pandey U (2002). Soil cyanobacteria from arable lands of southern Rajasthan. *Phykos*, 41: 7–11.

Passioura JB (1988). Water transport in and to roots. *Ann Rev Plant Physiol Plant Mol Biol*, 39: 245–265.

Piorreck M and Pohl P (1984). Formation of biomass, total protein, chlorophylls, lipids and fatty acids in green and blue green algae under different nitrogen regimes. *Phytochemistry*, 23: 207–216.

Rouser G, Kritchevisky G and Xama-moto (1976). Column chromatographic and associated procedures for separation and determination of phospholipids and glycolipids. In: *Lipid Chromatographic Analysis*, Vol 3, (Eds) Marinetti G V and Marcel D. New York, p. 713–776.

Rowell DL (1994). *Soil Science: Methods and Applications*. Longman Group UK Ltd, Harlow, UK.

Whitton BA (2000). Soils and rice fields. In: *The Ecology of Cyanobacteria*, (Eds) Whitton BA and Potts M. Kluwer Academic Publishers, The Neatherlands, p. 233–255.

Young I M and Ritz K (2000). Tillage: Habitat space and function of soil microbes. *Soil and Tillage Research*, 53: 201–213.

Soil Microflora, 2009 *Pages* **108–121**
Editor: **Rajan Kumar Gupta, Mukesh Kumar & Deepak Vyas**
Published by: **DAYA PUBLISHING HOUSE, NEW DELHI**

Chapter 11

Soil Denitrifying Bacteria and Environmental Factors Regulating Denitrification in Soil

Paromita Ghosh
G. B. Pant Institute of Himalayan Environment and Development, Garhwal Unit, Upper Bhaktiyana, P.O. Box. 92, Srinagar, Garhwal, Uttarakhand
E-mail:pghosh@gbpihed.nic.in/paroghosh@rediffmail.com

ABSTRACT

Denitrification involves the conversion of NO_3^- and NO_2^- to the gases N_2 and N_2O which escape from the soil. Denitrification is important to primary production, water quality and the chemistry and physics of the atmosphere at ecosystem, landscape regional and global scales. There is a concern that an increase in atmospheric N_2O from increased denitrification might cause a depletion of the ozone concentration allowing passage of more short wavelength ultraviolet radiation which is responsible for the occurrence of skin cancer and global warming. Increased denitrification might result from increases in the combined nitrogen inputs from either biological or industrial N_2 fixation. The present article describes important reported studies on major factors affecting denitrification in soil and soil denitrifying bacteria.

Keywords: *Denitrifying bacteria, Factors affecting denitrification in soil.*

Introduction

Denitrification is usually defined as the reduction of nitrate (NO_3^-) to a gaseous product (Gayon and Depetit, 1886; Payne, 1973a). The consequence of which is a loss of fixed nitrogen from a terrestrial or aquatic environment. The evolution of gaseous nitrogen as a result of microbiological action upon fixed compounds of nitrogen is known as biological denitrification. The reductive pathway is generally accepted to be as follows:

$$NO_3^- \text{---} 1 \text{---} NO_2^- \text{---} 2 \text{---} NO^- \text{---} 3 \text{---} N_2O \text{---} 4 \text{---} N_2$$

In principal each reductive step by itself should be able to support growth because each reduction is coupled to the generation of APT (Koike and Hattori, 1975b). Thus denitrifying bacteria can be found using the truncated pathway. The truncation may occur in one of the two ways (i) When organisms that are capable of carrying out every step of the pathway have available to them only intermediates of the pathway that are more reduced than NO_3^- and (ii) when organisms are genetically incapable of carrying out various steps of the pathway. Within the second category the biochemical variation of the denitrification pathway may be subcategorized as follows:

1. Organisms capable of reducing nitrate only to nitrite (*i.e.* lacking reactions 2, 3 and 4).
2. Organisms capable of reducing nitrate only to nitrous oxide (*i.e.* lacking reaction 4)
3. Organisms capable of reducing nitrite but not nitrate to dinitrogen (*i.e.* lacking reaction 1) and
4. Organisms capable of reducing nitrate to nitrite and nitric oxide to nitrous oxide (*i.e.* lacking reactions 2 and 4).

The prokaryotes of subcategory (1) are most numerous but they are not considered denitrifiers as their activities do not produce a gaseous product. Only organisms capable of reduction of nitrate to nitrogen or reduction of nitrogenous oxide through any of the truncated pathways 2, 3 and 4 just described are considered denitrifiers.

The Denitrifying Bacteria

Biological denitrification is apparently effected only by prokaryotes and even among them only a certain genera of bacteria is known to account for the most of the denitrifying activity in soil. Denitrification is present in strains contained in ten different prokaryotic families (1) Rhodospirillaceae; (2) Cytophagaceae; (3) Budding or appendaged bacteria; (4) Spirileaceae; (5) Pseudomonaceae; (6) rhizobiaceae; (7) Halobacteriaceae; (8) Neisseriaceae; (9) Nitrobacteraceae; (10) Bacillaceae. Three other groups of bacteria that are also known to denitrify are: (*a*) Obligate anaerobes; (*b*) Gram positive organisms other than bacillus and (*c*) The Enterobacteriaceae. The most important denitrifiers have been listed in Table 11.1.

Denitrification has classically been viewed as taking place only under anaerobic conditions but most of the bacteria are facultative in the sense that these responsible bacteria are commonly aerobic but they have become adapted to the utilization of nitrate or nitrite in environments containing little or no free oxygen. Under most conditions where denitrification is encountered in nature, a mixed flora is involved, *e.g.* of bacteria found in soil:

Obligate anaerobes: *Clostridium* (several species)

Facultative anaerobes: *E. coli*, *Citrobacter* sp., *Klebseilla* sp., *Enterobacter (Aerobacter) aerogenes*, *Erwinia carotovora.*

Aerobes; *Pseudomonas* (several strains), *Bacillus* (several strains).

The major genera with denitrifying species grouped according to distinctive physiological features are as follows:

Table 11.1: Organisms Capable of Denitrification (Ingraham, 1981)

Sl.No.		Remarks
A	**Phototrophic Bacteria**	
1.	*Rhodopseudomonas sphaeroids*	Denitrification is strain dependent, organism also capable of nitrogen fixation.
B.	**Gliding Bacteria**	
1.	*Cytophaga johnsonae*	Denitrification is strain dependent of 16 strains studied 4 reduce
2.	*Lysobacter antibioticus*	NO_3- to NO_2, of these one reduces NO_2- to gaseous product of 18
3.	*Simonsiella muelleri*	strains studied 9 reduce No_3- of these 4 reduce NO_2- with or
		without production of visible gas
C.	**Budding Bacteria**	
1.	Hypomicrobium spp.	Denitrifies as a methylotroph
D.	**Spiral and curved bacteria**	
1.	Aquaspirillum itersonii	Produce N_2O as an end product of denitrification
2.	*Aquaspirillum psychrophilum*	Produces visible gas as an end product of denitrification.
3.	*Aquaspirillum dispar*	Does not produce gas as an end product of denitrification.
4.	*Azospirillum lipoferum*	Denitrification is strain-dependent, some strains produce N_2O as an end product of denitrification, capable of nitrogen fixation.
5.	*Campylobacter sputorum*	
E.	**Gram-negative bacteria**	
1.	*Pseudomonas aeruginosa*	
2.	*Pseudomonas fluorescens*	Some strains produce N_2 as the end product of denitrification
3.	*Pseudomonas chloraphis*	Some strains produce N_2 as the end product of denitrification
4.	*Pseudomonas aureofaciens*	
5	*Pseudomonas stutzeri*	Some strains grown on N_2O as electron acceptor
6.	*Pseudomonas mendocina*	
7.	*Pseudomonas mallei*	An animal pathogen
8.	*Pseudomonas pseudomallei*	An animal pathogen
9.	*Pseudomonas caryophylli*	A plant pathogen
10.	*Pseudomonas lemoignei*	
11.	*Pseudomonas solanacearum*	A plant pathogen
12.	*Pseudomonas pickettii*	Some strains grow on N_2O as electron acceptor
13.	*Pseudomonas pseudoflava*	A facultative chemolithotroph (hydrogen bacterium)
14.	*Pseudomonas denitrificans*	
15.	*Pseudomonas perfectomarinus*	A marine bacterium
16.	*Pseudomonas nautica*	A marine bacterium
17.	*Gluconobacter* spp.	
18.	*Alcaligenes faecalis*	Some strains reduced NO_2- to gas but are unable to reduce NO_3- to NO_2-
19.	*Alcaligenes eutrophus*	A facultative chemolithotroph (hydrogen bacterium)
20.	*Agrobacterium tumefaciens*	A plant pathogen-some strains grow on N_2O as electron acceptor.
21.	*Agrobacterium radiobacter*	Some strains grow on N_2O as electron acceptor.

Contd...

Table 11.1–Contd...

Sl.No.		Remarks
F.	**Gram-negative facultatively anaerobic bacteria**	
1.	*Chromobacterium violaceum*	Visible gas usually not produced from NO_2- or NO_3-
2.	*Chromobacterium lividum*	Visible gas usually not produced from NO_2- or NO_3-
3.	*Flavobacterium* spp.	Some strains unable to reduce
G.	**Gram-negative cocci and coccobacilli**	
1.	*Neisseria sicca*	Able to reduce NO_2- but not NO_3-
2.	*Neisseria subflava*	Able to reduce NO_2- but not NO_3-
3.	*Neisseria flavescens*	Able to reduce NO_2- but not NO_3-
4.	*Neisseria mucosa*	Able to reduce NO_2-and NO_3-
5.	*Neisseria animalis*	Nitrate reduction is strain dependent
6.	*Neisseria cavie*	Nitrate reduction is strain dependent
7.	*Neisseria denitrificans*	Reduces NO_2- but not NO_3-
8.	*Branhamella catarrhalis*	Nitrate reduction is strain dependent, may or may not produce visible gas
9.	*Acinetobacter* spp.	
10.	*Kingella denitrificans*	
11.	*Paracoccus denitrificans*	N_2O and N_2 produced from nitrate, a facultative chemolithotroph (hydrogen bacterium)
12.	*Paracoccus halodenitrificans*	N_2O and N_2 produced from nitrate
H.	**Gram negative chemolithotrophic sulfur bacteria**	
1.	*Thiobacillus denitrificans*	Some strains reduce NO and N_2O to N_2
2.	*Thiomicrospira denitrificans*	Does not grow aerobically apparently an obligate denitrifier
3.	*Thermothrix thioparus*	No visible gas produced from NO_3- but reduced NO_2-, athermophilic facultative chemolithothroph
I.	**Gram positive spore forming bacteria**	
1.	*Bacillus licheniformis*	
2.	*Bacillus cereus*	
3.	*Bacillus polymyxa*	
4.	*Bacillus macerans*	
5.	*Bacillus stearothermophilus*	
6.	*Bacillus laterosporus*	
7.	*Bacillus pasteurii*	
8.	*Bacillus pantothenticus*	
9.	*Bacillus pulvifaciens*	
10.	*Bacillus nitrollens*	
11.	*Bacillus azotoformans*	Isolated by N_2O enrichment
J.	**Gram positive non-spore forming bacteria**	
1.	*Corynebacterium nephridii*	Only one strain known, produces N_2O as end product
2.	*Propionibacterium acidi propionici*	Denitrification is strain dependent
K.	**Others**	
1.	*Halobacterium* sp.	

Organotrophic (general aerobic): *Pseudomonas, Alcaligenes, Flavobacterium (Achromobacterium), Paracoccus (Cornybacterium), Acinetobacter, Cytophaga, Gluconobacter, Xanthomonas.*

Oligocarbophilic: *Hypomicrobium, Aquaspirillum.*

Fermentative: *Azospirillum, Chromobacterium, Bacillus, Wolinella*

Halophilic: *Halobacterium, Paracoccus*

Thermophilic: *Bacillus, Thermothrix*

Spore former: *Bacillus*

Magnetotactic: *Aquaspirrillum*

N_2 fixing: *Rhizobium, Bradyrhizobium, Azospirillum, Pseudomonas, Rhodopseudomonas, Agrobacterium*

Animal or pathogenic association: *Neisseria, Kingella, Moraxella, Wolinella.*

Phototrophic: *Rhodopseudomonas*

Lithotrophic 1. H_2 use: *Paracoccus, Alcaligenes, Bradyrhizobium, Pseudomonas.*

2. S use: *Thiobacillus, Thiomicrospira, Thiospaera, Thermothrix.*

3. NH_4^+ use: *Nitrosomonas.*

Environmental Factors Affecting Denitrification

pH

Optimum pH for denitrification varies with the organisms nitrate concentration. Denitrification is most rapid in the slightly alkaline range between pH 7 and 8, (Wiljer and Delwiche, 1954; Nommik, 1956), though the limits with respect to pH are quite wide ranging from pH 3.5 (Cady and Bartholomew, 1960) to pH 11.2 (Prakasam and Loehr, 1972). Generally more N_2O is produced in acidic condition (Cady and Barthlomew, 1960; Bollag *et al.*, 1973; Garcia, 1973). *P. aeruginosa* denitrifies in a pH range of 5.8 to 9.2 with an optimum range of 7.0 to 8.2. When the enzyme system from *P. aeruginosa* was tested in three different buffer system, denitrification was optimum at pH 7.4, 30 per cent at pH 6 and 50 per cent at pH 10 (Fewson and Nicholas, 1961). Wijler and Delwiche reported on the effect of pH on the distribution of various gases. Under the condition of their study with mixed flora, NO is a major dentrification product when the pH is below 5. Above pH 7 N_2O is produced but is subsequently reabsorbed and reduced to nitrogen gas. Below pH 7 some N_2O is formed but there is little reabsorption with further denitrification, this reaction is apparently differentially affected by acid conditions. NO is produced at low levels between pH 6.0 and 7.0. Under more acidic condition, NO is produced at higher levels amounting to approximately 20 per cent of total nitrogen evolved at pH5. At pH 4.9 N_2O and NO are evolved in almost equal volumes.

Temperature

Optimal rates of denitrification have been reported to occur at 65°C by Nommik (1956) who attributed this high optimum to the predominance of thermophilic *Bacillus* spp. Thermophilic denitrification is thought to be a combination of biological and chemical reactions in which nitrate respiring *Bacillus* species generated nitrite which reacted chemically and biologically to form gaseous products. Bailey (1976) found that denitrification was completely inhibited at 5°C although Schmid and Beauchamp (1976) indicate that denitrification may still occur at or near 0°C. Bailey (1976) found that the quantity of N_2 produced decreased as temperature decreased from 30° to 6° to 8°C. Focht and

Verstraete (1977) indicate, however, that the amount of N_2O or N_2 do not to get affected significantly by temperature. The direct effect of temperature on product formation seems to pertain only to environments that would favour accumulation of NO_2^- rather then NO_3^- (Payne, 1973). Broadbent and Clark (1965) stated that the relative proportions of N_2O and N_2 from denitrification depend upon temperature with N_2O predominant at lower temperatures and N_2 at higher. Rolston *et al.*, (1978) found that the denitrification gases contained a larger proportion of N_2O at 8 to 10°C than at 23°C in plots to which manure has been added. Much of the effect of temperature on the gases produced may be through the degree of anoxic developments. To some extent, indigenous denitrifiers may have temperature optima adapted to their climatic region (Gamble *et al.*, 1977). Denitrification appears to follow a seasonal trend with losses of N_2O and N_2 being markedly higher in summer than in winter (Rolston *et al.*, 1978; Bremmer *et al.*, 1980b). Several workers have observed marked diurnal variability in the rate of N_2O emission from soils that appears to be related to soil temperature (Ryden *et al.*, 1978; Denmead, 1979; Blackmer *et al.*, 1982; Lensi and Chalamet, 1982). Maximum rates generally occur in the afternoon and minimum rates during the night.

Oxygen

Aeration is the greatest and most variable factor that affects the denitrification process in soil. The oxygen concentration below which denitrification occurs ranges from 0.7 µg ml^{-1} (Goering and Cline, 1970) to 2 µg ml^{-1} (Wheatland *et al.*, 1959), although oxygen becomes limiting to respiration at a lower value of 0.1 µg ml^{-1} (Chance, 1957). It is generally accepted that dissimilatory nitrate reduction in most organisms becomes dominant only under anaerobic condition. If oxygen consumption exceeds O_2 diffusion, anoxic conditions will develop and denitrification may occur. It is notable that O_2 concentrations reported in the denitrification literature are in at least 10 different units; µg atoms L-1, µmol L^{-1}, mL L^{-1}, mg L^{-1}, per cent O_2 saturation, per cent air saturation atmospheres, Pascal, ppm and mmHg. This situation does not foster understanding. Although some of these units are easily derived from another others are not because of the effect of temperature, salt and pressure. µmol L^{-1} is the best alternative as it is the basic unit used to describe the substrate and inhibitor concentrations for cell metabolism.

Interactions

Oxygen and Temperature

Misra *et al.* (1974) noted a synergistic effect of oxygen and temperature upon denitrification. They found at 19.5°C the rate constant for nitrate reduction was reduced more than ten fold from a gaseous O_2 conc. of 0 to 20 per cent whilst a temperature of 34.5°C over the same O_2 conc. range affected decrease to only 75 per cent of the higher value. Higher temperatures cause a reduction in the soluble oxygen concentration as well as an increase in biological processes. This interaction may explain why denitrification in thermic semi arid soils is characterized as a rapid and transient processes confined to the uppermost 10 cm since this zone is subjected to the highest temperatures and most variable change in temperature and aeration upon irrigation (Focht, 1978; Focht and Stolzy, 1978; Focht *et al.*, 1979; Ryden *et al.*, 1979b).

Oxygen and Water Content

The diffusion rate of oxygen to pockets or zones within the soil depends largely on the water content of the system and the oxygen consumption rates of microorganisms and plant roots.

Water Content

Water content influences denitrification directly by providing an environment suitable for microorganisms. A dry soil cannot support a large population of microorganisms. Thus water content must be at a certain level to have an active microbial population. A more important effect of water content on denitrification, however occurs indirectly through blockage of the soil pores and reduction of O_2 transport to zones of high microbial activity, and respiring plant roots. The greatest potential for denitrification would be for completely water saturated soil. Denitrification occurs over a very narrow range of soil-water content or soil-water potential. Denitrification apparently does not occur when soils become drier than about -30 centibars of soil-water potential, as discussed by Focht and Verstraete, (1977). Pilot and Patrick, (1972) found that the critical soil-water potential for which soils change from a reducing environment for denitrification to an oxidizing environment were between -2 and -4 centibars in three different soils. Thus the effect of water content on denitrification is an integration of thickness of water films, the number of large pores that can conduct oxygen at any soil water potential value and the rate that oxygen is consumed at various depths within the soil profile. Total denitrification rates from irrigated cropped plots of Ryden *et al.*, (1979b) reached a maximum of 2 kg N ha^{-1} day^{-1} at the peak of the denitrification cycle after irrigation and less than 0.2 kg ha^{-1} day^{-1} at the minimum rate between irrigations.

Carbon Supply

Regardless of the amount of NO_3^- present or a lack of O_2 in the soil, denitrification requires carbon (C). Thus the rate and amount of denitrification will be largely influenced by the amount of C, its position in the profile and its position in relation to the NO_3^- of the soil. The carbon supply influences denitrification directly by supplying the necessary substrate for growth of denitrifiers and indirectly through the consumption of O_2 by other microorganisms that deplete O_2 in the soil.

Substrate Concentration (Nitrate, NO_3^-)

Dentrification rates in soils have been found by many workers to be independent of NO_3^- concentration over a fairly wide range. Another complicating factor affecting the order of the reaction kinetics, is the physical process of NO_3^- diffusion to zones or microsites where denitrification is occurring. Since diffusion is concentration dependent, the effect of diffusion would make the process appear to be first order. This phenomenon has been validated by Phillips *et al.*, (1978). The concentration of nitrate has an effect also on the proportion of volatile gases produced. At high nitrate concentration the predominant gas produced is N_2O as found by Nommik, 1956; Cooper and Smith, 1963; Blackmer and Bremmer, 1978b). As NO_3^- concentration become small the predominate gas produced is N_2.

Soil Profile

The zone of largest microbial activity and O_2 consumption is generally at the soil surface. Percentages of 'C' are generally highest near the soil surface, decreasing rapidly with depth in most soil. For denitrification to occur nitrate must be in the same position in the profile as the 'C'. For the soil profiles studied by Rolston *et al.* (1978) most denitrification occurred in the upper 30 to 45 cm of soil. The principle reason was that 'C' levels decreased very quickly with depth and O_2 levels increased below 60 cm of soil as a result of small amounts of root respiration and microbial activity below that depth and diffusion of O_2 from below. In soil both denitrifier density and organic carbon decreases with increase in depth.

Fertilizer Application

The total denitrification losses from fertilizers applied to soil are difficult to determine because insufficient data are collected in direct measurements of denitrification and many N⁻ balance studies have large errors associated with leaching, residual soil N and so on. Allison (1955; 1966) estimated that 10-30 per cent of fertilizer applied is denitrified. His analysis was based upon a review of many N⁻ balance experiments. Hauck, (1969) estimated that an average denitrification loss from cropped land was 16 kg N ha^{-1} yr^{-1}. Broadbent and Carlton, (1978) used ^{15}N depleted fertilizer over three years to attain an accurate N-balance. Their N⁻ balance is probably one of the best balances available because they used $^{15}N^{-}$depleted fertilizer over a large field area with very intensive measurements of the crop and soil. They estimated that 22 per cent of the fertilizer giving maximum yield (224 kg N/ha) was lost by denitrification. Another study determined that between 6 and 14 per cent of 330 kg N/ha applied over 3 years to grass and clover was lost by denitrification. Those losses occurred only if the 'N' was applied in February, whereas little loss occurred if the N was applied in October. Total denitrification loss from field experiments of Ryden *et al.*, (1979b) was 51 kg N/ha over a 4 month period from a fertilizer application of 335 kg N/ha. Crasswell and Martin, (1975) however found for an Australian wheat soil that 98 per cent of applied ^{15}N could be recovered from field experiments. Rolston *et al.*, (1976) directly measured both N_2O and N_2 evolution in a cropped field soil and found that 45 per cent of a single application of 300 kg N/ha was lost as N_2 and N_2O from a cropped plot maintained very close to saturation for a 30 day period.

Plants

Plant roots have several effects on the rhizosphere soil that may influence the potential for denitrification. First, roots release carbonaceous materials (organic substrate for denitrifiers) into the rhizosphere by excretion of soluble compounds, sloughing off root surface and root cap cells and production of mucigel polysaccharide (Warembourg and Billies, 1979). Thus large populations of denitrifiers frequently exist in the rhizosphere (Woldendrop, 1963a), where they may be 10 to 100 times more numerous than in the root-free soil (Netti, 1955). The metabolism of the carbonaceous material by the rhizosphere microflora will tend to deplete the soil of O_2, as, indeed, will root respiration. It is evident that if denitrification is limited by O_2 or C supply then the presence of plant roots will tend to stimulate denitrification. Many studies have confirmed that the presence of plant roots enhances the denitrification of added NO_3^- (Woldendrop, 1963b; Stefanson, 1972a;b;c; Brar, 1972; Garcia, 1975; Bailey, 1976; Volz *et al.*, 1976) and sometimes causes a decrease in the mole fraction of N_2O produced (Stefanson, 1972a;b). However Stefanson, (1976) observed that while wheat roots stimulated denitrification in soils of low organic matter content they had no effect in soils high in organic matter. The second major effect of roots is that they absorb NO_3^- and supply of NO_{3-} is limiting denitrification, then, plants tend to decrease the rate of denitrification (Guenzi *et al.*, 1978; Buresh *et al.*, 1981). Smith and Tiedje, (1979b) confirmed that when soil NO_3^- concentrations are high denitrification rates are increased in the rhizosphere, where as when NO_3^- concentrations are low denitrification rates are decreased in the presence of roots.

Animals

In grassland ecosystems the urine patches formed by grazing animals are recognized as focal points for the loss of N via NH_3 volatilization (Catchpoole *et al.*, 1983; Sherlock and Goh, 1984) and leaching (Ball *et al.*, 1979). However deposition of urine also results in an immediate release of N_2O which does not occur from addition of urea of equivalent 'N' content (Sherlock and Goh, 1983). This rapid production and release of N_2O may be due to stimulation of denitrification due to rapid onset of anaerobiosis caused by concomitant inputs of readily available 'C' and rapid urea hydrolysis (Sherlock and Goh, 1983).

Tillage Method

In comparison with conventional tillage, the lack of soil disturbance and the presence of surface mulch under zero tillage, result in increased bulk density, a reduction in large pores, reduced aeration, larger but less aerobic aggregates and generally higher moisture content in the surface soil (Dowdell *et al.*, 1979a; Lin and Doran, 1984). Levels of water soluble carbon can also be higher in surface soils under zero tillage (Lin and Doran, 1984). Such soil conditions obviously tend to favour the activity of denitrifiers and denitrifier populations are generally greater in the surface soil under zero rather than conventional tillage (Aulakh *et al.*, 1984a; Lin and Doran, 1984; Broder *et al.* (1984) studies have also shown that losses of N_2O (Burford *et al.*, 1981) or N_2O plus N_2 (Rice and Smith, 1982; Aulakh *et al.*, 1984a;b) are greater from zero tilled rather than conventionally tilled fields although the ratio of N_2O:N_2 emitted is not changed measurably. Aulakh *et al.* (1984a) estimated N_2O plus N_2 losses from cropped conventionally tilled and zero tilled fields as 3-7 and 12-16 kg N/ha/yr respectively. The generally lower rate of mineralization and therefore nitrification and the smaller populations of nitrifiers under zero tillage (Rice and Smith, 1982; Broder *et al.*, 1984) suggest that greater losses of N_2O under zero as compared to conventional tillage are the results of greater denitrification rather than nitrification.

Bacterial Populations and Micro Sites

Bacteria are not uniformly distributed throughout the soil, some commonly being clustered colonies. These populations obviously develop in micro sites where conditions are optimal, that is, where there is the highest likelihood of stimulatory and least likelihood of inhibitory pulses. Parkin recently advanced the concept that denitrification in soils is often found in "Hot Spots" which are created by decaying organic matter that generates the anaerobic micro site.

Time or Dynamic Changes in Rates of Denitrification

The rates measured by Rolston *et al.* (1978) varied from 0 to 70 kg N/ha/day. The denitrification rate of 70 kg N/ha/day was for a plot maintained nearly water saturated into which 34 metric tones of manure per hectare had been incorporated in the upper 10 cm of soil three weeks before the NO_3^- fertilizer was applied to the wet soil surface. Most of the denitrification occurred in the first five days after the fertilizer was applied. The soils were kept continuously wet (near saturation) during this period. No denitrification could be detected after about 20 days. This period of 20 days corresponded approximately to the time in which the NO_3^- would have been leached from the top 30 to 45 cm of soil. The maximum rate measured in a cropped field plot (perennial ryegrass) by Rolston *et al.* (1976) was about 10 kg N ha^{-1} day^{-1}. Again, most of the denitrification occurred over a 30 day period. For uncropped plots without added carbon maintained close to saturation, the maximum denitrification rate was about 2.5 kg N ha^{-1} day^{-1}.

Management Practices

Management of fertilizer 'N' applied to soils may be directed in either of two directions (Rolston, 1981): (*i*) one goal would be to minimize denitrification losses such as would be the objective of a grower applying fertilizer and wanting to maximize the use of that fertilizer. The other management objective would be to increase denitrification below the root zone to minimize the amount of NO_3^- leaching below disposal sites. The latter objective appears to be least conducive to management. To minimize denitrification, the major factors affecting denitrification should not occur at the same point in the soil or at the same time. For example, high concentration of NO_3^- should not be allowed in the profile along with high concentration of carbon and high water content. The absence of either NO_3^- or C or the presence of O_2 will result in very little or no denitrification. Thus management practices should be directed at managing the water, the C and the fertilizer such that they do not occur simultaneously in the soil profile. A proper management decision would be that fertilizer should not be applied to the surface soil (high C) at times when water contents are likely to be very close to saturation. Another management decision to minimize denitrification would be to ensure that fertilizer was placed at such a depth where plant roots could still obtain the needed 'N' yet not be in a zone of high carbon or potentially high water content. Application of manure along with fertilizer would be an unrealistic practice if one wanted to minimize denitrification. If water contents could not be maintained outside the range where denitrification occurs it may be possible to time fertilizer application so that NO_3^- will not occur in the zone containing high 'C' at a time when water contents would be close to saturation. For instance it may be possible to apply ammonium fertilizer immediately before a large irrigation for wetting the profile at the beginning of that season. The next irrigation would be timed and managed such that the NO_3^- could be moved out of the zone where most of the 'C' occurs yet remain within the zone where roots could obtain the 'N'.

Conclusion

The denitrification capacity is spread among a wide variety of physiologic and taxonomic groups. The most common denitrifiers in nature are species of *Pseudomonas* followed by the closely related *Alcaligenes.* These denitrifiers are responsible for the production of the greenhouse gas, nitrous oxide, which is now a significant contributor of global warming. Thus we need to have a detailed information through more extensive research and development about these denitrifiers which are now of much ecological significance, especially in relation to soil.

The phenomena of denitrification is known for a long time, but recently it has gained much importance in relation to tropical agricultural soil, as it accounts for significant loss of fertilizer 'N' from these soil. This fertilizer 'N' is lost to the atmosphere partly as nitrous oxide which leads to global warming. The rates and total amount of denitrification are the integrated results of all soil and environmental processes affecting denitrification. Hence we see that the denitrification rates and amounts would be expected to vary substantially with soil type, soil profile characteristics, irrigation or rainfall regime, cropping pattern, temperature, pH, carbon or manure inputs and fertilizer management. These factors play major role in denitrifying enzyme activity and the population of denitrifiers in soil.

References

Allison FE (1955). The enigma of soil nitrogen balance sheets. *Advan Agron,* 7: 213–250.

Allison FE (1966). The fate of nitrogen applied to soils. *Adv Agron,* 18: 219.

Aulakh MS, Rennie DA and Paul EA (1984 a). Acetylene and N-serve effects up upon N_2O emissions from NH_4^+ and NO_3^- treated soils under aerobic and anaerobic conditions. *Soil Biol Biochem,* 16: 351–356.

Aulakh MS, Rennie DA and Paul EA (1984b). Gaseous N losses from soil under zero till compared with conventional till management systems. *J Environ Qual,* 13: 130–136.

Bailey LC (1976). Effects of temperature and root on denitrification in soil. *Can J Soil Sci,* 56: 79–87.

Ball PR, Keeney DR, Theobald PW and Nes P (1979). Nitrogen Balance in urine affected areas of a New Zealand pasture. *Agronomy Journal,* 71: 309–314.

Blackmer AM and Bremner JM (1978b). Inhibitory effect of nitrate on reduction of N_2O to N_2 by soil microorganisms. *Soil Biol Biochem,* 10: 187–191.

Blackmer AM, Robbins SG and Bremner JM (1982). Diurnal variability in rate of emission of nitrous oxide from soils. *Soil Sci Soc Am J,* 46: 937–942.

Bollag JM, Drzymala S and Kardos LT (1973). Biological versus chemical nitrite decomposition in soil. *Soil Science,* 116: 44–50.

Brar S (1972). Influence of roots on denitrification. *Plant Soil,* 36: 713–715.

Bremner JM, Blackmer AM and Waring SA (1980b). Formation of nitrous oxide and dinitrogen by chemical decomposition of hydroxylamine in soil. *Soil Biol Biochem,* 12: 263–269.

Broadbent FE and Carlton AB (1978). Field trials with isotopically labelled nitrogen fertilizers. In: *Nitrogen in the Environment,* Vol 1, (Eds) Neilsen DR and MacDonald JG. Academic Press, New York, pp 1– 41.

Broadbent FE and Clark FE (1965). Denitrification. In: *Soil Nitrogen,* (Eds) Bartholomew WV and Clark FE. Am Soc of Agron, Madison, Wisconsin, p 344–359.

Broder MW, Doran JW, Peterson GA and Fenster CR (1984). Fallow tillage influence on spring populations of soil nitrifiers, denitrifiers and available nitrogen. *Soil Sci Soc Am J,* 48: 1060–1067.

Buresh RJ, DeLaune RD and Patrick Jr (1981). Influence of *Spartina Alterniflora* on nitrogen loss from marsh soil. *Soil Sci Soc Am J,* 45: 660–661.

Burford JR, Dowdell RJ and Crees R (1981). Emission of nitrous oxide to the atmosphere from direct-drilled and ploughed clay soils. *J Science of Food and Agric,* 32: 219–223.

Cady FB and Bartholomew W V (1960). Sequential products of anaerobic denitrification in Norfolk soil material. *Soil Sci Soc Am Proc,* 24: 477–482.

Catchpoole VR, Oxenham DJ and Harper LA (1983). Transformation and recovery of urea applied to a grass pasture in southeastern Queensland. *Aust J Exp Agric Husb,* 23: 80–86.

Chance B (1957). Cellular oxygen requirements. *Proc Fed Amer Soc Exp Biol,* 16: 671.

Cooper GS and Smith RL (1963). Sequence of products formed during denitrification in some diverse western soils. *Soil Sci Soc Am Proc,* 27: 659–662.

Crasswell ET and Martin AE (1975). Isotopic studies of the nitrogen balance in a cracking clay. II: Recovery of nitrate ^{15}N added to columns of packed soil and microplots growing wheat in the field. *Australian J of Soil Research,* 13 (1). 53–61.

Denmead OT(1979). Chamber systems for measuring nitrous oxide emission from soils in the field. *Soil Sci Soc Am Proc,* 43: 89–95.

Dowdell RJ, Crees R, Burford JR and Cannell RQ (1979a). Oxygen concentration in a clay soil after ploughing or direct drilling. *J Soil Sci,* 30: 239–245.

Fewson CA and Nicholas DJ (1961). Nitrate reductase from *Pseudomonas aeruginosa. Biochim Biophys Acta,* (13). 49: 335–349.

Focht DD (1978). Methods for analysis of denitrification in soils. In: *Nitrogen in the Environment*, Vol 2, (Eds) Neilson DR and MacDonald JG. Academic Press, New York, p 429–490.

Focht DD and Stolzy LH (1978). Long term denitrification studies in soils fertilized with $(^{15}NH_4)_2SO_4$. *Soil Sci Soc Am J*, 42: 894–898.

Focht DD and Verstraete W (1977). Biochemical ecology of nitrification and denitrification. In: *Advances in Microbial Ecology*, Vol 1, (Ed) Alexander M. Plenum Press, New York, p. 135–214.

Focht DD, Stolzy LH and Meek BD (1979). Sequential reduction of nitrate and nitrous oxide under field conditions as brought about by organic amendments and irrigation management. *Soil Biol Biochem*, 11: 37–46.

Gamble RN Betlach MR and Tiedje JM (1977). Numerically dominant denitrifying bacteria from world soil. *Appl and Environ Microbiol*, 33: 926–939.

Garcia JL (1973). Influence de la rhizosphere du riz sur e' activite' de'nitrifiante potentielle des sols de rizieres du Senegal. *Oecol Plant*, 8: 315–323.

Garcia JL (1975). effect rhizosphere du riz sur la denitrification. *Soil Biol Biochem*, 7: 139–141.

Gayon U and Dupetit AG (1886). Re-cherches sur la reduction des nitrates par les infiniment petits *Mem Soc Sci Phys Nat Bordeaux Ser*, 32: 201.

Goering JJ and Cline JD (1970). A note on denitrification in sea water. *Limnol Oceanogr*, 15: 306–309.

Guenzi WD, Beard WE, Watanabe FS, Olsen SR and Porter LK (1978). Nitrification and denitrification in cattle manure amended soil. *J Environ Qual*, 7: 196–202.

Hauck RD (1969). Quantitative estimates of N cycle processes: review and comments. Paper presented at a meeting on recent developments in the use of ^{15}N in soil plant studies. Sponsored by the *Joint FAO/IAEA Div of Atomic Energy in Food and Agriculture*. Sofia, Bulgaris, 21 pp.

Ingraham JL (1981). Microbiology and genetics of denitrifiers. In: *Denitrification, Nitrification and Atmospheric Nitrous Oxide*, (Ed) CC Delwiche. Wiley-Interscience Publication, pp 45–65.

Koike I and Hattori A (1975b). Energy yield of denitrifying bacterium, *Pseudomonas denitrificans*, under aerobic and denitrifying conditions. *J Gen Microbiol*, 88: 11–19.

Lensi R and Chalamet A (1982). Denitrification in waterlogged soils; *in situ* temperature dependent variations. *Soil Biol and Biochem*, 14: 51–55.

Linn DM and Doran JW (1984). Aerobic and anaerobic microbial populations in no till and plowed soils. *Soil Sci Soc Am J*, 48: 794–799.

Misra C, Neilsen DR and Biggar JW (1974). Nitrogen Transformations in soil during leaching I: Theoretical considerations. *Soil Sci Soc Am J*, 38: 289–293.

Netti IT (1955). Denitrifying bacteria of the oak rhizosphere. *Mikrobiologiya*, 24: 429–434.

Nommik H (1956). Investigations on denitrification in soil. *Acta Agric Scand*, 6: 195–228.

Payne, W J (1973a). Reduction of nitrogenous oxide by microorganisms. *Bacteriol Rev*, 37: 409–452.

Phillips RE, Reddy KR and Patrick Jr WH (1978). the role of nitrate diffusion in determining the order and rate of denitrification in flooded soil II theoretical analysis and interpretations. *Soil Sci Soc Am J*, 42: 272–278.

Pilot L and Patrick Jr WH (1972). Nitrate reduction in soils; effect of soil moisture tension. *Soil Sci,* 114: 312–316.

Prakasam TB and Loehr RC (1972). Microbial nitrification and denitrification in concentrated wastes. *Water Res,* 6: 859–869.

Rice CW and Smith MS (1982). Denitrification in no till and plowed soil. *Soil Sci Soc Am J,* 46: 1168–1173.

Rolston DE (1981). Nitrous oxide and nitrogen gas production in fertilizer loss. In: *Denitrification, Nitrification and Atmospheric Nitrous Oxide,* (Ed) CC Delwich. John Wiley and Sons, p. 127–149.

Rolston DE, Fried M and Goldhamer DA (1976). Denitrification measured directly from nitrogen and nitrous oxide gas fluxes. *Soil Sci Soc Am J,* 40: 259–266.

Rolston DE, Hoffman DL and Toy DW (1978). Field measurement of denitrification. I. Flux of N_2 and N_2O. *Soil Sci Soc Am J,* 42: 863–869.

Ryden JC, Lund LJ and Focht DD (1978). Direct in–field measurement of nitrous oxide flux from soil. *Soil Sci S . Am J,* 42: 731–737.

Ryden JC, Lund LJ, Letey J and Focht DD (1979b). Direct measurement of denitrification loss from soils: II Development and application of field methods. *Soil Sci Soc Am J,* 43: 110–118.

Schmid AE and Beauchamp EG (1976). Effects of temperature and organic matter on denitrification in soil. *Can J Soil Sci,* 56: 385–391.

Sherlock RR and Goh KM (1984). Dynamics of ammonia volatilization from simulated urine patches and aqueous urea applied to pasture I Field experiments. *Fertilizer Research,* 5: 181–195.

Sherlock RR and Goh KM (1983). Initial emission of nitrous oxide from sheep urine applied to pasture soil. Soil Biol. Biochem. 15: 615–617.

Smith MS and Tiedje JM (1979b). Phases of denitrification following oxygen depletion in soil. *Soil Biol Biochem,* 11: 261–267.

Stefanson RC (1972a). Soil denitrification in sealed soil-plant systems I: Effect of plants, soil, water content and soil organic matter content. *Plant Soil,* 37: 113.

Stefanson RC (1972b). Soil denitrification in sealed soil plant systems. II: Effect of soil water content and form of applied nitrogen. *Plant Soil,* 37: 129.

Stefanson RC (1972c). Soil denitrification in sealed soil-plant systems III: Effect of disturbed and undisturbed soil samples. *Plant Soil,* 27: 141.

Stefanson RC (1976). Denitrification from nitrogen fertilizer placed at various depths in the soil-plant system. *Soil Sci,* 121: 353–363.

Volz MG, Ardakani MS, Schulz RK, Stolzy LH and McLaren AD (1976). Soil nitrate loss during irrigation: Enhancement by plant roots. *Agron J,* 68: 621–627.

Warembourg FR and Billies G (1979). Estimating the carbon transfer in the rhizosphere In: *The Soil-Root Interface,* (Eds) Harley JL and R Scott-Russel, London p. 181.

Wheatland AB, Barrett MS and Bruge AM (1959). Some observations on denitrification in rivers and estuaries. *Inst Sewage Purif J Proc,* 2: 140–150.

Wijler J and Delwiche CC (1954). Investigations on the denitrifying process in soil. *Plant Soil,* 5: 155–169.

Woldendorp JW (1963a). Influence of living plants on denitrification (article in French). *Ann Inst Pasteur (Paris),* 105: 426–433.

Woldendorp JW (1963b). The influence of living plants on denitrification. Mededelingen Van de Landbouwwhogeschool Te Wageningen, Netherland, 63(13): 1–100.

Soil Microflora, 2009 *Pages* **122–138**
Editor: **Rajan Kumar Gupta, Mukesh Kumar & Deepak Vyas**
Published by: **DAYA PUBLISHING HOUSE, NEW DELHI**

Chapter 12

Molecular Mechanism of Nitrogen Fixation in *Rhizobium*

***Ravi Rajhans*[1] *and Rajan Kumar Gupta*[2]**
[1]*Department of Biotechnology, Modern Institute of Technology, Dhalwala, Rishikesh, Uttarakhand*
[2]*Department of Botany, Pt. L.M.S. Govt. P.G. College, Rishikesh – 249 201, Dehradun, Uttarakhand*

Nitrogen-Fixing Rhizobium

Rhizobium, Bradyrhizobium, and *Azorhizobium,* collectively referred as rhizobia. They are attached with leguminous hosts and form specialized organs, nodules. Root nodules develop to house nitrogen-fixing bacteria in a microaerobic environment.This process, a type of symbiotic nitrogen fixation, restricted to a limited number of bacterial groups, including the genera *Rhizobium, Mesorhizobium, Sinorhizobium, Bradyrhizobium,* and *Azorhizobium* (collectively referred as rhizobia) and Frankia. Frankia is a filamentous gram positive actinomycete that induces nodules on a variety of woody plants from the *Betulaceae, casuarinaceae, Myricaceae, Elaegnaceae, Rhamnaceae, Rosaceae, Coriariaceae,* and *Datisticaceae* families (Benson, and Clawson. 2000, Benson and Silvester. 1993). Rhizobia carry most of the genes specifically required for nodulation either on large (500 kbp to 1.5 Mbp) plasmids or on symbiosis islands (Barnett, M. J., *et al.*, 2001, Freiberg, *et al.*, 1997, Kaneko, *et al.*, 2000, Kaneko, T., *et al.*, 2002). It has been recently discovered that bacteria from outside the *Rhizobiaceae* can induce nodules on legumes. For rexample, a strain of *Methylobacterium,* an α-proteobacterium, can nodulate Crotalaria, and β-proteobacteria related to Burkholderia can nodulate *Machaerium lunatum* and *Aspalathus carnosa* (Moulin, *et al.*,. 2001, Sy, *et al.*, 2001).

Nodules induced by Rhizobia are two general kinds, determinate and indeterminate.These differ in a number of ways, one of the most important being that indeterminate nodules are elongatedand have a persistent meristem that continually gives rise to new nodule cells that are subsequently infected by rhizobia residing in the nodule. *M. truncatula* has recently become a favored model for studies focusing on the genetics and cell biology of indeterminate nodule formation because it is diploid, and can be readily inbred to form genetically homogenous lines (Cook, D., 2000). Determinate nodules lack a persistent meristem, are usually round and do not display an obvious developmental gradient as do indeterminate nodules. *L.japonicus*, like *M. truncatula*, has genetic characterstics that make it particularly suitable as a role model to study the formation of determinate-type nodules (Handberg, and Stougaard, 1992).

Taxonomy and Host Specificity of Rhizobium Species

The current taxonomic classification of the rhizobia is given in Table 12.1.

Table 12.1

Rhizobium	*Host Plant(s)*
R. meliloti	*Medicago, Melilotus,* and *Trigonella* spp.
R. leguminosarum bv.viciae	*Pisum, Vicia, Lathyrus,* and
Lens spp. bv. trifolii	*Trifolium spp.*
bv.*phaseoli*	*Phaseolus vulgaris*
R. loti	*Lotus* spp.
R. huakuii	*Astragalus sinicus*
R. ciceri	*Cicer arietinum*
Rhizobium sp. strain NGR234	*Tropical legumes, Parasponia* spp. (nonlegume)
R. tropici	*Phaseolus vulgaris, Leucaena* spp., *Macroptilium* spp.
R. etli	*Phaseolus vulgaris*
R.galegae	Galega officinalis, G. orientalis
R.fredii	*Glicine max, G. soja* and other legumes
B.japonicum	*Glycine max, G. soja* and Other legumes
B.elkanii	*Glycine max, G. soja* and Other legumes
Bradyrhizobium sp. *Strain Parasponia*	*Parasponia* spp.
A. caulinodans	*Sesbania* spp.(stem nodulating)

The species name of the microsymbionts reflects in most cases the corresponding host plant nodulated and suggests that symbiosis is a species-specific process. The situation is much more complex than can be reflected in Table 12.1, in which some of the host plants are matched up with the microsymbiont. It is quite clear that the degree of host specificity varies tremendously among rhizobia (Young, and Johnston, 1989). Some strains have a very narrow host range, for example *Rhizobium leguminosarum* bv. Trofolii, while others, like *Rhizobium* sp. Strain NGR234, have a very broad host range.

Molecular Mechanism of Nitrogen Fixation

The process of reduction of atmospheric gaseous N_2 to ammonia is called nitrogen fixation. The enzyme nitrogenase is responsible for the reduction of nitrogen to ammonia. Nitrogen reduction is expensive and requires a large ATP expenditure. At least 8 electrons and 16 ATP molecules, 4 ATPs per pair of electrons are required.

Nitrogenase is a complex system consisting of two major protein components. The enzyme system is composed of dinitrogenase (MoFe protein) (MW 220,000) and dinitrogenase reductase (Fe protein) (MW 64,000). Dinitrogenase has two dissimilar polypeptides $\alpha_2\beta_2$. The α polypeptide is encoded by *nif D* and the β polypeptides are encoded by *nif K* genes. The dinitrogenase protein contains two active metallo clusters: the P cluster containing 8 iron and 7-8 sulfur atoms ($Fe_8S_{7\text{-}8}$) and iron- molybdenum cofactor (FeMoco) containing 7 iron, 9 sulfur, 1 molybdenum atom and 1 molecule of homocitrate (Fe_7S_9Mo-homocitrate). The P cluster acts as an intermediate electron acceptor and probably transfer the electron to FeMoco cluster. The FeMoco cluster functions as the site of nitrogen reduction. The dinitrogenase reductase protein (Fe protein) consists of two identical polypeptides, $\gamma 2$ encoded by nif H gene. Each polypeptide contains two iron atoms. The four Fe atoms are organized into a Fe_4S_4 cluster. Ferredoxin first reduces the Fe protein, and then Fe protein binds to ATP (Figure 12.1). ATP binding changes the conformation of Fe protein and lowers its reduction potential, enabling it to reduce MoFe protein. ATP is hydrolyzed when this electron transfer occurs. Finally, reduced MoFe protein donates electrons to atomic nitrogen. Nitrogenase is sensitive to oxygen and must be protected from oxygen inactivation with in the cell.

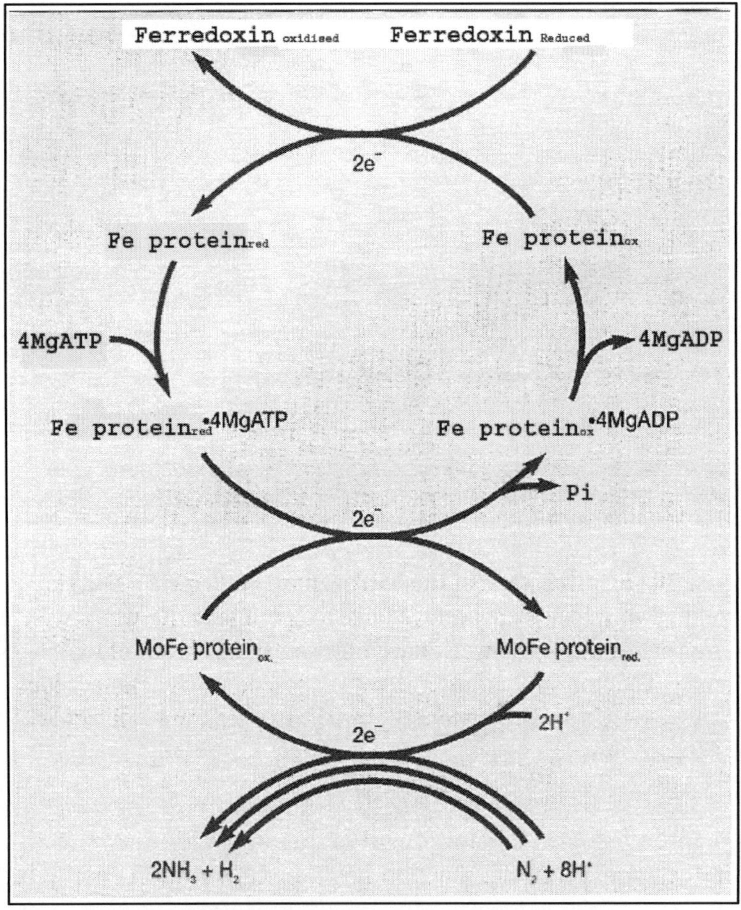

Figure 12.1: Mehanism of Nitrogenase Action

Nitrogen-Fixing Association Between Rhizobia and Legumes

Infection Process in Plants

Root infection by rhizobia is a multistep process that is initiated by preinfection events in the rhizosphere. During the growth in the rhizosphere of a host plan, rhizobia sense compounds such as flavonoids and betaines secreted by the host root and respond by inducing *nod* genes (Callaham and Torrey, 1981; Cooper and Rao, 1992; Chua, *et al.*, 1985). The *nod* genes encode approximately 25 proyeins required for the bacterial synthesis and export of nod Nod factor (Figure 12.2a). Rhizobia respond by positive chemotaxis to plant root exudates and move toward localized sites on the legume roots (Benson, and Silvester. 1993, -Anolles, Wrobel-Boerner, and Bauer. 1992, Dowling, and Broughton. 1986, Gaworzewska, and Carlile. 1982, Gulash, *et al.*, 1984). Both *Bradyrhizobium* and *Rhizobium* spp. are attracted by amino acids, dicarboxylic acids present in the exudates and very low concentration of excreted components, such as flavonoids, that may not have high nutritional value (Aguliar *et al.*, 1988; Armitage *et al.*, 1988; Caetano-Anolles *et al.*, 1988; Kape *et al.*, 1991; Peters and Verma, 1990).

Adhesion of Rhizobia to Root Hairs: Rhizobia are capable of binding to host root hairs. With *Rhizobium leguminosarum* this binding consists of two steps. The first is a weak, Ca+2-dependent binding step to root hairs that is mediated by a protein called rhicadhesin, which is thought to be present in most rhizobia (Lewin, *et al.*, 1990, Lewis-Henderson, and Djordjevic. 1991). Recently an R. *leguminosarum* bv.trifolii protein, RapA1, that has many of the properties attributed to rhicadhesin was described. Both are secreted proteins that can bind calcium, bind at bacterial cell poles, mediate calcium-dependent agglutination, and bind to root hairs. However, rhicadhesin and RapA1 most likely are not the same proteins, because they are different sizes and RapA1, being found only in *R.leguminosarum* and *Rhizobium etli,* does not appear to have the broad phylogenetic distribution of rhicadhesin. Following weak binding, a tight binding step that is mediated by the bacterial synthesis of cellulose fibrils is initiated (Lewin, *et al.*, 1987; Lewis-Henderson, and Djordjevic. 1991). The synthesis of these fibrils was shown to be required for *R.leguminosarum* to form biofilm-like caps on the tip of pea root hairs. Binding and capping may be needed for rhizobia to effectively colonize root hairs under natural conditions.

Lectins, plant protein with high affinity to carbohydrate moieties on the surface of appropriate rhizobial cells, have been identified as specific mediators of the attachment of the rhizobia to susceptible root hairs (Figures 12.2b, 12.3a). During the nodulation process, tryptophan secreted by the plant roots is metabolized to Indole acetic acid (IAA) by the rhizobia, and IAA together with unknown cofactor probably arising from the host plant roots, initiates hair curling or branching. The root hair grow around the bacterial cells (Figure 12.2c). Polygalacturonase, secreted by the rhizobia or possibly by the plant roots, depolymerizes the cell wall and allows bacteria to invade the soften plant tissues.

Extension of Infection Threads Through Root Hairs

After the penetration of the primary root hair wall, the infection proceeds by the development of an infection tube (thread) that is surrounded by cell membrane and a cellulosic wall (Figures 12.3b,c). It contains *rhizobium* cells lying end to end in a polysaccharide matrix. The infection tube (thread) penetrates through and between root cortex cells. As the infection thread grows, the cells enlarged nucleus moves and direct the development of infection thread (Figure 12.3d). With the formation of infection thread particular cortical cells divide to form a nodule primordium, and the infection thread grows toward these primordia (Libbenga and Harkes, 1973; Newcomb, 1981; Vasse and Truchetm, 1984; Wood and Newcomb, 1989). The root cortical cells through which infection threads will pass on their way to the nodule primordia change markedly before they are penetrated by an infection thread. The first cell of the developing nodule contains twice the normal number of chromosomes. These tetraploid cells give rise to the central nodule cells in which the rhizobia develop to produce nitrogen fixing tissue. Associated cells of normal ploidy give rise to uninfected supporting tissues that connect the nodule to the root vascular system.

Nodule Formation

The location of the nodule primordia in the root cortex depends on the type of nodule formed by a particular plant (Newcomb, 1981). In temperate legumes such as pea, vetch, and alfalfa, the primordium is formed from cells in the inner cortex (Dudley, *et al.* 1987, Libbenga, and Harkes, 1973). These legumes form indeterminate cylindrical nodules and have a persistent apical meristem (Newcomb, 1976). This persistent activity of the meristem ensures nodule elongation. The meristem is active, rhizobia are released from the infection thread into the plant cell cytoplasm (Brewin, 1991; Hirsch, 1992; Kijne, 1992; Newcomb, 1981). The differentiation of micro-and macrosymbiont leads to the establishment of a central zone of the nodule, in which nitrogen is reduced (Nap and Bisseling, 1990; Vasse, Billy *et al.*, 1990). Thus in indeterminate nodules, nodule growth and functioning occur simultaneously, and all intermediates in differentiation can be observed in a single longitudinal section of a nodule. On the other hand, in most tropical legumes example soybean and French bean, nodules have a determinate growth pattern (Newcomb, W., *et al.*, 1990; Turgeon and Bauer, 1985). A nodule meristem is induced in the root cortex and the bacteria are released into actively dividing meristematic cells, each daughter cell receiving rhizobia (Newcomb *et al.*, 1990; Newcomb, 1981).The invaded meristematic cells differentiate simultaneously to form the nitrogen-fixing central tissue (Newcomb, 1981). Determinate nodules do not elongate but enlarge.

Besides the formation of the infection threads through root hairs, which is widely studied, the rhizobia may enter through cracks in the epidermis. In legumes such as *Arachis hypogaea* (peanut) and *Stylosanthes* spp., microsymbionts infect their hosts by "crack entry" (Cook, 2000; Chandler, *et al.*, 1982). In the presence of rhizobia, cell divisions are induced in the cortex of an emerging lateral root. In both genera, no infection threads are formed, and rhizobia colonize the root apoplast presumably by cell wall digestion or, in *Stylosanthes spp.*, by progressive collapse of outer root cells. Continuous host cell divisions result in development of a uniformly infected central tissue resembling the determinate nodule type (Cook, 2000; Chandler, *et al.*, 1982; Sprent, and Faria, 1988). *Bradyrhizobium sp.* Infects *Parasponia andersonii (Ulmaceae)*, the only nonlegume plant genus to form nitrogen-fixing root nodules with rhizobia (Trinick, 1988). Upon inoculation, the first sign of root nodule initiation is the formation of swollen multicellular root hairs. Simultaneously, the colonizing bacteria stimulate cell division in the outer cortex. In due course, these cell divisions cause development of callus-like bumps which rupture the epidermis, especially at the base of multicellular root hairs, and infection follows through these wounds (crack entry). Another mode of infection is observed in Mimosa scabrella, a tropical tree, where the rhizobial infection sites are at junctions of epidermal cells (de faria, *et al.*, 1988). The bacteria penetrate the radial walls and proliferate intercellularly.

The route of infection is characterstic for the host, because the same bacteria can penetrate different host species by either crack entry, infection through intact epidermal cells, or root hair infection threads. Similarly, the structural and developmental characteristics of an efficient nodule are specified by the plant and not by the rhizobial strain, indicating that the host possesses the genetic information for symbiotic infection and nodulation and that the role of the bacteria is to switch on this plant developmental program (De'narie' *et al.*, 1992).

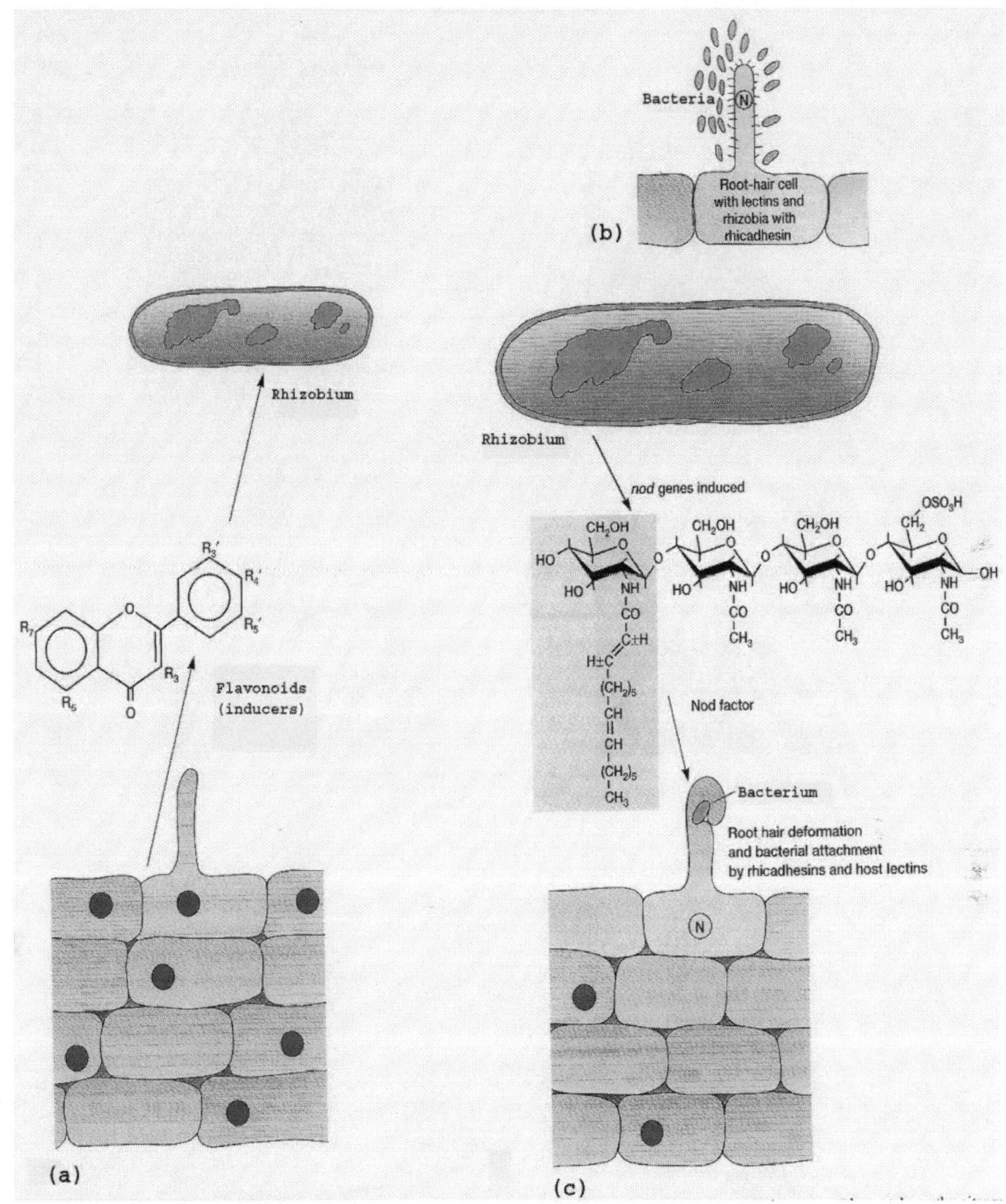

Figure 12.2: Root Nodule Formation by Rhizobium

(a) The plant root released flavonoids that stimulate production of various Nod metabolites by Rhizobium; (b) Attachement of Rhizobium to root hair involves specific bacterial protein called rhicadhesins and host plant lectins that effect the pattern of attachment and nod gene expression; (c) Structure of typical Nod factor that promote root hair curling and plant cortical cell division

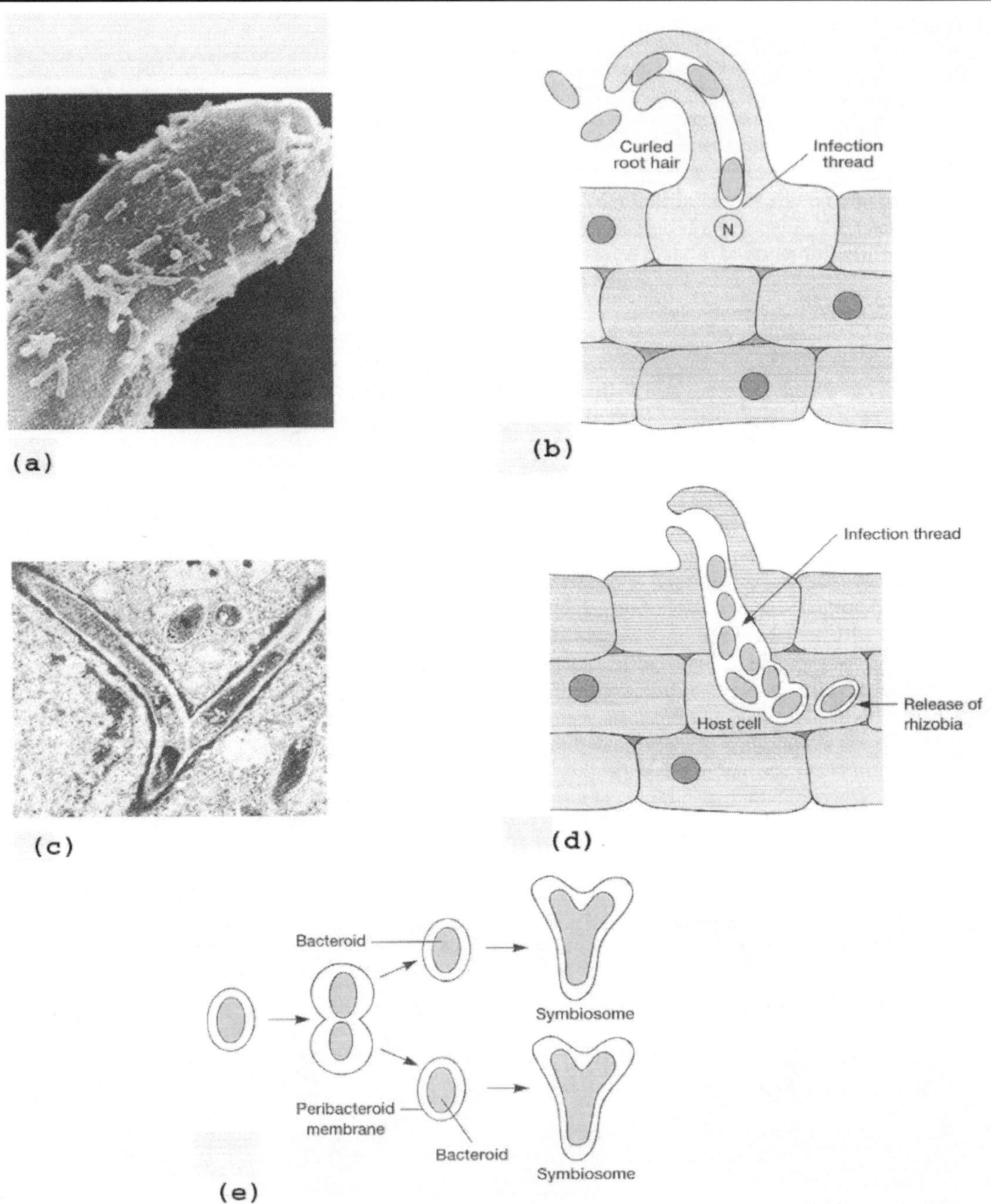

Figure 12.3: (a) A plant root hair covered with Rhizobium and under going curling; (b) Initiation of bacterial penetrations into root hair cells and infection thread growth coordinated by plant nucleus "N"; (c) A branched infection thread shown in an electron micrograph; (d)Cell to cell spread of Rhizobium through transcellular infection threads followed by release of rhizobia and infection of host cell; (e) Formation of bacteroids surrounded by plant derived peribacreroids membrane and differentiation of bacteroids into nitrogen fixing symbiosomes

Nitrogen Fixation

After invasion, a large proliferation of intercellular rhizobia occurs, which is associated with damage to the host cells. Rhizobia are not endocytotically released from these infection threads, but the infection thread will change in nature as rhizobia differentiate into the nitrogen-fixing bacteroid form (Lancelle, and Torrey, 1984; Trinick, 1979; Trinick, 1988). With in the infected tissue, rhizobia multiply, forming unusually shaped and some time grossly enlarged cells called bacteroids (Figure 12.3e). Interspersed with the bacteroid-filled cells of the nodule are uninfected vacuolated cells that may be involved in the transfer of metabolites between the plant and microbe tissues. During the transformation of normal rhizobial cells into bacteroids, the bacterial chromosome degenerate, eliminating the bacteroids capacity for independent multiplication. The bacteroid cells produced and contain active nitrogenase, but host plant tissue appears to play a role in the initiation and control of nitrogenase synthesis. The bacteriod with in the nodule carries out the fixation of the nitrogen. Under normal condition, neither free living rhizobia nor uninfected leguminous plants are able to bring about the fixation of the atmospheric nitrogen.

For active nitrogen fixation, the plant *rhizobium* association requires various organic and inorganic compounds.The trace element molybdenum is required and form an important part of the nitrogenase enzyme. Nitrogenase also contain high amount of sulfur and iron, which thus are requirement of active nitrogen fixation by nodules, cobalt and copper are also required in lower amount.

Nodule have a characteristic red brown color owing to the presence of leghemoglobin, which is a constant and prominent feature in the central tissue of all nitrogen-fixing leguminous nodules. The leghemoglobin serves as an electron carrier, supplying oxygen to the bacteroids for the production of the ATP and at the same time protecting the oxygen sensitive nitrogenase system. The heme protein of leghemoglobin is coded for by the rhizobia and the globulin protein by the plant. Leghemoglobin are unique for legume root nodules and occur no where else in the plant Kingdom. Specific expression of plant and bacterial genes accompanies the development of the rhizobial-plant symbiosis. The genes involved in the root nodule formation, which are collectively called *nodulin genes,* encode a series of Nod proteins that serve specific function in the establishment of nodules that permit symbiotic nitrogen fixation (Table 12.2).

The Rhizobium genes essential for infection and nodule formation can be divided into two classes. One class includes several sets of genes involved in the formation of the bacterial cell surface, such as genes determining the synthesis of exo-polysaccharides (exo genes), lipopolysaccharides (lps genes), capsular polysaccharides or K antigens, and β-1,2-glucans (ndv genes) (Breedveld and Miller, 1994; Gray and Rolfe, 1990; Leigh and Coplin, 1992; Noel, 1992; Reuhs *et al.*, 1993). The second class consists of the nodulation (nod or nol) genes. In most Rhizobium species studied to date, the nod genes reside on large symbiotic plasmids (pSym) that also carry the nif and fix nitrogen-fixing genes (166) (Table 12.2). In Rhizobium loti and *Bradyrhizobium and Azorhizobium spp.*, the symbiosis-related genes are localized on the chromosome (Chua *et al.*, 1985; Kaneko *et al.*, 2000). Most Rhizobium nod genes are not expressed in cultured cells but are induced in the presence of the plant. This induction is caused by flavonoids secreted by the plant and also requires the participation of the transcriptional-activator protein NodD. During the last few years, it has become clear that a major function of the nod genes is to ensure signal exchange between the two symbiotic partners. In the first step, flavonoids excreted by the plant induce, in conjunction with the Nod D protein, the transcription of bacterial nod genes (Fisher and Long, 1992; Schlaman *et al.*, 1992). In the second step, the bacterium, by means of the structural nod genes, produces lipooligosaccharide signals (Nod factors) (Demont, Debelle *et al.*, 1993; De'narie *et al.*, 1992; Spaink, 1992) that induce various root responses (Spaink *et al.*, 1991; Truchet, G., *et al.*, 1991). Mechanisms underlying host specificity depend on both the regulatory and the structural nod genes.

Table 12.2: some Feature of Nod Gene Products

Nod Protein	*Sequence Homology*
NodA	unknown
NodB	Deacetylase
NodC	Chitin synthases
NodD	Transcription activator,Lys R family
NodE	β-Ketoacyl synthase
NodF	Acyl carrier protein
NodG	Alcohol dehydrogenase, β-ketoacyl reductase
NodH	Sulfotransferase
NodJ	Capsular polysaccharide secretion protein
NodK	Unknown
NodL	Acetyltransferase
NodM	D-Glucosamine synthase
NodN	Unknwon
NodO	Hemolysin
NodP	ATP- sulfurylase
NodQ	ATP- sulfurylase and APS Kinase
NodS	Methyltrasnsferase
NodT	Transit sequence
NodU	Unknown
NodV	sensor two-compenent regulatory family
NodW	Regulator, two-compenent regulatory family
NodX	Acidic exopolysaccharide encoded by exo Z
NodY	Unknown
NodZ	Unknown

R. meliloti carries two extremely large plasmids (mega plasmids) of 1400 kilobase pair (Kb) of DNA (p Sym-a or mega plasmid 1) and 1700 kb (p Sym-b or mega plasmid 2). Both cluster I (*nif HDKE, nifN, fixABCX, nifA, nifB*) and cluster II (*fixLJ, fixK, fixNOQP, fixGHIS*) are located on 220 kb downstream of the *nif HDKE* operon and are transcribed in opposite orientation to it. A functional duplication of the region spanning *fixK* and *fix NOQP* is about 40 kb upstream of *nifHDKE*. A cluster of nod genes including the common nod genes (*nod ABC*) is located in the 30 kb region between *nif E* and *nif N*. Additional genes required for symbiosis are located on mega plasmid 2 and on the chromosome.

Table 12.3: Function Associated with Rhizobia Genes for Nitrogen Fixation

Gene	*Sequence Homology*
NifiH	Fe protein of nitrogenase
NifD	α subunit of Mo Fe protein of nitrogenase
NifK	β subunit of Mo Fe protein of nitrogenase
NifE	involved in Fe Mo cofactor biosynthesis
NifN	involved in Fe Mo cofactor biosynthesis
NifB	involved in Fe Mo cofactor biosynthesis Cystein desulfurase
NifS	Cystein desulfurase
NifW	Function unknown,require for full activity of FeMo protein
NifX	function unknown
NifA	positive regulator of nif,fix,and other genes
fixABCX	function unknown required for nitrogenase activity, Fix X show similarity with ferredoxins
fixNOQP	Membrane bound cytochrome oxidas
fixGHIS	Redox process coupled cation pump
fixLJ	Oxygen responsive two-component regulatory system involved in positive control of *fixK* and *nitA*
fixK/fixK2	positive regulator of *fixNOQP*
fixR	Function unknown

Control of the genes for nitrogenase and accessory functions necessary for nitrogen fixation involve a promoter type (with conserved sequences at -24 and -12 [-24/-12 promoter]); an RNA polymerase containing a unique σ factor (σ^{54}); and an activator protein (NifA).NifA is important in controlling expression of nitrogenase structural genes and genes encoding accessory function. A variety of mechanism has evolved to regulate *nifA* transcription with respect to the cellular oxygen conditions. In *R.meliloti* and *A.caulinodans* this control involves the fix LJ two component regulatory system. As a consequence, in *R.meliloti* and *A.caulinodans* oxygen control of nitrogen fixation genes is exerted at two levels (FixL,FixJ and NifA).The fixL and FixJ proteins are members of the ubiquitous two component regulatory system that enable bacteria to respond to environmental or cytoplasmic signals with specific cellular activities. Typically, signal sensing and transduction induce autophosphorylation at a conserved histidine residue in the C-terminal domain of the sensor protein and transfer of the phosphate to an aspartate residue in the N-terminal region of the response regulator protein.

A side reaction of the N_2-reduction process is the evolution of hydrogen gas. The evolution of H_2 waste photosynthetic energy and leads to lowered yields. Some *sym* plasmids also carry *hup* genes coding for hydrogenase activity. The hydrogenase oxidizes H_2 to water and recovers energy by chemiosmotic coupling with ATP synthesis. This beneficial process saves photosynthetic energy that would otherwise go to waste.

Isolation and Enumeration of Bacteria from Soil

Requirements

1. Martin's agar medium (Martin, 1950)

2. Soil sample
3. Pipettes 10 ml and 1 ml capacity
4. Erlenmeyer flasks (250 ml capacity)
5. Penicillin 30 mg
6. Streptomycin 30 mg
7. Incubator
8. Cotton blue + lactophenol

Procedure

1. Take 5 flasks, transfer 90 ml water in each, plug them, and label 1-5 and autoclave.
2. Serially make dilutions 10^{-1}, 10^{-2}, 10^{-3}, 10^{-4}, and 10^{-5} as described for isolation of soil fungi.
3. Transfer 1 ml soil suspension from 10^{-5} dilution into Thornton's agar plates in three replicates each containing 30 mg/l Nyastatin (supplemented after autoclaving just before pouring into plates).
4. Gently shake the plate so as to spread soil suspension uniformly on the medium.
5. Incubate the plates at 25±1°C for 24-48 hours.
6. Pick up each colony, prepare smear and perform Gram's staining fro differentiation.
7. Count CFUs of bacteria.
8. However, if a specific bacterium of interest is present, isolate it aseptically and perform further bacteriological tests for identification.

Isolation and Enumeration of Rhizosphere Microorganisms

Requirements

1. Martin's agar medium
2. Nutrient agar medium
3. Starch casein agar medium
4. Strepto-Penicillin (50 mg/l added to Martin's agar medium after autoclaving of medium)
5. Nyastatin (30 mg/l added to nutrient agar medium after autoclaving of medium)
6. Flasks
7. Pipette
8. Plant root system
9. Distilled water
10. Incubator

Procedure

1. Prepare Martin's agar plate (for fungi), nutrient agar plates (for bacteria) and starch casein plates (for actinomycetes) in triplicate for each.
2. Pour 90 ml distilled water in five flasks, label them 1-5, plug properly and autoclave at 15 lb/inch^2 for 30 minutes.

3. Collect plant root system by gentle removing with the help of sterile spatula.
4. Cut the root in small pieces (2 cm long, around 10 g) with adhering soil particles and transfer into flasks 1 containing 100 ml sterile distilled water as described earlier for soil microorganisms.
5. Gently shake the flasks using magnetic stirrer or electric shaker for 10 minutes so as to get proper soil suspension.
6. Serially dilute soil suspension by transferring 10ml suspension from flask 1 to 2, 2 to 3, 3 to 4 and 4 to 5 subsequently.
7. Transfer aliquots of 1ml soil suspension from dilution 10^{-3} on Martins agar plates, 1ml from dilution 10^{-4} on starch casein plates, and 1ml from dilution $10^{-}4/10^{-}5$ on Thornton's agar plates in three replicates.
8. Incubate the Thornton's agar at 25±1°C for 24-48 hours and record the result as described earlier.
9. Incubate Martin's agar plates at 25±1°C for 5-6 days and record the result.
10. Incubate the starch-casein agar plates at 30±1°C for 10-15 days and record the result.

Isolation of Rhizobium from Root Nodules

Requirements

1. Root nodules
2. YEM agar medium
3. Test tube with nylon mesh
4. Petri dishes
5. 0.1 per cent acidified $HgCl^-$ (1 g $HgCl^-$, 5 ml conc. HCl, 1 litre distilled water).
6. Sterile tap water
7. Nichrome blade
8. Plates containing YEM agar medium

Procedure

1. Procure healthy root nodules of a young leguminous plant by cutting with a blade.
2. Wash the nylon mesh under aseptic conditions so as to remove contaminants and adhering soil particles.
3. Thereafter, immerse them in 0.1 per cent acidified $HgCl^-$ for 5 minutes.
4. Transfer nodules in a sterile beaker containing 10 ml of 95 per cent ethanol and wait for 2-3 minutes.
5. Wash the nodules thoroughly for 5 times with sterile tap water, and blot dry by using sterile blotting paper.
6. Ascetically crush the nodules with glass rod or dissect the nodules by using nichrome blade and prepare dilutions.
7. Pour 1 ml suspension on YEM agar plates.

8. Incubate the inoculated plates at 28°C for 48 hours. Thereafter, observe the bacterial colonies which are gummy, translucent or white opaque.
9. Pick up a discrete colony into a test tube containing 9 ml sterile distilled water. Serially dilute it and then pour 1ml bacterial suspension onto YEM agar plates.
10. Incubate the inoculated plates. Colonies ensure the presence of rhizobia.
11. Pick up a single colony and transfer on YEM agar slants for preservation.
12. Identify by Gram's staining and other molecular techniques.

Isolation of Azotobacter from Garden Soil

Requirements

1. N_2-free glucose medium
2. Garden soil
3. Spatula
4. Incubator
5. Medium
6. K_2HPO_2: 1.000 g
7. $MgSO_2.7H_2O$: 0.200 g
8. $FeSO_2.7H_2O$: 0.050 g
9. $CaCl_2.2H_2O$: 0.100 g
10. $Na_2MoO_2.2H_2O$: 0.001 g
11. Glucose: 10.000 g
12. Water: 1 litre

Procedure

1. Add small amount of garden soil to a bottle having nitrogen-free medium containing glucose as a carbon source.
2. Incubate it at 30°C for 7 days.
3. After incubation, prepare the wet mount of slide from the surface.
4. If organisms are visible, prepare an agar plate of the same medium and streak the agar surface with surface growing culture.
5. Incubate the plate for another 7 days.
6. Prepare the wet mount of the culture and observe under microscope preferably with phase contrast.

Result

Large, mucoid, Gram-negative, rods or cocci, motile forming gummy colonies are visible. These characters confirm the presence of Azotobacter spp. The rod shape cells are arranged singly or in pairs and in irregular clumps. Azotobacter is quite similar to Azomonas aegilis but the late does not form cysts.

References

Aguliar JM, Ashby AM, Richards JM, Loake GJ, Watson MD and Shaw CH (1988). Chemotaxis of *Rhizobium leguminosarum* biovar phaseoli towards flavonoid inducers of the symbiotic nodulation genes. *J Gen Microbiol*, 134: 2741–2746.

Armitage J P, Gallagher A and Johnston AWB (1988). Comparison of the chemotactic behavior of *Rhizobium leguminosarum* with and without the nodulation plasmid. *Mol Microbiol*, 2: 743–748.

Atlas Bartha (2005). *Microbial Ecology: Fundamentals and Applications, 4th edn in Interaction between Microorganism and Plants*, p. 118–125.

Bakhuizen R (1988). The plant cytoskeleton in the *Rhizobium*-legume symbiosis. *PhD Thesis*, University of Leiden, Leiden, The Netherlands.

Barbour WM, Hatterman DR and Stacey G (1991). Chemotaxis of *Bradyrhizobuim japonicum* to soybean exudates. *Appl Environ Microbiol*, 57: 2635–2639.

Barnett MJ, Fisher RF, Jones T, Komp C, Abola AP, Barloy-Hubler F, Bowser L, Capela D, Galibert F, Gouzy J, Gurjal M, Hong A, Huizar L, Hyman RW, Kahn D, Kahn ML, Kalman S, Keating DH, Palm C, Peck MC, Surzycki R, Wells DH, Yeh KC, Davis RW, Federspiel NA and Long, SR (2001). Nucleotide sequence and predicted functions of the entire *Sinorhizobium meliloti* pSymA megaplasmid. *Proc Natl Acad Sci*, USA, 98: 9883–9888.

Benson DR and Clawson ML (2000). Evolution of the actinorhizal plant nitrogen-fixing symbiosis. In: *Prokaryotic Nitrogen Fixation: A Model System for the Analysis of a Biological Process*, (Eds) E. Triplett. Horizon Scientific Press, Wymondham, England, p. 207–224.

Benson DR and Silvester WB (1993). Biology of *Frankia* strains, actinomycete symbionts of actinorhizal plants. *Microbiol Rev*, 57: 293–319.

Breedveld MW and Miller KJ (1994). Cyclic β-glucan of members of the family of *Rhizobiaceae. Micobiol Rev*, 58: 145–161.

Brewin NJ (1991). Develoment of the legume root nodule. *Annu Rev Cell Biol*, 7: 191–226.

Caetan-Anolles G, Wrobel-Boerner E and Bauer WD (1992). Growth and movement of spot inoculated *Rhizobium meliloti* on the root surface of alfalfa. *Plant Physiol*, 98: 1181–1189.

Caetano-Anolles G, Crist-Estes DK and Bauer WD (1988). Chemotaxis of *Rhizobium melioloti* to the plant flavones leteolin requries functional nodulation genes. *J Baceriol*, 170: 3164–3169.

Callaham DA and Torrey JG (1981). The structural basis for the infection of root hairs in *Trifolium repens* by *Rhizobium. Can J Bot*, 59: 1647–1664.

Chandler MR (1978). Some observations on the infection of *Arachis lypogaea* L. by *Rhizobium. J Exp Bot*, 29: 749–755.

Chandler MR, Date RA and Roughley RJ (1982). Infection and root nodule development in *Stylosanthes* species by *Rhizobium. J Exp Bot*, 33: 47–57.

Chua YK, Pankhurst CE, Macdonald PE, Hopcroft D, Jaarvis BDW and Scott DB (1985). Isolation and characterization of Tn5-induced symbiosis mutants of *Rzhizobium loti. J Bacteriol*, 162: 335–343.

Cook D (2000). *Medicago truncatula*: A model in making! *Curr Opin Plant Biol*, 2: 301–304.

Cooper EJ and Rao JR (1992). Localized changes in flvonoid biosynthesis in roots of *Lotus pedunculatus* after infection by *Rhizobium loti. Plant Physiol,* 100: 444–450.

Davis EO, Evans IJ and Johnston AWB (1988). Indentification of *nod*X, a gene that allows *Rhizobium leguminosarum* bv. *Phaseoli. Mol Microbiol,* 4: 933–941.

de faria SM, Hay HT and Sprent JI (1988). Entry of rhizobia into roots of *Mimosa scabrella* Bentham occurs between epidermal cells. *J Gen Microbiol,* 134: 2291–2296.

Denarie J, Debelle F and Rosnberg C (1992). Signaling and host range variation in nodulation. *Annu Rev Microbiol,* 46: 497–531.

Demont NF, Debelle H Aurelle JC Prome (1993). Role of the *Rhizobium meliloti nod*F and nodE genes in the biosynthesis of lipo–oligosaccharide nodulation factors. *J Biols Chem,* 268: 20143–20142.

Dowling DN and Broughton WJ (1986). Competition for nodulation of legumes. *Annu Rev Microbiol,* 7: 131–157.

Dudley ME, Jacobs TW and Long SR (1987). Microscopy studies of cell division induced alfalfa roots by *Rhizobium meliloti. Planta,* 171: 289–301.

Fisher RF and Long SR (1992). *Rhizobium*: Plant signal exchange. *Nature (London),* 357: 655–660.

Freiberg C, Fellay R, Bairoch A, Broughton WJ, Rosenthal A and Perret X (1997). Molecular basis of symbiosis between *Rhizobium* and legumes. *Nature,* 387: 394–401.

Gaworzewska ET and Carlile MJ (1982). Positive chemotaxis of *Rhizobium leguminosarum* and other bacteria towards roots exudates from legumes and other plants. *J Gen Microbiol,* 128: 789–798.

Gray JX and Rolfe BG (1990). Exopolysaccharide production in *Rhizobium* and its role in invasion. *Mol Microbiol,* 4: 1425–1431.

Gulash M, Ames P, LaRosiliere RC and Bergman K (1984). Rhizobia are attracted to localized sites on legumes roots. *Appl Environ Microbiol,* 48: 149–152.

Handberg K and Stougaard JS (1992). *Lotus japonicas,* an autogamous, diploid legume species for classical and molecular genetics. *Plant J* 2: 487–496.

Hirsch AM (1992). Developmental biology of legume nodulation. *New Phytol,* 122: 211–237.

Kaneko T, Nakamura Y, Sato S, Asamizu E, Kato T, Sasamoto S, Watanabe A, Idesawa K, Ishikawa A, Kawashima K, Kimura T, Kishida Y, Kiyokawa C, Kohara M, Matsumoto M, Matsuno A, Mochizuki Y, Nakayama S, Nakazaki N, Shimpo S, Sugimoto M, Takeuchi C, Yamada M and Tabata S (2000). Complete genome structure of the nitrogen-fixing symbiotic bacterium *Mesorhizobium loti. DNA Res,* 7: 38–406.

Kaneko T, Nakamura Y, Sato S, Minamisawa K, Uchiumi T, Sasamoto S, Watanabe A, Idesawa K, Iriguchi M, Matsumoto M, Shimpo S, Tsuroka H, Wada T, Yamada M and Tabata S (2002). Complete genome structure of the nitrogen–fixing symbiotic bacterium *Bradyrhizobium japonicum* USDA110. *DNA Res,* 9: 189–197.

Kape R, Parniske M and Werner D (1991). Chemotaxis and *nod* gene activity of *Bradyrhizobium japonicum* in response to hydrocinnamic acids and isoflavonoids. *Appl Environ Microbiol,* 57: 1631–1632.

Kijne JW (1992). The fine structure of pea root nodule. 1. Vacuolar changes after endocytotic host cell infection by *Rhizobium leguminosarum*. *Physiol Plant Pathol,* 5: 75–79.

Lancelle SA and Torrey JG (1984). Early development of R–induced root nodules of *Parasponia rigida*. I. Infection and early nodule initiation. *Protoplasma,* 123: 26–37.

Leigh JA and Coplin DL (1992). Exopolysaccharides in plant-bacterial interactions. *Annu Rev Microbiol,* 46: 307–346.

Lewin A, Rosenberg C, Meyer ZAH, Wong CH, Nelson L, Manen JF, Stanley J, Downing DN, De'narie J and Broughton WJ (1987). Multiple host-specificity loci of the broad host-range *Rhizobium* sp. NGR234 selected using the widely compatible legume *Vigna unguiculata. Plant Mol Biol,* 8: 447–459.

Lewin A, Cervante's E, Wong CH and Broughton WJ (1990). *nod*SU, two new *nod* genes of the broad host–range *Rhizobium* strain NGR234 encode host-specific nodulation of the tropical tree *Leucaena leucocephala. Mol Plant-Microbe Interract,* 3: 317–326.

Lewis-Henderson WR and Djordjevic MA (1991). A cultivar-specific interaction between *Rhizobium leguminosarum* by *trifolii* and subterranean clover is controlled by *nod*M, other bacterial cultivar specificity genes, and a single recessive host gene. *J Baceriol,* 173: 2791–2799.

Libbenga KR and Harkes PAA (1973). Initial proliferation of cortical cells in the formation root nodules in *Pisum sativum. Planta,* 114: 17–28.

Martinez E, Romero D and Palacios R (1990). The *Rhizobium* genome. *Crit Rev Plant Sci,* 9: 59–93.

Mellor HY, Glenn AR, Arwas R and Dilworth MJ (1987). Symbiotic and competitive properties of motility mutants of *Rhizobium trifolii* TA1. *Arch Microbiol,* 148: 34–39.

Moulin L, Munive A, Dreyfus B and Boivin-Masson C (2001). Nodulation of legumes by members of the β-subclass of Proteobacteria. *Nature,* 411: 948–950.

Nap JP and Bisseling T (1990). Developmental biology of a plant-prokaryote symbiosis: The legume root nodule. *Science,* 250: 948–954.

Newcomb W (1976). A correlated light electron microscopic study of symbiotic growth and differentiation in *Pisum sativum* root hair nodules. *Can J Bot,* 54: 2163–2186.

Newcomb W (1981). Nodule morphogenesis and differentiation. *Int Rev Cytol Suppl,* 13: 246–298.

Newcomb W, Spippell D and Peterson RL (1979). The early morphogenesis of *Glycine max* and *Pisum sativum* root nodules. *Can J Bot,* 57: 2603–2616.

Noel KD (1992). Rhizobial polysaccharides required in symbiosis with legumes. In: *Molecular Signals in Plant-Microbe Communication,* (Ed) DPS Verma. CRC Press, Inc, Boca Raton, Fla, p. 341–358.

Peters NK and Verma DPS (1990). Phenolic compounds as regulators of gene espression in plant-microbe interactions. *Mol Plant-Microbe Interact,* 5: 33–37.

Prescott, Harley, Klein (2003). *Microbiology,* 5th edn. McGraw Hills Publication in metabolism: The use of energy in biosynthesis, p. 212–214.

Reuhs BL, Carlson RW and Kim JS (1993). *Rhizobium fredii* and *Rhizobium meliloti* produce 3-deoxy-D-mannose-2-oculosonic acid containing polysaccharides that are structurally analogous to group-II-K-antigens (capsular polysaccharides) found in *Escherichia coli. J Bacteriol,* 175: 3570–3580.

Schlaman HRM, Okker RJH and Lugtenberg BJJ (1992). Regulation of nodulation gene expression by NodD in rhizobia. *J Bacteriol,* 174: 5177–5182.

Spaink HP (1992). Rhizobial lipo-oligosaccharides: answer and questions. *Plant Mol Biol,* 20: 977–987.

Spaink HP, Sheeley, DM, van Brussel AAN, Gkushka J, York WS, Tak T, Geiger O, Kennedy EP, Reinhold VN and Lugtenberg BJJ (1991). A novel highly unsaturated fatty acid moiety of lipo–oligosaccharide signals determines host specificity of *Rhizobium. Nature (London),* 354: 125–130.

Sprent JL and de Faria SM (1988). Mechanisms of infection of plants by nitrogen fixing organisms. *Plant Soil,* 110: 157–165.

Sy A, Giraud E, Jourand P, Garcia N, Willems A, de Lajudie P, Prin Y, Neyra M, Gillis M, Boivin-Masson C and Dreyfus B (2001). Methylotrophic *Methylobacterium* bacteria nodulate and fix nitrogen in symbiosis with legumes. *J Bacteriol,* 183: 214–220.

Trinick MJ (1979). Structure of nitrogen fixing nodules formed by *Rhizobium* on roots of *Parasponia andersonni* Planck. *Can J Microbiol,* 25: 565–578.

Trinick MJ (1988). Biology of the *Parasponia-Bradyrhizobium* symbiosis. *Plant Soil,* 110: 177–185.

Truchet G, Roche P, Lerouge P, Vasse J, Camut S, de Billy F, Prome JC and De'narie J (1991). Sulphated lipo-oligosaccharide signals of *R.meliloti* elicit root nodule organogenesis in alfalfa. *Nature (London),* 351: 670–673.

Turgeon BG and Bauer WD (1982). Early events in the infection of the soybean by *Rhizobium japonicum.* Time course cytology of the initial infection process. *Can J Bot,* 60: 152–161.

Turgeon BG and Bauer WD (1985). Ultrastructure of infection thread development during the infection of soybean by *Rhizobium japonicum. Planta,* 163: 328–349.

Van Rhijin P and Vanderleyden J (1995).The rhizobium: Plant symbiosis.*Microbial Review,* 59: 124–142

Vasse J and Truchet G (1984). The *Rhizobium*-legume symbiosis: observation of root infection by bright-field microscopy after with methylene blue. *Planta,* 161: 487–489.

Vasse J, de Billy F, Camut S and Truchet G (1990). Correlation between ultrastructural differentiation of bacteroids and nitrogen fixation in alfalfa nodules. *J Bacteriol,* 172: 4296–4306.

Wood SM and Newcomb W (1989). Nodule morphogenesis the early infection of alfalfa (*Medicago saiva*) root hairs by *Rhizobium meliloti. Can J Bot,* 67: 3108–3122.

Young JP and Johnston AWB (1989). The evolution of specificity in the legume–*Rhizobium* symbiosis. *Trends Ecol,* 4: 331–349.

Soil Microflora, 2009 *Pages* **139–147**
Editor: **Rajan Kumar Gupta, Mukesh Kumar & Deepak Vyas**
Published by: **DAYA PUBLISHING HOUSE, NEW DELHI**

Chapter 13

Role of Soil Microorganisms in Nutrition and Health of Higher Plants

Dheeraj Mohan*, Preetesh Kumari, Kaushal Pratap Singh and Anuradha Chauhan

Microbiology Research Laboratory, Department of Botany, Raja Balwant Singh College, Agra, U.P.

The root system of a plant is very active metabolically and provides a continuous source of food for soil microorganisms in the form of secretions of organic compounds and sloughed-off dead cells and cell debris. Since the zone where roots and soils meet is a special environment, it has been named the rhizosphere (root zone). This zone comprises several poorly defined, heterogeneous regions in which microorganisms are particularly active.

Although activity in the rhizosphere is of great importance to the plant, it affects only a small fraction (about 5 percent) of the root surface. Some microorganisms are loosely associated with roots, but others develop on the root surface and many can invade root tissue, with effects that can be beneficial or harmful certain soil-inhabiting microorganisms produce diseases of great significance to agriculture and forestry. Others are beneficial-they partly inhibit the growth of disease organisms or kill them. The vast majority of the pathogens that infect roots are fungi, and they are exceptionally difficult to control or eradicate.

In some cases, the invasion of roots by microorganisms is desirable. This is true for the root-nodule bacteria of the genus Rhizobium that fix nitrogen, as well as for mycorrhizal fungi, which assist roots in accumulating phosphate and other essential minerals. Rhizosphere microorganisms can affect plant welfare in a number of ways. The processes of nutrient cycling, growth stimulation or

* E-mail: dheerajmsinghal6@yahoo.co.in, Phone: +91-9412426474

inhibition, and diseases are of great significance, but they are very complex population effects rather than the result of simple interactions between roots and known microorganisms. The rhizosphere is also influenced by external factors such as soil moisture and even the intensity of light reaching the plant. No single microorganism may be essential to the process, but the combined effect of the rhizosphere population can be profound.

The sum of the various interrelationships of rhizosphere microorganisms and roots can benefit plant growth by influencing the availability of essential nutrients, by producing plant growth regulators, and by suppressing root pathogens.

Mineral Cycling by Soil Microorganisms

By decomposing plant and animal residues, soil microorganisms release carbon, nitrogen, sulfur, phosphorus, and trace elements from organic materials in forms that can be absorbed by plants.

This process, known as mineralization, is the primary source of atmospheric carbon dioxide. Without mineralization of organic carbon, the carbon dioxide content of the air, which is essential for plant photosynthesis, would be progressively reduced and plant production would ultimately cease. Maintenance of the carbon cycle, therefore, is one of the most important biological processes on earth.

Microbial activities similar to those responsible for the carbon cycle also transform soil nitrogen, sulfur, and phosphorus, and to a lesser extent are instrumental in the conversion of other elements. Although particular attention has been directed to microbial transformations of nitrogen, plants also have a nutritional need for sulfur. The microbial transformations of nitrogen and sulfur are much alike because both elements can be oxidized and reduced. Sulfur reduction is necessary for the synthesis of sulfur-containing amino acids. Under anaerobic conditions, sulfur reduction may produce hydrogen sulfide, which can be harmful to plants. It accumulates in very wet soils, like rice paddies, and can cause straight head disease of rice and other physiological plant disorders. At low concentrations, however, it can supply the sulfur requirements of some plants.

The oxidation of sulfur by bacteria of the genus *Thiobacillus* is also of potential significance in agriculture. The product of this transformation is sulfuric acid, which can dissolve minerals that otherwise would not be available for plant growth. Farmers who add elemental sulfur to rock phosphate find that phosphorus is liberated more rapidly and in greater amounts than if the sulfur is omitted.

Microorganisms are also able to promote phosphorus solubilization by the production of chelators, which form complexes with metal ions and increase their solubility. Solvent action by microorganisms is not restricted to a few species; it is characteristic of many members of the rhizosphere population and can be accomplished in part by plant roots as well.

Microorganisms require many of the same nutrient elements that are essential to plants for their growth. When nitrogen, sulphur, or phosphorus is in short supply, the rhizosphere population will compete with roots for nourishment. Because of their abundance, small size, and relatively large surface area, and because they surround the absorbing part of the root, microbes will absorb nutrients at the expense of the plant. Ultimately, plants will display signs of nutrient deficiency and crop yields may decrease.

There has been speculation that in the rhizosphere oxygen consumption occurs more rapidly than diffusion, so that anaerobic sites may form in places in the root; such reduced conditions could be important in making ferrous ions from ferric, for instance, which increases iron solubility. Wheat roots have high populations of denitrifying bacteria, so oxygen-free conditions must exist in their presence.

Microorganisms are prolific producers of vitamins, amino acids, hormones, and other growth-regulating substances. Many bacteria and fungi isolated from soil are able to synthesize compounds that provoke a growth response in plant tissue. Some produce indole acetic acid or gibberellins, which are hormones that control plant growth, while others produce Vitamins. Many may also produce unidentified growth factors. Rhizosphere microorganisms are variously credited with promoting increased rates of seed germination, root elongation, root-hair development, nutrient uptake, and plant growth.

Inoculation

Various bacterial fertilizers have been marketed at different times, but commercial preparations known as azotobacterin and phosphobacterin have received the most attention. Azotobacterin is composed of cells of *Azotobacter chroococcum*, a bacterium able, under some conditions, to fix atmospheric nitrogen. Phosphobacterin contains the bacterium *Bacillus megaterium* var. *phosphaticum*, which mineralizes organic phosphorus compounds. Treated plants are favorably affected, but growth is not increased by more than 10 per cent. Moreover, the effect is said to be due not to nitrogen fixation or phosphorus solubilization, but to plant hormones.

Since the benefits are minimal and depend on conditions difficult to control in the field, and the results are unpredictable, bacterial fertilizers are not recommended for general use.

Mycorrhizal Fungi

Most plants, both wild and cultivated, have roots infected with fungi that increase nutrient and water uptake and may also protect the root from certain diseases. These infected roots are called mycorrhizae. Although the mycorrhizal fungi probably increase uptake of all the essential elements, they are usually most important in improving phosphorus nutrition. Phosphate is generally present in the soil in low concentrations and it is also highly immobile. Strands of fungal hyphae grow out from mycorrhizae and greatly increase the volume of soil from which phosphorus is obtained. So mycorrhizal plants, in general, can grow and thrive in soils much lower in phosphate and other essential nutrients than a comparable non-mycorrhizal plant. Many plants are so dependent on mycorrhizal fungi for nutrient uptake that they may starve if these fungi are absent. There are a number of types of mycorrhizae. The two that occur on the most economically important crops, the endomycorrhizae and the ectomycorrhizae, are discussed below:

1. Endomycorrhizae of Crop Plants and Forest Trees

Endomycorrhizae of the arbuscular type occur on neady all crop plants (plants in a few families such as the Cruciferae [cabbage, mustard, etc.] and Chenopodiaceae [beets, spinach, etc.] may be non-mycorrhizal). They also occur on many trees in temperate regions and on the majority of tree species native to the sub tropics and tropics.

A. Arbuscular mycorrhizal fungi are present in almost all soils and they are not host-specific. Thus, the same fungus producing. A. mycorrhizae on trees will form mycorrhizae on plants after land is cleared and planted to agricultural crops.

B. The mycorrhizal condition is normal for most plants, and at absence or scarcity of mycorrhizal fungi can greatly limit plant growth (Figure 13.1). Introduction of mycorrhizal fungi to soil environments lacking or with low populations of such organisms, such as biocide-treated soils, can enhance plant growth.

C. A. mycorrhizae are particularly important for many legumes in that they stimulate nodulation by Rhizobium, thereby increasing nitrogen fixation. Improved phosphorus nutrition of the mycorrhizal legume is responsible for increased nodulation.

D. A. mycorrhizal fungi survive in soil as resting spores. They obtain their food from the plant roots and they are unable to grow independently in soil. It is unlikely that they obtain much, if any, organic nutrient from soil.

E. A M fungi have not been grown in pure culture, which presents an obstacle to artificial inoculation. However, these fungi produce the largest spores of any known fungi, some being 0.5 mm or more in diameter. The spores can be easily extracted from the soil with sieves and then propagated on the roots of living plants. The infected roots can also be used for inoculation. Heavily infected palm roots collected in the wild have been used as a source of inoculum. If field-collected inoculum is used, it is important that it be free of dangerous pathogens.

There are situations in which inoculation with V A fungi is highly beneficial. If soil is treated with steam or fumigants to kill pathogens, V A fungi are also killed, and considerable time is required for them to become reestablished naturally. The nutrient deficiencies and associated stunting that often result may be prevented by inoculating the soil with V A fungi rather than by applying excessive rates of fertilizer.

The greatest opportunity for the use of V A fungi is in soils low in available phosphorus, which includes many untreated soils in tropical regions. There is evidence as well that many of these soils also contain less than the optimum number of spores of V A fungi. In such soils inoculation may enable the use of inexpensive rock phosphate fertilizer instead of the more expensive super and triple phosphates.

2. Ectomycorrhizae of Forest Trees

Ectomycorrhiza is the second most common type of mycorrhizae. It occurs on roots of pine, spruce, fir, larch, hemlock, willow, poplar, hickory, pecan, oak, birch, beech, and eucalyptus (Figure 13.2). The fungi that form ectomycorrhizae produce mushrooms and puffballs as their reproductive structures (fruit bodies). The fungi are spread in nature by millions of microscopic spores, finer than dust, which are released from fruiting bodies and moved great distances by winds.

Many forest trees, such as pines, cannot grow beyond the first year without an appreciable number of ectomycorrhizae. Ectomycorrhizae benefit trees by: increasing nutrient and water absorption from soil; increasing the tolerance of the tree to drought and extremes of soil conditions (acid levels, toxins, etc.); increasing the length of the feeder root system; and protecting the fine feeder roots from certain harmful soil fungi.

Ectomycorrhizal fungi cannot grow and reproduce unless they are in association with the roots of a tree host. These fungi obtain all their essential sugars, vitamins, amino acids, and other foods from their hosts. It is unlikely that these fungi as a group are directly involved in any significant decomposition of forest litter.

Certain forest trees, then, must have ectomycorrhizae to survive and grow, and the ectomycorrhizal fungi need their tree hosts to exist. This means that the introduction of tree species as exotics into regions where the appropriate mycorrhizal fungi are absent must be· accompanied by the introduction of their natural ectomycorrhizal fungi. In the past, this introduction has been accomplished mainly by

Figure 13.1: Endomycorrhizal (Left) and Nonmycorrhizal (Right) *Acacia nilotica* Plants

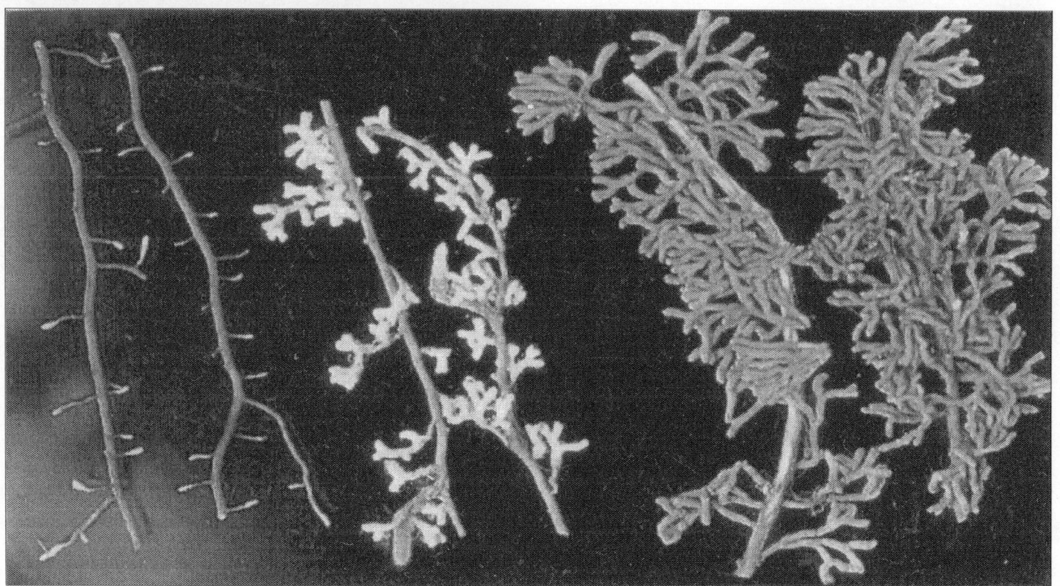

Figure 13.2: Examples of Pine Ectomycorrhizal
Each different ectomycorrhiza is formed by a different species of fungus.

using soil collected from under healthy trees with ectomycorrhizae, which is mixed into the upper layer of soil in nurseries. The seedlings planted in this soil usually form abundant ectomycorrhizae in one growing season, and they are then transplanted to the field. Unfortunately, this method is not without risk, since pathogens can be present in the introduced soils and cause serious damage to the trees. The logistics of transporting large volumes of soil great distances is an added problem. By far the most biologically sound method of correcting an ectomycorrhizal deficiency is by the use of pure cultures of selected species of ectomycorrhizal fungi.

In recent years, techniques have been devised to inoculate soil with pure vegetative and spore cultures of *Pisolithus tinctorius* in the United States and to introduce spore cultures of *Rhizopogon luteolus* onto pine seed and into soil in Australia. These two puffball-producing fungi form ectomycorrhizae on many commercially important forest trees. Thus far, *P. tinctorius* appears to enhance growth more than other ectomycorrhizal fungi, and it can be used to tailor-make seedlings to improve the performance of trees even in areas where other ectomycorrhizal fungi are present.

Pines with *Pisolithus* ectomycorrhizae formed in nurseries and planted in forest sites in the southern United State have not only survived, but have grown to twice the heights of comparable pines with naturally occurring ectomycorrhizae. The selection and use of specific ectomycorrhizal fungi may well determine whether a productive forest becomes established.

The puffball fungi usually produce an abundance of easily extractable spores. For example, one fruit body of *P. tinctorius* may contain 75 grams of spores, and there are more than one billion spores in a gram. Spores of *P. tinctorius* when kept dry and cool have been stored for more than 4 years without losing their ability to form ectomycorrhizae. The spores can be mixed into nursery soil and the seed of

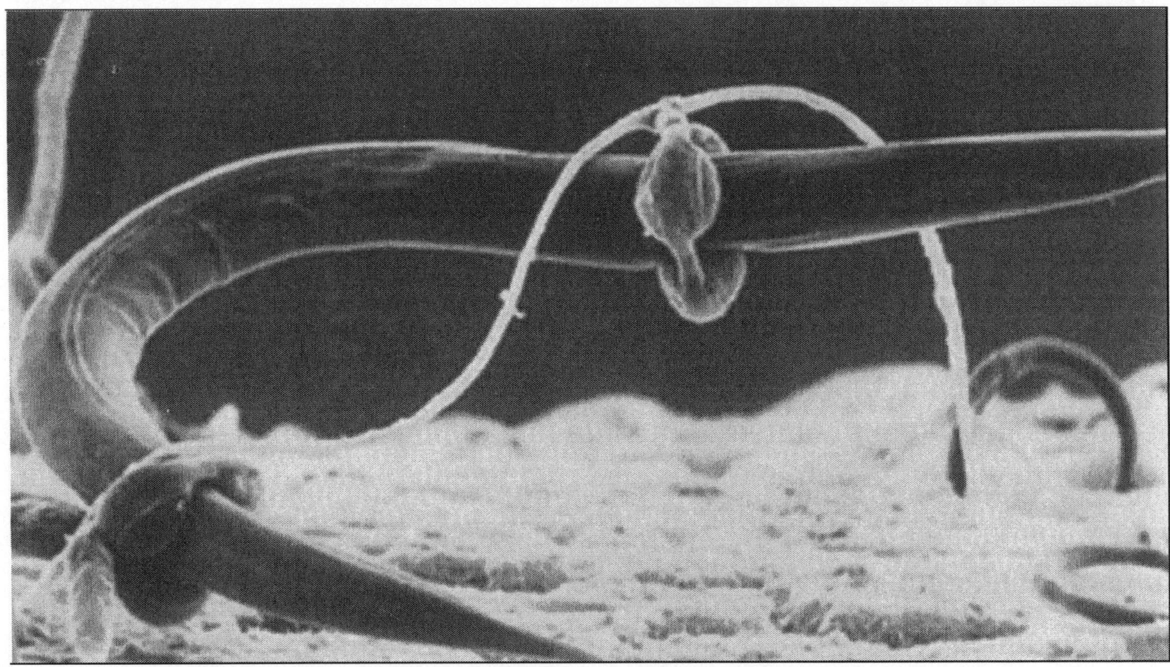

Figure 13.3: A Nematode-Trapping Fungus, which Captures Prey on Adhesive Knobs (*Dactylella drechsleri*)

the desired tree species planted. Moderate levels of fertility and at least 2-4 per cent organic matter should be maintained in the soil. Usually, the seedlings will have adequate ectomycorrhizae in 6-7 months after seed germination.

The production of vegetative cultures of ectomycorrhizal fungi requires aseptic culture technique. This means that the substrate on which the fungi are grown must be sterilized, usually by autoclaving, and maintained free of other microbial contamination for at least several weeks, or until the fungus has produced sufficient growth to overcome contamination. After it has been leached with water, this inoculum can then be added to soil.

Some species of ectomycorrhizal fungi are more beneficial to tree growth and development under different soil conditions than others. It is important to select and test different species to determine which are best suited to specific locations.

Biological Control of Soil–Borne Pathogens

Microorganisms that cause root diseases are sometimes suppressed by other microorganisms in the soil. In many instances disease-causing organisms may be present, but because of naturally occurring biological control, little or no disease results. The prevalence of pathogens may be reduced by crop rotation using a non-host crop, which often starves the pathogen and prevents it from reproducing.

It is also possible to increase the level of organisms that are antagonistic to soil-borne plant pathogens. This is generally done not by adding antagonistic microorganisms directly to the soil, but by the use of various organic amendments such as manure or plant residues. These amendments provide a source of food for soil-borne microorganisms that can inhibit the development of plant pathogens.

There have been many attempts to control pathogens in soil by the addition of specific microorganisms. In general, these attempts have failed to increase the level of naturally occurring biological control. For example, soil contains species of fungi that trap and feed on plant parasitic nematodes (Figure 13.3). However, application of additional nematode-trapping fungi failed to protect plants under field conditions. It is likely that unless the soil is altered in some way, it naturally contains the maximum number of nematode-trapping fungi that it can support.

There are a few examples where disease has been reduced by applying a hyper virulent strain or a mutant strain of a pathogen that is incapable of producing disease. Such strains may prevent development of the pathogenic strains. A nonpathogenic strain of the crown gall bacterium will thus protect plants from attack by a pathogenic strain.

In soil, most pathogenic fungi must pass through the rhizosphere or must live within this zone. Their success in colonizing or infecting plant roots depends upon their ability to compete with other rhizosphere microorganisms. The chemical and microbiological environment in this zone may be altered slightly, but significantly, to effect changes in the inoculum potential of pathogens, either by selections of plant genotypes that produce such changes or by careful regulation of nitrification in soil.

References

Alexander M (1977). *Introduction to Soil Microbiology*, 2nd Edition. John Wiley and Sons, New York.

Alexander M (1974). *Microbial Ecology*. John Wiley and Sons, New York.

Baker KF and Cook RJ (1974). *Biological Control of Plant Pathogens*. W. H. Freeman and Company, San Francisco.

Barber DA and Martin JK (1976). The release of organic substances by cereal roots into soil. *The New Phytologist*, 76: 69–80.

Barber DS (1968). Microorganisms and the inorganic nutrition of higher plants. *Annual Review of Plant Physiology*, 19: 71–88.

Barron GL (1977). The nematode-destroying fungi. *Topics in Mycobiology* No. 1. Guelph, Ontario: Canadian Biological Publications Ltd.

Brown ME (1974). Seed and root bacterization. *Annual Review of Phytopathology*, 12: 181–197.

Brown ME, Hornby D and Pearson V (1973). Microbial populations and nitrogen in soil growing consecutive cereal crops infected with take-all. *Journal of Soil Science*, 24(3): 296–310.

Carson EW (1974). *The Plant Root and its Environment*. The University Press of Virginia, Charlottesville.

Cook RJ (1977). Management of the associated micro biota. In: *Plant Disease: An Advanced Treatise in how Disease is Managed*, (Eds) JG Horsfall and EB Cowling. Academic Press, New York, pp. 145–166.

Cook RJ (1976). Interaction of soil-borne plant pathogens and other microorganisms: An introduction. *Soil Biology and Biochemistry*, 8: 267.

Doetsch RN and Cook TM (1976). *Introduction to bacteria and their ecobiology*. University Park Press, Baltimore, Maryland.

Garrett SD (1956). *Biology of Root-Infecting Fungi*. Cambridge University Press, New York, p. 11.

Garrett SD (1970). *Pathogenic Root-Infecting Fungi*. Cambridge University Press, Cambridge. England.

Gerdemann JW (1975). Vesicular-arbuscular mycorrhizae. In: *The Development and Function of Roots*, (Eds) JG Torrey and DT Clarkson. Academic Press, New York, pp. 575–591.

Gray TP and Parkinson D (1968). *The Ecology of Soil Bacteria*. Liverpool University Press, Liverpool.

Gray TP and Williams ST (1975). *Soil Microorganisms*. Longman, New York.

Harley JL (1979). *Proceedings of the Soil-Root Interface Symposium*. Academic Press, London.

Henis Y and Chet I (1975). Microbiological control of plant pathogens. *Advances in Applied Microbiology*, 19: 85.

Hornby D (1978). Microbial antagonisms in the rhizosphere. *Annals of Applied Biology*, 89: 97–100.

Joshi MM and Hollis JP (1977). Interactions of Beggiatoa and rice plants: detoxificaation of hydrogen sulfide in the rice rhizosphere. *Science*, 197: 179–180.

Kleinschmidt GD and Gerdemann JW (1972). Stunting of citrus seedlings in fumigated nursery soils related to the absence of endomycorrhizae. *Phytopathology*, 62: 447–1453.

Krasilnikov NA (1958). *Soil Microorganisms and Higher Plants*. Moscow: Academy of Sciences USSR. English translation, by Y. Halperin, 1961. Jerusalem: Israel Program for Scientific Translations, Ltd.

Marks GC and Kozlowski RT (1973). *Ectomycorrhizae: Their Ecology and Physiology*. Academic Press, New York.

Marx DH (1977). The role of mycorrhizae in forest production. In: *Proceedings of the TAPP (Technical Association of the Pulp and Paper Industry) Annual Meeting*, February 14–16,held in Atlanta, Georgia, Atlanta: TAPPI, pp. 151–161.

Mosse B (1977). Plant growth responses to vesicular-arbuscular mycorrhiza: X. Responses or stylosanthes and maize to inoculation in unsterile soils. *New Phytologist*, 78: 277–288.

Mosse B (1977). The role of mycorrhiza in legume nutrition on marginal soils. In: *Exploiting the Legume-Rhizobium Symbiosis in Tropical Agriculture: Proceedings of a Workshop*, University of Hawaii, August 1976, College of Tropical Agriculture Miscellaneous Publication No. 54, pp. 275–292. Honolulu: University of Hawaii.

Rovira AD, Newman FI, Bowen HJ and Campbell R (1974). Quantitative assessment of the rhizoplane microflora by direct microscopy. *Soil Biology and Biochemistry*, 6: 211–216.

Sanders FE, Mosse B and Tinker PB (Eds) (1974). Endomycorrhizas. *Proceedings of Symposium on Endomycorrhiza*. July, University of Leeds, Academic Press, New York.

Shipton PJ (1977). Monoculture and soil borne pathogens. *Annual Review of Phytopathology*, 15: 387 – 407.

Smith AM (1976). Availability of plant nutrients in reduced microsites in soil. *Annual Review of Phytopathology*, 14: 53–73.

Tansey MR (1977). Microbial facilitation of plant mineral nutrition. In: *Microorganisms and Minerals*, (Ed) ED Weinberg. Marcel Dekker, Inc, New York, pp. 343–385.

United Nations Educational, Scientific, and Cultural Organization (1969). *Soil Biology: Review on Research*. Natural Resources Research, Series No. 9. Paris: United Nations Educational, Scientific, and Cultural Organization. Distributed in the United States by UNIPUB. New York.

Soil Microflora, 2009 *Pages* **148–164**
Editor: **Rajan Kumar Gupta, Mukesh Kumar & Deepak Vyas**
Published by: **DAYA PUBLISHING HOUSE, NEW DELHI**

Chapter 14

Soil Microflora and their Impact on Soil Health

***Narendra Kumar*[1], *Pawan Kumar*[1] *and Surendra Singh*[2]**
[1]*School of Studies in Microbiology, Jiwaji University, Gwalior – 474 011, M.P., India*
[2]*Centre of Advanced Study in Botany, Banaras Hindu University, Varanasi – 221 005, U.P., India*
E-mail: surendrasingh.bhu@gmail.com, surendrasinghg@yahoo.co.in

ABSTRACT

Microorganisms are ubiquitous in the environment, where they have a variety of essential functions to play. Microorganisms live almost everywhere on earth where there is liquid water, including hot springs on the ocean floor and deep inside rocks within the earth's crust. Microorganisms are vital to humans and the environment, as they participate in the earth's elemental cycles such as the carbon and nitrogen cycle, as well as fulfilling other vital roles in virtually all ecosystems, such as recycling other organisms' dead remains and waste products through decomposition. Microbes also have an important place in most higher-order multicellular organisms as symbionts. Soil is our most fundamental terrestrial asset. It provides, along with sunlight and water, the basis for all terrestrial life: the biodiversity around us, the field crops that we harvest for food and fiber, and animal products. Healthy soils provide us with a range of 'ecosystem services' they support healthy plant growth, resist erosion, receive and store water, retain nutrients and act as an environmental buffer in the landscape. Soils supply nutrients, water and oxygen to plants, and are populated by soil biota which are essential for decomposition and recycling processes.

Many microbes are uniquely adapted to specific environmental niches, such as those that inhabit the Dead Sea *(Halobacterium),* the bacteria and cyanobacteria that inhabit the boiling water springs in Yellowstone National Park, and *Chlamydomonas nivalis,* the green alga that causes "pink snow." Microbes also play an essential role in the natural recycling of living materials. All naturally produced substances are biodegradable, which means that they can be broken down by living things, such as bacteria or fungi. Composting is biodegradation at its best. An examination of conditions that faster or impede composting gives insight to growth conditions of microorganisms as well as the proper functioning of the ecosystem.

Soils are the lifeblood of our country and our society, but many of us don't know the first thing about soil structure, nutrient ratios or humus. Somewhat behind the scenes, soil erosion, nutrient depletion, water scarcity, salinity and the disruption of biological cycles have led to widespread land degradation, a fundamental and persistent problem. Healthy soils deliver productive agricultural yields, support ecosystems and biodiversity, stabilize climate through carbon storage, and are essential for maintaining environmental, economic and political stability. Applying compost to our soils is a critical step in restoring and maintaining the health of our soils. Soil is our most fundamental terrestrial asset. It provides, along with sunlight and water, the basis for all terrestrial life: the biodiversity around us, the field crops that we harvest for food and fiber, and animal products (such as meat, milk, eggs, wool). Healthy soils provide us with a range of 'ecosystem services'–they support healthy plant growth, resist erosion, receive and store water, retain nutrients and act as an environmental buffer in the landscape. Soils supply nutrients, water and oxygen to plants, and are populated by soil biota which are essential for decomposition and recycling processes.

Keywords: *Azospirillum, Azotobacter, Soil, cyanobacteria, Mycorrhiza.*

Introduction

The uniqueness of microorganisms and their often unpredictable nature and biosynthetic capabilities, has made them likely candidates for solving particularly difficult problems in the life sciences and other fields as well. The various ways in which microorganisms have been used over the past 50 years to advance medical technology, human and animal health, food processing, food safety and quality, genetic engineering, environmental protection, agricultural biotechnology, and more effective treatment of agricultural and municipal wastes provide a most impressive record of achievement. Many of these technological advances would not have been possible using straight forward chemical and physical engineering methods, or if they were, they would not have been practically or economically feasible.

Nevertheless, while microbial technologies have been applied to various agricultural and environmental problems with considerable success in recent years, they have not been widely accepted by the scientific community because it is often difficult to consistently reproduce their beneficial effects. Microorganisms are effective only when they are presented with suitable and optimum conditions for metabolizing their substrates Including available water, oxygen (depending on whether the microorganisms are obligate aerobes or facultative anaerobes), pH and temperature of their environment. Meanwhile, the various types of microbial cultures and inoculants available in the market today have rapidly increased because of these new technologies. Significant achievements are being made in systems where technical guidance is coordinated with the marketing of microbial products. Since, microorganisms are useful in eliminating problems associated with the use of chemical fertilizers and pesticides, they are now widely applied in nature farming and organic agriculture.

India is an agricultural country. Nearly 70 per cent of the populations thrive in rural areas, engaged in agriculture making the backbone of our economy. To meet the tremendous increase in food production, chemical fertilizers are used to enhance the crop production. However, chemical fertilizers and chemical pesticides have changed the agriculture scenario in the world.

Microbes play a vital role as organic fertilizers in facilitating uptake of nutrients in a crop. Rhizobia and blue-green algae fix atmospheric nitrogen and reduce dependency upon artificial application of chemical fertilizers.

Microorganisms appear in many different forms, the most popular types being fungi and bacteria. These microorganisms along with the macroorganisms such as worms and beetles are all essentially responsible for "trash patrol" *i.e.* they decay organic field residue so that the plants can use the nutrients to grow healthier. Microorganisms are single celled organisms and one of the simplest known to man. They multiply very rapidly and this is vital to the soil. Bacteria come in many different forms. For example they can be round, spiral shape or rod shaped. Bacteria also contribute to how plants grow by secreting enzymes that are helpful to dissolve the material the plant can not use. The results are improved soil stability and structure. It is important to remember that plants use other things to grow as well such as sunlight, carbon dioxide and different chemicals that the earth provides to them. Although microorganisms play an important role in the growth of plants, they can still thrive without them.

Environmental pollution, caused by excessive soil erosion and the associated transport of sediment, chemical fertilizers and pesticides to surface and groundwater, and improper treatment of human and animal wastes, has caused serious environmental and social problems throughout the world. Often engineers have attempted to solve these problems using established chemical and physical methods. However, they have usually found that such problems cannot be solved without using microbial methods and technologies in coordination with agricultural production.

For many years, soil microbiologists and microbial ecologists have tended to differentiate soil microorganisms as beneficial or harmful according to their functions and how they affect soil quality, plant growth and yield, and plant health. Beneficial microorganisms are those that can fix atmospheric nitrogen, decompose organic wastes and residues, detoxify pesticides, suppress plant diseases and soil-borne pathogens, enhance nutrient cycling, and produce bioactive compounds such as vitamins, hormones and enzymes that stimulate plant growth. Harmful microorganisms are those that can induce plant diseases, stimulate soil-borne pathogens, immobilize nutrients, and produce toxic and putrescent substances that adversely affect plant growth and health.

Soil Health

Soil health is defined as "the capacity of a soil to function within ecosystem boundaries to sustain biological productivity, maintain environmental health, and promote plant and animal health." In some cases, the term "soil quality" may be used; the two terms have the same meaning. Soil properties that determine soil health include soil texture, depth of soil, infiltration, bulk density, water-holding capacity, soil organic matter, pH, electrical conductivity, microbial biomass, carbon and nitrogen, potentially mineralizable nitrogen, and soil respiration.

It is useful to compare agricultural soil with nearby soils under permanent vegetation such as trees. Soils with poor health often have inferior tilth, lower organic matter contents, few living organisms, and show signs of soil erosion, crusting, and soil compaction. Eventually, poor soil health results in problems with crop establishment, root growth, and crop yields. Increasing amounts of fertilizers, pesticides, and tillage are needed to maintain yields on poor-quality soil. In this chapter we are going to discuss some important soil properties that determine soil health. Several definitions of soil health have been proposed during the last decades. The term soil quality described the status of soil as related to agricultural productivity or fertility (Singer *et al.*, 2000). In the 1990s, it was proposed that soil quality was not limited to soil productivity but instead expanded to encompass interactions with the surrounding environment, including the implications for human and animal health. In this regard, several examples of definitions of soil quality have been suggested (Doran *et al.*, 1994). In the mid-1990s, the term soil health was introduced. For example, a programme to assess and monitor soil health used the terms quality and health synonymously to describe the ability of soil to support crop growth without becoming degraded or otherwise harming the environment (Acton *et al.*, 1995). Others broadened the definition of soil health to capture the ecological attributes of soil, and went beyond its capacity to simply produce particular crops. These attributes are chiefly associated with biodiversity, food web structure and functional measures. Doran *et al.*, (1997) proposed the following definition of soil health:

"The continued capacity of soil to function as a vital living system, within ecosystem and land-use boundaries, to sustain biological productivity, promote the quality of air and water environments, and maintain plant, animal and human health". Soil health thus focuses on the continued capacity of a soil to sustain plant growth and maintain its functions regardless of the fitness for any certain purposes. Examples of dynamic soil properties could be organic matter content, the number or diversity of organisms, and microbial constituents or products (Singer *et al.*, 2000).

Soil is a finite and non-renewable resource because regeneration of soil through chemical and biological weathering of underlying rock requires geological time (Huber *et al.*, 2001). Deterioration of soil, and thereby soil health, is of major concern for human, animal and plant health because air, groundwater and surface water consumed by humans can be adversely affected by mismanaged and contaminated soil (Singer *et al.*, 2000). As such, deteriorating soil health and the benefits of soil management has become of political concern. A healthy soil functions to buffer nutrients as well as contaminants and other solutes via sorption to or incorporation into clay particles and organic materials. The soil itself thus serves as an environmental filter for removing undesirable solid and gaseous constituents from air and water (Parr *et al.*, 1992). The extent to which a soil immobilizes or chemically alters substances that are toxic, thus effectively detoxifying them, reflects the degree of soil health in the sense that humans or other biological components of the system are protected from harm (Singer *et al.*, 2000).

Biological Indicators Used for Determining Soil Health

Indicators of soil health have been defined as measurable surrogates for environmental processes that collectively tell us whether the soil is functioning normally (Pankhurst *et al.*, 1997).

The concept of soil health refers to the biological, chemical, and physical features necessary for long-term, sustainable agricultural productivity with minimal environmental impact. Thus, soil health provides an overall picture of soil functionality. Healthy soils maintain a diverse community of soil organisms that help to: (*i*) control plant diseases as well as insect and weed pests; (*ii*) form beneficial symbiotic associations with plant roots (*e.g.* nitrogen-fixing bacteria and mycorrhizal fungi); (*iii*) recycle plant nutrients; (*iv*) improve soil structure with positive repercussions for its water and nutrient-holding capacity; and (*v*) improve crop production. One of the most important objectives in assessing the health of a soil is the establishment of indicators for evaluating its current status.

A microbial parameter that represents properties of the environment or impacts to the environment, which can be interpreted beyond the information, that the measured or observed parameter represents by itself (Christensen, 1992).

Soil health encompasses a diverse set of microbial measurements due to bacteria, fungi and protozoan indicators. They are grouped according to the different soil health parameters of the ecosystem that is biodiversity, carbon cycling, nitrogen cycling, biomass, microbial activity, key species and bioavailability. The indicators relate to the ecosystem (*e.g.* processes), community (*e.g.* biomass and biodiversity) or population (*e.g.* species or functions) levels and this relationship is noted together with relations to policy-relevant end point. Microorganisms possess the ability to give an integrated measure of soil health, an aspect that cannot be obtained with physical/chemical measures and/or analyses of diversity of higher organisms. Microorganisms respond quickly to changes, hence they rapidly adapt to environmental conditions. The microorganisms that are best adapted will be the ones that flourish. This adaptation potentially allows microbial analyses to be discriminating in soil health assessment, and changes in microbial populations and activities may therefore function as an excellent indicator of change in soil health (Kennedy *et al.*, 1995). Microorganisms also respond quickly to environmental stress compared to higher organisms, as they have intimate relationship with their surroundings due to their high surface to volume ratio. In some instances, changes in microbial populations or activity can precede detectable changes in soil physical and chemical properties, thereby providing an early sign of soil improvement or an early warning of soil degradation.

The bioavailability of chemicals, *e.g.* heavy metals or pesticides, is also an important issue of soil health because of its connection with microbial activities. The impact of such chemicals on soil health is dependent on microbial activities. It has been shown that the bioavailability of poly-aromatic hydrocarbons was lower in autumn compared to early spring due to a higher microbial activity after the growing season. Therefore, the total content of chemicals in soil is not a reliable indicator of its bioavailability (Logan, 2000) and thereby soil health. Instead, bioavailability has to be measured in relation to bioassays and specific microbial processes.

Soil Microorganisms

Soil is the outer most covering of the earth's crust, which consists of loosely arranged layers of materials composed of inorganic and organic constituents in different stages of organization. It is a natural medium in which plants grow, multiply and die thus providing a perennial source of organic matter, which could be recycled for plant nutrition. It provides the physical support for the anchorage of root system and also serves as a reservoir of air, water and nutrients, which are essentially required for growth of the plants. The portion of earth beneath the soil is known as bed rock and it does not contribute directly to the growth of plants (Subba Rao, 1986).

Over the last billion and more years, millions of biological species have evolved on the soil and earth and there exists certain specific relationship between soil, plants and microorganisms (Rangaswami, 1988). However, integrated knowledge of different aspects of soil physics, chemistry, biology and topography is not very wide spread. As a consequence, India is facing soil abuse for one or the other reason. The care and maintenance of soil health is therefore, essential for safeguarding the future of our agriculture.

Furthermore, there is a considerable pressure on land because of the accelerated developmental needs and rapid increase in population, which is growing at an alarming rate of 18 per cent per year. It is expected to reach a billion by 2025 and 9.4 billion by 2050 (Husle, 1995; Fisher and Helig, 1997; Litvin, 1998). It is therefore imperative that in India having a very high man to land and animal to land ratio, attention should be given for the reclamation of the less fertile and problem soils that are at present lying fallow.

In addition, excessive use of inorganic fertilizer and plant protection chemicals for maximizing crop yield and change in traditional cultivation practices, resulted in deterioration of physical, chemical and biological health of cultivated land. Green revolution in India during 1970's brought about an increase in agriculture production of various crop. The introduction of various cultivation practices with the intensive use of fertilizers, pesticides and other inputs had greatly polluted the soil, water and environment. Consequently, this has directly or indirectly affected the human health. The contamination and degradation coupled with other activities have increased the salt affected area.

Soil eco-system, a major resource of sustaining agriculture procedure is the ecologically vulnerable resource and to fulfill the demand of food supply for rapidly growing population, the soil/land system is under great stress. Salt affected soils affects the economy and livelihood security of people in more than hundred countries as they occupy about 831.4 million hectare around the world, of this 397 million hectare (47.8 per cent) is saline while 434.3 million hectare (52.2 per cent) is sodic (Tyagi, 2004).

Besides sustaining agriculture, soil is a natural habitat for all types of microorganisms beneficial as well as harmful to plants and animals. The soil microorganisms beneficial for crop production, protection and maintenance of soil fertility are known for centuries and soils differing in physical and chemical attitudes support diverse forms of microorganisms.

The soil microflora include bacteria, fungi, protozoa, algae and viruses. The microbial population is found in the upper layer of soil, top soil (10-30 cm) in depth. The number of microorganisms decreases with the depth of the soil and after about 150 cm very little or no microorganisms are found. Fertile loam soil of the upper layer may contain anywhere from 10,00,00 to a billion live microorganisms. Each soil is distinct in the composition of micro-flora, even though microflora of the soil of the world are broadly similar.

Among the different group of microorganisms inhabiting the soil, bacteria are found to be most abundant. There exists a great nutritional and physiological versatility in soil bacteria.

In addition, soil also contains a large population of actinomycetes, which are filamentous bacteria known to produce various antibiotics. The bacteria along with other microorganisms in the soil play an important role in the bio-geochemical cycling and soil fertility.

Bacteria in Soil

Many types of bacteria are found in the soil. Of these some are autotrophic and utilize inorganic compounds for their energy and growth. The majority of bacteria, however, are heterotrophic and utilize large amounts of organic matter. These belong to the order eubacleriales and actinomycetales. The later group is most frequently represented by organisms of the genera *Streptomyces, Nocardia and Micromonospora*. The important group includes;

Nitrogen-fixing Bacteria

Biological nitrogen-fixation is estimated to contribute 180×10^6 metric tons of fixed nitrogen/year globally of which 80 per cent comes from symbiotic associations and the rest from free-living or associative systems (Postgate, 1998). The ability to reduce and siphon out such appreciable amounts of nitrogen from the atmospheric reservoir and enrich the soil is confined to bacteria and archaea (Young, 1992). These include, a) symbiotic nitrogen fixing (N_2-fixing) forms, *viz. Rhizobium,* the obligate symbionts in leguminous plants and *Frankia* in non-leguminous trees, and b) Non-symbiotic (free-living, associative or endophytic) N_2-fixing forms such as cyanobacteria, *Azospirillum, Azotobacter, Acetobacter diazotrophicus, Azoarcus,* etc.

Symbiotic Nitrogen Fixers

Microorganisms that form the symbiotic association with other organisms or plants and fix the atmospheric nitrogen (N_2) are known as symbiotic nitrogen fixers. Two groups of nitrogen fixing bacteria, *i.e.* rhizobia and *Frankia* have been studied extensively. *Frankia* forms vesicles (the site for aerobic N_2-fixation) on more than 280 species of woody plants from 8 different families (Schwintzer and Tjepkema, 1990), however, its symbiotic relationship is not yet well understood. Species of *Alnus* and *Casuarina* are globally known to form effective symbiosis with *Frankia* (Dommergues and Marco-Bosco, 1998). In India, a technique for isolation of *Frankia* by single spore culture technique was developed, and PCR-RFLP markers were identified for screening actinorrhizal symbionts (Varghese *et al.*, 2003). In the context of rhizobia, considerable change in taxonomic status has come out during the last years. Sahgal and Johri (2003) outlined the current status of rhizobial taxonomy and enlisted 36 species distributed among seven genera (*Allorhizobium, Azorhizobium, Bradyrhizobium, Mesorhizobium, Methylobacterium, Rhizobium* and *Sinorhizobium*) based on the polyphasic taxonomic approach.

Although most *Rhizobium* isolates can nodulate more than one host species and also several different bacterial species are often isolated from a single legume, it is only from a few legumes that the symbionts have, so far, been investigated thoroughly (Sprent *et al.*, 1987). The family Fabaceae (formerly Leguminoseae) is important both ecologically and agriculturally, since it is a major source of biological nitrogen-fixation (Akkermans and Van Dijk, 1981). Species of *Parasponia* and *Tremma* are the only non-legumes that form an effective symbiosis with *Rhizobium* or *Bradyrhizobium*. There appears a common evolutionary origin, as on the basis of chloroplast genome sequence data they all form a single clad within the angiosperms (Soltis *et al.*, 1995). A few aquatic legumes bear stem nodules in addition to the normal root nodules. This peculiarity is restricted to 15 of the 250 species of *Aeschynomene*, and 1 out of 15 species of *Neptunia i.e. N. oleracea*.

Non-symbiotic Nitrogen Fixers

Non-symbiotic nitrogen fixation is known to be of great agronomic significance. The main limitation to non-symbiotic nitrogen-fixation is the availability of carbon and energy source for the energy intensive nitrogen fixation process. This limitation can be compensated by moving closer to or inside the plants *viz.* in diazotrophs present in rhizosphere, rhizoplane or those growing endophytically. Some important nonsymbiotic nitrogen-fixing bacteria include, *Achromobacter, Acetobacter, Alcaligenes, Arthrobacter, Azospirillum, Azotobacter, Azomonas, Bacillus, Beijerinckia, Clostridium, Corynebacterium, Derxia, Enterobacter, Herbaspirillum, Klebsiella, Pseudomonas, Rhodospirillum, Rhodopseudomonas* and *Xanthobacter* (Saxena and Tilak, 1998). Some of them are:

Azotobacter

The family azotobacteriaceae comprises of two genera (Tchan, 1984) namely, *Azomonas* (non-cyst forming) with three species (*A. agilis, A. insignis* and *A. macrocytogenes*) and *Azotobacter* (cyst forming) comprising of 6 species (Tchan,1984), namely, *A. chroococcum, A. vinelandii, A. beijerinckii, A. nigricans, A. armeniacus* and *A. paspali*. *Azotobacter* is generally regarded as a free-living aerobic nitrogen-fixer. *Azotobacter paspali* which was first described by Dobereiner and Pedrosa, (1975), has been isolated from the rhizosphere of *Paspalum notatum*, a tetraploid subtropical grass, and is highly host specific. Various crops in India have been inoculated with diazotrophs particularly *Azotobacter* and *Azospirillum* (Saxena and Tilak, 1999). Application of *Azotobacter* and *Azospirillum* has been reported to improve the yields of both annual and perennial grasses. Saikia and Bezbaruah (1995) reported increased seed germination of *Cicer arietinum, Phaseolus mungo, Vigna catjung* and *Zea mays*. However, yield improvement is attributed more to the ability of *Azotobacter* to produce plant growth promoting substances such as phytohormone Indole acetic acid (IAA) and siderophore azotobactin, rather than to diazotrophic activity.

Azospirillum

Members of the genus *Azospirillum* fix nitrogen under microaerophilic conditions, and are frequently associated with root and rhizosphere of a large number of agriculturally important crops and cereals. Due to their frequent occurrence in the rhizosphere, these are known as associative diazotrophs. Sen (1929) made one of the earliest suggestions that the nitrogen nutrition of cereal crops could be met by the activity of associated nitrogen fixing bacteria such as *Azospirillum*. This organism came into focus with the work of Dobereiner and associates from Brazil (Dobereiner *et al.*, 1976), followed closely by reports from India (Tilak and Murthy, 1981). After establishing in the rhizosphere, azospirilla usually, but not always, promote the growth of plants. Despite their N_2-fixing capability (~1–10 kg N/ha), the increase in yield is mainly attributed to improved root development due to the production of growth promoting substances and consequently increased rates of water and mineral uptake (Fallik *et al.*, 1994).

Acetobacter

The family acetobacteriaceae includes genera, *Acetobacter, Gluconobacter, Gluconoacetobacter* and *Acidomonas* (Yamada *et al.*, 1997). Based on 16S rRNA sequence analysis, the name *Acetobacter diazotrophicus* has been changed to *Gluconoacetobacter diazotrophicus* (Yamada *et al.*, 1998). In addition to *G. diazotrophicus*, two more diazotrophs, *G. johannae* and *G. azotocaptans* have been included in the list (Fuentez-Ramirez *et al.*, 1997). The genetic diversity of *G. diazotrophicus* isolated from various sources does not exhibit much variation (Cabellaro-Mellado and Martinez Romero, 1994). However, Suman *et al.*, (2001) found that the diversity of the isolates of *G. diazotrophicus* by RAPD analysis was more conspicuous than that reported on the basis of morphological and biochemical characters. The SDS-PAGE and multilocus enzyme electrophoresis analysis also revealed certain differences among strains of *G. diazotrophicus* suggesting genotypic differences (Vandamme *et al.*, 1996). Investigations of isolates of *G. diazotrophicus* from pineapple suggested that only certain genetically related groups of this bacterium or its ancestors have acquired the capability of colonizing plants by themselves or with the aid of the vectors such as insects or fungi (Tapia-Hernandez *et al.*, 2002). *G. diazotrophicus* has been found to harbour plasmids of 2–170 kb.

Azoarcus

Azoarcus, an aerobic/microaerophilic nitrogen-fixing bacterium was isolated from surface-sterilized tissues of kaller grass (*Leptochloa fusca* (L.) Kunth) (Reinhold *et al.*, 1986), and can infect roots of rice plants as well. Kallar grass is a salt-tolerant grass used as a pioneer plant in Pakistan on salt-affected low fertility soils. Repeated isolation of one group of diazotrophic rods (Reinhold *et al.*, 1993) from kallar grass roots and the results of polyphasic taxonomy led to the identification of genus *Azoarcus*, with two species, *A. indigens* and *A. communis*. Nitrogen-fixation by *Azoarcus* is extremely efficient (specific nitrogenase activity, one order of magnitude higher than those found for bacteroids). Such hyper-induced cells contain tubular arrays of internal membrane stacks that can cover a large proportion of the intercellular volume. These structures are considered as vital for high efficiency N_2-fixation (Reinhold *et al.*, 1986).

Phosphate Solubilizing Microorganisms

Phosphorus (P) is major essential macronutrients for biological growth and development. P in soils is immobilized or becomes less soluble either by absorption, chemical precipitation, or both. A survey of Indian soils revealed that 98 per cent of these need phosphorus fertilization either in the form of chemical or biological fertilizer. Although P content in an average soil is 0.05 per cent, only 0.1

per cent of the total P present is available to the plants because of its chemical fixation and low solubility. Application of chemical phosphatic fertilizers is practiced, though a majority of the soil P reaction products are only sparingly soluble. Under such conditions, microorganisms offer a biological rescue system capable of solubilizing the insoluble inorganic P of soil and make it available to the plants.

Phosphate solubilizing microorganisms (PSM) include largely bacteria and fungi, which can grow in media containing tricalcium, iron and aluminium phosphate, hydroxyapatite, bonemeal, rock phosphate and similar insoluble phosphate compounds as the sole phosphate source. Such microbes not only assimilate P but a large portion of soluble phosphate is released in quantities in excess of their own requirement (Gaur, 1990). The most efficient PSM belong to genera *Bacillus* and *Pseudomonas* amongst bacteria and *Aspergillus* and *Penicillium* amongst fungi. The reported bacilli include, *B. brevis, B. cereus, B. circulans, B. firmus, B. licheniformis, B. megaterium, B. mesentricus, B. mycoides, B. polymyxa, B. pumilis, B. pulvifaciens* and *B. subtilis* from the rhizosphere of legumes, cereals (rice and maize), arecanut palm, oat, jute and chilli (Kole and Hajra, 1998). *Pseudomonas striata, P. cissicola, P. fluorescens, P. pinophillum, P. putida, P. syringae, P. aeruginosa, P. putrefaciens* and *P. stutzeri* have been isolated from rhizosphere of *Brassica*, chickpea, maize, soybean and other crops, desert soils and Antarctica lake (Gupta *et al.*, 1998). In addition, *Escherichia freundii, E. intermedia, Serratia phosphaticum* and species of *Achromobacter, Brevibacterium, Corynebacterium, Erwinia, Micrococus, Sarcina* and *Xanthomonas* are active in solubilizing insoluble phosphates. Cyanobacteria, *viz. Anabaena* sp., *Calothrix brauni, Nostoc* sp., *Scytonema* sp. and *Tolypothrix ceylonica* can also solubilze phosphate (Gupta *et al.*, 1998). Among phosphate solubilizing fungi, *Aspergillus niger, A. flavus, A. nidulans, A. awamori, A. carbonum, A. fumigatus, A. terreus* and *A. wentii* have been reported from the rhizosphere of maize, soybean, chilli, tista soils, acidic lateritic soils and compost (Prerna *et al.*, 1997). *Paeciliomyces fusisporus, Penicillium digitatum, P. simplicissimum, P. aurantiogriseum, Sclerotium rolfisii* and species of *Cephalosporium, Alternaria, Cylindrocladium, Fusarium* and *Rhizoctonia* are other solubilizers of insoluble phosphate. Amongst yeasts, *Torula thermophila, Saccharomyces cerevisiae* and *Rhodotorula minuta* can solubilize inorganic phosphate.

Other Plant Growth Promoting Rhizobacteria

Other microorganisms that are known to be beneficial to plants are the plant growth promoting rhizobacteria (PGPR). In addition to supplying combined nitrogen by biological nitrogen-fixation, certain bacteria affect the development and function of roots by improving mineral (NO_3^-, PO_3^{3-} and K^+) and water uptake. Considerable research is underway globally to exploit the potential of one such group of bacteria that belong to fluorescent pseudomonad (FLPs). FLPs help in maintenance of soil health, protect crop from pathogens and are metabolically and functionally most active (Lata *et al.*, 2002). *P. corrugata*, a form that grows at 4°C under laboratory conditions (Pandey and Palni, 1998), produces antifungals such as diacetylphloroglucinol and/or phenazine compounds that aid in phosphate solubilization.

Cyanobacteria

Cyanobacteria evolved very early in the history of life, and share some of the characteristics of gliding bacteria on one hand and those of higher plants on the other. Cyanobacteria can photosynthesize and fix nitrogen, and these abilities, together with great adaptability to various soil types, make them ubiquitous. Cyanobacteria are the organisms that are very important in the formation of biological soil crusts. They are photosynthetic, and live within the first ten inches of top soil. Cyanobacteria help to reduce erosion by helping in binding the particles of soil together. When conditions are wet they become active and move through the soil, leaving behind a sticky filaments to which particles of soil cling to. When the filaments become wet, they absorb water and swell up to ten times to their original size which helps to store moisture within the upper layer of soil where many plants root systems and other organisms live. Cyanobacteria also play a more direct role in aiding plant survival and growth. Cyanobacteria are important because they help to convert atmospheric nitrogen into an assimilable form (NH_3) that can be utilized by plants. Cyanobacteria form symbiotic relationships ranging from algae to angiosperms. They form microbiotic crusts in intimate association with surface soil, which contribute significantly to the stabilisation of soil towards erosion (Eldridge *et al.*, 1994). Cyanobacteria have mainly been used as indicators of heavy metals contamination (*e.g.* from sewage sludge application) in soil. Most experiments have shown a negative correlation between the number of cyanobacteria or nitrogenase activity and the concentration of heavy metals (Brookes, 1995; Lorenz *et al.*, 1992; Dahlin *et al.*, 1997; Scherr *et al.*, 2001). It has also been noted that cyanobacteria may be too sensitive to experimental conditions to provide a robust indicator of heavy metal contamination (Brookes 1995; Lorenz *et al.* 1992).

Cyanobacteria also have a unique potential to contribute to productivity in a variety of agricultural and ecological situations. Cyanobacteria have been reported from a wide range of soils, thriving both on and below the surface. Most paddy soils have a natural population of cyanobacteria which provides a potential source of nitrogen fixation at no cost. Due to this important characteristic of nitrogen fixation, the utility of cyanobacteria in agriculture to enhance production is beyond doubt. Many studies have been reported on the use of dried cyanobacteria to inoculate soils as a means of aiding fertility, and the effect of adding cyanobacteria to soil on rice yield was first studied in the 1950s in Japan. The term 'algalization' is now applied to the use of a defined mixture of cyanobacterial species to inoculate soil, and research on algalization is going on in all major rice producing countries.

In recent years people throughout the world are focusing their attention towards cyanobacteria for their possible use for the photobiological production of biofuel, as biofertilizer, ammonia, amino acids, various metabolites, vitamins, toxins, therapeutic substances, aqua or animal feed. Various chemicals including restriction enzymes, pharmacological probes and labeled compounds for research, as well as fluorescent probes for clinical diagnostics are now commercially available (Rathore *et al.*, 2007).

Mycorrhiza

Majority of higher plants exist in natural symbiosis with mycorrhizal fungi. The group of mycorrhizal fungi includes ectomycorrhizal (mainly forest trees), arbuscular mycorrhizal (terrestrial plants) and ericoid mycorrhizal fungi. They colonise plant roots and provide the plant with nutrients, especially phosphorus, due to the increased nutrient availability caused by the extra-radical mycelium. Furthermore, mycorrhizal associations can have a positive influence on plant diversity (Allen *et al.*, 1995), plant stress and disease tolerance, and on soil aggregation. Colonization by AM has been shown to be highly dependent on the presence of host plants, on land use and on soil management practices (Kling *et al.*, 1998). Spore abundance and diversity have been shown to discriminate between extensively and intensively managed soils (Oehl *et al.*, 2001) and AM diversity has been reported to be sensitive to heavy metal contamination, organic pollutants and atmospheric deposition (Siciliano *et al.*, 1999; Cairney *et al.*, 1999; Egli *et al.*, 2001; Egerton-Warburton *et al.*, 2000).

Plant Pathogens

The presence of plant pathogens (*e.g.* fungi) in soil may indicate the existence of other soil health problems, *e.g.* nutrient imbalance (Hornby *et al.*, 1997). A suppressive soil is able to suppress specific plant diseases by inherent biotic and abiotic factors. The suppressiveness of a certain soil may thus be an indicator of plant health. Several methods are available for determining soil suppressiveness as reviewed by Van Bruggen and Grunwald (1996). It can be determined by inoculation of target-plant seeds directly into the test soil or into a pathogen-infested test soil. The frequency of diseased plants and/or pathogenic propagules in soil is scored after incubation for about 3 to 4 weeks and compared to a reference soil. The plant bioassay is a conventional technique and a positive correlation between the plant bioassay and the actual field measurement has been shown for suppressiveness of pea root rot (Persson *et al.*, 1999). A specific test plant system has to be selected for a monitoring programme and the correlation between bioassay and field measurements has to be confirmed on a diverse set of soils.

Human Pathogens

Human pathogens can enter into the agricultural soils through amendment with manure and sewage sludge. The presence of human pathogenic bacteria in soil is an indicator of potential human infection and as such an indicator of human health. Presence of *E. coli*, has traditionally been used as an indicator of faecal contamination (*e.g.* coastal waters) and hence as an indicator of the possible presence of other more pathogenic bacteria (Rhodes *et al.*, 1988). Since, the ability of the pathogenic bacteria to survive in the environment may not necessarily be equal to that of *E. coli* (Morales *et al.*, 1996), it would be advantageous if the pathogens were enumerated directly.

Microorganisms and their Role in Soil

Microorganisms, especially algae and lichen, are pioneer colonizers of barren rock surfaces. Colonization by these organisms begins the process of soil formation necessary for the growth of higher plants. After plants have been established, decomposition by microorganisms recycles the energy, carbon, and nutrients in dead plant and animal tissues into forms usable by plants. Therefore, microorganisms have a key role in the processing of materials that maintain life on the earth. The transformations of elements between forms are described conceptually as the elemental cycles.

In the carbon cycle, microorganisms transform plant and animal residues into carbon dioxide and the soil organic matter known as humus. Humus improves the water-holding capacity of soil, supplies plant nutrients, and contributes to soil aggregation. Microorganisms may also directly affect soil aggregation. The extent of soil aggregation determines the workability or tilth of the soil. A soil with good tilth is suitable for plant growth because it is permeable to water, air, and roots. which returns nitrogen to the atmosphere by transforming $NO_3^?$ to N_2 or nitrous oxide (N_2O) gas. Microorganisms are crucial to the cycling of sulfur, phosphorus, iron, and many micronutrient trace elements.

In addition to the elemental cycles, there are several interactions between plants and microbes which are detrimental or beneficial to plant growth. Some soil microorganisms are pathogenic to plants and cause plant diseases such as root rots and wilts. The region of soil surrounding plant roots, the rhizosphere, may contain beneficial microorganisms which protect the plant root from pathogens or supply stimulating growth factors. The interactions between plant roots and soil microorganisms is an area of active research in soil microbiology and needs greater attention by the soil microbiologists.

The numerous natural substances that are used by microorganisms indicate that soil microorganisms have diverse mechanisms for degrading a variety of compounds. Human activity has polluted the environment with a wide variety of synthetic or processed compounds. Many of these hazardous or toxic substances can be degraded by soil microorganisms. This is the basis for the treatment of contaminated soils by bioremediation, the use of microorganisms or microbial processes to detoxify and degrade environmental contaminants. Soil microbiologists study the microorganisms, the metabolic pathways, and the controlling environmental conditions that can be used to eliminate pollutants from the soil environment.

The biological activity in soil is largely concentrated in the top soil, the depth of which may vary from a few to 30 cm. In top soil, the biological components occupy a tiny fraction (<0.5 per cent) of the total soil volume and make up less than 10 per cent of the total organic matter in soil. These biological components consist mainly of soil organisms, especially microorganisms. Despite their small volume in soil, microorganisms are key players in the cycling of nitrogen, sulphur, and phosphorus, and the decomposition of organic residues. Thereby they affect nutrient and carbon cycling on a global scale (Pankhurst *et al.*, 1997). That is, the energy input into the soil ecosystems is derived from the microbial decomposition of dead plant and animal organic matter. The organic residues are, in this way, converted to biomass or mineralized to CO_2, H_2O, mineral nitrogen, phosphorus, and other nutrients. Mineral nutrients immobilized in microbial biomass are subsequently released when microbes are grazed by microbivores such as protozoa and nematodes (Bloem *et al.*, 1997). Microorganisms are further associated with the transformation and degradation of waste materials and synthetic organic compounds (Torstensson *et al.*, 1998).

In addition to the effect on nutrient cycling, microorganisms also affect the physical properties of soil. Production of extra-cellular polysaccharides and other cellular debris by microorganisms help in maintaining soil structure, as these materials function as cementing agents that stabilize soil aggregates. Thereby, they also affect water holding capacity, infiltration rate, crusting, erodibility, and susceptibility to compaction (Elliott *et al.*, 1996).

Methane Oxidation

Methane (CH_4) is found extensively in nature and is a greenhouse gas in the atmosphere. Methane is produced in anoxic environments by methanogenic archaea and consumed by aerobic methane oxidizing bacteria, the methanotrophs (Ritchie *et al.*, 1997). Important terrestrial sites for methane oxidation are wetland areas receiving a high input of organic material. Furthermore, landfills containing high amounts of organic wastes are a source of methane and the habitat of many methanotrophs (Ritchie *et al.*, 1997). Net production of methane can be considered as an indicator of greenhouse gas emission and may further be linked to monitoring of the atmospheric balance.

Methanogens involves a complex mixture of anaerobic bacteria, which convert up to 90 per cent of the combustible energy of the degradable matter to CH_4 and CO_2. Approximately 65 per cent of the methane released to the atmosphere is produced by methanogens.

Nitrogen Cycling

Soil microorganisms play key roles in the nitrogen cycle. The atmosphere is approximately 78 per cent nitrogen gas (N_2), a form of nitrogen that is available to plants only when it is transformed to ammonia (NH_3) by either soil bacteria (N_2 fixation) or by humans (manufacture of fertilizers). Soil bacteria also mediate denitrification, the mineralization of soil organic nitrogen through nitrate to gaseous N_2 by soil microorganisms, is a very important process in global N^-cycling. This cycle includes nitrogen mineralization, nitrification, denitrification and N_2^-fixation. Ammonium is subsequently either immobilized by soil microorganisms (that is, assimilated into new biomass) or oxidized to nitrite (NO_2^-) and subsequently to nitrate (NO_3^-) by aerobic nitrification. Chemoautotrophic bacteria, the nitrifier population, carry out this process. At this step, leaching of N to the groundwater may occur due to the negative charge of the nitrate ion. Under normal circumstances, however, nitrate is subsequently reduced to gaseous nitrogen (N_2) via nitrous oxide (N_2O) by anaerobic denitrification. Denitrification is represented by a variety of soil bacteria (Zumpft, 1992). Nitrification and denitrification together lead to losses of bioavailable N since nitrous oxide and gaseous N_2 may be lost to the atmosphere. N_2 can be re-fixed into the soil by N_2-fixing microorganisms. Nitrous oxide is a greenhouse gas when lost to the atmosphere.

The denitrifying capacity is a widespread feature among soil bacteria and therefore denitrification can be used as a representative for microbial biomass (Stenberg, 1999). Since, denitrification is an anaerobic process, the amount of denitrification found in soil is very much dependent on abiotic factors such as precipitation and soil compaction. Thus, soil management practices readily influence the amount of denitrification found in agricultural fields. Denitrification measurements together with nitrification measurements may indicate the deposition of ammonia in N-limited habitats. Denitrification is considered to be a wasteful process in agricultural context, since more than 50 per cent of chemical fertilizers are denitrified before it reaches to the crop plants.

References

Acton DF and Gregorich EG (1995). Executive summary. In: *The Health of our Soils Towards Sustainable Agriculture in Canada* (Eds) Acton DF and Gregorich EG. Centre for Land and Biological Resources Research, Research Branch Agriculture and Agri-food, Canada, p. 111–120.

Akkermans ADL and Van Dijk C (1981). Non-leguminous root nodule symbioses with actinomycetes and rhizobia. In: *Nitrogen Fixation*, (Ed) WJ Broughton. Cambridge University Press, Cambridge, p. 57–75.

Allen EB, Allen MF, Helm DJ, Trappe JM, Molina R, and Rincon E (1995). Patterns and regulation of mycorrhizal plant and fungal diversity. *Plant and Soil*, 170: 47–62.

Bloem J, de Ruiter P and Bouwman LA (1997). Food webs and nutrient cycling in agro-ecosystems. In: *Modern Soil Microbiology*, (Eds) Van Elsas JD, Trevors JT and Wellington EMH. Marcel Dekker Inc, New York, p. 245–278.

Brookes PC (1995). The use of microbial parameters in monitoring soil pollution by heavy metals. *Biology Fertility Soils*, 19: 269–279.

Cabellaro-Mellado J and Martinez-Romero E (1994). Limited genetic diversity in the endophytic sugarcane bacterium *Acetobacter diazotrophicus*. *Appl Environ Microbiol*, 60: 1532–1537.

Cairney JWG and Meharg AA (1999). Influences of anthropogenic pollution on mycorrhizal fungal communities. *Environ Pollution*, 106: 169–182.

Christensen H, Griffiths B and Christensen S (1992). Bacterial incorporation of tritiated-thymidine and populations of bacteriophagous fauna in the rhizosphere of wheat. *Soil Biol Biochem*, 24: 703–709.

Dahlin S, Witter E, Mart A, Turner A and Baath E (1997). Where's the limit? Changes in the microbiological properties of agricultural soils at low levels of metal contamination. *Soil Biol Biochem*, 29: 1405–1415.

Dobereiner J, Marriel JE and Nery M (1976). Ecological distribution of *Spirillum lipoferum* Beijerinck. *Can J Microbiol*, 22: 1464–1473.

Dommergues YR and Marco-Bosco (1998). The contribution of N_2 fixing trees to soil productivity and rehabilitation in tropical, subtropical and Mediterranean regions. In: *Microbial Interactions in*

Agriculture and Forestry, (Eds) Subba Rao NS and Dommergues YR. Oxford and IBH, New Delhi, p. 65–96.

Doran JW and Parkin TB (1994). Defining and assessing soil quality. In: *Defining Soil Quality for a Sustainable Environment,* (Eds) Doran JW, Coleman DC, Bezdicek DF and Stewart BA. Soil Sci Soc America, Inc, Madison, p. 3–21.

Doran JW and Safley M (1997). Defining and assessing soil health and sustainable productivity. In: *Biological Indicators of Soil Health,* (Eds) Pankhurst CE, Doube BM and Gupta VVSR. CAB International, p. 1–28.

Egerton-Warburton LM and Allen EB (2000). Shifts in arbuscular mycorrhizal communities along an anthropogenic nitrogen deposition gradient. *Ecol Applications,* 10: 84–496.

Egli S and Mozafar A (2001). Eine standard methode zur Erfassung des Mykkorrhiza- fektionspotenzials in Land wirtschaftsböden. *VBB Bulletin,* 5: 6–7.

Eldridge DJ and Greene RSB (1994). Microbiotic soil crusts: A review of their roles in soil and ecological processes in the rangelands of Australia. *Australian J Soil Res,* 32: 389–415.

Elliott LF, Lynch JM and Papendick RI (1996). The microbial component of soil quality. In: *Soil Biochemistry,* (Eds) Stotzky G and Bollag JM. Marcel Dekker Inc, New York, p. 1–21.

Fallik E, Sarig S and Okon Y (1994). Morphology and physiology of plant roots associated with *Azospirillum.* In: *Azospirillum.* Plant Associations, (Ed) Okon Y. CRC Press, Boca Raton, Florida, p. 77–84.

Fisher G and Helig GK (1997). *Population momentum* and the demand on land and water resources. *Phil Trans Roy Soc London,* Series B, 352: 869–889.

Fuentez-Ramirez LE, Cabellaro-Mellado J, Sepulveda- Sanchez J and Martinez-Romero E (1997). Location of *Acetobacter diazotrophicus* in inoculated sugarcane by GUS detection. In: *11th International Congress on Nitrogen Fixation,* Abs, p. 18.

Gaur AC (1990). *Phosphate Solubilizing Microorganisms as Biofertilizers.* Omega Scientific Publishers, New Delhi, p. 176.

Gupta RP, Vyas MK and Pandher MS (1998). Role of phosphorus solubilizing microorganisms in P–economy and crop yield. In: *Soil-Plant-Microbe Interaction in Relation to Nutrient Management,* (Ed) Kaushik BD. Venus Printers and Publishers, New Delhi, p. 95–101.

Hornby D and Bateman GL (1997). Potential use of plant root pathogens as bioindicators of soil health. In: *Biological Indicators of Soil Health,* (Eds) Pankhurst CE, Doube BM and Gupta VVSR. CAB International, p. 179–200.

Husle JH (1995). *Science Agriculture and Food Security.* NRC Press, Ottawa, Canada.

Huber S, Syed B, Freudenschuss A, Ernstsen V and Loveland P (2001). Proposal for a European soil monitoring and assessment framework. Technical report no. 61, European Environment Agency, Copenhagen, Denmark.

Kennedy AC and Papendick RI (1995). Microbial characteristics of soil quality. *J Soil Water Conservation,* 50: 243–248.

Kling M and Jakobsen I (1998). Arbuscular mycorrhiza in soil quality assessment. *Ambio,* 27: 29–34.

Kole SC and Hajra JN (1998). Occurrence and acidity of tricalcium phosphate and rock phosphate solubilizing microorganisms in mechanical compost plants of Calcutta and an alluvial soil of West Bengal. *Environ Ecol,* 16(2): 344–349.

Lata, Saxena AK and Tilak KVBR (2002). Biofertilizers to augment soil fertility and crop production. In: *Soil Fertility and Crop Production,* (Ed) Krishna KR. Science Publishers, USA, p. 279–312.

Litvin D (1998). Drit poor. *Economist,* 21: 3–16.

Logan TJ (2000). Soils and environmental quality. In: *Handbook of Soil Science,* (Ed) Sumner M E. CRC Press, Boca Raton, Florida, p. G155–G169.

Lorenz SE, McGrath SP and Giller KE (1992). Assessment of free-living nitrogen-fixation activity as a biological indicator of heavy-metal toxicity in soil. *Soil Biol Biochem,* 24: 601–606.

Morales A, Garland JL and Lim DV (1996). Survival of potentially pathogenic human-associated bacteria in the rhizosphere of hydroponically grown wheat. *FEMS Microbiol Ecol,* 20: 155–162.

Oehl F, Ineichen K, Mader P and Wiemken A (2001). Einfluss der landwirtschaftlichen Nutzungsintensität auf die Diversität der arbuskulären Mykorrhizapilze. *VBB Bulletin,* 5: 10–12.

Pandey A and Palni LMS (1998). Isolation of *Pseudomonas corrugata* from Sikkim, Himalaya. *World J Microbiol Biotechnol,* 14: 411–413.

Pankhurst CE, Doube BM and Gupta VVSR (1997). Biological indicators of soil health-a synthesis. In: *Biological Indicators of Soil Health,* (Eds) Pankhurst CE, Doube BM and Gupta VVSR. CAB International, p. 419–435.

Parr JF, Papendick RI, Hornick SB and Meyer RE (1992). Soil quality: Attributes and relationship to alternative and sustainable agriculture. *American J Altern Agriculture,* 7: 5–11.

Persson L, Larsson-Wikstrom M and Gerhardson B (1999). Assessment of soil suppressiveness to *Aphanomyces* root rot of pea. *Plant Disease,* 83: 1108– 1112.

Postgate, J (1998). *Nitrogen Fixation,* 3rd Edition. Cambridge University Press, Cambridge UK.

Prerna A, Kapoor KK and Akhaury P (1997). Solubilization of insoluble phosphate by fungi isolated from compost and soil. *Environ Ecol,* 15(3): 24–527.

Rangaswami G (1988). Soil plant-microbe interrelationship. *Indian Phytopath,* 41(2): 165–172.

Rathore NK, Raghuvanshi R and Singh S (2007). Nutraceutical, pharmaceutical and bioactive potential of cyanobacteria. In: *Advances in Applied Phycology,* (Eds) Gupta R and Pandey VD. Daya Publishing House, New Delhi, p. 275–293.

Reinhold B, Hurek T, Niemann EG and Fendrik I (1986). Close association of *Azospirillum* and diazotrophic rods with different root zones of Kallar grass. *Appl Environ Microbiol,* 52: 520.

Reinhold B–Hurek, Hurek T, Gillis M, Hoste B, Vancanneyt M, Kersters K and DeLey J (1993). *Azoarcus* gen. nov. nitrogen fixing proteobacteria associated with roots of Kallar grass [*Leptochloa fusca* (L.) Kunth] and description of two species, *Azoarcus indigens* sp. nov. and *Azoarcus communis* sp. nov. *Int J Syst Bacteriol,* 43: 574–584.

Rhodes MW and Kator H (1988). Survival of *Escherichia coli* and *Salmonella* spp. in estuarine environments. *Appl Environ Microbiol,* 54: 2902–2907.

Ritchie DA, EdwardsC, McDonald IR and Murrell JC (1997). Detection of methanogens and methanotrophs in natural environments. *Global Change Biol,* 3: 339–350.

Sahgal M and Johri BN (2003). The changing face of rhizobial systematics. *Curr Sci,* 84: 43–48.

Saikia N and Brezbaruah B (1995). Iron-dependent plant pathogen inhibition through *Azotobacter* RRLJ 203 isolated from iron-rich acid soils. *Indian J Exp Biol,* 33: 571–575.

Saxena AK and Tilak KVBR (1998). Free-living nitrogen fixers: Its role in crop production. In: *Microbes for Health, Wealth and Sustainable Environment,* (Ed) Verma AK. Malhotra Publishing Company, New Delhi, p. 25–64.

Saxena AK and Tilak KVBR (1999). Potentials and prospects of *Rhizobium* biofertilizer. In: *Current Trends in Life Sciences: Agromicrobes,* (Ed) Jha MN. Today and Tomorrow Printers and Publishers, New Delhi, p. 51–78.

Scherr C, Fliessbach A and Mäder P (2001). Photoautotrophe Bodenmikroorganismen als Bioindikatoren fur Schwermetalle. *VBB Bulletin,* 5: 7–9.

Schwintzer R and Tjepkema JD (1990). *The Biology of Frankia and Actinorrhizal Plants.* Academic Press Inc, San Diego, USA.

Sen J (1929). The role of associated nitrogen-fixing bacteria on nitrogen nutrition of cereal crops. *Agric J India,* 24: 967–980.

Siciliano SD and Roy RD (1999). The role of soil microbial tests in ecological risk assessment: Differentiating between exposure and effects. *Human Ecol Risk Assess,* 5: 671–682.

Singer MJ and Ewing S (2000). Soil quality. In: *Handbook of Soil Science,* (Ed) Sumner ME. CRC Press, Boca Raton, Florida, p. G271–G298.

Soltis DE, Soltis PS, Morgon DR, Swensen SM, Mullin BC, Dowd JM and MG (1995). Chloroplast gene sequence data suggest a single origin of the predisposition for symbiotic nitrogen fixation in angiosperms. *Proc Natl Acad Sci USA,* 92: 2647–2651.

Sprent JI, Sutherland J and de Faria SM (1987). Some aspects of the biology of nitrogen-fixing organisms. *Phil Trans Royal Soc London,* B 317: 11–119.

Stenberg B (1999). Monitoring soil quality of arable land: Microbiological indicators. *Acta Agri Scandinavia,* 49: 1–24.

Subba Rao NS (1986). *Soil Microbiology, 4th edn: Soil Microorganism and Plant Growth.* Oxford and IBH Publication Company Pvt Ltd, New Delhi.

Suman A, Shasany AK, Singh M, Shahi HN, Gaur A and Khanuja SPS (2001). Molecular assessment of diversity in endophytic diazotrophs of sub-tropical Indian sugarcane. *World J Microbiol Biotechnol,* 17: 39–45.

Tapia-Hernandez A, Bustillos-Cristales MR, Jimenezsalgado T, Cabellaro-Mellado J and Fuentes-Ramirez LE (2002). Endophytic *nifH* gene diversity in African sweet potato. *Can J Microbiol,* 44: 162–167.

Tchan YT (1984). Azotobacteriaceae. In: *Bergey's Manual of Systematic Bacteriology,* (Eds) Krieg NR and Holt JG. Williams and Wilkins company, Baltimore, 1: 219–234.

Tilak KVBR and Murthy BN (1981). Occurrence of *Azospirillum* in association with roots and stems of different cultivars of barley (*Hordeum vulgare* L). *Curr Sci,* 50: 496–498.

Torstensson L, Pell M and Stenberg B (1998). Need of a strategy for evaluation of arable soil quality. *Ambio,* 27: 4–8.

Tyagi NK (2004). Soil sodicity: A global overview. In: *Proceeding International Conference on Sustainable Management of Sodic Land,* U.P. Council of Agriculture Research, Lucknow, p. 1–15.

Van Bruggen AHC and Grünwald NJ (1996). Test for risk assessment of root infection by plant pathogens. In: *Methods for Assessing Soil Quality,* (Eds) Doran JW and Jones AJ. Soil Sci Soc America, Madison, Wisconsin, p. 293–310.

Vandamme P, Pot B, Gillis M, DeVos P, Kersters K and Swings J (1996). Polyphasic taxonomy: A consensus approach to bacterial systematics. *Microbial Rev,* 60: 407–438.

Varghese R, Chauhan VS and Mishra AK (2003). Hypervariable spacer regions are good sites for developing specific PCR–RFLP markers and PCR primers for screening actinorrhizal symbionts *J Biosci,* 28: 437–442.

Yamada Y, Hoshino K and Ishikawa T (1998). Taxonomic studies of acetic acid bacteria and allied organisms. XII, The phylogeny of acetic acid bacteria based on the partial sequences of 16S ribosomal RNA. *Int J Syst Bacteriol,* 48: 3270–3280.

Yamada Y, Hoshino K and Ishikawa T (1997). The phylogeny of acetic acid bacteria based on the partial sequences of 16S ribosomal RNA: The elevation of the subgenus *Gluconoacetobacter* to generic level. *Biosci Biotechnol Biochem,* 61: 1244–1251.

Young JPW (1992). Phylogenetic classification of nitrogen–fixing organisms. In: *Biological Nitrogen Fixation,* (Eds) Stacey G, Burris RH and Evans HJ. Chapman and Hall, New York, p. 43–86.

Zumpft, W G (1992). The denitrifying prokaryotes. In: *The Prokaryotes: A Handbook on the Biology of Bacteria: Ecophysiology, Isolation, Identification, Applications,* (Eds) Balows, A, Truper, H G, Dworkin, M, Harder, W, and Schleifer, KH. Springer-Verlag, New York, 1: 554–582.

Soil Microflora, 2009 *Pages* **165–191**
Editor: **Rajan Kumar Gupta, Mukesh Kumar & Deepak Vyas**
Published by: **DAYA PUBLISHING HOUSE, NEW DELHI**

Chapter 15

Soil Microbes and their Importance

***Anjana K. Vala*[1] *and Anita Suresh Kumar*[2]**
[1]*Department of Bioinformatics, Bhavnagar University, Bhavnagar – 364 002, Gujarat*
E mail: anjana_vala@yahoo.co.in
[2]*A-12, Everest Flats, Waghawadi Road, Bhavnagar – 362 002, Gujarat*

ABSTRACT

Much of the knowledge earlier available on soil science was in relation to plant growth and was passed on from generation to generation. Soil microbiology has been an area of focus for several researchers after the initiatives of Antonie Lavoisier (1794) and J B Bossingault (1834). Microbes not only play vital ecological roles but also possess significant traits which can be commercially exploited. This review provides a glimpse of the significant diverse roles played by soil microbiota.

Keywords: *Bioferitilisers, Bioinsecticides, Bioremediation, Diversity, Soil microbes.*

Introduction

Soils are made up of four basic components: minerals, air, water, and organic matter. The mineral portion consists of three distinct particle sizes classified as sand, silt, or clay. Sand is the largest particle that is generally referred to as soil. Soil organic matter contains dead organisms, plant matter, and other organic materials in various phases of decomposition. The soil can be viewed as a living community rather than an inert body. The oldest fossil of a living land creature which has been reported is that of a soil organism about a billion years old. Soil organisms are a complex mosaic of living organisms–algae, cyanobacteria (blue-green algae), bacteria, lichens, mosses, liverworts, and fungi–that grow on or just below the soil surface. An acre of living topsoil contains approximately 900 pounds of earthworms, 2,400 pounds of fungi, 1,500 pounds of bacteria, 133 pounds of protozoa, 890 pounds of arthropods and algae, and even small mammals in some cases (Pimentel *et al.*, 1995).

Unlike soil science, whose ancient origin can be traced back to the Roman and Aryan times, soil microbiology emerged as a distinct branch in 1838, when the French agricultural chemist JB Boussingault showed the phenomena by which legumes can obtain nitrogen from air (Subbarao, 1995). Since then, this field has remained the centre of attraction for several researchers. A time line of noteworthy contributions in soil microbiology has been shown in Table 15.1. (Subba Rao, 1995; Rangaswami and Bagyaraj, 2004; http://www.nap.edu/html/biomems/hbarker.html; http://www.asm.org)

Table 15.1: Timeline of Soil Microbiology

Scientist	*Year*	*Contribution*
Matcura	1961	Further pursued work on rhizosphere
Alexander	1961	Started school of soil microbiology at Cornell University
Carnahan *et al.*	1960	Biological N fixation
Rovira	1956	Further pursued work on rhizosphere
Ruinen	1956	Concept of phylosphere
Umbreit	1947	Studied problems related to autotrophy
Katznelson	1946	Further pursued work on rhizosphere
Starkey	1945	Transformations of iron bacteria
Waksman	1944	Discovered Streptomycin
Barker	1944	Anaerobic fermentation by methane bacteria
Lochhead	1940	Further pursued work on rhizosphere
Allen and Allen	1940	Root nodule bacteria
Garrett	1936	Established a school on soil fungi and their ecological classification
Van Niel	1931	Chemoautotrophic soil bacteria and bacterial photosynthesis
Cholodny	1930	Contact Slide method for soil microorganisms
Starkey	1929	Further pursued work on rhizosphere
Fleming	1929	Discovered Penicillin
Rossi	1927	Contact Slide method for soil microorganisms
Walksman		Penned a book entitled Soil Microbiology
Rayner and Melin	1921-1927	Intensive study of Mycorrhiza
Conn	1918	Direct Soil Examination of soil microorganisms
Russel and Hutchinson	1909	Enunciated importance of protozoa in controlling bacterial population and activity in soil
Hiltner	1904	Initiated work on rhizosphere
Koning	1904	Suggested that fungi play an important role in the decomposition of organic matter and the formation of humus.
Lipman and Brown	1903	Ammonification of organic nitrogenous substances by soil microorganisms and developed the tumbler or beaker method for studies on different types of transformations in soil

Contd...

Table 15.1–Contd...

Scientist	*Year*	*Contribution*
Omeliansky	1902	Anaerobic soil bacteria degrade cellulose
Beijerinck	1893 and 1901	Studied transformation of nitrogen in nature is largely due to activities of soil organisms
Winogradsky	1891	
Winogradsky	1895	Isolated the first free-living nitrogen-fixing organism, *Clostridum pasteurianum*
Schloesing and Laurent	1892	Reported that amount of nitrogen absorbed from air by bacteria approximately equals the gain by the plant and soil
Winogradsky	1890	Established the role of bacteria in nitrification and is one of the pioneers in soil Bacteriology
Koch	1882	Discovered gelatin plate technique that initiated examination of specific types of soil bacteria
Warington	1878	Demonstrated that nitrification in soil was a microbial process taking place in two steps
Schloesing and Muntz	1877	Evidenced the role of bacteria in nitrification while studying sewage purification
Woronin	1866	Demonstrated nodules in legumes were formed by a specific group of bacteria
Leishman	1858	
Leibig	1856	Showed that nitrates were formed in soil due to addition of nitrogenous fertilizers
Boussingault	1838	Showed that legumes could obtain nitrogen from air when grown in soil which was not heated.
Lavoisier	1794	Chemical elements in plants came from soil and air

Soil Microbial Diversity

Interest in microbial diversity has grown rapidly in the scientific community (Wilson, 1988; Franklin, 1993; Benizri *et al.*, 2002). Soil is a natural medium in which microbes live, multiply and die. Biological communities fertilize the soil, breaking down dead organisms and releasing nutrients for use by living plants. Soil-dwelling organisms release bound minerals, converting them into plant-available forms that are then taken up by plants. The organisms recycle nutrients again and again with the death and decay of each new generation of plants. These organisms are referred to as soil livestock (http://www.attra.org/attra-pub/soilmgmt.html). They are the unsung, unheralded and unrecognized heroes of the ecosystem. However, the microbes are not uniformly distributed through the soil. Each species and group exists where they can find appropriate space, nutrients, and moisture. They occur mostly in the top few inches of soil although microbes have been found as deep as 10 miles (16 km) in oil wells. Topsoil is the most biologically diverse part of the earth (U.S. Department of Agriculture, 1998). Soil microorganisms can be described as a 'black box' (Paul and Clark, 1989). Over the years, there has been a growing interest in microbes as fertility of soil depends not only on its chemical composition, but also on the qualitative and quantitative nature of microbiota inhabiting it. Maintenance of viable, diverse populations and functioning microbial communities in the soil is essential for sustainable agriculture (Beare *et al.*, 1995; Benizri *et al.*, 2002).

In soil, microorganisms are closely associated with soil particles, mainly clay–organic matter complexes (Foster, 1988). They are mainly distributed around roots, on humus, in litter. Interior as well as exterior surfaces of soil aggregates are microhabitats for soil microorganisms. Microbes are found as single cells or as micro colonies embedded in a matrix of polysaccharides (Smiles, 1988; Wood, 1989). Soil microbial activity and interaction with other microbes, macroorganisms and with soil particles depend largely on conditions at the microhabitat level that may differ among microhabitats even over very small distances (Wieland *et al.*, 2001).

Considering the cell structure and function as criteria, there are three groups of cellular organisms: eukaryotes, eubacteria, and archaea. Microbes belonging to all these groups have been reported from various kinds of soils.

Eubacteria

Eubacteria are recognized as the most dominant group of microorganisms in various kinds of soil (Liesack and Stackebrandt, 1992; Visscher *et al.*, 1992; Borneman *et al.*, 1996). They are present in all types of soil, but their population decreases with increase in the depth of soil (Duineveld *et al.*, 2001; Wieland *et al.*, 2001). Soil rich in organic matter has been reported to consist of more microorganisms (Bruns and Slatar, 1982; Subba Rao, 1997).

Autotrophic and heterotrophic bacteria are present in a wide range of soils (Tate, 1995). Autotrophic bacteria (purple and green bacteria) synthesize their own organic matter from carbon dioxide or inorganic carbon sources, whereas heterotrophic bacteria depend on pre-formed organic matter for their nutrition and energy support. Photoautotrophs derive their energy from sunlight that they catch and transform into chemical energy through the bacteriochlorophyll pigment. Chemoautotrophs oxidize inorganic materials to derive energy and at the same time, they gain carbon from CO_2 (Tate, 1995). There is a group of bacteria known as obligate chemoautotrophs. Within this group, *Nitrobacter* utilizes nitrite and *Nitrosomonas* ammonium, while *Thiobacillus* converts inorganic sulfur compounds to sulfate and *Ferrobacillus* converts ferrous ions to ferric ions (Alexander and Clark, 1965; Baudoin *et al.*, 2002).

Among various morphological forms of bacteria, bacilli are common in soil, where as spirilli are very rare in natural environments (Baudoin *et al.*, 2001, 2002). Bacteria have been classified into two broad categories. The autochthonous or indigenous population are more uniform and constant in soil, since their nutrition is derived from native soil organic or mineral matter (*Arthrobacteria* and *Nocardia)* (Herman *et al.*, 1993). Zymogenous bacteria require the input of an external substrate, and their activity in soils is variable. They often produce resting propagules (*Pseudomonas* and *Bacillus*). When specific substrates are added to soil, the number of zymogenous bacteria increases and gradually declines when the added substrate is exhausted (cellulose decomposers, nitrogen utilizing bacteria) (Giri *et al.*, 2005).

The most common soil bacteria belong to genera *Pseudomonas*, *Arthrobacter*, *Achromobacter*, *Bacillus*, *Clostridium*, *Micrococcus*, *Flavobacterium*, *Corynebacterium*, *Sarcina*, *Azosprillium*, and *Mycobacteria* (Loper *et al.*, 1985; Bruck, 1987; Lynch, 1987a,b). *Escherichia* is encountered rarely in soils except as a contaminant from sewage, where as *Aerobacter* is frequently encountered and is probably a normal inhabitant of certain soils (SubbaRao, 1997). Another group of bacteria common in soil is the Myxobacteria belonging to the genera *Myxococcus*, *Chondrococcus*, *Archangium*, *Polyangium*, *Cytophaga* and *Sporocytophaga*. The latter two genera are cellulolytic and, hence, are dominant in cellulose-rich environments (Slater, 1988; Benizri *et al.*, 2001). Bacteria can withstand extreme climates, although temperature and moisture influence their population (Woese, 1987; Benizri *et al.*, 2002).

On the other hand, the cyanobacteria are a structurally diverse assembly of Gram-negative eubacteria characterized by their ability to perform oxygenic photosynthesis. They have characteristics common to bacteria and algae and are therefore, often named "blue-green algae"(the characteristic blue–green color is due to pigments phycocyanin and chlorophyll). Some cyanobacteria also possess heterocysts, which are involved in nitrogen fixation. The rice fields are a good habitat for the development of certain cyanobacteria where they fix atmospheric nitrogen (Prescott *et al.*, 1996). The dominant soil cyanobacteria belong to the genera *Chrococcus, Aphanocapsa, Lyngbya, Oscillatoria, Phormidium, Microcoleus, Cylindrospermum, Anabaena, Nostoc,* Cytonema, and *Fischerella* (Subba Rao 1997; Benizri *et al.*, 2002).

Actinomycetes (confined to the order *Actinomycetales)*, bear certain similarities to Fungi Imperfecti in the branching of the aerial mycelium, which profusely sporulate, and in the formation of distinct clumps or pellets in liquid cultures (Benson, 1988). The number of actinomycetes increases in the presence of decomposing organic matter. They are intolerant to acidity and their numbers decline below pH 5.0. The most conducive range of pH is between 6.5 and 8.0. Waterlogging of soil is unfavorable for the growth of actinomycetes, whereas desert soils of arid and semi-arid zones sustain sizeable populations of the same, probably due to the resistance of spores to desiccation. The percentage of actinomycetes in the total microbial populations increases with the depth of soil. *Streptomyces* constitutes the most common genus of soil actinomycetes (nearly 70 per cent), where as, *Nocardia* and *Micromonospora, Actinomyces, Actinoplanes* and *Streptosporangium* are encountered occasionally (Prescott *et al.*, 1996; Subba Rao, 1997). Temperatures between 25 and 30°C are conducive for the growth of actinomycetes although thermophilic cultures growing at 55 and 65°C are common in compost heaps where they are numerically extensive and belong mostly to the genera *Thermoactinomyces* and *Streptomyces* (Giri *et al.*, 2005).

Archaebacteria

Archaebacteria or ancient bacteria, are primitive prokaryotes, or the earliest organisms to have appeared on the earth (Giri *et al.*, 2005). Instead of peptidoglycan, their cell wall contains proteins and non-cellulosic polysaccharides. Their cell membrane contains branched chain lipids that enable them to bear extreme temperatures and pHs. Their rRNA nucleotides are quite different from those of other organisms (DeLong and Pace, 2001; Huber *et al.*, 2002). Diverse groups of archaea are enlisted in Table 15.2.

Fungi

They dominate all types of soils and represent the greatest diversity among soil microbes. Distribution of fungal flora in soils is influenced by the same environmental factors that apply for the distribution of bacteria. However, as fungi are heterotrophs, quality and quantity of organic matter have a direct bearing on fungal numbers in soils. Fungi are dominant in acid soils, where they have monopoly for utilization of organic substrates as an acidic environment is not conducive to the existence of either bacteria or actinomycetes (Bolton *et al.*, 1993). They are also present in neutral or alkaline soils and some can tolerate a pH over 9.0. Arable soils contain abundant fungi since they are strictly aerobic and an excess of soil moisture decreases their numbers. Fungi exhibit a selective preference for various soil depths. Species common in lower depths are rarely found on the surface. This specific distribution is ruled by the availability of organic matter and by the ratio between oxygen and carbon dioxide in the soil atmosphere at various depths. Farm practices including crop rotation and fertilizer or pesticide applications influence the nature and dominance of fungal species (Hawksworth, 1991a,b). Many fungi, which are commonly isolated from soils, come under the class Fungi Imperfecti by virtue of the fact that they produce abundant asexual spores, but lack sexual stages (Lynch, 1987a,b).

Table 15.2: Diversity of Archaea

Archaea	*Genera*	*Traits*
Methanogens	*Methanococcus* and *Methanospirillum, Methanobrevibacter, Methanobacterium, Methanosarcina*	Strict anoxybionts, occur in marshy area, produce methne from CO_2 or fumaric acid
Methane generating thermophiles	*Methanothermus*	Grow at ~ 100°C, found near hydrothermal vents
Extreme halophiles	*Halobacterium, Halorubrum, Natrinobacterium, Natronococcus*	Occur in highly saline environments, produce pigments
Sulfur and sufate reducing hyperthermophiles	*Thermococcus, Archaeoglobus, Thermoproteus, Pyrodictium, Pyrolobus*	Occur in high temperature environments (~80°C, pH 2), reduce sulfur to H_2S in absence of oxygen
Sulfur oxidizers	*Sulfolobus*	Oxidise sulfur consuming CO_2/under aerobic condition oxidize sulfur to sulfuric acid
Thermophilic extreme acidophiles	*Thermophilus, Picrophilus*	Grow only in extremely hot acid environments

The following genera of fungi are most commonly encountered in soils: *Acrostalagmus, Aspergillus, Alternaria, Botrytis, Cephalosporium, Gliocladium, Monilia, Penicillium, Scopulariopsis, Spicaria, Trichoderma, Trichothecium, Verticillium, Cladosporium, Pillularia, Cylindrocarpon,* and *Fusarium, Absidia, Cunninghamella,Mortierella,Mucor, Rhizopus, Zygorynchus, Pythium, Chaetomium,* and *Rhizoctonia* (Newman, 1985; Hawksworth, 1991a; Subba Rao, 1997). Filamentous fungi in soil degrade organic matter and help in soil aggregation. Certain fungi like *Alternaria, Aspergillus, Cladosporium, Dematium, Gliocladium, Helminthosporium, Humicola,* and *Metarhizium* produce substances similar to humic substance in soil and, hence, may be important in the maintenance of soil organic matter (Hawksworth, 1991b). Many yeasts belonging to true Ascomycetes such as *Saccharomyces* and those belonging to Fungi Imperfecti such as *Candida* have been isolated from soils. However, their number in soil is relatively low.

Algae

Algae are photoautotrophic organisms by virtue of chlorophyll present in their cells. Soil algae are ubiquitous where moisture and sunlight are available. They are less abundant in soil as compared to fungi (Metting, 1988). Diatoms have also been found in soils. The algae, which are dominant in soils, are members of the class Chlorophyceae. Some of the common green algae occurring in most soils belong to the genera *Chlorella, Chlamydomonas, Chlrococcum, Oedogonium, Chlorochytrium,* and *Protosiphone* (Metting, 1988; Lynch, 1990).These microbes are observed in the form of green scum on the surface of soils where as some algae are microscopic. They may be unicellular (*Chlamydomonas*) or filamentous (*Spirogyra, Ulothrix*). Algae have also been found below the surface of the soil and beyond the reach of sunlight. However, their number here is low compared to that of algae inhabiting the surface of soil (Metting, 1988; Subba Rao, 1997).

Protozoa

Soil protozoa are unicellular, lack chlorophyll (with few exceptions), and are characterized by a cyst stage which help them to withstand adverse soil condition. They are abundant in the upper layer of the soil and their numbers are directly dependent on the bacterial population. Protozoa live in soil at the cost of bacteria. Protozoa perform diverse functions in soils. They: (*i*) mineralize nutrients and make them available to plants and other soil organisms (*ii*) help regulating bacterial populations and (*iii*) provide food source for other soil organisms and help to suppress disease by competing with or feeding on pathogens. The flagellated protozoans predominate soil and are classified under the class (*a*) Mastigophora (*e.g. Allantion, Bodo, Cercobodo, Cercomonas, Entosiphon, Heteromita, Monas, Oikomonas, Spiromonas, Spongomonas* and *Tetramitus*. (*b*) Sarcodina (comprising of protozoa which move with the help of pseudopodia) including *Amoeba, Biomyxa, Difflugia, Euglypha, Hartmanella, Lecythium, Nuclearia* and *Trinema* and (*c*) Cilliata, distinguished by cillia which help in locomotion (*e.g.Balantiophorus, Colpidium, Colpoda, Enchelys, Gastrostyla, Oxytricha, Uroleptus* and *Vorticella*) (Subba Rao, 1995).

Viruses

They are the smallest inhabitants of soil and are known to attack bacteria (bacteriophages), actinomycetes (actinophages), cyanobacteria (cyanophages) and fungi (mycoviruses). The most extensively studied system is mycovirus of *Penicillium chrysogenum*. Reports are also available revealing viruses infecting *Penicillium, Aspergillus, Periconia* and *Ustilago* (Subba Rao, 1995).

Ecological Tasks of Soil Microbes

Soil microbes carry out diversified functions. They convert ambient nitrogen into plant available forms (nitrate and ammonia), defend plants against pathogens by overcoming pathogens for food, break down leaf litter to usable nutrients, increase soil porosity by gluing soil particles together, increase water infiltration, etc.

Soil bacteria are mainly involved in controlling atmospheric composition and recycling of biomass, where as archaea produce and consume low molecular weight compounds, and aid bacteria in recycling dead biomass. Amongst the eukaryotes, fungi may have either a single cell form (yeast) or multicellular form (mould). They assist in recycling biomass and also stimulate plant growth. Algae possessing a rigid cell wall and photosynthetic ability form important primary producers. In addition, protozoa play a significant role in mineralizing nutrients, making them available for plants and other soil organisms. Following are some of the important roles played by soil microbes.

Microbial Compounds in Humification and Mineralization

Microorganisms mineralize not only all primary resources to inorganic end products, but utilize most plant-derived materials to resynthesize their own biomass. Their cell walls can be categorized into (*i*) *fungal*: constituting chitin, glucan or cellulose and (*ii*) *bacterial*: constituting glycolipids, peptidoglycans, proteoglycans, and glycoproteins (Gleixner *et al.*, 2001).

Microbial cell wall materials are an important source of soil organic material. Large quantities of easily degradable amino sugars are released upon hydrolysis of soil, this suggests that these amino sugars are probably well stabilized within the highly cross-linked peptidoglycan after death of microbes (Guggenberger *et al.*, 1999).

Apart from the transient products in the decomposition of organic matter, a dark coloured fairly stable soil organic matter called humus with known and unknown physical and chemical properties is also an integral part of the organic matter complex in soil. Humus can be defined as a ligno-protein complex or an amino acid-lignin complex containing approximately 45 per cent lignin compounds, 35 per cent amino acids, 11 per cent carbohydrates, 4 per cent cellulose, 7 per cent hemicellulose, 3 per cent fats, waxes and resins and 6 per cent other miscellaneous substances including plant growth substances and inhibitors. While soil microorganisms in general take part in humus formation few fungi and bacteria also possess melanin pigments (black to brown in color) in addition to the above mentioned compounds. Melanins are precursors of humic substances in soil comprising a polymeric core of phenolic, indolic, quinone, hydro-quinone and semi-quinone monomers and are non-hydrolyzable (Butler and Day 1998; Saiz-Jimenez, 1996). Spore extracts of fungi like *Aspergillus niger* possess properties similar to humic acids (SubbaRao, 1995). Thus, microbial components serve as structural units for the synthesis of humic substances.

Soil Aggregation

Role of microorganisms in soil aggregation has been well documented by several researchers. The presence of fungal hyphae provides physical binding for the particles, but it cannot wholly explain the formation and stabilization of aggregates. Andrade *et al.* (1998) showed involvement mycorrhizal fungus (*Glomus mossae*) in formation of soil aggregates. Rillig and Steinberg (2002) reported the growth of mycorrhizal fungus (*Glomus intraradices*) and its metabolites to be involved in formation and stabilization of soil aggregates. Although these studies illustrate the effect of fungi, there are reports stating the limit of the effect of fungi on aggregation. Bossuyt *et al.* (2001) reported that beyond a certain limit, fungal growth has no further effect on aggregation.

Soil aggregation can be attributed to the nature of the metabolites produced by soil microbes. A few of them are briefly described here.

Glomalin is a glycoprotein found on mycorrhizal hyphae as well as on surface of aggregates (Wright, 2000). It is a N-linked oligosaccharide with tightly bound iron. It is insoluble and possibly hydrophobic in its native state (Wright *et al.*, 1996). A number of studies have confirmed the role of this fungal metabolite in the formation and stabilization of aggregates (Wright and Upadhyaya, 1998; Bird *et al.*, 2002; Rillig and Steinberg, 2002). Rillig *et al.* (2002) reported that the production of glomalin by the hyphae had a stronger effect on aggregation than the direct action of enmeshing.

The importance of polysaccharides in the aggregation process has been well recognized for several years (Robert and Chenu, 1992). Bacterial metabolites and exudates from roots are the most active polysachcharides involved in soil aggregation. The effectiveness of polysaccharides stems from their macromolecular structure which enables them to be adsorbed on clay particles and also to glue particles together. Puget *et al.* (1999) showed the predominance of plant carbohydrates in the largest aggregates. Moreover, they also highlighted the role of high concentrations of microbial carbohydrates in formation of stable aggregates. These results confirm the hypothesis that the stability of the aggregates is partly due to the polysaccharides produced by microorganisms decomposing the vegetal residues trapped in the aggregates.

In addition to glomalin and polysaccharides, the role of lipids in soil aggregation has also been studied. Owing to their hydrophobic nature, the presence of lipids in soil aggregates improves their resistance to slaking (Paré *et al.*, 1999). However, compared to other compounds, the role of lipids in the aggregation processes is less studied (Capriel *et al.*, 1990; Dinel *et al.*, 1991).

Weathering Down of Rocks

Wearing down of rocks is caused by many microbes. Amongst these are Epilithic lichens which cause detachment of grains and particles. These lichens also are responsible for the partial protection *i.e.* they form a protective crust (Gehrmann *et al.*, 1988; Dornieden *et al.*, 1997, 2000; Banfield *et al.*, 1999). On the other hand epilithic free living fungi and cyanobacteria are mainly involved in chemical wear down of rocks and also photosynthesis induced alkalinization of its surfaces. Chasmolithic and endolithic cyanobacteria, algae, chemoorganotrophic bacteria and fungi may produce acid polysaccharides or extracellular polymeric substances and may also be responsible for differential heating and heat transfer through protective pigments. These organisms are responsible for exfoliation, chipping, pitting and increase in porosity of rocks. The rhizines, hyphae and microcolonies of endolithic lichens, fungi, bacteria (rarely algae) penetrate and propogate on rocks and may excrete organic acids such as oxalic acid (Dornieden *et al.*, 1997, 2000; Sterflinger and Krumbein, 1997). This causes pores and cracks in rocks. If this penetration is deep, this may allow penetration of surface water and chemicals. Organic acid production also plays a vital role in mineral deterioration, affecting both biogenic chemical weathering and soil formation (Gadd, 1999).

Biogeochemical Role of Soil Microbiota

Soil microbes serve as agents for conversion of complex organic compounds into simple inorganic compounds or constituent elements. This process is generally termed as mineralization and provides continuity of elements as nutrients for plants and animals.

The role of soil microbes with respect to transformations of nitrogen, carbon, sulphur, phosphorus and other compounds have been illustrated in Table 15.3.

Table 15.3: Microorganisms and Element Transformations*

Element	*Process*	*Microorganisms*
Nitrogen	Proteolysis (Proteins → peptides → amino acids)	*Clostridium histolyticum* and *C. sporogenes*, *Proteus* spp., *Pseudomonas* spp., *Bacillus* spp.
	Amino acid degradation: Ammonification (Alanine+1/2O_2 → Pyruvic acid+NH_3)	Most of the microbes
	Nitrification $2NH_3+3O_2 \rightarrow 2HNO_2+2H_2O$ $HNO_2+1/2O_2 \rightarrow HNO_3$	Ammonia oxidizers (*e.g. Nitrosomonas europaea, Nitrosovibrio tenuis*, Nitrosococcus nitrosus, *Nitrosococcus oceanus* Nitrate oxidizers (*Nitrobacter winogradskyi* and *Nitrospina gracilis*)
	Reduction of Nitrate to ammonia ($HNO_3 + 4H_2 \rightarrow NH_3 + 3H_2O$)	Several heterotrophic bacteria
	Denitrification ($2NO_3^- \rightarrow 2NO_2^- \rightarrow 2NO \rightarrow N_2O \rightarrow N_2$	Denitrifying bacteria (*e.g. Achromobacter*, *Agrobacterium*, *Alcaligenes*, *Bacillus*, *Chromobacterium*, *Flavobacterium*, *Hyphomicrobium*, *Pseudomonas*, *Thiobacillus* and *Vibrio*)
	Nitrogen fixation	Nitrogen fixing bacteria
	(*a*) Non symbiotic	
	(*b*) Symbiotic	
Carbon	CO_2 fixation	CO_2 fixing bacteria
	(*a*) $CO_2 + 4H \rightarrow (CH_2O)x + H_2O$	Autotrophic bacteria Heterotrophic microbes
	(*b*) Pyruvic acid + CO_2 → oxaloacetic acid	

Contd...

Table 15.3: Microorganisms and Element Transformations*

Element	*Process*	*Microorganisms*
	Organic carbon compound degradation (*e.g.* cellulose → cellobiose → glucose → CO_2, water and other end products)	Many microbes
Sulfur	Oxidizing sulfur to sulfate	Bacteria (*Thiobacillus, Beggiatoa, Thiothrix* and *Thioploca* spp.), fungi and actinomycetes (*Aspergillus, Penicillium* and *Microsporium*)
	Assimilation	Many heterotrophs
	Reduction of sulfate to H_2S	*Desulfatomaculum* spp., *Desulfovibrio desulfuricans*
	Oxidation of H_2S to elemental S	Variety of pigmented sulfur bacteria
Phosphorus	Phosphate solubilization	Phosphate soubilizing microorganisms (*e.g. Aspergillus, Penicillium, Bacillus* and *Pseudomonas*)
Iron	Oxidation (ferrous to ferric ion)	*Gallionella, Siderophacus, Siderocapsa, Siderophaera, Ferribacterium, Naumanniella, Ochrobium, Sideromonas, Sideronema, Siderobacter, Siderococcus, Ferrobacillus, Leptothrix, Sphaerotilus, Toxothrix, Crenothrix, Colnothrix, Thiobacillus ferrooxidans* and *Ferrobacillus ferrooxidans*

*: Pelczar, 1993.

Besides the elements mentioned in Table 15.3, plants as well as microorganisms require traces of Fe, Mn, Cu, Zn, Mo and Co, non-availability of which may cause manifestation of specific symptoms on different plant parts. Some of these elements play a vital role in cellular enzymatic activities. Heavy metals in soil are associated with a number of soil components which determine their behavior in the soil and influence their bioavailability (Boruvka and Drabek, 2004). Microbes bring about transformation of metals by various mechanisms. Microorganisms can mobilize metals through autotrophic and heterotrophic leaching, chelation by microbial metabolites and siderophores, and methylation, which can result in volatilization (Gadd, 2005). Such processes can lead to dissolution of insoluble metal compounds and minerals, including oxides, phosphates, sulphides and more complex mineral ores, and desorption of metal species from exchange sites on, *e.g.* clay minerals or organic matter. Microorganisms can also mobilize metals, metalloids and organometallic compounds by reduction and oxidation processes (Gadd, 1993a; Gharieb *et al.*, 1999; Lovley, 2000). A number of processes lead to immobilization of metals. Although immobilization reduces the external free metal species, it may also promote solubilization in some circumstances by shifting the equilibrium to release more metal into solution. Following are some examples of microbial involvement in metal transformations. The conversion of manganous (Mn^{2+}) to manganic ion (Mn^{4+}) is a microbiological process involving bacteria (*Azotobacter chroococcum, Pseudomonas fluorescens, P. trifolii, Leptothrix* spp., *Aerobacter* spp. *Proteus* spp. *Corynebacterium* sp., *Flavobacterium* sp., *Chromobacterium* sp., *Metallogenium* spp., *Hyphomicrobium*, etc.), yeast (*Cryptococcus albidus*) and moulds (*Cladosporium, Curvularia, Helminthosporium* and *Cephalosporium*). Deficiency of Cu in higher plants depends on the extent of fixation of this elements in soil. Cu is precipitated by H_2S producing microbes. Apart from *D. desulfuricans*, which reduces sulfate, bacteria like *Clostridium lentoputrescens, Proteus vulgaris* and *E. coli* may also produce H_2S from sulfur containing amino compounds under anaerobic conditions (SubbaRao, 1995).

Restoration of Polluted Soils and Bioremediation

One of the very important roles of microbes is in remediation and restoration of polluted sites. It is imperative to combat the large number of toxic compounds being released into the environment because of increased anthropogenic, industrial and/or agricultural activities. These contaminants can be broadly categorized as: Organic contaminants (*e.g.* pesticides, fuels, solvents, alkanes, polycyclic aromatic hydrocarbons (PAHs), nitrogen and phosphorus compounds, explosives, and dyes) and inorganic contaminants (*e.g.* toxic heavy metals). Microbial potentials are harnessed for environmental clean up.

Bioremediation, either as a spontaneous or as a managed strategy, is usually considered a safer and cleaner methodology than the conventional techniques for the clean-up of polluted systems. The purpose of soil bioremediation is "not only to enhance the timely degradation, transformation, remediation or detoxification of pollutants by biological means, but also to protect soil quality" (Adriano *et al.*, 1999). A multidisciplinary approach is required for a successful implementation of a bioremediation technology (Blackburn and Hafker, 1993; Boopathy, 2000).

Microbes, plants and plant-microbe associations are mainly involved in bioremediation processes (Bumpus, 1993; Dec and Bollag, 1994; Harvey *et al.*, 2002; Korda *et al.*, 1997; Liu and Sulfita, 1993; Lynch, 2002; Pointing, 2001; Roper *et al.*, 1996; Siciliano and Germida, 1998; Walton *et al.*, 1994). These agents can effectively transform the organic pollutants, extensively modify structure and toxicological properties of contaminants or completely mineralize the organic molecules into innocuous inorganic end products because of their powerful enzymatic machinery. The enzymes can even carry out processes for which no efficient chemical transformations could have been devised (Gianfreda *et al.*, 2006).

Margesin *et al.* (1999) have found that monitoring of soil microbial lipase activity is a valuable indicator of diesel oil biodegradation in freshly contaminated, unfertilized and fertilized soils. Fungal species can be used to degrade oil spills in the coastal environment, which may enhance ecorestoration as well as in the enzymatic oil processing in industries (Gopinath, 1998).

Microbes employ diverse mechanisms to respond to inorganic contaminants like toxic heavy metals. All major microbial groups have roles in metal immobilization and mineral formation. Microorganisms produce range of metal-binding compounds. Non-specific metal-binding compounds range from simple organic acids and alcohols to macromolecules such as polysaccharides, humic and fulvic acids (Birch and Bachofen, 1990; Beech and Cheung, 1995; Bridge *et al.*, 1999; Sayer and Gadd, 2001). Both live and dead biomass have metal binding potentials. Metal species can be accumulated within cells via membrane transport systems of varying affinity and specificity. The fate of a metal depends on the type of element and the organism, binding, localization, precipitation, or transportation to specific structures (Gadd, 2005). Among various microbial components, peptidoglycan carboxyl group of Gram-positive bacteria, phosphate groups of Gram–negative bacteria, chitins of fungal cell walls are some of the effective binding sites for metals. Fungal phenolic polymers and melanins possess many potential metal-binding sites with oxygen-containing groups including carboxyl, phenolic and alcoholic hydroxyl, carbonyl and methoxyl groups being particularly important (Gadd, 1993). Interestingly, it has been reported that the fungal component of soil can immobilize the total Chernobyl radiocaesium fallout received in upland grasslands (Dighton *et al.*, 1991).

Mycorrhizal fungi also have ability to sequester potentially toxic elements. Since plants growing on metalliferous soils are generally mycorrhizal, an important ecological role for the fungus has frequently been postulated (although such a role, *e.g.* phytoprotection, is often difficult to establish experimentally). Mycorrhizal fungi exhibit "constitutive and adaptive resistance" to metals (Meharg and Cairney, 2000). Microbes also aid plants in phytoremediation, *i.e.*, the use of plants to remove or detoxify environmental pollutants (Baker and Brooks, 1989; Salt *et al.*, 1998).

Thus, microbial processes have a potential application in treating contaminated lands. In the bioremediation context, production of sulphuric acid by *Thiobacillus* species has been used to solubilize metals from sewage sludge, thus enabling separation from the sludge which can then be used as a fertilizer (Sreekrishnan and Tyagi, 1994). Autotrophic leaching has been used to remediate other metal-contaminated solid materials including soil and red mud, the main waste product of Al extraction from bauxite (Vachon *et al.*, 1994). Although some processes could be used *in situ* (*e.g.* leaching using S-oxidising bacteria), many are probably most suitable for *ex situ* use in bioreactors, where the mobilized or immobilized metal can be separated from soil components (White *et al.*, 1998). Living or dead fungal and bacterial biomass and metabolites have been used to remove metals and metalloids from solution by biosorption or chelation (Macaskie, 1991; Gadd, 2001).

An integrated process of bacterial sulfate-reduction with bioleaching by sulfur-oxidizing bacteria is reported to remove contaminating toxic metals from soils. Sulfur- and iron-oxidising bacteria liberated metals from soils in the form of an acid sulphate solution that enabled almost all the metals to be removed by bacterial sulphate reduction (White *et al.*, 1998). Large scale bioreactors have been developed using bacterial sulphate-reduction for treating metal-contaminated waters (Barnes *et al.*, 1992; Gadd, 1992). Many filamentous fungi can sorb trace elements and are used in their commercial biosorbents (Morley and Gadd, 1995).

Commercial Potentialities of Soil Microbes

Besides playing an important ecological role, soil microbes have marketable potentials also. They can be harnessed as sources of biofertilizers, bioinsecticides, antibiotics, industrially important enzymes, bioremediation agents, etc. A few of them are already in use while efforts are under way for commercial exploitation of certain microbial potentialities. These diverse roles of a few groups of soil microbes are discussed below.

Microbial Biofertilizers

Biofertilizers are biologically active products or microbial inoculants of bacteria, algae and fungi, that aid plant growth. Besides, chemical fertilizers (like fine limeflour) whose main action is to stimulate soil biological activity, also are considered as biofertilizers. In contrast to their chemical counterparts, judicious use of biofertilizers is ecofriendly and more economical over long term (http://www.urbancreeks.org).

Microbial biofertilizers either contain single species (often selected strains of microbes) or a wide range of microorganisms to gain potentially wider and synergistic effects. Some products may simply contain microbial extracts that include growth promoting factors.

Generally, aerobes are considered as better microbial inoculant as compared to anaerobes. Though this is largely true, benefits of many anaerobic organisms should not be neglected. In fact there always exist copious anaerobic pockets in the soil which are best being filled with beneficial anaerobic organisms rather than harmful or wasteful microbes.

Bacteria

Different types of soil bacteria are used either as live inocula or providers of biofertilizer extracts. Among these, fluorescent pseudomonads have been extensively studied for their growth promoting effects. Few bacteria used as biofertilizers are detailed below.

Cyanobacteria

These primitive oxygenic, photosynthetic bacteria (blue green algae or BGA) are significant in many ways. Many forms of cyanobacteria are capable of nitrogen fixation (free living in the soil or associated with plant roots and nodules). Some of the free living nitrogen fixing cyanobacterial genera include *Aulosira, Anabaena, Anabaenopsis, Calothrix, Campylonema, Cylindrospermum, Fischerella, Hapalosiphon, Michrochaete, Nostoc* and *Tolypothrix*, etc. Besides atmospheric nitrogen fixation, they synthesize and excrete several vitamins and growth substances (vitamin B_{12}, auxin and ascorbic acid) that contribute towards better plant growth.

On the other hand, some cyanobacteria exist in association with fungi, liverworts ferns and flowering plants and fix atmospheric nitrogen. Blue green algae and *Azolla* (an aquatic fern) constitute a system which is the main source of algal biofertilizer in South and South East Asia, particularly for low land paddy. *Anabaena azollae* occurs as an endophyte in the vegetative leaf of *Azolla* and fixes atmospheric nitrogen (Gupta, 2003; Subba Rao, 1995).

Additionally, cyanobacteria also provide oxygen that aids plant root health and nutrient uptakes. Soil biological activity is supported by the provision of oxygen, chelating agents (for improving nutrient uptake) and growth stimulating factors including hormones and simple proteins. BGA also abet in improving soil structure and water holding capacity through production of sticky polysaccharides. A few forms of cyanobacteria *e.g. Anabaena, Nostoc, Tolypothrix, Aulosira* and *Anacystis* are capable of solubilising phosphorus from its mineral form making it available to microbes and plants (Natesan and Shanmugasundaram, 1989; Bose *et al.*, 1971).

Indian Agricultural Research Institute (New Delhi), Madurai Kamraj University (Madurai) and Bharathidasan University (Trichi) are amongst the leading institutions actively involved in promoting BGA biofertilizers technologies in India.

Pseudomonads

Pseudomonads due to their vital role as PGPR (plant growth promoting rhizobacteria) have received worldwide attention. Fluorescent pseudomonads (strains exhibiting fluorescence under uv light) because of their versatility in growth and nutrient absorption 'mop up' nutrients in the rhizosphere (Subba Rao, 1995). They produce siderophores (low molecular weight iron chelating agents) and sequester the limited supply of iron and thus controls the growth of pathogens by making iron unavailable to them. Strains from the genera *Pseudomonas, Bacillus* and *Rhizobium* are among the most powerful solubilizers of phosphorus (Rodriguez and Fraga, 1999). *Pseudomonas aeruginosa, P. putida, Pseudomonas* sp., etc. are amongst the ones well documented. PGPR promote mycorrhizal functioning also. Villegas and Fortin (2001) showed an interesting specific synergistic interaction between the P solubilizing bacterium *Pseudomonas aeruginosa* and the AM fungus *Glomus intraradices.* Many Indian industries are involved in commercialization of *Pseudomonas* biofertilisers (http://www.indianindustry.com).

Bacilli

Among these, members of genus *Bacillus* have a significant contribution as phosphate solubilizing microorganisms (PSMs). They secrete organic acids that dissolve insoluble phosphate and thus make it available to plants. *Bacillus subtilis, B. coagulans* and *B. megaterium* are amongst the promising phosphate solubilizers. Bacterization of soil using a commercial preparation 'phosphobacterin' containing phosphate solubilizing *Bacillus megaterium* was first initiated and widely used in the

U.S.S.R. (Subba Rao, 1995). Application of *Bacillus megatherium* var. phosphaticum as biofertilizer under the commercial name of phosphorein has also been reported (El-Zeiny, 2007).

Rhizobia

These are Gram-negative, nitrogen-fixing bacteria that form nodules on host plants. They are able to enter into symbiotic relationship with legumes. Currently, there are 44 recognized species of nodule bacteria on legumes within 11 genera, 9 belonging to α- proteobacteria, *Allorhizobium, Azorhizobium, Bradyrhizobium, Devosia, Mesorhizobium, Methylobacterium, Ochrobactrum, Rhizobium* and *Sinorhizobium* (Sahgal and Johri, 2006). They fix atmospheric nitrogen and thus, not only increase the production of the inoculated crops, but also leave a fair amount of nitrogen in the soil, which benefits the subsequent crop. Rhizobia can also be expoited for restoration of degraded lands, *e.g.* R8 for saline soils and R36 and R73 for acidic soils (Gupta, 2003). Moreover, strains of rhizobia are also potential phosphate solubilizers (Villar-Igea *et al.*, 2007). First commercial *Rhizobium* biofertilizer as 'Nitragin' was produced in the USA in 1895. (http://www.fao.org/ag/agl/agll/ipns/index en.jsp?term=r095)

In India, mass production technology of Rhizobial biofertilizers was transferred by DBT to entrepreneurs M/S Javery Agro, Amrawati and M/S Pratistha Industries Ltd., Secunderabad. However, the product is yet to be marketed. (http://dbtindia.nic.in)

Azotobacters and Azospirilla

These are potential asymbiotic nitrogen fixers which also synthesize growth promoting antibiotic substances helpful to the plants. Efficient strains of *Azotobacter* are reported to fix 30 kg of nitrogen from 1000 kg of organic matter. Similarly, *Azospirillum* with farmyard manure, is reported to save 15-25 kg equivalent of nitrogen per hectare in crops like *Sorghum* and other millets (Subba Rao, 1995). According El-Zeiny (2007), *Azotobacter* spp., commercialized as "microbien", is a promising biofertilizer.

Lactic Acid Bacteria

In the soil system they can be beneficial by aiding the fermentative decomposition of organic matter and producing growth promoting substances that benefit many microorganisms and plants. They also produce bacteriocins and other natural toxins that restrict the growth and activity of other soil microbes particularly harmful ones.

Actinomycetes

Actinomycetes, a range of hardy bacteria, are responsible for smell of "good" soil. Some species of actinomycetes have been used to provide extracts useful as biofertilizers with high levels of growth promoters. There are about 160 species of angiosperms, which are known to form nitrogen fixing root nodules with actinomycetes belonging to the genus *Frankia*. This genus helps in nitrogen fixation in non-leguminous plant species and therefore, can be used for land reclamation through reforestation due to high biomass production with out the need of expensive nitrogen fertilizers (Gupta, 2003). They not only fulfill the nitrogen requirement of trees but also help in fixing nitrogen in the virgin soil thus improving the fertility of degraded lands. They improve soil conditions and also suppress pathogens by excreting phenolic or other organic compounds. Few actinomycetes can produce ionophore antibiotics which interestingly increase the uptake of nutrients including cations (to the detriment of susceptible organisms). In a biofertilizer, the presence or stimulation of organisms that antagonize harmful microorganisms may be one of the modes of action that help obtain better plant yields.

Fungi

Mycorrhiza and yeast are two of the main types of fungi used in biofertilizer production. Other fungi such as *Aspergillus oryzae* and *A. niger* may also be used, often for their extracts which contain growth promoters.

Mycorrhizal Fungi

Mycorrhizal fungi form a bridge between the roots and the soil, gathering nutrients from the soil and giving them to the roots. Most of the crops are capable of indulging in a relationship with fungi partly inhabiting ("infecting") the plant roots and partly growing out into the soil.

The mycorrhizal fungi can be categorized into two major types: Ectomycorrhizal Fungi (EM) and Endomycorrhizal Fungi (AM). While both types penetrate the plant roots, ectomycorrhizae spread their hyphae between root cells, while endomycorrhizae hyphae penetrate root cells. Ectomycorrhizae hosts include members of the Pine, Oak and Beech families, as well as few others in scattered families. Endomycorrhizae are the most common, and are found in grasses, shrubs, some trees, and many other plants. EM fungi are usually specific to a certain host species, but most species of endomycorrhizae will form relationships with almost any AM host plant, and is, therefore, much easier to specify.

Mycorrhizae improve survival and growth of the seedling by enhancing uptake of nutrients (especially phosphorus) and water, by lengthening root life and providing protection against pathogens. Mycorrhizae also benefit plants indirectly by improving structure of the soil. AM hyphae excrete gluey, sugar-based compounds called Glomalin, that helps to bind soil particles, and makes stable soil aggregates. This gives the soil structure, improves air and water infiltration, and enhances carbon and nutrient storage (Peters, 2002). In particular, different forms of mycorrhiza have great importance for the P nutrition of higher plants. Most agricultural crops are potential host plants for arbuscular mycorrhizal (AM) fungi. In addition to an exudation of carboxylates, phosphatases and plant hormones, mycorrhiza increase the exploitation of the soil volume by the hyphal network, which increases the active adsorption surface and spreads beyond the phosphate depletion zone (Lange Ness and Vlek, 2000; Martin *et al.*, 2001). Mycorrhizal hyphae have a higher affinity for phosphate(Lange Ness and Vlek, 2000). AM fungi store phosphate in the form of orthophosphate, polyphosphate and organic phosphate in their vacuoles and transfer it to the roots of the host plant (Ezawa *et al.*, 2002). *Glomus*, *Gigaspora*, *Sclerocystis*, *Endogone*, etc. are a few of the important fungal genera forming mycorrhizal associations.

Mycorrhizae is most frequently applied via hand seeding, seed drilling, hydroseeding, broadcast and tilling, planting, or as a nursery medium. While installing container plants, packets of mycorrhizae are planted along with the plant (1 packet per foot of plant height or container width) (RTI, 2003). Better establishment and growth of seedling may be achieved by supplementing the fungal inoculants with *Rhizobium*/*Azotobacter* and phosphate solubilizing microorganisms.

In India, technology for mass production of AM fungi suitable for majority of crops (including plantation crops) has been developed by TERI, New Delhi. The technology has been transfered to two industries M/s Cadila Pharamaceutical Ltd., Ahemdabad, and KCP Sugar and Industries (Pvt.) Ltd., Chennai, who have launched the products under the brand name 'Josh–a root booster', and 'Ecorrhiza-VAM' and 'Nurserrhiza-VAM', respectively. (http://dbtindia.nic.in)

Aspergilli and penicillia

Aspergillus niger, *Aspergillus awamori*, *Penicillium oxalicum*, *P. bilaiae*, *P. citrinum*, *Penicillium albidum* are capable of solubilizing insoluble P and releasing it into soil, thus making it available to plants

(Morales *et al.*, 2007; Takeda and Knight, 2006; Wang *et al.*, 2005). Members of these genera also exhibit siderophore production potential. Vala *et al.* (2000) reported production of hydroxamate siderophores by *P. citrinum* isolated from marine sediments.

Trichoderma spp.

Strains of various *Trichoderma* spp. dominate soil and plant habitats and control plant disease. They also stimulate plant growth especially root production. Many members are commercially available (Subba Rao, 1995).

Yeasts

Many yeasts provide growth promoters and are used as use in biofertilizers. One of them include low alcohol strains of beer yeast (*Saccharomyces cerevisiae*) which are sometimes used in the manufacture of biofertilizers and are also included in animal probiotics.

The microbes have potential benefits but results obtained by using them as biofertilizers are characteristically variable. This is perhaps due to the vacillation in product quality, environment and the vagaries of not understanding the actual modes of action.

Bioinsecticides/Biopesticides

Over the centuries, as microorganisms have evolved in their environment, the metabolites they produce have provided a competitive advantage; consequently, the isolation of these compounds continues to be a fruitful agricultural research area for the suppression of both arthropod pests and plant pathogens. Microorganisms are endowed naturally with ability to produce metabolic byproducts that are toxic to many organisms. Metabolic byproducts of some microbes prove to be toxic against arthropod pests. Instead of relying on the microbe to produce arthropod-active toxins in the field, the microbe can be cultured in a fermentor and the resulting metabolites can be harvested, purified, formulated and used effectively against major pests. Biodegradable, non-toxic and cost effective nature of bioinsecticides make them superior to chemical insecticides (Godfrey *et al.*, 2005).

The sources of commercially exploited microbes that produce bioinsecticides are bacteria, fungi and also viruses.

Bacteria

In general, Gram positive bacteria are the main resources of bacterial bioinsecticides. The natural epizootics of insect pathogens occur commonly in native and managed systems, significantly assisting pest management. However, except for *Bacillus thuringiensis* (Bt), the application of microorganisms for pest control in agricultural systems is extremely limited (Flint, 1992). The reasons for this include: (1) the high cost of *in vitro* or *in vivo* production; (2) limited persistence and efficacy due to UV light degradation, high humidity requirements or temperature sensitivity in the field; (3) slow speed of kill; (4) poor shelf-life or special handling needs; and (5) high levels of specificity. The latter can preserve populations of natural enemies following application, but also makes it difficult to balance market size with registration costs. In addition, microorganisms generally only have one mode of entry into the host. Bacteria, viruses and protozoa must be ingested to cause an infection, whereas fungi cause an infection when the conidium (spore) attaches to and penetrates the insect cuticle.

Bacilli

One additional way of protecting a microbial-derived toxin and efficiently delivering it to the target pest is through genetically modified plants that express an insecticidal protein, such as *Bacillus thuringiensis* (Bt). Although there is considerable controversy worldwide regarding the applicability and sustainability of this technology, it does undeniably represent an effective way to deliver a toxic dose to the pest (Shelton *et al.*, 2002).

Bacillus thuringiensis (Bt) products are especially important for pest control in organic cropping systems. Bt strains are active against particular groups of insects (Tanada and Kaya, 1993). In 1938, *Bacillus thuringiensis* was first used as a microbial insecticide. Bt *kurstaki* and Bt *aizawai*, are both active against lepidopterous larvae while, Bt *tenebrionis* is active against certain beetles and Bt *israelensis* is active against mosquitoes and black flies.

Other bacilli such as *Paenibacillus* (formerly *Bacillus*) *popilliae* and *B. sphaericus*, have insecticidal activity with efficacy against Coleoptera (white grubs) and Diptera (mosquitoes), respectively. The first US registration of microbial pesticides containing *B. papillae* and *B. tentimorhus* against Japanese beetle larvae was done in 1948. Despite their commercial potential, these bacteria currently have a limited market share.

Actinomycetes

Actinomycetes are significant contributors in production of bioinsecticides. The soil environment where actinomycetes flourish is an extremely complicated and diverse system.

1. *Saccharopolyspora spinosa*: This soil-dwelling actinomycete discovered from a Caribbean soil sample in 1982, produces several metabolites called spinosyns, of which two biologically active compounds form the basis for the insecticide Spinosad, the first product in the naturalyte class of pest management tools (Boek *et al.*, 1994; Boucher, 1999). More than 30 different spinosyns have been isolated from *S. spinosa* alone, and these are being evaluated for pesticidal properties.

   Spinosad is selectively active on insects in the Orders: Lepidoptera, Diptera, Hysanoptera, and some Coleoptera and Hymenoptera. Targeted crops are cotton, vegetables, tree fruits, and nuts at use rates from 50-180 grams AI/Ha. Registrations are in place on almonds, stone and pome fruits, citrus and cole crops. Spinosad kills susceptible species by causing rapid excitation of the insect nervous system, leading to involuntary muscle contractions, tremors and paralysis. For insecticidal action, ingestion of spinosad is essential. Therefore, it has little effect on sucking insects and most nontarget predatory insects (it is highly toxic to syrphid fly larvae). Spinosad is relatively fast-acting the insect usually dies 1 to 2 days after ingesting the active ingredient and there appears to be no recovery (Godfrey *et al.*, 2005).

2. *Streptomyces avermitilis*: Abamectin (a mixture of avermectins) is an insecticide and acaricide produced by *Streptomyces avermitilis* (isolated from a soil sample collected in Shizuoka Prefecture, Japan) (Hotson, 1982). Abamectin acts on insects by interfering with neural and neuromuscular transmission and paralyzes the arthropods, resulting in the cessation of feeding and death 3 to 4 days after exposure. Abamectin is most effective when ingested by target arthropods, but also works on contact. A unique attribute of abamectin is that its acaricidal properties are coupled with activity on a few other insect species, especially dipterous and lepidopterous leafminers (*Liriomyza* and *Gracillariidae*, respectively), the lepidopterous tomato pinworm *(Keiferia lycopersicella)*, citrus thrips *(Scirtothrips citri)*, fire ants in turf and the homopterous pear psylla *(Cacopsylla pyricola)*.

   Abamectin penetrates leaf tissue and provides long-term (3 to 5 weeks) control of various mite species in field-grown roses and other ornamentals, strawberries, citrus, cotton and pears. As with synthetic insecticides, the development of resistance is a concern; however, this has been largely avoided to date, except in greenhouse systems. With the exception of predatory mites, abamectin is fairly nontoxic to natural enemies. In addition to the ones mentioned above, numerous other microbial species may also produce compounds useful as insecticides (Godfrey *et al.*, 2005; http://www.bioscience.ws/encyclopedia/index.php?title= Abamectin).

Fungi

Fungi which produce spores or toxins can be mass produced for bioinsecticide production. Among the different entomopathogenic fungi, the ones successfully used are: *Entomophthora thaxteriana* and *Aschersonia aleyrodis*, *Beuvaria bassiana*, *Metarrhizium anisopliae* and *Nomuraea rileyi*. *Beuvaria bassiana* and *M. anisopliae* have been harnessed for commercial pordcution of Biotrol FBB and Biotrol FMA, respectively (Nutrilite products, USA). Apart from these, *Verticillium lecanii* and *Hirsutella thompsonii* are commercially exploited for production of microbial insecticides in Europe and USA, respectively. *Beuvaria bassiana* and *Verticillium lecanii* are exploited in India too. Fungi like *Phytophthora palmivora*, *Colletotricum gloeospriodes* and *Cercospora rodmanii*, exhibiting herbicidal properties are either commercialized or experimented (Subba Rao, 1995).

Viruses

Nucleopolyhedroviruses (NPV) can potentially be used against lepidopterous larvae. NPV is effective against insects like *Lymantria*, *Mamestra*, *Neodiprion*, *Orgyia*, *Pieris*, *Prodenia*, *Spodoptera* and *Trichoplusia* (Subba Rao, 1995). A number of NPVs are registered in the United States, with two of particular significance to California agriculture. The NPV from the beet armyworm *(Spodoptera exigua)* is registered in the United States with a provisional registration for use in California in 2005, for use in field, vegetable and floriculture crops. Similar products containing NPV from corn earworm (*Helicoverpa zea*) and tobacco budworm (*Heliothis virescens*) have also been reported (Godfrey *et al.*, 2005). NPV is commercialized under the trade name Elcar by Sandoz Inc. (Subba Rao, 1995). In India cotton crops alone demand 50 per cent of the insecticide supply, hence NPV can be a promising insecticide alternative (Gupta, 2003).

Antibiotics

Broadly defined, antibiotics include a chemically heterogeneous group of small organic molecules of microbial origin that, at low concentrations, are deleterious to the growth or metabolic activities of other microorganisms (Thomashow *et al.*, 1995). After the discovery of penicillin by Alexander Fleming earlier in twentieth century, a massive screening effort followed and thousands of different antibiotic compounds have been identified, most produced by soil bacteria and fungi (Wiener, 1996). Antibiotic sales mount to a total of more than $8 billion worldwide each year (http://www.actionbioscience.org).

Most species of the genus *Streptomyces* are significantly active in producing a variety of antibiotics (Table 15.4). Over 55 per cent of the antibiotics detected between 1945 and 1978 originated from the genus *Streptomyces* alone, representing a total of more than 5,000 compounds (Berdy, 1980). The genus *Bacillus* also has a significant contribution in antibiotic production (Table 15.4). Their antibiotics share a full range of antimicrobial activity: *e.g.* bacitracin and gramicidin are effective against Gram-positive bacteria; colistin and polymyxin are anti-Gram-negative; difficidin is broad spectrum; and mycobacillin and zwittermicin are anti-fungal. (http://www.textbookofbacteriology.net/Bacillus.html)

Table 15.4: Some Microbes and Antibiotics Produced by them

Microorganism	Antibiotic(s)
Streptomyces griseus	Streptomycin and cyclohexamide
S. antibioticus	Mitomycin
S. virginae	Virginiamycin
S. erythrius	Erythromycin
S. caespitosus	Actinomycin
Streptomyces noursei	nystatin
Nocardia uniformis	Nocardins
Penicillium sp.	Penicillin
Bacillus licheniformis	Bacitracin
B. subtilis	Polymyxin, difficidin, subtilin, mycobacillin
Bacillus cereus	Cerexin, zwittermicin
Brevibacillus brevis	gramicidin, tyrothricin
Paenibacillus polymyxa	Polymyxin, colistin
Pseudomonas acidophila	Sulfazecin
P.mesoacidophila	Isosulfazocin
S. sannanensis	Sannamycin

Enzymes

There is renewed interest in the study of enzymes, mainly due to the recognition that they not only play an indispensable role in the cellular metabolic processes but have also gained considerable attention in the industrial community. Enzymes have found broad-based application in a variety of commercially relevant areas (http://www.wiley-vch.de/books/biopoly/pdf v07/vol07 04.pdf). The global market for industrial enzymes is expected to increase to over \$2.7 billion by 2012 (http://www.bccresearch.com/pressroom/RBIO030E.htm).

Microbial enzymes are often more useful than those derived from plants or animals due to the following reasons: (*i*) great variety of catalytic activities available, (*ii*) possiblility of high yields and stability (*iii*) ease of genetic manipulation, (*iv*) convenient and safer production (v) rapid growth of microorganisms on inexpensive media, and (*vi*) production is independent of seasonal fluctuations (Wiseman, 1995).

The industrial enzyme market is divided into three application segments: technical enzymes, food enzymes, and animal feed enzymes (Hasan *et al.*, 2006). Microbes have a spectacular contribution in all the application segments. One of the most important and profitable applications for enzymes is in detergents, where the total global market size was 0.6 billion USD in 2000 (Novozymes data). Among technical enzymes carbohydrases and proteases are the principal enzymes used in food and animal feed. Proteases, amylases, lipases, and cellulases are used in cleaning compounds including; laundry detergents, dishwashing detergents, and other cleaners. Protease and amylase lead the market with current shares of 25 per cent and 20 per cent, respectively (Hasan *et al.*, 2006). Many laundry-detergent products contain at least a protease, and many contain cocktails of enzymes including proteases, amylases, cellulases, and lipases. (http://www.wiley-vch.de/books/biopoly/pdf v07/vol07_04.pdf)

Nearly all commercial proteases are so called serine proteases originating from the *Bacillus* family and contain the catalytic triad of amino acids (*i.e.,* aspartic acid, histidine, serine) in their active sites. Most commercial amylases are derived from either genus *Bacillus* or *Aspergillus*. Mannaway (a mannase), which originates from *Bacillus,* is reported to effectively cleave the β-1,4-linkages between mannose units in guar, thereby dramatically reduce the reappearing stain phenomenon in fabrics (http://www.wiley-vch.de/books/biopoly/pdf v07/vol07 04.pdf). Soils contain a diverse population of *Bacillus* family, which may provide enzymes with new–fangled applications.

Though cellulose is considered to be highly recalcitrant, soil microbes possess expertise in degradation of cellulosic materials as they are constantly exposed to the same. Thus, they also prove to be a potential source of cellulases. Commercial cellulases come from both bacterial and fungal sources. *Trichoderma reesei* is one of the most common examples of cellulose degraders (Subba Rao, 1995).

Novo Nordisk, (Denmark), Solvay Enzymes (Germany), Godo Shusei (Japan) and Enzyme Development (USA) have exploited the enzymatic prospective of *Bacillus licheniformis* in detergent, silk degumming, feed, denture cleaners, waste degradation and food applications. In addition, Gist-Brocades (the Netherlands), Solvay Enzymes (Germany), Nagase Biochemicals (Japan) and Rohm (Germany) have commercialized enzymes from *B. subtilis* for various cosmetic, pharmaceuticals, food, detergent, cleaning; research, alcohol, baking, brewing, feed, leather, photographic waste management applications. Apart from the above mentioned bacilli, Novo Nordisk, (Denmark), Gist-Brocades (the Netherlands), Amano Pharmaceuticals (Japan), Enzyme Development (USA), Wuxi Synder Bioproducts (China) and Advanced Biochemical (India) have explored the enzymes produced by *Bacillus* sp. for their utilization in detergents, food textiles and other industrial purposes. Other bacteria such as *Pseudomonas aeruginosa* and *Clostridium* sp. have been used for production of various enzymes for applications in research and technical fields by Nagase Biochemicals and Amano Pharmaceuticals (Japan) (Gupta *et al.,* 2002). Unichem International (Spain) has launched the products for cosmetics using *Rhizomucor meihei* lipase as a biocatalyst and claims much higher quality of product, requiring minimum downstream refining (http://www.au-kbc.org/beta/bioproj2/uses.html.). Lipases from *A. niger* and *R. oryzae* have applications in dietary supplements, foods and beverages (http://www.bio-cat.com/products.phb per cent 3Fsortby=application per cent 26application). Besides these examples, there are many more enzymes originating from a range of microbiota and having sizable commercial utility.

There is an increasing demand of industrially important enzymes and so far only about 2 per cent of the world's microorganisms have been tested as enzyme sources (Hasan *et al.,* 2006). Explorations of soil organisms and applications of metagenomics to isolate novel biocatalysts from the uncultured microbiota in the environment (Yun and Ryu, 2005) would lead to development of promising products.

Apart from the multifarious roles mentioned, soil microbiota exhibit a variety of functions. Products having better commercial value can be envisaged by exploring soil microflora using a polyphasic approach. In depth research on the utilities of soil microbiota, in addition to their role in the environment they exist and their interaction with other forms of life, may unveil a lot more hidden information.

Acknowledgements

The authors are grateful to Prof. RV Upadhyay, Chief Coordinator, Department of Bioinformatics, Bhavnagar University, for support and encouragement.

References

Adriano DC, Bollag JM, Frankenberger WT Jr and Sims RC (1999). Biodegradation of Contaminated Soils. Agronomy Monograph 37, American Society of Agronomy. *Crop Science of America, Soil Science Society of America*, 772 pp.

Alexander M and Clark FE (1965). Nitrifying bacteria. In: *Methods of Soil Analysis, Part 2: Chemical and Microbiological Properties* (Eds) CA Black. American Society of Agronomy, Madison, Wisconsin, USA, pp. 1477–1483.

Andrade G, Mihara KL, Linderman RG and Bethlenfalvay GJ (1998). Soil aggregation status and rhizobacteria in the mycorrhizosphere. *Plant Soil*, 202: 89–96.

Baker AJM and Brooks RR (1989). Terrestrial higher plants which accumulate metallic elements: A review of their distribution, ecology and phytochemistry. *Biorecovery*, 1: 81–126.

Banfield JF, Barker WW, Welch SA, Taunton, A. (1999). Biological impact on mineral dissolution: Application of the lichen model to understanding mineral weathering in the rhizosphere. *Proc Nat Acad Sci USA*, 96: 3403–3411.

Barnes LJ, Janssen FJ, Sherren J, Versteegh JH, Koch RO and Scheeren PJH (1992). Simultaneous removal of microbial sulphate and heavy metals from wastewater. *Trans Inst Mining Metall*, 101: 183–190.

Baudoin E, Benizri E and Guckert A (2001). Metabolic structure of bacterial communities from distinct maize rhizosphere compartments. *Eur J Soil Biol*, 37: 85–93.

Baudoin E, Benizri E and Guckert A (2002). Impact of growth stages on bacterial community structure along maize roots by metabolic and genetic fingerprinting. *Appl Soil Ecol*, 19: 135–145.

Beare MH, Coleman DC, Crossley DA Jr, Hendrix PF and Odum EP (1995). A hierarchical approach to evaluating the significance of soil biodiversity to biogeochemical cycling. *Plant Soil*, 170: 5–22.

Beech IB and Cheung CWS (1995). Interactions of exopolymers produced by sulphate-reducing bacteria with metal ions. *Int Biodeter Biodeg*, 35: 59–72.

Benizri E, Baudin E and Guckert A (2001). Root colonization by plant growth promoting rhizobacteria. *Biocont Sci Technol*, 5(11): 557–574.

Benizri E, Dedourge O, Di Battista-Leboeuf C, Nguyen CS, Piutti and Guckert A (2002). Effect of maize rhizodeposits on soil microbial community structure. *Appl Soil Ecol*, 21: 261–265.

Benson DR (1988). The genus *Frankia*: actinomycetes symbionts of plants. *Microb Sci*, 5: 9–12.

Berdy J (1980). Recent advances in and prospects of antibiotic research. *Process Biochem*, 15: 28–36.

Birch L and Bachofen R (1990). Complexing agents from microorganisms. *Experientia*, 46: 827–834.

Bird SB, Herrick JE, Wander MM, Wright SF (2002). Spatial heterogeneity of aggregate stability and soil carbon in semi-arid rangeland. *Environ Pollut*, 116: 445–455.

Blackburn JW, Hafker WR (1993). The impact of biochemistry, bioavailability and bioactivity on the selection of bioremediation techniques. *Trends in Biotech*, 11: 328–333.

Boek LD, Hang C, Eaton TE, Godfrey OW, Michel KH, Nakatsukasa WM, and Yao RC (1994). Process for producing A83543 compounds. US Patent No. 5,362,634. Assigned to DowElanco.

Bolton H Jr, Fredrikson JK and Elliot LE (1993). Microbiology of the rhizosphere. In: *Soil Microbial Ecology*, (Ed) FB Jr Metting. Dekker, New York, p. 27–63.

Boopathy R. (2000). Factors limiting bioremediation technologies. *Biores Technol*, 74: 63–67.

Borneman J, Skroach PW, O'Sullivan EW, Palus JA, Rumjanek NG, Jansen JL, Nienhuis J, and Triplett EW (1996). Molecular microbial diversity of an agricultural soil in Wisconsin. *Appl Environ Microbiol*, 62: 1935–1943.

Boruvka L, Drabek O. (2004). Heavy metal distribution between fractions of humic substances in heavy polluted soils. *Plant Soil Environ*, 50: 339–345.

Bose P, Nagpal US, Venkataraman GS and Goyal SK (1971). *Curr Sci*, 40: 165.

Bossuyt H, Denef K, Six J, Frey SD, Merckx R and Paustian K (2001). Influence of microbial populations and residues quality on aggregate stability. *Appl Soil Ecol*, 16: 195–208.

Boucher, TJ (1999). Spinosad: The first selective, broad-spectrum insecticide. Storrs: University of Connecticut Cooperative Extension. www.hort.uconn.edu/ipm/general/htms/spinosad.htm

Bridge TAM, White C, Gadd GM (1999). Extracellular metal-binding activity of the sulphate reducing bacterium *Desulfococcus multivorans*. *Microbiol*, 145: 2987–2995.

Bruck TD (1987). The study of microorganisms in situ: progress and problems. In: *Ecology of Microbial Communities*, (Eds) Fletcher M, Gray TRG, Jones, JG. SGM symposium 41.Cambridge University Press, Cambridge, pp 1–17.

Bruns RG and Slatar JH (1982). *Experimental Microbial Ecology*. Blackwell, Oxford, 683 pp.

Bumpus JA (1993). White-rot fungi and their potential use in soil bioremediation processes. In: *Soil Biochemistry*, (Eds) J.M Bollag and G. Stotzky. Marcel Dekker, New York, p. 65–100.

Butler MJ and Day AW (1998). Fungal melanins: A review. *Can J Microbiol*, 44: 1115–1136

Capriel P, Beck T, Borchet H, Harter P (1990). Relationship between soil aliphatic fraction extracted with supercritical hexane, soil microbial biomass, and soil aggregate stability. *Soil Sci Soc Am Proc*, 54: 415–420.

Dec J and Bollag JM. (1994). Use of plant material for the decontamination of water polluted with phenols. *Biotechnol Bioeng*, 44: 1132–1139.

Dighton J, Clint GM, Poskitt J (1991). Uptake and accumulation of 137Cs by upland grassland soil fungi: a potential pool of Cs immobilization. *Mycol Res*, 95: 1052–1056.

Dinel H, Schnitzer M, Mehuys GR (1991). Soil lipids: origin, nature, content, decomposition and effect on soil physical properties. In: *Soil Biochemistry*,Vol 6, (Ed) Stotzky G. Dekker, New York, pp 397–430.

Dornieden T, Gorbushina AA, Krumbein WE (1997). Änderungen der physikalischen Eigenschaften von Marmor durch Pilzbewuchs. *Int J Restor Build Monu*, 3: 441–456.

Dornieden T, Gorbushina AA and Krumbein WE (2000). Biodecay of mural paintings and stone monuments as a space/time related ecological situation: An evaluation of a series of studies. *Int Biodeter Biodegrad*, 46: 261–270.

Duineveld BM, Kowalchuk GA, Keijzer A, van Elsas JD and van Veen JA (2001). Analysis of bacterial communities in the rhizosphere of *Chrysanthemum* via denaturing gradient gel electrophoresis of

PCR–amplified 16S rRNA as well as DNA fragment coding for 16S rRNA. *Appl Environ Microbiol,* 67: 172–178.

El-Zeiny, O.A.H. (2007). Effect of Biofertilizers and Root Exudates of Two Weeds as a Source of Natural Growth Regulators on Growth and Productivity of Bean Plants (*Phaseolus vulgaris* L.). *Research Journal of Agriculture and Biological Sciences,* 3(5): 440–446.

Ezawa T, Smith SE and Smith FA (2002). P metabolism and transport in AM fungi. *Plant Soil,* 244: 221–230.

Flint ML. (1992). Biological approaches to the management of arthropods. In: *Beyond Pesticides: Biological Approaches to Pest Management in California.* Oakland, CA. DANR Pub 3354. p 2–30.

Foster RC (1988). Microenvironment of soil microorganisms. *Biol Fertil Soils,* 6: 189–203.

Franklin JF (1993). Preserving biodiversity: species, ecosystems, or landscapes? *Ecol Appl,* 3: 200–205.

Gadd GM (1992). Microbial control of heavy metal pollution. In: *Microbial Control of Pollution,* (Eds) Fry JC, Gadd GM, Herbert RA, Jones CW, Watson-Craik I. Cambridge University Press, Cambridge, pp 59–88.

Gadd GM (1993). Microbial formation and transformation of organometallic and organometalloid compounds. *FEMS Microbiol Rev,* 11: 297–316.

Gadd GM (1999). Fungal production of citric and oxalic acid: importance in metal speciation, physiology and biogeochemical processes. *Adv Microb Physiol,* 41: 47–92.

Gadd GM (2001). Accumulation and transformation of metals by microorganisms. In: *Biotechnology: A Multi-Volume Comprehensive Treatise, Vol 10: Special Processes*l (Eds) Rehm H-J, Reed G, Puhler A, Stadler P. Wiley-VCH,Weinheim, pp 225–264.

Gadd GM (2005). Microorganisms in toxic metal-polluted soils. In: *Microorganisms in Soils: Roles in Genesis and Functions,* (Eds) Buscot F and A Varma. Springer-Verlag, Berlin, Heidelberg, p. 325–356.

Gehrmann C, Krumbein WE and Petersen K (1988). Lichen weathering activities on mineral and rock surfaces. *Stud Geobotan,* 8: 33–45.

Gharieb MM, Kierans M and Gadd GM (1999). Transformation and tolerance of tellurite by filamentous fungi: accumulation, reduction and volatilization. *Mycol Res,* 103: 299–305.

Gianfreda L, Mora ML and Diez MC (2006). Restoration of polluted soils by means of microbial and enzymatic processes. *RC Suelo Nutr Veg,* 6(1): 20–40.

Giri B, Huong Giang P, Rina Kumari, Ram Prasad, Ajit Varma (2005). Microbial Diversity in Soils. In: *Microorganisms in Soils: Roles in Genesis and Functions,* (Eds) Buscot F and A Varma. Springer-Verlag, Berlin, Heidelberg, p. 19–55.

Gleixner G, Czimczik C, Kramer C, Lühker BM and Schmidt MWI (2001). Plant compounds and their turnover and stability as soil organic matter. In: *Global Biogeochemical Cycles in the Climate System,* (Eds) Schulze E-D, Heimann M, Harrison S, Holland E, Lloyd J, Prentice IC, Schimel D. Academic Press, San Diego, p. 201–216.

Godfrey L, Grafton-Cardwell E, Kaya H and Chaney W (2005). Microorganisms and their byproducts, nematodes, oils and particle films have important agricultural uses. *California Agriculture,* 59 (1): 35–40.

Gopinath S, Hilda A, Ramesh VM (1998). Detection of biodegradability of oils and related substances. *J Environ Biol,* 19: 157–165.

Guggenberger G, Frey SD, Six J, Paustian K and Elliott ET (1999). Bacterial and fungal cell-wall residues in conventional and no-tillage agroecosystems. *Soil Sci Soc Am J,* 63: 1188–1198.

Gupta PK (2003). *Elements of Biotechnology.* Rastogi Publ, Merut, 602 pp.

Gupta R, Beg QK and Lorenz P. (2002). Bacterial alkaline proteases: molecular approaches and industrial applications. *Appl Microbiol Biotechnol,* 59: 15–32.

Hasan F, Shah AA, Hameed A (2006). Industrial applications of microbial lipases. *Enzyme Microbial Technol,* 39 (2): 235–251.

Harvey PJ, Xiang M, and Palmer JM (2002). Extracellular enzymes in the rhizosphere. In: *Proc. Inter-Cost Workshop on Soil-microbe-Root Interactions: Maximising Phytoremediation/Bioremediation,* Grainau, Germany, p. 23–25.

Hawksworth DL (1991a). *The Biodiversity of Microorganisms and Invertibrates: Its Role in Sustainable Agriculture.* CAB International/Redwood Press, Melksham, UK, 302 pp.

Hawksworth DL (1991b). The fungal dimension of diversity: magnitude, significance, and conservation. *Mycol Res,* 95: 641–655.

Herman RP, Provencio KR, Torrez RJ and Seager GM (1993). Effect of water and nitrogen additions on free-living nitrogen fixer populations in desert grass root zones. *Appl Environ Microbiol,* 59: 3021–3026.

Hotson IK (1982). The avermectins: A new family of antiparasitic agents. *J S Afr Vet Assoc.,* 53(2): 87–90.

Korda A, Santas P, Tenente A and Santas R (1997). Petroleum hydrocarbon bioremediation: Sampling and analytical techniques, *in situ* treatments and commercial microorganisms currently used. *Appl Microbiol Biotechnol,* 48: 677–686.

Lange Ness RL, Vlek PLG (2000). Mechanism of calcium and phosphate release from hydroxy-apatite by mycorrhizal hyphae. *Soil Sci Am J,* 64: 949–955.

Liesack W, Stackebrandt E (1992). Occurrence of novel groups of the domain bacteria as revealed by analysis of genetic material isolated from an Australian terrestrial environment. *J Bacteriol,* 174: 5072–5078.

Loper JE, Haack C and Schroth MN (1985). Population dynamics of soil pseudomonads in the rhizosphere of potato (*Solanum tuberosum* L.). *Appl Environ Microbiol,* 49: 416–422.

Liu S and Sulfita JM (1993). Ecology and evolution of microbial populations for bioremediation. *Trends in Biotech,* 11: 344–352.

Lovley DR (ed). (2000). *Environmental Microbe-Metal Interactions.* ASM Press,Washington, DC.

Lynch JM (1987a). Microbial interactions in the rhizosphere. *Soil Microorg,* 30: 33–41.

Lynch JM (1987b). Soil biology: Accomplishments and potential. *Soil Sci Soc Am J,* 51: 1409–1412.

Lynch JM (1990). *The Rhizosphere.* Wiley, New York.

Margesin R, Zimmerbauer G and Schinner F (1999). Soil lipase activity: A useful indicator of oil biodegradation. *Biotechnol Tech,* 13: 313–333.

Martin FM, Perotto S and Bonfante P (2001). Mycorrhizal fungi: A fungal community at the interface between soil and roots. In: *The Rhizosphere,* (Eds) Pinton R, Varanini Z, Nannipieri P. Dekker, New York, p. 263–296.

Meharg AA and Cairney JWG (2000). Co-evolution of mycorrhizal symbionts and their hosts to metal-contaminated environments. *Adv Ecol Res,* 30: 69–112.

Metting B (1988). Micro-algae in agriculture. In: *Micro-algal Biotechnology,* (Eds) Borowitzka MA, Borowitzka LA. Cambridge Univ Press, Cambridge, p. 288–304.

Morales A, Alvear M, Valenzuela E, Rubio R and Borie F (2007). Effect of inoculation with *Penicillium albidum*, a phosphate-solubilizing fungus, on the growth of *Trifolium pratense* cropped in a volcanic soil. *J Basic Microbiol,* 42(3): 275–280.

Moreno J, Gonsalez Loper J, Vela GR (1986). Survival of *Azotobacter* spp. in dry soils. *Appl Environ Microbiol,* 51: 123–125.

Morley GF, Gadd GM (1995). Sorption of toxic metals by fungi and clay minerals. *Mycol Res,* 99: 1429–1438.

Natesan R. and Shanmugasundaram S. 1989. Extracellular phosphate solubilization by the cyanobacterium *Anabaena* ARM310. *J Biosci,* 14(3): 203–208.

Newman EI (1985). The Rhizosphere: carbon sources and microbial populations. In: *Ecological Interactions in Soil, Plants, Microbes and Animals,* (Eds) Fitter AH, Atkinson D, Read DJ, Usher MB. Blackwell, Oxford, p. 107–121.

Paul EA and Clark FE (1989). *Soil Microbiology and Biochemistry*. Academic Press, San Diego.

Par´e T, Dinel H, Moulin AP and Townley-Smith L (1999). Organic matter quality and structural stability of a Black Chernozemic soil under different manure and tillage practices. *Geoderma,* 91: 311–326.

Pelczar MJ, Chan ECS, Krieg, NR. (1993). *Microbiology*. Tata McGraw Hill Publication, 918 pp.

Peters S. 2002. *Mycorrhiza 101*. Reforestation Technologies International, Salinas, CA.

Pimentel D *et al.* (1995). Environmental and economic costs of soil erosion and conservation benefits. *Science,* 267(24): 1117–1122.

Prescott LM, Harley JP, Klein DA (1996). The diversity of the microbial world. In: *Microbiology,* (Eds) Prescott LM, Harley JP, Klein DA. WCB Publishers, Dubuque, Iowa.

Puget P, Angers DA and Chenu C (1999). Nature of carbohydrates associated with water-stable aggregates of two cultivated soils. *Soil Biol Biochem,* 31: 55–63.

Rangaswami G and Bagyaraj DG (2004). *Agricultural Microbiology,* Prentice Hall of India, 440 pp.

Reforestation Technologies International (2003). *MycoPak Product Information.*

Rillig MC and Steinberg PD (2002). Glomalin production by an arbuscularmycorrhizal fungus: a mechanism of habitat modification? *Soil Biol Biochem,* 34: 1371–1374.

Rillig MC, Wright SF and Eviner T (2002). The role of arbuscularmycorrhizal fungi and glomalin in soil aggregation: comparing effects of five plant species. *Plant Soil,* 238: 325–333.

Robert M and Chenu C (1992). Interactions between soil minerals and microorganisms. In: *Soil Biochemistry,* Vol 7, (Eds) Stotzky G, Bollag JM. Dekker, New York, p. 307–404.

Rodriguez H and Fraga R (1999). Phosphate solubilizing bacteria and their role in plant growth promotion. *Biotech Adv*, 17: 319–339.

Sahgal M and Johri B (2006). Taxonomy of rhizobia: Current status. *Curr Sci*, 90(4): 486–487

Saiz-Jimenez C (1996). The chemical structure of humic substances: recent advances. In: *Humic Substances in Terrestrial Ecosystems* (Ed) Piccolo A . Elsevier, Amsterdam, p. 1–44.

Salt DE, Smith RD and Raskin I (1998). Phytoremediation. *Annu Rev Plant Physiol Plant Mol Biol*, 49: 643–668.

Sayer JA and GaddGM (2001). Binding of cobalt and zinc by organic acids and culture filtrates of *Aspergillus niger* grown in the absence or presence of insoluble cobalt or zinc phosphate. *Mycol Res*, 105: 1261–1267.

Siciliano SD and Germida JJ (1998). Mechanism of phytoremediation: biochemical and ecological interactions between plants and bacteria. *Environm Review*, 6: 65–79.

Slater JH (1988). Microbial population and community dynamics. In: *Microorganisms in Action: Concepts and Application in Microbial Ecology*, (Eds) Lunch JM, Hobbie JB. Blackwell, Oxford, p. 51–74.

Smiles DE (1988). Aspects of the physical environment of soil organisms. *Biol Fertil Soils*, 6: 204–215.

Soil Biology Primer [online]. Available: soils.usda.gov/sqi/concepts/soil_biology/biology.html [access date14–4].

Sreekrishnan TR and Tyagi RD (1994). Heavy metal leaching from sewage sludges: a technoeconomic evaluation of the process options. *Environ Technol*, 15: 531–543.

Sterflinger K, Krumbein WE (1997). Dematiaceous fungi as a major agent for biopitting on Mediterranean marbles and limestones. *Geomicrobiol J*, 14: 219–230.

Subba Rao NS (1995). *Soil Microorganisms and Plant Growth*. Oxford and IBH Publ, New Delhi

Subba Rao NS (1997). *Soil Microbiology*. IBH Publ, Oxford.

Takeda M and Knight JD (2006). Enhanced solubilization of rock phosphate by *Penicillium bilaiae* in pH–buffered solution culture. *Can J Microbiol*, 52(11): 1121–1129.

Tanada Y and Kaya HK. (1993). *Insect Pathology*. Academic Press, San Diego, 666 pp.

Tate RL III (1995). *Soil Microbiology*. Wiley, New York.

Thomashow LS and Weller DM (1995). Current concepts in the use of introduced bacteria for biological disease control: Mechanisms and antifungal metabolites. In: *Plant–Microbe Interactions*, Vol 1, (Eds) G. Stacey and N. Keen. Chapman and Hall, New York, pp. 187–235.

US Department of Agriculture (1998). *Soil Biodiversity*. Soil Quality Information Sheet, Soil Quality Resource Concerns. January. 2 p.

Villar-Igea M, Velázquez E, Rivas R, Willems A, van Berkum P., Trujillo ME, Mateos PF, Gillis M and Martínez-Molina E (2007). Phosphate solubilizing rhizobia originating from Medicago, Melilotus and Trigonella grown in a Spanish soil. In: *First International Meeting on Microbial Phosphate Solubilization Series: Developments in Plant and Soil Sciences*, (Eds) Velazquez E, Rodriguez-Barrueco C. Springer, Netherlands, 102: 149–156.

Vachon RPD, Tyagi J, Auclair C and Wilkinson KJ (1994). Chemical and biological leaching of aluminium from red mud. *Environ Sci Technol*, 28: 26–30.

Vala AK, Vaidya SY and Dube HC (2000). Siderophore production by facultative marine fungi. *Indian J Mar Sci,* 29: 339–340.

Villegas J and Fortin JA (2001). Phosphorus solubilization and pH changes as a result of the interaction between soil bacteria and arbuscular mycorrhizal fungi on a medium containing NH_4^+ as nitrogen source. *Can J Bot,* 79: 865–870.

Visscher PT, Vandenede FP, Schaub BEM and van Gemerden H (1992). Competition between anoxygenic phototrophic bacteria and colorless sulfur bacteria in a microbialmat. *FEMS Microbial Ecol,* 101: 51–58.

Walton BT, Hoylman AM, Perez MM, Anderson TA, Johnson TR, Guthrie EA and Christmas RF (1994). Rizhosphere microbical community as a plant defense against toxic substances in soils. In: *Bioremediation through Rizhosphere Technology,* (Eds) Anderson TA, Coats JR. American Chemical Society: Washington DC, p. 82–92.

Wang G H, Zhou DR, Yang Q, Jin J and Liu XB (2005). Solubilization of rock phosphate in liquid culture by fungal isolates from rhizosphere soil. *Pedosphere,* 15: 532–538.

White C, Sharman K, and Gadd GM (1998). An integrated microbial process for the bioremediation of soil contaminated with toxic metals. *Nature Biotechnol,* 16: 572–575.

Wieland G, Neumann R, Backhaus H (2001). Variation of microbial communities in soil, rhizosphere, and rhizosphere in response to crop species, soil type, and crop development. *Appl Environ Microbiol,* 67: 5849–5854.

Wiener P (1996). Experimental studies on the ecological role of antibiotic production in bacteria. *Evol Ecol,* 10: 405–421.

Wilson EO (1988). *Biodiversity.* National Academy Press, Washington, DC.

Wiseman A (1995). Introduction to principles. In: *Handbook of Enzyme Biotechnology,* 3rd edn (Ed) A Wiseman. Padstow, Cornwall, UK: Ellis Horwood Ltd, TJ Press Ltd, p. 3–8.

Woese CR (1987). Bacterial evolution. *Microbiol Rev,* 51: 221–271.

Wood M (1989). *Soil Biology.* Chapman and Hall, London.

Wright SF (2000). A fluorescent antibody assay for hyphae and glomalin from arbuscular mycorrhizal fungi. *Plant Soil,* 226: 171–177.

Wright SF and Upadhyaya A (1998). A survey of soils for aggregate stability and glomalin, a glycoprotein produced by hyphae of arbuscular mycorrhizal fungi. *Plant Soil,* 198: 97–107.

Wright SF, Franke-Snyder M,Morton JB and Upadhyaya A (1996). Time-course study and partial characterization of a protein on hyphae of arbuscular mycorrhizal fungi during active colonization of roots. *Plant Soil,* 181: 193–203.

Yun J and Ryu S (2005). Screening for novel enzymes from metagenome and SIGEX, as a way to improve it. *Microbial Cell Factories,* 4: 8 (doi: 10.1186/1475–2859–4–8).

Soil Microflora, 2009 *Pages* **192–200**
Editor: **Rajan Kumar Gupta, Mukesh Kumar & Deepak Vyas**
Published by: **DAYA PUBLISHING HOUSE, NEW DELHI**

Chapter 16

Freeze Recovery and Nitrogenase Activity in Antarctic Cyanobacterium *Nostoc commune*

Rajan Kumar Gupta[1] and Mukesh Kumar[2]
[1]*Department of Botany, Pt. L.M.S. Government P.G. College, Rishikesh – 249 201, Uttarakhand*
E-mail: rajankgupta1@rediffmail.com
[2]*Department of Botany, Sahu Jain (P.G.) College, Najibabad, U.P.*

ABSTRACT

Nostoc commune was cultured and propagated on sand with aqueous N-free BG-11 medium. Laboratory experiments were conducted to characterize the *in vivo* freeze recovery physiology of nitrogenase activity. Nitrogenease activity was monitored by the acetylene reduction activity. Frozen *Nostoc* mats were thawed and warmed to 2,4,6,8,10,20, or 25°C, nitrogenase activity was detected within 6h after thawing. Optimum thawing temperature with respect to the recovery of nitrogenase activity was 20°C. In Subsiquent experiments, laboratory grown *Nostoc* mats were used along with the following conditions: prefreezing treatment of 3d of exposure to light or darkness, freezing and then thawing to 20°C in light or darkness, with or without DCMU or chloramphenicol. Approximately 25 per cent of the energy in the initial recovery of nitrogenase activity (up to 12 h after thawing) appeard to be supplied via the utilization of carbon compounds stored before freezing. Photosynthetic condition (*i.e.,* light and without DCMU) were necessary for maximum recovery of nitrogenase activity. In the presence of protein synthesis inhibitor chlormphenicol, nitrogenase activity was still detected at 12 to 48 h after thawing. Although damage may occur to nitrogenase, some of the enzyme was capable of surviving the freeze-thaw period *in vivo*. However, complete recovery of nitrogenase activity (equal to prefreezing activity) may entail some de novo synthesis of nitrogenase.

Introduction

Cyanobacteria, especially terrestrial forms, are exposed frequently to alternating freeze-thaw cycles during the early spring and late fall seasons in Antarctic regions. However, the question of how cyanobacteria respond to freezing has received little attention. Ono and Murata (1981 a, b) investigated the effect of chilling on the photosynthetic activities of *Anacystis nidulans*. They noted that both the dark reactions and primary photochemical reaction are damaged by chilling treatment. They correlated the chilling susceptibility with the fluidity of membrane lipids. Increased membrane permeability, as measured by leakage of potassium ions and other ions from the cytoplasm to the surrounding medium, was attributed to an alterd state of cytoplasmic lipids at chilling temperatures (Ono and Murata 1982). This loss of ions then resulted in an inactivation of photosynthesis.

The effects of freezing on nitrogenase activity has been only studied with nitrogenase extract preparation. Haystead *et al.* (1970) showed that nitrogenase activity in extracts of *Anabaena cylindrica* reapidly decreased when incubated at 0°C. After 12d at 0°C, there was 61.7 per cent inhibition of nitrogenase activity when compared with control samples (20°C). In initial studies on crude extracts of *Clostridium* sp, Dua and Burris (1963) reported an inactivation of approximately 85 per cent after 12 at 0°. Zumft and Mortenson (1975) have the data on the cold lability of nitrogenae (specifically the azoferodoxin protein component). They postulated the possibility of structural changes in protein at lower temperature. There have been no reports describing the effects of freezing on *in vivo* nitrogenase activity in cyanobacteria or the ability of in vivo nitrgenase activity to recover from freezing.

In this study, we examind the recovery response of *in vivo* nitrogenase activity after a freeze period as to the energy source(s) involved, recovery of preexisting enzyme versus *de novo* systhesis, and the effect of prefreezing conditions.

Material and Methods

Culture Conditions

The cultures of *N. commune* was propagated on sand with aqueous modified BG-11 medium (N-free) in glass petri dishes (80 mm diameter). Several attempts were made to isolate heterotrophic diazotrophs associated with the colonies by plating the colonies on various media selective for heterotrophic diazotrophs. All these attempts proved negative. Furthermore no sustained nitrogenase activity occurred in the dark with the algal mats. From this we concluded that no contaminating heterotrophic diazotrophs were present. the cultures were grown on Petri dishes placed under a light bank of continuous cool day light fluorescent tubes (50 $\mu E/m^2/s$). Sand was used as a substratum because it is biologically inert material. Also, using sand, an aquesous medium could be added as needed and therefore eliminate the need for successive transfers (as with the use of an agar medium). The use of an aqueous medium in a simulated terrestrial system also facilitated the use of metabolic inhibitors.

Experimental Conditions

All light treatments were performed in an incubator continuously illuminated with cool day light fluorescent tubes (50 $\mu E/m^2/s$) maintained at 14±1°C. Dark treatments were performed in the same incubator in Petri dishes which were wrapped with alumunium foil. The plates were exposed to either light or darkness or 3 d before freezing. The algal mats were then frozen for 3–7 days at –14°C. Preliminary experiments showed that the duration of freezing between 1 and 20 d had no effect on the recovery of *in vivo* nitrogenase activity. Also our study was focussed on freeze recovery and not chilling injury. thus we wanted complete freezing of the *N. commune* mats. At the beginning of thawing, colonies were exposed to light or darkness alone or with exposure to metabolic inhibitor. The thawing time equilibration to ambient temperature was approximately 20 min.

Metabolic Inhibitor

3-(3,4 dichlorophenyl)-1,1-dimethyl urea (DCMU) was used to inhibit photosynthesis. DCMU was added to experimental cultures to yield a final concentration of 2 µM. Protein synthesis was inhibited with chlormaphenicol added to the experimental cultures to yield a final concentration of 50 µg/ml.

Acetylene Reduction Assay for Nitrogenase Activity

Colonies were separated from substratum, rinsed in the medium and placed in vacutainer tubes (7.5 ml). The tubes were fitted with rubber stopper and injected with acetylene to produce 10-12 per cent acetylene atmosphere in the head space. The tubes were incubated in light at indicated temperatures (see results) for 1 h. Ethylene produced was quantitified with Tracor model 540 gas chromatograph with dual columns (2 mm by 2 m) of Porapak R and flame ionization detector Column injector and detector temperature were maintained at 50 °C.

Results

Freeze Recovery of Nitrogenase Activity

The purpose of the experimentation (Figure 16.1) was to determine at what temperature the greatest recovery of nitrognease activity could be achieved. the experiments were performed for maximum nitrogenase activity in subsequent experimentation. The amount of nitrogenase activity was greatest when colonies were thawed to 20 °C. On the basis of these results, the mats were thawed to 20 °C in all subsequent experiments.

Exposure to light during the thawing period enhances the recovery of nitrogenase activity (Figure 16.2). Colonies thawed during the light recovered detectable nitrogenase activity by 6 h, followed by a rapid increase in activity. As a control for each experiment shown in Figure 16.2 each set was analyzed for nitrogenase activity just before freezing (*i.e.*, after the light or dark prefreeze treatment) to determine the time involved for freeze recovery nitrogenase activity to achieve the level of prefreeze nitrogenae activity. The control for the cultures pretreated in the light and thawed in light showed prefreezing nitrogenase activity of 1.6 µmol of C_2H_4/mg chl a/h. Freeze recovery nitrogenase activity equaled this prefreeze activity by 24-36 h after thawing.

Cyanobacterial mat thawed in the dark (with a light prefreeze treatment) displayed nitrogenase activity at 12h (Figure 16.2) which was equal to 25 per cent of the nitrogenase activity at 12h of mats thawed in light. Negligible nitrogenae activity was measured from 24-72 h after thawing. After exposure to light at 72 h, nitrogenase activity began to increase rapidly. The prefreeze (control) nitrognease activity of these cultures was 1.63 µmol of C_2H_4/µg chl a/h. Freeze recovery nitrogenase activity equaled this prefreeze activity by ca 36 h after exposure to light.

The above procedure was repeated except that colonies were kept in the dark for 72 h before freezing. This dark period was utilized to deplete the cells of stored carbon compounds (Lex and Stewart, 1973), normally used as an energy or reductant source. After the 72 h dark prefreeze treatment, nitrogenase activity in the colonies was undetectable. Upon thawing, recovery of nitrogenase activity was much lower (Figure 16.2) than when cultures were exposed to light 3 d before freezing and thawed in light.

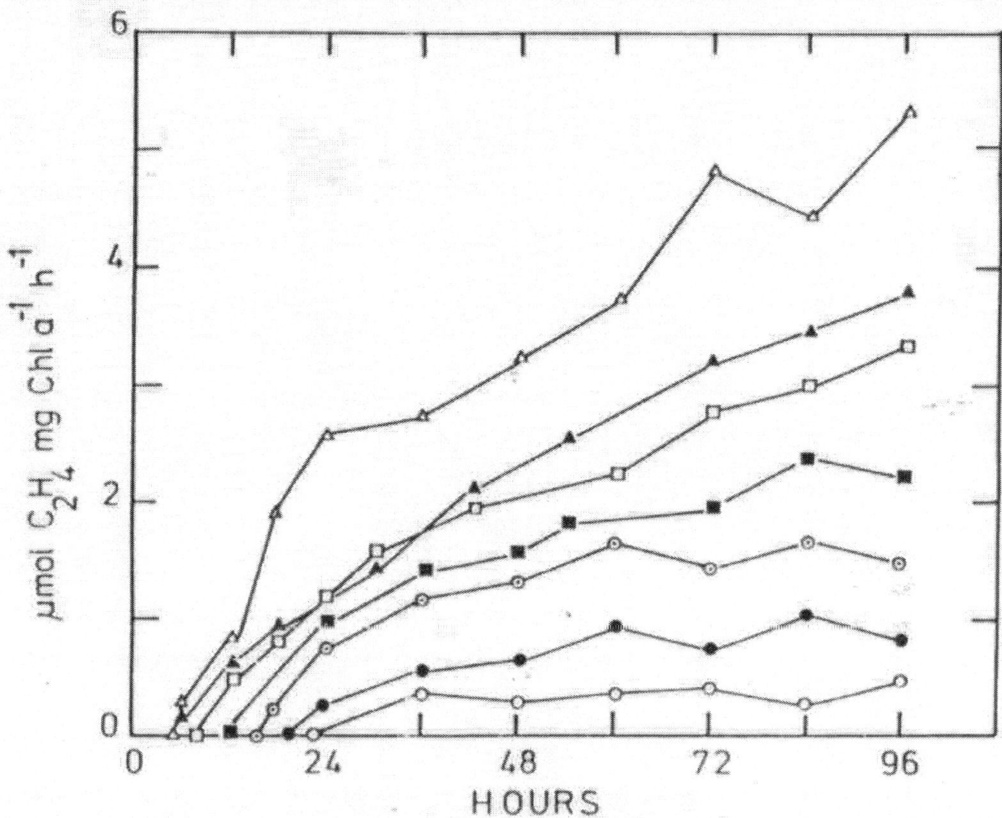

Figure 16.1: Recovery of Nitrogenase Activity from Freezing by *Nostoc commune* Thawed to 2° (O), 4° (●), 6° (①), 8° (■), 10° (□), 20° (Δ) and 25° (▲) Centrigate

Effect of Metabolic Inhibitor on Recovery of Nitrigenase Activity

Figure 16.3 shows the effects of the metabolic inhibitors on the nitrogenase activity of unfrozen *Nostoc* mat. DCMU (an inhibitor of photosynthesis) was effective at 3 µM on *Anbaena cylindrica* (Lex and Stewart, 1973) and also appeared to be effective at this concentration on *Nostoc*. Chlormaphenicol (a translational inhibitor of protein synthesis in bacteria) is effective at 50 µg/ml on *Anabaena cylindrica* (Murray and Benemann, 1979) and also appeared to be effective at this concentration on *Nostoc*.

After a prefreezing treatment of 3 d in the dark, followed by 3 d of freezing, mats were thawed in light in the presence of DCMU. At 48 h. After thawing the DCMU was removed by rinsing the colonies six times with sterile distilled water. Within 12 h after missing, nitrogenase activity was detected.

Nitrogenase activity through 96 h after thawing, increase to more than double its rate of activity from 48 to 96 h after thawing

Table 16.1 shows the effect of chloramphenicol on the recovery of nitrogenase activity from freezing under various prefreezing and thawing conditions. When algal mats were thawed in the presence of

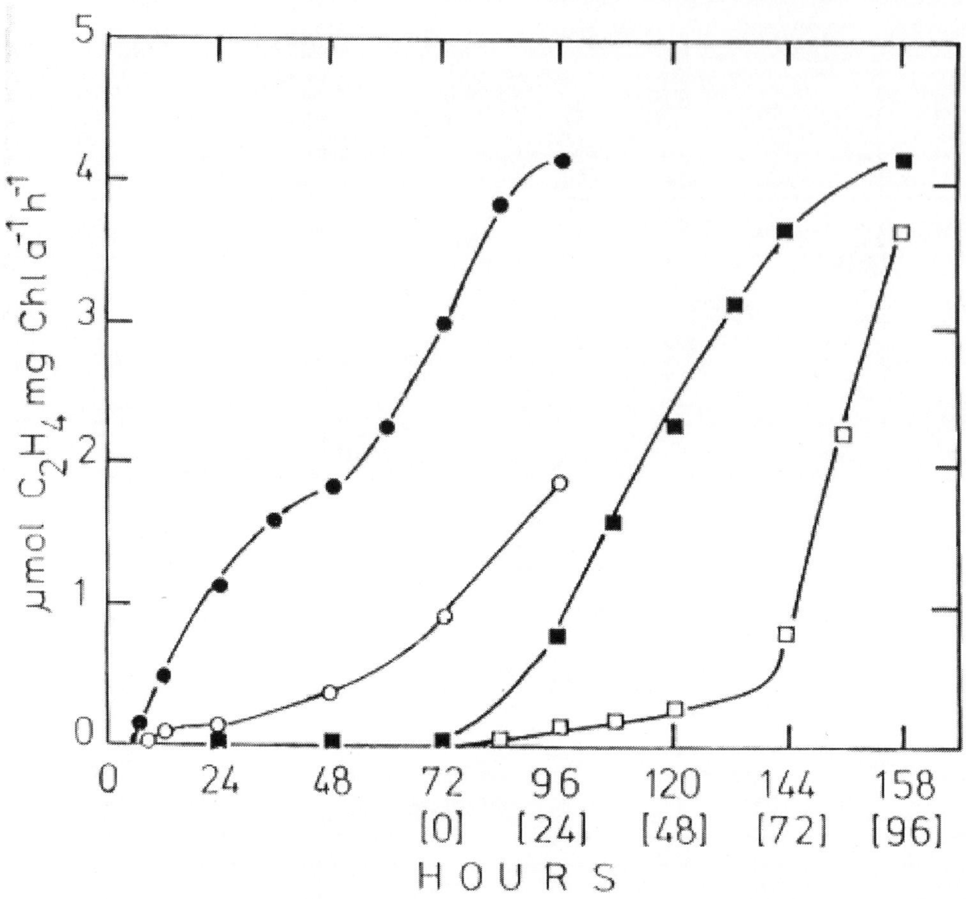

Figure 16.2: Recovery of Nitrogenase Activity from Freezing by *Nostoc commune* Grown Under Laboratory Conditions and Thawed to 20°C

Symbols: ●: **Cultures exposed to light for 3 d before freeznig and thawed in light; ■: Cultures exposed to light for 3 d before freezing and thawed in dark followed by exposure to light at 72h; O: Cultures kept in dark for 3 d before freezing and thawed in light; ❑: Cultures kept in dark for 3 d before freezing and thawed in dark followed by exposure to light at 72 h**

chloramphenicol, the response of nitrogenase activity was similar for the various prefreezing and thawing treatments. When mats were exposed to light before freezing and thawed in the light with chloramphenicol, nitrogenase activity was detected 48 h after thawing. However, the activity was much lower than that of controls. The mats thawed in dark with chloramphenicol showed nitrogenase activity similar to that of controls, although the peak in activity was at 24 h rather than at 12 h after thawing as in controls. By 48 h, nitrogenase activity had declined to equal that of controls, both of which were detected at very low levels. Mats kept in dark before freezing and thawed in the light again showed nitrogenase activity to 48 h after 48 h thawing, and this activity was also much less than that of controls. Nitrogenase activity was detected in these colonies again at 96 h after thawing.

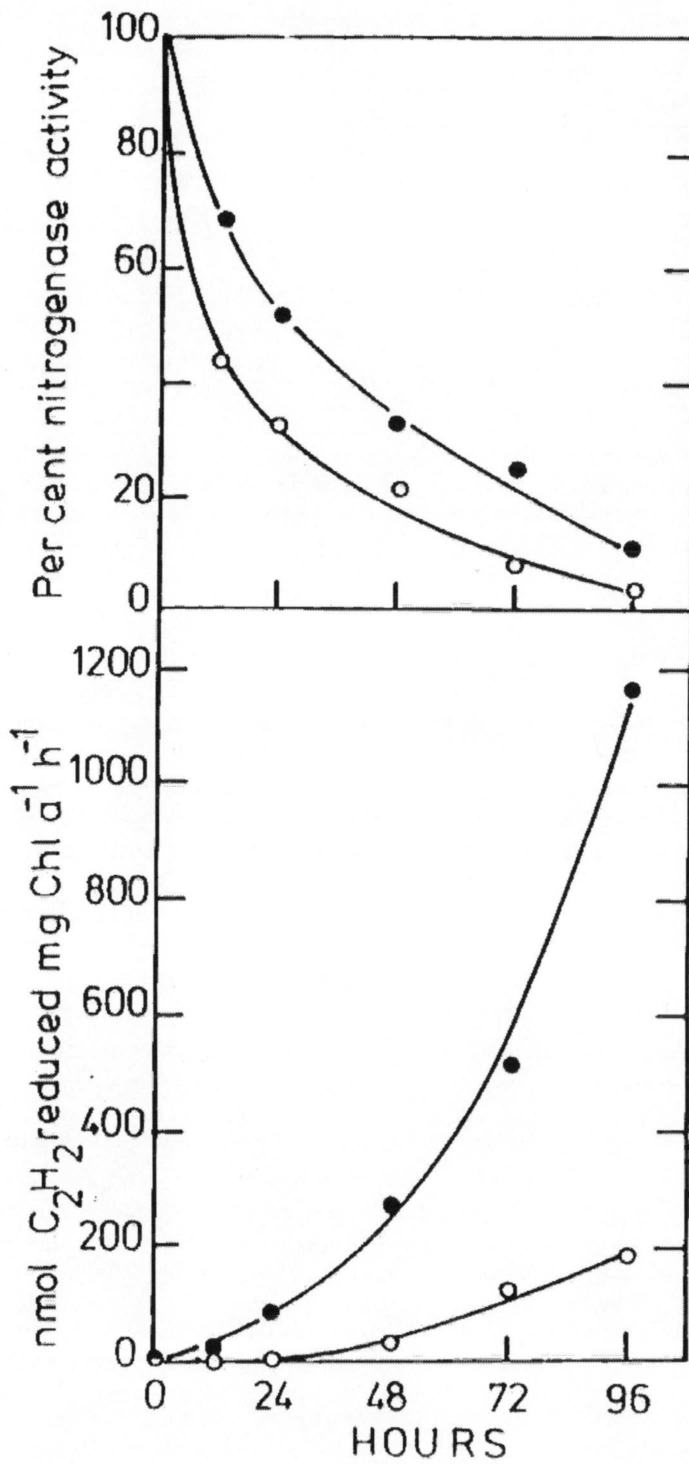

Figure 16.3: Effect of DCMU (●) or Chloramphenicol (O) on the Nitrogenase Activity of Unfrozen Laboratory Grown *Nostoc commune* Mats. Nitrogenase activity in controls were 480 and 3,600 nmol of acetylene reduced per mg of chlorophyll *a* per h for (●) and (O) respectively.

Figure 16.4: Efect of DCMU (●) on the Recovery of Nitrogenase Activity from Freezing in Laboratory Grown *Nostoc commune* Mats. (Control :●)

Table 16.1: Effect of Chloramphenicol (50 □ g/ml) on the Recovery of Nitrogenase Activity from Freezing in Laboratory Grown *Nostoc commune* Mat[a]

Time (h)	Nitrogenase Activity[b]					
	Light[c]				Dark[c] then Light[d]	
	Light[d]		Dark[c]			
	Control	Cm	Control	Cm	Control	Cm
0	0	0	0	0	0	0
12	630	20	190	40	20	50
24	1040	110	20	120	70	10
48	1460	110	20	30	260	60
72	1990	0	20	0	510	0
96	2790	0	20	0	1220	110

(a) *Nostoc* mat was exposed to light or kept dark before freezing and were thawed to 20°C in the light or dark. Colonies kept dark before freezing and thawad in dark showed no detectable nitrogenase activity in controls or with chloramphenicol (cm) treatment; (b) Nitrogenase activity expressed as nmoles of acetylene reduced/mg chl *a* per hour; (c) Prefreezing treatment; (d) Thawing treatment.

Discussion

Preefreezing conditions (light or dark) appear to have a major impact on the freeze recovery of nitrogenase activity. Exposure to light before freezing promotes a more rapid rate of recovery of nitrogenase activity after thawing (Figure 16.2). When mats are kept dark before freezing. The rate of recovery is much slower. The light pretreatment would increase the potential for storage of carbon compounds (*i.e.* carbohydrates, cyanophycean granules), whereas the dark pretreatment would promote the utilization of stored carbon compounds (Lex and Stewart, 1973) Many reports have shown that freeze protection in higher plants can afforded by carbohydrates (Lineberger and Steponkus 1980; Santarius, 1971), amino acids (Heber *et al.*, 1971), organic acids (Santarius, 1973), and proteins (Volger and Heber, 1975). This storage of arganic compounds may raise osmotic potential and hence play an important role in freeze protection. This study shows that conditions enabling cells to store carbon compounds before freezing (assuming compound storage during light pretreatment) have a beneficial effect on the freeze recovery rate of nitrogenase activity in *Nostoc commune.*

The ATP requirement of nitrogenase activity in cyanobacteria is normally supplied by photophosphorylation (Lex and Stewart, 1973) and possibly by oxidative phosphorylation. The oxidation of pyruvate via Krebs cycle activity, usually supplies the majority of reductant (Eady and Postgate, 1974; Lex and Stewart, 1973). Reductant may be supplied by the pentose phosphate pathway (Haystead and Stewart, 1972) involving a glucose -6- phosphate-$NADP^+$ complex. Energy requirements during freeze recovery of nitrogenase activity appear to be met via photosynthetic activity. Utilization of stored carbon compounds (under dark conditions) can supply energy for nitrogenase activity during the first 12 h after thawing (Figure 16.2). However, this source of energy can only support nitrogenase activity equal to about 25 per cent of that in mats exposed to light upon thawing.

When mats were treated with chloramphenicol immediately upon thawing. Nitrogenase activity was still detected 12 to 48 h after thawing (Table 16.1). Murray and Benemann (1979) showed a decrease in nitrogenase activity within 1 h after the addition of chloramphenicol to air grown cultures of *Anabaena cylindrica*. They attributed this decrease to an oxygen inactivation of existing nitrogenase in the aerobically grown cultures and no *de novo* synthesis of nitrogenase to replace inactivated enzyme. When air grown *Nostoc commune* mats were treated with chloramphenicol, nitrogenase activity decreased substantially after 24 h and continued to decrease through 72 h. On the basis of the results of Murray and Benemann (1979), this decrease in nitrogenase activity of *Nostoc* mats may be due to oxygen inactivation or normal turnover of existing nitrogenase activity of *Nostoc* mats may be due to oxygen inavtivation or normal turnover of existing nitrogenase. In the present study, low rates of nitrogenase activity were detected between 12 and 48 h after thawing in the presence of chloramphenicol. This implies that some nitrogenase is capable of surviving a freeze period *in vivo*. However, *de novo* synthesis of nitrogenase or other critical proteins is required for complete recovery of nitrogenase activity from freezing.

In the field, although temperature become limiting from late fall until early spring, *Nostoc commune* mats can recover from freezing with little physiological damage. When condition are favourable it appears that *N. commune* can resume growth and nitrogen fixation activity in existing colonies rather than relying solely on the formation of new colonies. This tolerance to freezing allows the mats to persist and possibly contribute substantial levels of nitrogen to localized field sites.

Acknowledgements

The author wish to thank the Head of Department, Centre of Advanced study in Botany, Prof. A.K. Kashyap for providing the laboratory facilities and all necessary help and UGC, New Delhi for financial support.

References

Dua RD and Burris RH (1963). Stability of nitrogen fixing enzymes and the reactivation of a cold labile enzyme. *Proc Natl Acad Sci USA*, 50: 169–174.

Eady RR and Postgate JR (1974). Nitrogenase. *Nature (London)*, 249: 805–810.

Haystead A and Stewart WDP (1972). Characteristic of the nitrogenase enzyme of blue-green alga *Anabaena cylindrica*. *Arch Microbiol*, 82: 325–336.

Haystead A, Robinson R and Stewart WDP (1970). Nitrogenase activity in extracts of heterocystous and non-heterocystous blue-green algae. *Arch Microbiol*, 74: 235–243.

Heber U, Tyankova L and Santarius KA (1971). Stabilization and inactivation of biological membranes during freezing in the presence of amino acids. *Biochim Biophys Acta*, 241: 587–592.

Lex M and Stewart WDP (1973). Algal nitrogenase, reductant pools and photosytem I activity. *Biochim Biophys Acta*, 292: 436–443.

Lineberger RD and Steponkus PL (1980). Cryoprotection by glucose, sucrose and reffinose to chloroplast thylakoids. *Plant Physiol*, 65: 298–304.

Murrey MA and Benemann JR (1979). Nitrogenase regulation in *Anabaena cylindrica*. *Plant Cell Physiol*, 20: 1391–1401.

Ono T and Murata N (1981a). Chilling susceptibility of the blue green alga *Anacystis nidulans*. I. Effect of growth temperature. *Plant Physiol*, 67: 176–181.

Ono T and Murata N (1981b). Chilling susceptibility of the blue green alga *Anacystis nidulans*. II. Effect of growth temperature. *Plant Physiol*, 67: 182–187.

Ono T and murata N (1982). Chilling susceptibility of the blue green alga *Anacystis nidulans*. III. Effect of growth temperature. *Plant Physiol*, 6: 25–129.

Santarius KA (1971). The effects of freezing on thylakoid membranes in the presence of organic acids. *Plant Physiol*, 48: 156–162.

Santarius KA (1973). The protective effect of sugars on chloroplast membranes during temperature and water stress and its relationship to frost, desiccation and heat resistance. *Planta*, 113: 105–114.

Volger HG and Heber U (1975). Cryoprotective leaf proteins. *Biochim Biophys Acta*, 412: 335–345.

Zumft WG and Mortenson LE (1975). The nitrogen fixing complex of bacteria. *Biochim Biophys Acta*, 416: 1–52

Soil Microflora, 2009 *Pages* **201–211**
Editor: **Rajan Kumar Gupta, Mukesh Kumar & Deepak Vyas**
Published by: **DAYA PUBLISHING HOUSE, NEW DELHI**

Chapter 17

Seasonal Variation in Root Colonization and Rhizosphere Soil Spore Population of Mycorrhiza Species in Various Plants Growing in Alkali Soil

Kaushal Pratap Singh[1*], *Dheeraj Mohan*[1], *Rekha Yadav*[1], *Seema Bhadauria*[1] *and Chatar Singh*[2]

[1]*Microbiology Research Laboratory, Department of Botany, Raja Balwant Singh College, Agra, U.P.*
[2]*Department of Agroforestry, Institute of Agriculture Sciences, Bundelkhand University, Jhansi, U.P.*

ABSTRACT

Nineteen plant species were taken for the study of AM fungi at experimental sites *i.e.* Dannahar, Nouner, Ishwarpur and Ujhaia (Mainpuri district). All weeds were found to have mycorrhizal associations. However, both qualitative and quantitative variations regarding AM association infections level were observed five species of Glomus *viz. Glomus fasciculatum, G. aggregatum, G. constrictum, G. mossease, G. macrocarpus*, two identified species of *Gigaspora*, one *Acaulospora laevis* and *Sclerocystis* were frequently observed. *G. fasciculatum* seems to be a predominant species of AM in *A. nilotica* followed *G. aggregatum*. Among different weeds, Cenchrus observed to be most heavily infected with mycorrhizal fungi showing 88 per cent root colonization along with abundant mycelium, arbuscules and spores but without vesicles. Environmental conditions had significant effects on the distribution, density and composition of AM fungi. In the present investigation AM fungal sporulation was maximum in winter season and minimum in rainy season. During rainy season the quantity of soil nitrogen was higher, which suppressed

* Corresponding Author: E-mail: kaushalmpi1978@yahoo.com, Phone: +91-9412660984

the AM fungi colonization and favoured the sporulation resulting in an increase in spore population during winter season. During rainy season, not only the number of spores decreased but percentage of root infection was also decreased.

***Keywords**: Seasonal variation, Root colonization, Mycorrhizal spore population, Plants, Alkali soil.*

Introduction

Mycorrhiza terms denote the symbiotic association between plant roots and fungal mycelia. Frank in 1885, a German pathologist, first introduced the word mycorrhiza. This symbiotic relationship is for the mutual benefit of both the partners. Such associations have been known for over 100 years, but only in the past 30 years their significance to plant growth and health was realized. AM fungi are geographically ubiquitous and occur over a broad ecological range. They are commonly found in association with agricultural crops, most shrubs, most tropical tree species and some temperate tree species. AM fungi have been observed in 1000 genera of plants representing some 200 families. There are at least 3,00,000 receptive hosts in the world flora (Kendrick and Berch, 1985) and there are 120 species.

Arbuscular mycorrhizas occur in most angiosperms as well as in some gymnosperms, pteridophytes and bryophytes. Meyer (1973) estimated that most flowering plants have endomycorrhiza (almost entirely of the AM type) in contrast to only 3 per cent with ectomycorrhiza. As for known Arbuscular Mycorrhiza is absent only from a few plant families, mainly those that form only ectomycorrhiza (Pinaceae, Betulaceae) or the two other specific types of endomycorrhiza (Ericales, Orchidaceae). Biological boundaries are never rigid, however, and some plant families and even species can form both ecto and Arbuscular Mycorrhizal association, *e.g.* oak, hazel, juniper, sweetgum, poplar, leptosermum, tulip and some eucalyptus.

Most plant species in the two families of major economic importance, the Fabaceae and Poaceae, are normally mycorrhizal. Yet even here some species are more prone to AM infection than others. The root systems of many fodder legumes, *e.g. Stylosanthes* sps. and *Trifolium* sps., usually have dense AM infections where as lupins have little or none (Morley and Mosse, 1976). Rye may be less mycorrhizal than other cereals and maize is usually heavily infected (Gerdemann, 1968). In other crops, where considerable AM infection have been reported include sorghum, barley, wheat, upland rice, grapevine, citrus, tobacco, cotton, pineapple, onion etc.

AM fungi are virtually ubiquitous, being present in tropical, temperate and arctic regions. Over 81 per cent of the 89 shrub species in arid and semi arid areas of the world are endomycorrhizal (Lindsey, 1984). Mycorrhization of forest tree species and its effect on biomass production is very well known (Kormanik, 1980; Bhagyaraj *et al.*, 1989; Reena and Bhagyaraj, 1990; Thapar *et al.*, 1993) Durga and Gupta, 1995; Rahangdale and Gupta, 1998; Shanmughavel *et al.*, 2001; Chouhan and Pokhriyal, 2002).

Mycorrhization in tree species like *Acacia nilotica, A. arabica, Leuceana leucocephala, Prosopis juliflora, Albizzia lebbeck, Parkinsonia aculata, Tamarindus indica, Cassia siammea, Dalbergia sissoo, Zizyphus* sp. growing in alkaline region is also reported by Bhadauria and Yadav (1999).

In the present article, the effect of different seasons on the percent root colonization and rhizosphere soil spore population of mycorrhizal species in various plants growing in alkali soil shall be discussed.

Material and Methods

Collection of Arbuscular Mycorrhizal Spores from Rhizosphere

The most common method of processing a sample, after it has been thoroughly mixed, is the wet sieving and decanting technique. Originally developed by nematologists, this technique was adapted for AM studies by Gerdemann and Nicolson (1963). The details of this method is as follows:

1. Mixed a measured portion of soil (50 g is suitable for quantitative studies) by hand in luke-warm water (200 ml is a convenient volume) in large beaker until all soil aggregates have dispersed to leave a uniform suspension.
2. Decanted most of the suspension through a 710 µm sieve (supported in a funnel) in to 1 litre graduated cylinder.
3. Resuspended the residue in more water and decanted, repeating this four or five times to give about 700 ml in the cylinder and leaving only grit, sand and heavy organic particles in the beaker.
4. Washed roots and other organic matter on the sieve with a fine jet of water, *e.g.* from a squeeze bottle and collected washings in the cylinder.

Infestation

Roots were washed under running tap water to quantify mycorrhizal association properly washed roots were cleared in boiling 10 per cent KOH aqueous solution for 48 hours and stained in tryphan blue following several washing in distilled water to drain out KOH (Phillips and Hayman, 1970). Stained roots were cut in to 1 cm segments and 100 such segments were randomly picked up and examined under microscope. Arbuscular mycorrhizal colonization was determined by Nicolson's formula (1955):

$$\text{Per cent colonization} = \frac{\text{Number of root segments colonized}}{\text{Total number of segments examined}} \times 100$$

Identification of AM Fungi

The identification of spores was done with the help of their characters described in standard keys *viz.* Gerdemann and Bakshi (1976), Hall (1977), Amess and Schneder (1979), Schenck (1979), Schenck and Kinlock (1980), Schenck and Smith (1982), Spain *et al.* (1990), Mukherji *et al.* (1992) and Mehrotra (1995).

Results

Maximum percent root colonization was found in the summer season in Cenchrus grass *i.e.* 88 per cent followed by rain 82 per cent and winter 79 per cent. The spore population per 100 g of soil was found maximum in winter (270/100 g of soil) followed by summer (197/100 g of soil) and rain (181/100 g of soil). AM species, which were found in this grass, were *Glomus fasciculatum*, *Glomus mosseae* and *Glomus macrocarpus*.

In case of trees, the maximum percent root colonization was found in Acacia nilotica in summer season followed by rain and winter season *i.e.* (82 per cent, 80 per cent and 79 per cent respectively) and the spore population in Acacia nilotica per 100 g of soil was maximum in winter followed by summer and rain (269, 177, 174/100 g of soil). The AM fungi found in this tree were *Glomus fasciculatum*, *Glomus aggregatum* and *Gigaspora margarita*.

Table 17.1a: Seasonal Variation in Per cent Root Colonization and Rhizosphere Soil Spore Population of 19 Plant Species at Study Site

Sl.No.	Name of Plant Species	Summer		Rain		Winter		AM Species
		% Infection	Spore Pop. /100 g soil	% Infection	Spore Pop. /100 g soil	% Infection	Spore Pop. /100 g soil	
				[A] Trees				
1.	Acacia fomesiana	61	145	57	101	52	196	Gf, Gm
2.	*A. nilotica*	82	177	80	174	76	269	Gm, Ga, Gf
3.	*A. tortilus*	67	144	59	107	57	200	Gm, Gmos
4.	*L. leucocephala*	55	100	53	98	50	196	Gf, Gma
5.	*Prosopis juliflora*	79	171	77	167	75	268	Gf, Gm
6.	*Dalbergia sissoo*	75	170	72	161	71	257	Gm, Ss
7.	*Perkinsonia aculata*	33	62	30	51	27	149	Gf, Gm
8.	*Emblica officinalis*	60	146	57	140	55	238	Al, Gm
9.	*Eucalyptus hybrid*	67	149	61	142	58	245	Gmos, Gm
10.	*Albizia lebbeck*	10	20	8	15	7	93	Gmos, Ga
				[B] Grasses				
11.	*Sporobolus diander*	50	108	47	102	35	199	Gf, Gc
12.	*Solanum xanthium*	42	107	38	103	32	198	Al, Gm
13.	*Xanthium xanthocarpus*	35	72	33	91	31	185	Gc, Sc
14.	*Achyranthes* sp.	21	65	18	62	16	150	Sc, Gcs
15.	*Cenchrus* sp.	88	197	82	181	79	270	Gf, Gmos, Gm
16.	*Zizypus* sp.	60	146	57	109	54	200	Gf, Gmos
17.	*Rumex dentalis*	15	57	12	51	10	46	Gf, Sc
18.	*Sporobolus munja*	39	75	37	69	36	162	Gf, G. sps
19.	*Comopus didymus*	36	72	34	68	32	160	Gcs

Gf: *Glomus fasciculatum*; Ga: *Glomus aggregatum;* Gm: *Gigaspora margarita;* Gcs: *Glomus constrictum*; Al: *Acaulospora laevis*; Gc: *Gigospora calospora;* Gmos: *Glomus mosseae*; Sc: *Sclerocystis coremoides*; Gma: *Gigospora macrocarpus.*

Table 17.1b: Seasonal Variation in Per cent Root Colonization and Rhizosphere Soil Spore Population of 19 Plant Species at Study Site

Analysis of Variance (% Root Colonization)					*Analysis of Variance (% Spore population)*				
Source	*DF*	*SS*	*MS*	*F*	*Source*	*DF*	*SS*	*MS*	*F*
P	18	134184.01	7454.67	2944.67	P	18	693342.25	38519.01	14765.20
S	2	1801.40	900.70	355.79	S	2	451222.20	225611.10	86481.70
P x S	36	525.80	14.61	5.77	P x S	36	44811.27	1244.76	477.14
Error	228	577.20	2.53		Error	228	594.80	2.61	

Contd...

Table 17.1b–Contd...

CD for Significant Treatments				CD for Significant Treatments			
Treatment	S. E. (M)	t-val at 5%	CD at 5%	Treatment	S. E. (M)	t-val at 5%	CD at 5%
P	0.41082	1.96	1.139	P	0.41703	1.96	1.156
S	0.16324	1.96	0.452	S	0.16571	1.96	0.459
P x S	0.71156	1.96	1.972	P x S	0.72233	1.96	2.002
C. V.		3.32 per cent		C. V.		1.17 per cent	

Prosopis juliflora showed 79 per cent root colonization in summer season followed by rain (77 per cent) and winter (75 per cent). The spore population per 100 g of soil in Prosopis juliflora showed maximum number in winter season *i.e.* 171 and 167/100 g soil respectively. The AM species associated with *P. juliflora* were *Glomus fasciculatum* and *Gigaspora margarita.*

In case of Dalbergia sissoo, the maximum percent root colonization was found in summer season followed by rain and winter (*i.e.* 75 per cent, 72 per cent and 71 per cent respectively) while the spore population per 100 g of soil was found maximum in winter followed by summer and rainy seasons *i.e.* 257, 170 and 161/100 g of soil. The AM species that were associated with Dalbergia sissoo were *Gigaspora margarita* and *Sclerocystis coremoides.*

Like wise in the case of Eucalyptus hybrid maximum percent root colonization was found in summer season followed by rain and winter seasons (67, 61 and 58 per cent respectively) while spore population per 100 g of soil were maximum in winter season followed by summer and rainy season (*i.e.* 245, 149 and 142 spores/100 g of soil). The AM species associated with E. hybrid were *Glomus mosseae* and *Glomus aggregatum.*

Acacia tortilus and *Acacia fornesiana* were also showed maximum percent root colonization in summer season (62 per cent and 61 per cent) followed by rain (59 per cent and 57 per cent) and least percent root colonization was found in winter season (57 per cent and 52 per cent respectively). The spore populations per 100 g of soil were found maximum in winter (200 and 196 spores/100 g soil) respectively followed by summer (144 and 145/100 g soil) and the least spore population were found in rain *i.e.* 107 and 101/100 g soil respectively. The AM species found in *Acacia tortilus* and *Acacia farnesiana* were *Gigaspora margarita, Glomus mosseae* and *Glomus fasciculatum* respectively.

Emblica officinalis, Leucaena leucocephala and *Zizyphus* sps. Also showed good root percent colonization in summer (*i.e.* 60 per cent, 55 per cent and 60 per cent respectively) followed by rain (57 per cent, 53 per cent and 57 per cent respectively) and winter (55 per cent, 50 per cent and 54 per cent respectively). However, spore population/100 g soil was found maximum in winter season (*i.e.* 238, 196 and 200 spores/100 g soil respectively). The AM species that were associated with these plants were *Acaulospora laevis, Gigaspora margarita, Glomus fasciculatum, Glomus macrocarpus, Glomus mosseae* respectively.

In case of *Sporobolus diander, Solanum xanthium* and *Sporobolus munja* the percent root colonization were found maximum in summer season (*i.e.* 50 per cent, 42 per cent and 39 per cent respectively) followed by rain (*i.e.* 47 per cent, 38 per cent and 37 per cent respectively) and winter seasons (*i.e.* 35 per cent, 32 per cent and 36 per cent respectively). The spore populations per 100 g soil were found maximum in winter season (*i.e.* 199, 198 and 162 spores/100 g soil) followed by summer (*i.e.* 108, 107 and 75 spores/100 g soil respectively) and rainy seasons (*i.e.* 102, 103 and 69 spores/100 g soil). The

Table 17.2: Morphology of AM Fungi Isolated from Rhizosphere Soils of Study Site

S. No.	*AM Species*	*Spore Shape*	*Spore Size (m)*	*Nature of Spore*	*Thickness of Compound (m)*	*Number of Wall Layer*	*Wall Layers Morphology*	*Size of Hyphae (m)*	*Nature of Hyphae*
1.	*Acaulospora laevis*	Subglobose	87.51	Deep yellow	8.9	6 in 4 group	Outer wall yellow laminated	26.77	Subglobose
2.	*Entrophosphora columiana*	Subglobose	79.64	Light golden brown colour	3.83	7 group	Outer wall hyaline	26.44	Subglobose
3.	*Glomus fasciculatum*	Globose	40	Contains oil globules	4	3	Outer wall thin yellowish	4	Cylindrical
4.	*Glomus mosseae*	Globose	12	Smooth white	5	3	Outer wall thin yellowish	4	Cylindrical
5.	*Glomus macrocarpus*	Globose	100	In loose clusters	–	–	–	–	Substending hyphae
6.	*Glomus aggregatum*	Globose	40	Yellow	4	4	Outer wall thicker yellowish	3	Recurve sharp at spore base
7.	*Gigaspora margarita*	Spherical	30	Dull white with cream coloured tinct	10	8	–	–	–
8.	*Gigaspora calospora*	Spherical	140	Bright orange	10	5	Outer wall thin	30	Septate
9.	*Sclerocystis coremoides*	Subglobose pulvinate flattened at base	340	Peridium inter oven 20–70 thick dull	35	50	Tightly grouped brown	2	–

Figure 17.1: The growth of *Prosopis juliflora* and Grass in Alkaline Soil

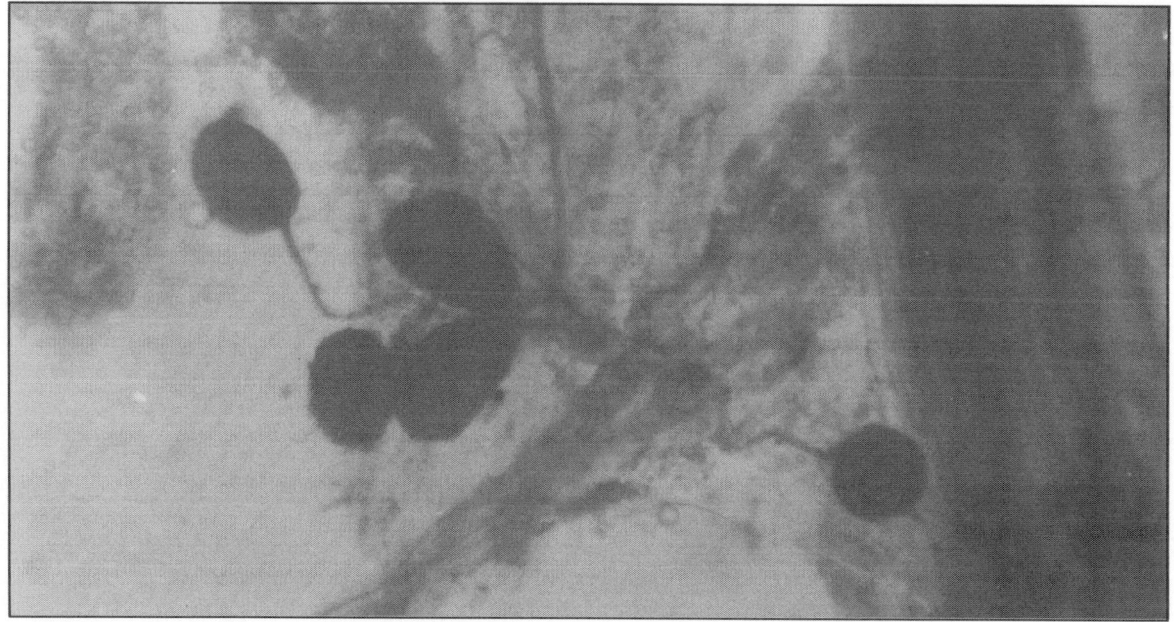

Figure 17.2: Spores of *Glomus fasciculatum* Attached to a Mycorrhizal *Acacia nilotica* Root, Stout, Permanent Hyphae, Finally Branbched, Ephemeral Lateral Hyphae. X50.

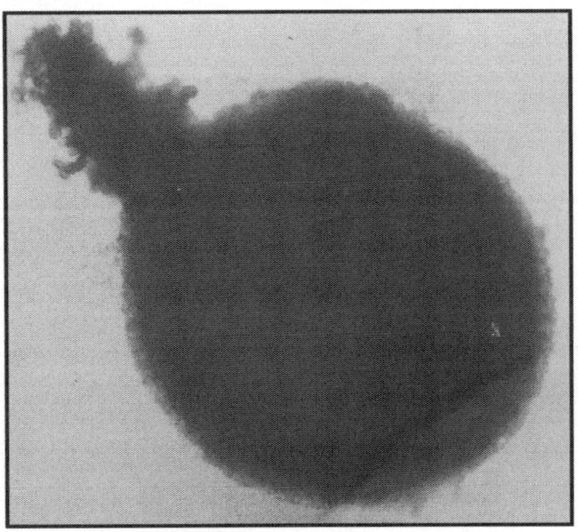

Figure 17.3: Spore of *Acaulospora laevis*

Figure 17.4: Spore of *Glomus constrictum*

AM species associated with *Sporobolus diander*, *Solanum xanthium* and *Sporobolus munja* were *Acaulospora laevis*, *Gigaspora margarita*, *Gigaspora calospora*, *Sclerocystis coremoides* and *Glomus fasciculatum* respectively. *Cornopus didymus*, *Xanthium xanthocarpum* and *Achyranthus aspera* showed maximum percent root colonization in summer season *i.e.* 36 per cent, 35 per cent and 21 per cent followed by rainy season 34 per cent, 33 per cent and 18 per cent respectively. The least colonization was found in winter season *i.e.* 32 per cent, 31 per cent and 16 per cent respectively. However, spore population/100 g soil were found maximum in winter season *i.e.* 165, 160 and 150 spores/100 g of soil followed by summer season *i.e.* 72, 72 and 65 spores/100 g soil and least spore population in *Cornopus didymus*, *Xanthium xanthocarpum* and *Achyranthus aspera* were found in rainy season *i.e.* 68, 91 and 62 spores/ 100 g soil respectively. The AM species associated with these plants were *Glomus constrictum* and *Sclerocystis coremoides* and *Glomus aggregatum* respectively.

In case of *Albizia lebbeck* and *Rumex dentalis*, the maximum percent root colonization was found in summer season *i.e.* 10 per cent and 15 per cent followed by winter season *i.e.* 8 per cent and 12 per cent respectively. The least percent root colonization was found in winter season *i.e.* 7 per cent and 10 per cent. The spore population in *Albizia lebbeck* and *Rumex dentalis* recorded maximum spore population in winter season 93 and 46 spores/100 g soil and the least spore population were recorded in rainy season *i.e.* 15 and 12 spores/100 g soil. The AM species associated with these plants were *Glomus fasciculatum*, *Gigaspora calospora* and *G. calospora* and *Sclerocystis coremoides* respectively.

Discussion

Nineteen plant species were taken for the study of AM fungi at experimental sites *i.e.* Dannahar, Nouner, Ishwarpur and Ujhaia (Mainpuri district). All weeds were found to have mycorrhizal associations. However, both qualitative and quantitative variations regarding AM association infections level were observed five species of Glomus *viz. Glomus fasciculatum*, *G. aggregatum*, *G. constrictum*, *G. mossease*, *G. macrocarpus*, two identified species of *Gigaspora*, one *Acaulospora laevis* and *Sclerocystis* were frequently observed. *G. fasciculatum* seems to be a predominant species of AM in *A. nilotica* followed *G. aggregatum*. Among different weeds, Cenchrus observed to be most heavily infected with mycorrhizal fungi showing 88 per cent root colonization along with abundant mycelium, arbuscules and spores but without vesicles. Further in most of weeds more than one species of Glomus was found to be associated, although. Brundrett and Kendrick (1990) has indicated that members of certain families including Chenopodiaceae and Caryophyllaceae rarely or mycorrhizal association but in the present studies the weed belonging to these families were found to readily form mycorrhizal infection status of the weeds with respect to the soil types and tree species.

In case of trees, all trees were found to have mycorrhizal associations. However, both quantitative and qualitative variations regarding AM association level were observed, 5 species of *Glomus viz. Glomus fasciculatum, G. mosseae, G. constrictum, G. macrocarpus, G. aggregatum*, 2 species of *Gigaspora viz. Gigaspora margarita* and *Gigaspora calospora*, one *Acaulospora laevis*, one *Sclerocystis coremoides* and one unidentified *Sclerocystis* sps. (Bhadauria *et al.*, 1998). Percent infection and spore population in the mycorhizosphere of all the nineteen plant species varied seasonally. It may be due to variation in environmental factors, such as total precipitation, temperature, and relative humidity etc. This observation is in conformity with that of Bhadauria and Yadav (1999).

Mycorrhizal activity is defined as a total of the actions and the interaction of host, AM fungi, rhizosphere, microorganisms and the environment. Environmental conditions had significant effects on the distribution, density and composition of AM fungi. In the present investigation AM fungal sporulation was maximum in winter season and minimum in rainy season. Lussenhop (1996) was also found similar results. During rainy season the quantity of soil nitrogen was higher, which suppressed the AM fungi colonization and favoured the sporulation resulting in an increase in spore population during winter season. During rainy season, not only the number of spores decreased but percentage of root infection was also decreased. Thimn and Larink (1995) have reported that wet conditions in the soil usually retarded the AM fungal infection. It may be also due to substrate competition with other rhizosphere microorganisms. Hass and Menge (1990); Ross and Ruttencutter (1977) have noted the microbial suppression of AM fungal sporulation favoured by suitable environmental conditions, which increased the microbial competition with AM fungi.

References

Amess RN and Schneder RW (1979). Entrophosphora: A new genus in the Endogonaceae. *Mycotaxon*. 8: 347–352.

Bhadauria S, Yadav R and Singh R (1998). Survey for detection of native high pH tolerant VAM fungi in alkaline soil. *Proc. Nat. Acad. Sic. India*, 68(B), II.

Bhadauria S and Yadav R (1999). Vesicular Arbuscular mycorrhizal association in fuel wood trees growing in alkaline soil. *Mycorrhiza News*, 10(4): 14.

Bhagyaraj DJ, Byra Reddy MS and Nalini PA (1989). Selection of an efficient inoculant VA–Mycorrhizal fungus for Leucaena. *For Eco Manage*, 27: 81–85.

Brundrett M and Kendrick B (1990). The roots and mycorrhizas of herbaceous woodland plant structural aspects of morphology. *New Phytol*, 114(3): 469–479.

Chauhan YS and Pokhriyal TC (2002). Effects of nitrogen and Rhizobium inoculation treatments on some growth parameters in *Albizia lebbek* (L) Benth seedling. *Indian Forester*, 27(4): 316–322.

Durga VVK and Gupta S (1995). Effect of vesicular Arbuscular mycorrizae on the growth and miniral nutrient of teak (*Tectona grandis* Linn. F.) *Indian Forester*, 121: 518–527.

Frank AB (1885). Uber die auf Wurzei Symbiose beruhende Evanahurg Gewiisser baume durch Unterirdische. *Pilze Ber Dtsch Bot Ges*, 3: 128–145.

Gerdemann JW (1968). Vesicular Arbuscular mycorrhizae and plant growth. *Ann Rev Phytopathol*, 6: 397–418.

Gerdemann JW and Bakshi BK (1976). Enzymatic study of the metabolism of vesicular arbuscular mycorrhiza. I. Effect of mycorrhiza formation and phosphorus nutrition on soluble phosphate activities in onion roots. *Physiol Veg*, 14: 833–841.

Gerdemann JW and Nicolson TH (1963). Spores of mycorrhizal endogone species extracted from soil by wet sieving and decanting. *Brt Mycol Soc*, 4–6: 235–244.

Hall IR (1977). Species and Mycorrhizal infection of Newzealand Endogonaceae. *Trans Brit Mycol Soc*, 68: 341–356.

Hass TUI and Menge JA (1990). VAM fungi and soil characteristics in Avocado (Persea Americana mill.) orchard soils. *Plant and Soil*, pp 127–212.

Kendrick B and Berch S (1985). Mycorrhiza application on agriculture and forestry. In: *Comprehensive Biotechnology*, (Eds) Robinson CW and Howel JA, (Ed-in-Chief) Moo Young. Pergamon Press, Oxford, 4: 109–152.

Kormanik PP (1980). Effect of nursery practices on VAM development and hardwood seedling production. In: *Proc Tree Nursery Conf*, September 2–4, Lake Barkely, USA.

Lindsey DL (1984). The role of vesicular Arbuscular mycorrhizae in shrub establishment: VA mycorrhizae and reclamation of arid and semiarid lands. In: *Proc Conf on Mycorrhizae and Reclamation of Arid and Semiarid Lands*, Dubios, Wyomina, p. 53–68.

Lussenhop J (1996). Collembola as mediators of microbial symbiont effects upon soybean. *Soil Biology and Biochemistry*, 28 (3): 363–369.

Mehrotra VS (1995). Arbuscular mycorrhizal association in plant colonizing over burden soil at an open cast coalmine city. Mycorrhiza biofertilizers for the future. In: *Proc. of the sSecond National Conf on Mycorrhiza*, (Eds) A Adholiya and S Singh. TERI, New Delhi, pp 22–28.

Meyer FH (1973). Distribution of ectomycorrhizae in native and man made forests. In: *Ectomycorrhizae: Their Ecology and Physiology*, (Eds) GC Marks and TT Kozlowski. Academic Press, New York, pp. 79–105.

Morley CD and Mosse B (1976). Abnormal vesicular Arbuscular mycorrhizal infection in white clover induced by Lupin. *Trans Brit Mycol Soc*, 67: 510–513.

Mukherji KG, Sharma MM, Kaushik A and Raut P (1992). Taxonomy of the VAM fungi. In: *Mycorrhizae: An Asian Overview*. TERI, New Delhi, pp 12–22.

Nicolson TH (1955). The mycotrophic habit in grasses. *PhD Thesis*, University of Nottingham.

Phillips JM and Hyaman DS (1970). Improved procedures for clearing roots and staining parasitic and vesicular clearing roots and staining parasitic and vesicular Arbuscular mycorrhizal fungi. *Trans Brit Myco Soc*, 55: 158–161.

Rahangdale R and Gupta N (1998). Selection of VAM inoculant for some forest tree species. *Indian Forester*, 23(4): 331.

Reena S and Bhagyaraj DJ (1990). Growth stimulation of *Tamarindus indica* by selected VAM fungi. *World J Micobial and Biotechnical*, 6: 59–63.

Ross JP and Ruttencuter R (1977). Population dynamic of two vesicular arbuscular mycorrhizal fungi and the role of hyperparasitic fungi. *Phytopathology*, 61: 490–496.

Schcnbeck F (1979). Endomycorrhizae in relation to plant diseases. In: *Soil Borne Plant Pathogens* (Eds) Scheppers B and Gams W. Academic Press, p. 271.

Schenck NC and Kinlock RA (1980). Incidence of mycorrhizal fungi on six field crops in monoculture on newly cleared woodland site. *Mycologia*, 72: 445–456.

Schenck NC and Smith GS (1982). Additional new and unreported species of mycorrhizal fungi (Endogonaceae) from Florida. *Mycologia*, 74: 77–82.

Shanmughavel P, Thangavel P and Peddappaiah RS (2001). Growth performance and economic return of Leucaena leucocephala in agroforestry. *Indian Journal of Forestry*, 24(4): 480–483.

Spain JL, Silverding EL and Schenck NC (1990). *Gigospora ramisporaphora*: A new species with novel sporophores from Brazil. *Mycotaxon*, 34: 667–677.

Thapar HS, Vijayan AK and Uniyal K (1993). Vesicular Arbuscular mycorrhizal association and root colonization in some important tree species. *Indian Forester*, 18(3): 207–212

Thimn T and Larink O (1995). Grazing preferences of some Collembola for endomycorrhizal fungi. *Biology and Fertility of Soil*, 19(2–3): 266–268.

Soil Microflora, 2009 *Pages* **212–228**
Editor: **Rajan Kumar Gupta, Mukesh Kumar & Deepak Vyas**
Published by: **DAYA PUBLISHING HOUSE, NEW DELHI**

Chapter 18

Nitrogen Fixing Soil Microflora

Yashveer Singh
Department of Botany, Dr. S.P. Mukherjee Govt. Degree College, Bhadohi, U.P.
E-mail: yashveersingh_16@yahoo.co.in

ABSTRACT

The requirement of nitrogenous compounds of plants is supplied as fertilizers which are rather expensive and beyond the reach of small and poor farmers in developing countries. Organic nitrogen is a major constituent of all living organisms. Utilization of free-living nitrogen-fixing cyanobacteria (blue-green algae) and symbiotic fern *Azolla* and bacterium *Rhizobium* in agriculture and economy has important role in many countries. The nitrogen-fixing soil microflora, cyanobacteria, as biofertilizer are being used in paddy fields. The pot trials suggested that the heterocystous cyanobacterial N_2-fixation rates are as much as 70 kg. N ha^{-1} in six weeks and rice grains yield of 33 per cent. Nitrogen-fixation and grain yields from the field trials have been judged against controls without added inorganic nitrogen in the field and an average of 30 kg. N ha^{-1} is fixed per rice cycle. Cyanobacterial inocula can be achieved from a small pond or pit in 4-5 weeks. Small poly-packs of 100-200 gm of dry algal flakes of selected species can be used and sold for algalization of paddy field.

The association between cyanobacterium *Anabaena azolla* and water fern *Azolla* is one of the most efficient N_2-fixing system. As a green manure, this system can fix about 100-150 kg. of nitrogen per hectare per year in approximately 50-60 tonnes of biomass. One crop of *Azolla* applied to the paddy field is roughly equivalent to the addition of 30 kg. of urea per hectare. There is about 18 per cent greater rice yield in fields with *Azolla* compared to field without the fern as being common for China. The heterocystous frequency of *A. azollae* is 25 per cent compared to 5-10 per cent in non-symbiotic species

Undoubtedly the best known N_2-fixing symbiotic soil microflora is the bacterium *Rhizobium* which grows aerobically at 15-30°C at pH around 7.0. Species of *Rhizobium* live in root nodules of leguminous plants. The growth of legumes is greatly increased by this symbiosis especially on nitrogen poor soils. For the symbiotic association, the plant roots and rhizobial cells must come in contact in the soil and the plant and rhizobia have to recognize each other. The repeated use of rhizobial inoculated seeds on soils for a few years builds up the population of the derived rhizobial strain and leads to optimal and sustained N_2-fixation.

Keywords: *Azolla, Cyanobacteria, Nitrogen-fixation, Rhizobium, Soil microflora.*

Introduction

Nitrogen fixation and photosynthesis are undoubtedly the two most important physiological processes of plants and microbes on earth. The former involves the fixation of nitrogen, and the latter that of carbon. The atmosphere contains only about 360 ppm of CO_2 but it has about 80 per cent N_2. This nitrogen is, however, not in a form that most plants or animals can use. Most plants require nitrogenous compounds such as nitrate or ammonium and these are usually supplied to them as fertilizers which are rather expensive and beyond the reach of small and poor farmers in developing tropical countries. In the long run excessive use of chemical fertilizers may reduce the soil fertility. Organic nitrogen is a major constituent of all living organisms. Proteins, nucleic acids, vitamins and several other molecules all contain nitrogen.

Utilization of nitrogen-fixing cyanobacteria (blue-green algae), *Azolla* (a fern) and *Rhizobium* (a bacterium) in agriculture has a long history. Dao and Tran (1979) suggested that *Azolla* was profusely used to fertilize paddy crop in Vietnam at least from the eleventh century and it was introduced in China by early Ming dynasty in fourteenth century. A description of *Azolla* use from 540 B.C. also may refer to international cultivation of the plant (Lumpkin and Plucknett, 1982). Microalgae technologies applicable to agriculture include *in situ* biological nitrogen fixation by free living cyanobacteria and *Azolla*. To date, these technologies are largely restricted to rice cultivation in the tropical countries. Eukaryotic, palmelloid, mucilaginous microalgal flora have potential as soil conditioners. Probably the best known nitrogen fixing microbe is the bacterium *Rhizobium*. Species of this genus live in root nodules of leguminous plants.

The two atoms in the dinitrogen molecule are held together by a tight triple (N≡N) bond. When this bond is broken and the single atoms incorporated into another usuable molecule (NH_3), the process is called nitrogen fixation. It requires a lot of energy. Ammonia is used by plants and microbes as a building block for the synthesis of amino acids and of other nitrogenous compounds. In fertilizer manufacture factories, nitrogen is fixed industrially, by means of the Haber-Bosch process requiring hydrogen gas, high temperatures and enormous energy. Certain groups of many microbes, called the nitrogen fixers, can accomplish the same at physiologically normal temperatures. The conversion of atmospheric N_2 into NH_3 by the nitrogen fixing microbes, mostly certain bacteria and cyanobacteria is called biological nitrogen fixation. The blue-green algae which are oxygenic, photosynthetic and the first oxygen evolving microbes are now usually called cyanobacteria.

Nitrogen-Fixing Microbes

Current evidences suggest that only prokaryotic organisms fix nitrogen. Nitrogen fixing representatives occur among the archaebacteria, actinomycetes, cyanobacteria and bacteria (Table 18.1). Many of the bacteria do not fix nitrogen by themselves but do so only when in symbiosis with a higher plant. The energy, as fixed carbon, needed for breaking the strong triple bond between the two atoms of the N_2 molecules has to come from the higher plant. There is a great symbiosis between nitrogen fixing microbes and other organisms which may even be animals. Thus, wood-devouring termites contain in their guts population of nitrogen fixing bacteria that help the termites overcome the nitrogen deficiencies of their staple diet. Another typical example of symbiotic association is between *Anabaena azollae* and the aquatic fern *Azolla* which is widely used as a source of nitrogen fertilizer in rice paddies of India, China and Vietnam.

Table 18.1: Selected Nitrogen Fixing Microbes

Sl.No.	Species	Property
1.	***Archaebacteria***	
	Methanococcus spp.	It grows diazotrophically upto 64°C.
2.	***Actinomycetes***	
	Frankia spp.	It is gram-positive and fixes nitrogen in root nodules of several shrubs and trees.
3.	***Cyanobacteria* (BGA)**	
	Nostoc spp.	Filamentous, heterocystous blue-green alga and fixes nitrogen in rice fields. May be gram-negative.
	Anabaena spp.	Filamentous, heterocystous blue-green alga and fixes nitrogen. Some species form symbiosis with *Azolla* and *Cycas*.
	Aulosira fertilissima	Filamentous, heterocystous blue-green alga and fixes nitrogen in rice fields as a bio-fertilizer.
4.	**Bacteria (Eubacteria)**	
	Azotobacter vinelandii	Gram-negative, fixes N_2 in air, has a protein which protects nitrogenase from O_2 damage.
	Azospirillum spp.	Gram-negative, associated with roots of some cereals and grasses.
	Clostridium spp.	Gram-positive and obligate anaerobe.
	Klebsiella pneumoniae	Gram-negative and well studied for nitrogen fixing genes.
	Rhizobium spp.	Gram-negative, fixes N_2 in root nodules of legumes.
	Rhodopseudomonas spp.	Gram-negative, purple-green an oxygenic photosynthetic bacteria.

Free-Living Cyanobacteria

The concept of cyanobacterial biofertilizers is inherently attractive. Many free living cyanobacteria (Blue-green algae) fix nitrogen and being photosynthetic, do not compete either with crops or the heterotrophic soil microflora for carbon or energy (Kumar and Kumar, 1988). Their importance lies in the fact that they can fix nitrogen in aerobic environments not with standing the fact that the nitrogenase enzyme is inactivated by oxygen. Cyanobacteria are morphologically the most diverse and complex of the prokaryotes (Stanier and Cohen-Baziere, 1977). Species with biofertilizer potential are the heterocystous filamentous forms (Figure 18.1) belonging to the orders Nostocales and Stigonematales. The biofertilizer potential of nitrogen-fixing cyanobacteria with traditional methods of their biomass production and use including their sources to obtain in India have been given by many workers (Singh, 2002; Singh, 2003).

Distribution

The cyanobacteria are unique in the microbial world for their ability to simultaneously fix nitrogen in aerobic habitats and carbon by the oxygenic eukaryotic plant mechanism (Haselkorn, 1986). Species

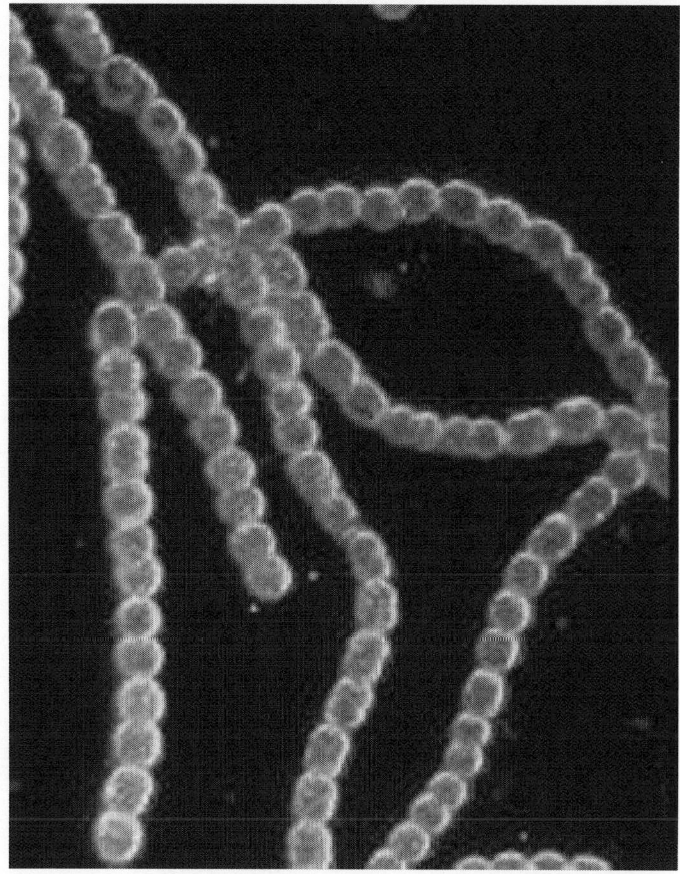

Figure 18.1: Microphotograph of a Cyanobacterium *Anabaena variabilis* Showing Vegetative Cells and Heterocysts (X approx. 400)

of *Aulosira, Nostoc, Anabaena, Gloeocapsa, Scytonema, Gloeotrachia, Cylindrospermum, Westiellopsis, Camptylonema* and other genera are wide spread in Indian soils and rice fields and contribute significantly to their fertility. Unfortunately, there are no reliable methods to accurately estimate the abundance, biomass or productivity of edaphic cyanobacteria and this introduces an element of serious error in any quantitative estimates of the contribution of cyanobacteria to soil nitrogen. Abundance has been determined by direct observation, various planting techniques, pigment extraction and most probable numbers (MPN) methods. The best methods for calculating abundance of filamentous cyanobacteria are currently that of Renaud and Laloe (1985). Cyanobacteria are ubiquitous members of the soil microflora. Species diversity may be large in both temperate and tropical habitats (Metting 1981; Roger and Reynaud 1982). Some studies suggest that species distribution is discontinuous within regions. Venkataraman (1975) reported that only 33 per cent of 2213 Indian soil samples yielded heterocystous cyanobacteria from enrichment culture and that the distribution among and within regions was highly variable. On the other hand, Renaud (1982) recorded nitrogen fixing cyanobacteria in 86 out of 89 rice paddy soil samples in Senegal. Roger (1989) estimated that standing crops of nitrogen-fixing cyanobacteria in rice fields can range from few kg to 0.5 tonnes dry wt. per hectare. There appears to be a potential of about 25-30 kg nitrogen per hectare per crop (Whitton and Roger, 1989).

The biochemistry and molecular genetics of N_2-fixation have been studied intensively in the soil bacteria *Azotobactor, Klebsiella* and *Rhizobium*. Relatively less is known about cyanobacteria. Unlike heterotrophic bacteria, reducing equivalents and energy for nitrogen fixation in cyanobacteria are derived from photosynthesis. Added to these cost is the energy required to maintain anaerobic conditions in the heterocysts where nitrogen fixation occurs. Together, these needs are satisfied by expenditure of upto 20 ATP for each molecules of dinitrogen reduced (Bothe, 1982). Combined with energy required for growth and reproduction of inocula on soil or in flood waters, these basic metabolic needs ultimately define the potential of these organisms as biofertilizers.

Physiology and Ecology

The growth and nitrogen fixation of cyanobacteria in soil or in flooded paddy fields is affected by several physical, chemical and biological factors such as light, temperature wetting-drying cycles (Roger and Renaud, 1979). Many heterocystous cyanobacteria tend to be adversely affected by the high light intensities which can reach 1,10,000 luxin tropical areas, though they can escape high irradiances by moving vertically in the water column. Availability of light influences succession of eukaryotic microalgae and cyanobacteria in the soil algal community.

Cyanobacteria generally grow and fix nitrogen optimally between 30°–35°C and thus temperature is not the growth limiting factor in the tropics. Although evidence from the native grassland communities in Canada suggests that cyanobacteria fix significant amount of nitrogen during episodes of snowmelt (Coxson and Kershaw, 1983), temperature probably will be an important consideration in any attempt to introduce a biofertilizer technology to temperate agriculture.

Among chemical properties, pH is the most important factor influencing species composition, growth and nitrogen fixation of cyanobacteria. Soil cyanobacteria grows best under neutral to alkaline conditions as illustrated by correlations among water pH in flooded paddies, cyanobacterial diversity and abundance of cyanobacterial spores during the dry season between rice cycles (Roger and Renaud, 1982). A pH range of around 7 to 8.5 is preferred by many cyanobacteria. Because N_2 fixation is an anaerobic process, soil and paddy oxidation-reduction (redox) potential may be important. Under reduced conditions less energy would be expended to maintain suitable environment within the heterocysts (the cells which fix nitrogen). Watanabe and Roger (1984) describe four microhabitats in the flooded paddy ecosystem within which cyanobacteria can grow and fix nitrogen: (1) the submerged soil surface, (2) the water column, (3) the free-floating state, and (4) epiphytic state on rice and aquatic weeds. Certainly the redox potential varies among these habitats and influences N_2-fixation by cyanobacteria through out the rice cycle.

Nutrients and agrichemicals also influence the activities and growth of cyanobacteria (Metting, Rayburn and Reynaud, 1988). The abundance of heterocystous cyanobacteria in many rice fields tends to be positively correlated with pH and available phosphorus content of the soil. Phosphorus is often unavailable or prohibitively expensive in developing countries. This might be the most difficult hurdle to overcome for expansion of algalization. Addition of lime ($CaCO_3$) to rice fields stimulates growth of cyanobacteria. The response could be due to a combination of increased alkalinity and increased availability of some nutrients (K, Mo). Work done in India and Egypt has demonstrated that the inoculation of rice fields with cyanobacteria may be equivalent to the addition of around 20-45 kg nitrogen per hectare annually, with the additional advantage that algalization also increase the organic matter content of the soil thereby enhancing its fertility.

The major limitation in a widespread adaptation of the cyanobacterial inoculation technology in developing countries is that most soil fields suffer from deficiency of P and to lesser extent of Fe and Mo. Amelioration of P-deficiency can greatly stimulate growth and nitrogen fixation by cyanobacteria.

Cyanobacteria are tolerant to extended periods without water and variations among strains within species appear to be related to habitats of origin (Roger and Reynaud, 1982). Species morphology also influences tolerance to desiccation. For example, *Cylindrospermum* species forming globose mucilaginous aggregates tend to dominate paddies at the end of the cultivation cycle after fields are drained. Effects of herbicides and other agrichemicals on growth and N_2-fixation by free living cyanobacteria are variable and differ widely among strains (Pipe and Shubert, 1984). In general, herbicides that interfere with photosynthesis are detrimental to cyanobacteria at field application rates in liquid culture but less so in soil. Cyanobacteria are generally more tolerant of such herbicides than vascular plants. Most hormone-analogue herbicides (2, 4-D and Phenoxys) as well as fungicides and insecticides are rarely harmful to cyanobacteria. Parasitism, antagonism, competition and grazing have all been implicated as important biotic factors in algalization studies. Antibiotic activity among cyanobacteria and eukaryotic microalgae are well documented from laboratory studies (Metting and Pyne, 1986) but have not been studied in soil. Invertibrates (*Daphnia*, *Cyclops*, etc.) from many phyla and protozoa eats cyanobacteria in rice paddies (Grant and Seegers, 1985). Studies suggest that paddy soil vary widely in standing crop and diversity of cyanobacteria. Dominant species and successional trends change seasonally according to the interaction of the factors as mentioned earlier.

Cyanobacterial Inoculation Technology

Biofertilization of rice fields by inoculation with free-living cyanobacteria is termed algalization (Singh, 1961; Venkataraman, 1972). Algalization technology (inoculation methods) was developed and devised by Indian farmers. Efforts in the early years were carried out by the All India Coordinated Project on Algae (AICPA, 1979). Technical methods for production of cyanobacterial biomass were first developed in Japan for *Tolypothrix tenuis*. Similar village-level technologies were also under development in China. In India, during 1970s Indian farmers prepared their own inocula in outdoor soil culture as recommended by AICPA. Usually a mixed culture of *Aulosira*, *Anabaena*, *Nostoc* and *Tolypothrix* is grown in small (2 sq. meter) ponds for two months. Dry algal flakes are taken from these ponds and about 8-10 Kg. of these are inoculated per hectare one weak after transplanting rice seedlings. This technology has been applied to over 3 million hectares in India, 4 lakh hectares in Burma and small areas in Sri Lanka and Nepal but the results have been rather erratic, inconsistent and limited in some trials. In most algalization trials the emphasis has been placed only on the grain yield; other parameters have usually been overlooked. Few, if any, experiments have been conducted on the effects of algalization in more than one cropping cycle. These and other shortcomings prompted by Roger and Watanabe (1986) to consider algalization as an unproven technology of doubtful value. Metting (1988) suggested that for many Indian rice fields already having large numbers of heterocystous cyanobacteria, algalization might never prove worthwhile unless the inoculum size is very large. Roger (1989) concluded that the effects of inoculation of rice fields by free-living cyanobacteria are erratic and irreproducible and there is no clear-cut example of a statistically significant increased yield following algalization.

Most reports of algalization trials are from laboratory experiments with individual rice plants in pots. Extrapolation from pot trials suggests N_2-fixation rates of as much as 70 kg nitrogen per hectare in six weeks and increased grain yields of 33 per cent. Stewart (1980) suggested that increased water retention and perhaps other factors favorable to cyanobacterial growth in well-tended pots might enhance N_2-fixation. Generally, nitrogen fixation and grain yields judged against controls without added inorganic nitrogen in field trials have been about half the average for experiments.

Most algalization trials have used inocula developed from mixed laboratory cultures. There appears to exist no data to support the claim made by Agrawal (1979) that the cyanobacteria so introduced actually establish themselves in the field even after repeated inoculations. The only quantitative study to determine the fate of the inoculated strains has been done by Reddy and Roger (1988); they found that out of five species of heterocystous cyanobacteria, only one, namely *Aulosira fertilissima* was able to establish itself significantly, the other four tending to be eliminated. Two approaches with some future potential appear to be to test the fate of fast-growing strains following inoculation and to develop pesticide-resistant strains for use as inoculants. Any success with such strains might prove useful in algalization. A wider exploitation of the cyanobacterial inoculation technology will require a more thorough understanding of the ecology of cyanobacteria in soil, especially when reliable methods for quantitative estimation of their biomass have been developed. As part of the research efforts, the potential for use of cyanobacterial biofertilizers in temperate, irrigated agriculture should be assessed. Documented field studies under temperate conditions are limited. Inoculated pots in England sown to wheat with two species of *Nostoc* and an *Anabaena* species and measured increased acetylene reduction activity but could not detect a crop response. In another experiments inoculated pots sown to maize with an *Anabaena* sp. and *Tolypothrix tenuis*. Both cyanobacteria began to colonize the irrigated soil surface but disappeared after ten weeks and significant acetylene reduction activity was not detected.

Microalgal plant growth regulators are naturally occurring chemicals that exhibit positive or negatives on plant growth, development and metabolism. Plant growth regulators include antibiotics, herbicides and other compounds with hormone like activities. The plant growth regulators production by cyanobacteria are well documented especially antibacterial, algicidal, antifungal and antiprotozoan compounds are well identified. The extracts of a strain of *Nostoc muscorum* inhibits mycelial development of the soil-borne phytopathogen *Cunninghamella blakesleana*. The cyanobacterial products exert positive influence on other microflora. For example the aqueous extract of *Microcystis aeruginosa* enhanced the growth of *Chlorella vulgaris*. Cyanobacteria have an additive effect on yield of algalization in the presence of full compliments of inorganic nitrogen fertilizer (Roger and Kulasooriya, 1980). Cyanobacteria have been shown to release phosphorus from minerals. Cyanobacterial plant growth regulator effects have been attributed to production of amino acids, vitamins and hormones. A material with vitamin B_{12}-like activities has been produced by a *Cylindrospermum* species. The plant growth regulator effects of *Calothrix* spp., *Anabaena* spp. and *Nostoc* spp. suggest varied auxin-like responses. Gibberellin-like enhanced root development is reported for extracts of *Phormidium* and *Aulosira fertilissima* (Singh and Trehan, 1973). Cytokinin may be the active ingredient for effects of cyanobacterial extract on growth of vegetables.

Various soil surfaces are commonly consolidated by cryptogamic communities, of which microalgal crusts are important constituents. Excluding basic and applied research with cyanobacteria and rice, the literature on soil crusts is the most extensive dealing with terrestrial algae (Metting, 1987). Functions of microalgal crusts include consolidation of the crust, mediation of infiltration and retention of water and nitrogen fixation. Desert and steppe surfaces that support algal crusts are consolidated by the combined aggregating effects of mucilaginous sheath and filamentous nature of dominant cyanobacteria. A crust of *Scytonema* reduces loss of soil moisture to the air nearly 18-fold. Besides as microalgal crust, cyanobacteria are early colonizers of primary and secondary substrates. They even quickly colonize the volcanic materials (Metting, 1981) and their importance in secondary plant community succession has already been established. There is precedent from use of natural blooms of certain cyanobacteria in our country for reclamation of soil. Singh (1961) described a method for reclaiming saline-alkaline soils, called "Usar" soils by construction of artificial impoundments within which cyanobacterial blooms are grown during the rainy season. The biomass subsequently incorporated into the soil to improve physical properties that influence infiltration and drainage (Kaushik, Krishna Murti and Venkataraman, 1981). When the method is employed and gypsum ($CaSO_4$) is also incorporated into a sodic soil, exchangeable Na^+ and electrical conductivity are lowered and hydraulic conductivity is increased. The ability to wash away accumulated salts and exchange divalent cations for Na are keys to reclamation of saline and sodic soils.

Methods of Cyanobacterial Use

Cyanobacterial incoulum may be used in the rice fields containing 8-10 cm water, after 7 days of rice seedlings transplantation. Cyanobacterial incoulum may be used at the rate of 8-12 kg per hectare. Maintain the water level in the field for 5-6 days. Small amount of super phosphate may be used at the time of crop transplantation. Pesticides or fungicides may be used if necessary. Some precautions must be considered at the time of cyanobacterial inoculation in the paddy fields-

1. Always maintain 8-10 cm water level in the field.
2. Use cyanobacterial cultures in paddy field after 7-8 days of transplantation.
3. Maintain the soil pH in between 6.5-8.5 and temperatures 25-40°C.
4. Spray 0.05 per cent $CuSO_4$ solution to check the growth of green algae.

Azolla-Anabaena Symbiosis

In contrast to free-living cyanobacteria, the *Azolla-Anabaena* symbiosis is widely employed as an appropriate technology in rice production. The association between the water fern *Azolla* and the nitrogen fixing cyanobacteria *Anabaena azollae* is one of the most efficient N_2-fixing systems from the energy point of view because it enables harvesting of solar energy through photosynthesis and its utilization for the nitrogen fixation. The cyanobacterium *Anabaena azollae* is harbored in a dorsal lobe leaf cavity. The heterocyst frequency of *Anabaena azollae* is 25 per cent compared to 5-10 per cent in non-symbiotic species. Regulation of heterocyst development and nitrogen-fixing genes (NIF) activity in *Anabaena* are related to glutamine synthetase (GS) activity and synthesis (Haselkorn, 1986).

For centuries, the association of *Azolla–Anabaena azollae* has been used as green manure and forage in China, India and other Asian countries. This system can fix about 100-150 kg. of nitrogen per hectare per year in approximately 40-60 tonnes of biomass, if multiple crop growth cycle of the association is practiced over a year. This biomass yield is achievable in view of the fact that *Azolla* plants tend to double in about 5 days under optimal conditions. Further, when *Azolla* dies, it releases over 50 per cent of nitrogenous compounds in about 3 weeks and nitrogen so released becomes available to the rice crop. One crop of *Azolla* applied to the paddy field is roughly equivalent to the addition of 30 kg. urea per hectare. *Azolla* is quite suitable for double cropping of rice in irrigated areas, but its application is labour intensive. Its high protein content makes it a good feed supplement, in China as high as 50 per cent of the diet for pigs is made of *Azolla.* One hectare of green *Azolla* can provide enough fodder for upto 200 pigs. Liu (1979) reported 18 per cent greater rice yield in fields with *Azolla* compared to field without the fern as being common for China.

Important environmental parameters are temperature, light, nutrients, pH, salinity, humidity, desiccation, wind, predators and pests. Lumpkin (1987) reviewed the effects of environmental factors on *Azolla* utilization. Successful growth of *Azolla* appears to be more demanding than that of cyanobacteria. Survival of *Azolla* is greatly affected in drying paddy fields whereas that of cyanobacteria is not so affected as they can readily withstand several dry days between intermittent rains. Sporocarp formation in *Azolla* can eliminate this constraint as the sporocarps can withstand the dry spell but we know very little about the factors which lead to the development of sporocarps. As *Azolla* has a high requirement for light, its growth starts being affected by the shading from the elongating rice plants a few weeks after transplanting. In many places in India, temperature is not a limiting factor for *Azolla*, but phosphorus is. However, *Azolla* plants can accumulate high concentrations of phosphorus (luxury storage); thus the use of phosphatic fertilizer in ponds for growing inocula permits them to store large quantities for subsequent utilization when transferred to the P-deficient paddy fields, atleast for a few days.

In *Azolla* calcium deficiency is indicated by yellow, erect leaves and reduced root growth whereas purple, bullet shaped leaves and excessively long roots characterize a shortage of phosphorus. Maximum growth and N_2-fixation by *Azolla* requires greater phosphorus fertility than rice.

China, India, West Africa and Vietnam are the only countries with a long tradition of *Azolla* propagation. In China, *Azolla* is usually cultivated from May to June in the North and from March to April in the South China. *Azolla* is grown as an annual fern from July to December in our country. Substantial agronomic research with *Azolla* began in 1970s. The International Network on Soil Fertility and Fertilizer Evaluation for Rice (INSFFER) has coordinated field trials since 1979 at 19 sites in 9 countries. Watanabe (1985) summarized the results of the first four years as follows:

1. Incorporation of one *Azolla* crop into soil before or after transplanting rice is equal to about 30 kg. of nitrogen per hectare.
2. Incorporation of two crops of *Azolla* grown before and after transplanting is equivalent to a split application of 60 kg. of nitrogen per hectare.
3. Yield response of rice to *Azolla* is roughly proportional to fertilizer additions.
4. The growth of *Azolla* is not significantly affected by the spacing of rice plants.

The current *Azolla* cultivation technology is very labour-intensive. Methods for propagation have not changed because the fern must be grown in the rice fields themselves. This is one of the problem that is being addressed scientifically before the technology which can be improved and used on a larger scale. *Azolla* is maintained in standing water and because varieties are locally adapted, the central mass culture of quality inocula is not feasible but, if the sporulation can be induced and if methods for harvesting and storage are developed, the potential for *Azolla* use will be more significant.

Rhizobium

Probably the best known nitrogen-fixing soil microflora is the bacterium *Rhizobium*. It grows aerobically at 15-30°C at a pH around 7.0. Calcium (Krishnan *et al.*, 2007), phosphorus, potassium, some micronutrients and small amounts of biotin (Guillen-Navarro *et al.*, 2005) are necessary requirements for its growth. Species of *Rhizobium* live in root nodules of leguminous plants. They provide the plants with fixed nitrogen, while the plants in turn provide energy in the form of photosynthates to the bacteria. The growth of legumes is greatly increased by this symbiosis, especially on nitrogen-poor soils. However, some requirements must be met before an optimal symbiosis can be established: Firstly, the plant roots and rhizobial cells must come in contact in the soil. If the soil lacks rhizobia, there can be no symbiosis. Secondly, the plant and rhizobia have to 'fit' or recognize each other. Although most plants can be infected by several strains of rhizobia, and most rhizobia can infect several plant species, yet only very few or even a single rhizobial strain can form the optimal symbiosis with a specific plant. Finally, rhizobia must occur in large numbers to produce enough nitrogen or to be able to out-compete the non-effective rhizobial strains.

In those areas where the legume is native, compatible rhizobia are naturally available in adequate amounts. In areas where the plant is introduced, chances of presence of the compatible rhizobial strain would be very low. The best way to tackle this shortcoming is to produce the derived rhizobial strains in the laboratory and then add them to the soil or coat the legume seeds with these bacteria. This process is called inoculation. The repeated use of inoculated seeds or soils for a few years builds up the population of the derived rhizobial strain and leads to optimal and sustained N_2-fixation. Plant-interacting microorganisms can establish either mutualistic or pathogenic associations. Although the outcome is completely different, common molecular mechanisms that mediate communication between the interacting partners seem to be involved. Specially, nitrogen-fixing bacterial symbionts collectively called rhizobia and phytopathogenic bacteria have adopted similar strategies and genetic traits to colonize, invade and establish a chronic infection in the plant host. A great deal of research effort has gone into the rhizobial inoculation procedures, infection and invasion of roots by rhizobia during nodulation (Voisin *et al.*, 2003; Gage, 2004). Inoculating the host legume with its specific symbiont is usually done by mixing a culture of the particular strain with presterilized peat. However, since peat varies in composition, certain synthetic carriers of constant quality, such as polysaccharide or alginate, are now being evaluated as inoculant carriers (Dreyfus *et al.*, 1988). In Australia, the inoculated seeds are pelleted with calcium carbonate or rock phosphate in order to protect the symbiont against acidity.

Inoculant performance and plant response are affected by availability of fixed nitrogen in the soil, indigenous rhizobial populations, and other edaphic, climatic and management factors. Phosphorus is a particularly critical factor in this context. If the soil is deficient in phosphorus, inoculation will not prove as beneficial as it would in phosphorus-rich soil. Similarly, if the soil has been heavily fertilized with nitrogenous fertilizer, inoculation will have little or no effect. Infact, inoculant applications will only increase legume yield if the fixed nitrogen content already in the soil is not sufficient to meet the crop's needs. Any environmental stress that restricts growth reduces demand for nitrogen by the crop. It is really the balance between nitrogen supply from soil sources and demand that controls the potential increase in yield.

Significant practical successes with inoculation have been noticed when a cultivated legume is introduced into an area for the first time, and also with those annual legume crops for which the numbers of soil rhizobia decline sharply in between crops (Henzell, 1988). Data from worldwide trials have revealed that out of all the legume crops, soybeans tend to give the most consistent response to inoculation of *Rhizobium* (Graham, 1985).

Technological advances during the last few decades have generated the ability to identify which strains of *Rhizobium* form nodules in the field (Soto *et al.*, 2006; Suzuki *et al.*, 2007; Laguerre *et al.*, 2007). This has become possible through use of serological methods, differential sensitivity to antibiotics and isoenzyme patterns. There have also been notable improvements in the production and storage of rhizobial cultures, and strategies used by rhizobia to increase nodulation and the inoculation methodology (Thompson, 1980; Perret *et al.*, 2000; Ma *et al.*, 2002). However, a very common observation has been that one major problem of biological nitrogen fixation application in agriculture is that inoculant rhizobial strains fail to continue to nodulate plants in the yield. Even when they are initially successful, they tend to be soon replaced by indigenously-occurring strains.

Soil bacteria belonging to various genera of the order Rhizobials are able to invade legume roots in nitrogen-limiting environments, leading to the formation of a highly specialized organ called root nodule. Nodule formation is a complex process that requires a continuous and adequate signal exchange between the plant and the bacteria, of which we only have a fragmentary knowledge. Rhizobia are attracted by root the exudates and colonize plant root surface. Flavonoids present in the exudates activate the expression of the bacterial nodulation genes involved in the synthesis and secretion of nodulation factors, lipochito-oligosaccharides that are recognized by the plants. Nodulation factors together with additional microbial signals such as polysaccharides and secreted proteins allow bacteria attached to root hairs to penetrate the root through a tubular structure called the infection thread, which grows towards the root cortex where the root primordium is developing (Soto *et al.*, 2006). Cells of *Rhizobium* are rod-shaped which usually do not fix nitrogen when grown in culture. But they have the unique capacity to recognize and invade particular legumes and induce a coordinator response in the host; this response includes cell-division and the synthesis of several proteins. The bacterium normally infects the tip of a growing root hair which then undergoes curling, branching or some other change. The host plant then makes a tube or infection thread within the root hair and the bacteria enter the host through the tube. Thereafter the root cells divide and redivide to form a nodule (Figure 18.2). At the same time the bacterial cells within the infection tube become pinched off, covered by a plant-specified membrane, and then released into the cytoplasm of the nodule cells. In this state the bacterial cells are called "bacteroids". They actively fix nitrogen and excrete the ammonia into the host cells. The host then utilizes this ammonia by combining it with glutamic acid to form the amide glutamine which then gives and disseminates the fixed nitrogen to the rest of the host plant. In return, the bacteria are supplied with carbohydrates in a protected environment.

Legumes and actinorrhizal plants capable of fixing nitrogen can grow on nitrogen-deficient soil. If the actinorrhizal plant is also deficient in phosphorus, then mycorrhizal infection of the plant is additionally beneficial as the mycorrhizal fungi provide phosphorus to the plants. In nature, roots of most legumes and actinorrhizal trees are infected with vesicular-arbuscular mycorrhizal (VAM) fungi which play an important role in the growth and nutrition of the plant. More work is needed to improve the production of VAM fungi for use as inoculants. The root nodules form several specific proteins called nodulins. The functions of only a few of these are known. In the *Rhizobium*-legume symbiosis, particular legume species are nodulated only by certain species or strains (Table 18.2). Rhizobia under stress (nutritional or environmental) conditions showed highly effectiveness for symbiotic nitrogen-fixation with *Phaseolus vulgaris* (Krouma *et al.*, 2006; Mnasri *et al.*, 2007). There are reports that pesticides also reduce symbiotic efficiency of N_2-fixing rhizobia and host plants (Fox, 2007). The effect of a heavy metal, Cadmium, on nodulation and nitrogen-fixation of soybean and in other legumes have been documented by Chen *et al.* (2003).

Associative Nitrogen Fixation

Associative N_2-fixing organisms such as *Azospirillum* do not form the highly developed symbiotic system with the root nodules. But it may aid in the growth of non-leguminous plants by supplying fixed nitrogen to them. The free-living bacteria (*e.g.* species of *Azospirillum*, *Herbaspirillum*, etc.) associate with the root zone of tropical pasture grasses and cereals, and probably also benefit these plants by producing certain growth hormones. There are some reports that associative nitrogen fixers may be responsible for fixing up to about 10 kg. nitrogen per hectare per month in sugarcane fields during the rainy season. In Pakistan, associative biological nitrogen fixation by *Klebsiella* in the rhizosphere of the "kallar" grass growing on highly saline wastelands can provide up to 25 per cent of the total

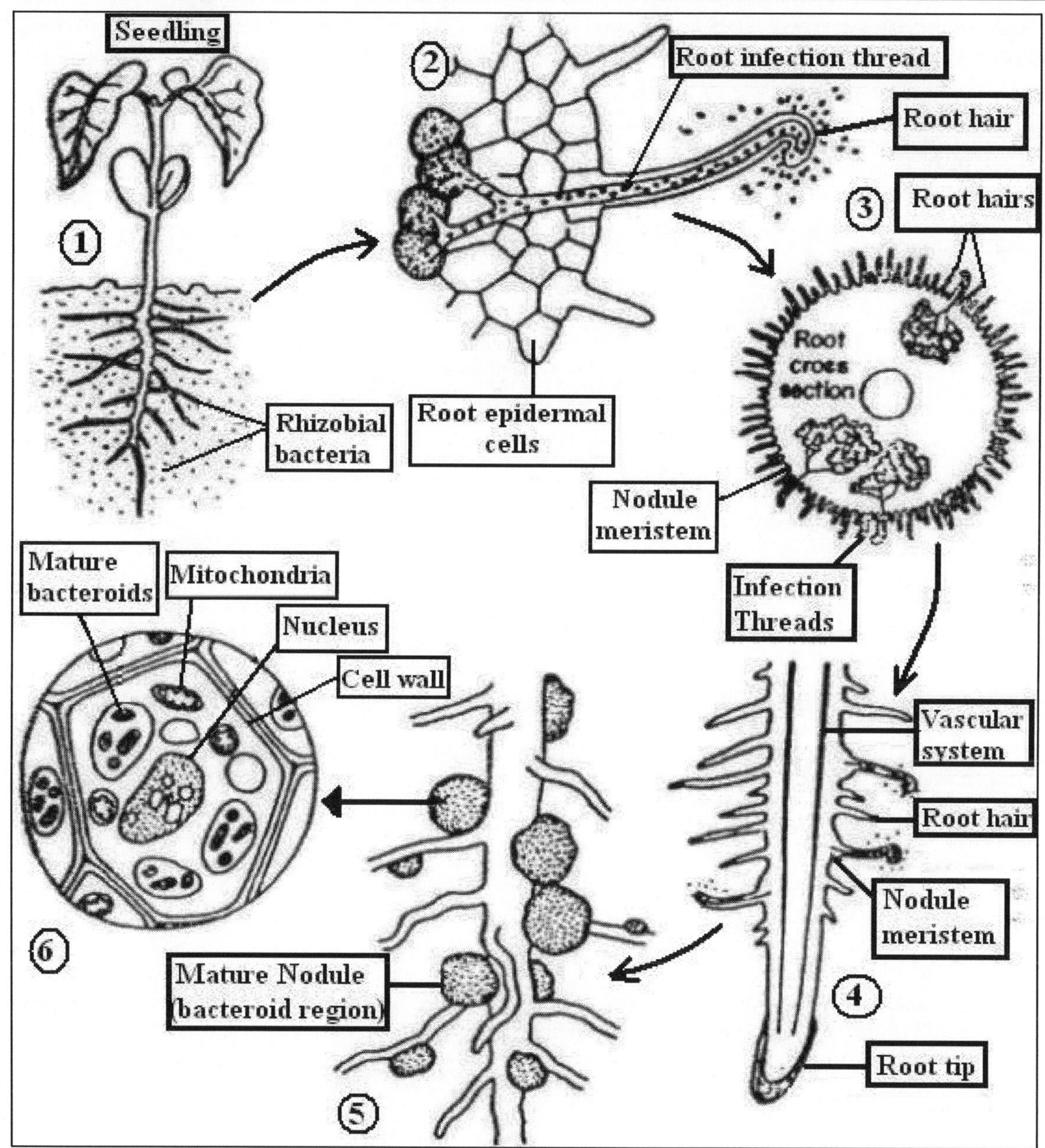

Figure 18.2: Developmental Stages of a Legume Root Nodule

nitrogen. Establishment of the "Kallar" grass has transformed large areas of saline-sodic land and the grass serves as good fodder and green manure.

Unlike *Rhizobium* which forms symbiotic associations with several agriculturally important legumes such as soybeans, peas, beans, lentils and chickpeas (pigeon peas), the actinomycete *Frankia* is associated with a variety of trees and shrubs which have no importance in agriculture though they frequently occur as pioneer species that help to reclaim derelict land.

Table 18.2: The Hosts of Some Species of *Rhizobium* and Related Genera

Sl.No.	Species	Host	Remarks
1	*R. leguminosarum*	Pea, Vicia, Lentils, Clover	Species are closely related to each other.
	R. trifolli		
	R. phaseoli	*Phaseolus*	
	R. meliloti	Alfalfa	
2	*R. sesbania*	*Sesbania*	Induces both stem and root nodules; can also fix nitrogen in free living culture.
3	*R. fredii*	Soybean	Induces non-fixing root nodules on some strains of soybean.
4	*Bradyrhizobium japonicum*	Soybean	Slow growing, induces nitrogen fixing root nodules.

Biochemistry

The work on biochemistry of nitrogen fixation was reviewed by Nicholas (1986) and three discoveries in this area are particularly noteworthy: (i) Irrespective of its source, the enzyme nitrogenase is usually made of the same two proteins; (ii) Leghaemoglobin mediates the flow of oxygen to the vigorously respiring bacteroids in nodules in such a way that the oxygen tension is kept low so as not to inactivate the nitrogenase (nitrogenase is damaged by oxygen), and (iii) During the reduction of nitrogen to ammonia, bound intermediates are produced. Further, the routes for assimilation of the fixed nitrogen (*i.e.* ammonia) into other nitrogenous compounds of the host legume have also been elucidated.

Despite the above discoveries, however, the biochemical mechanism of N_2-fixation is still incompletely known. There are gaps in our knowledge of the structure of the Fe-Mo protein of the enzyme nitrogenase. How does oxygen inactivate nitrogenase? What intermediates are produced during the conversion of nitrogen to ammonia?

Stem Nodules

Whereas most legumes form nodules on their roots, a few form them on stem. Stem-nodulated species occur in the genera *Sesbania*, *Aeschynomene* and *Neptunia*. A few good examples are *Aeschynomene nilotica* and *Sesbania rostrata*. The importance of stem-nodulated plants lies in the fact that they combine the capacity to utilize fixed nitrogen through root absorption (as in most rooted plants) with nitrogen fixation in their nodules. This enables them to fix nitrogen in habitats rich in nitrogenous compounds (habitats in which nitrogen fixation by most nitrogen fixers is inhibited). With a few to developing plants whose nitrogen fixation will not be inhibited in nitrogen-rich habitats, this unique property of stem-nodulated legumes has catalyzed interest in the prospect of transferring their stem-nodulation genes into other legumes lacking stem nodules.

Some New Symbiotic Associations

Cyanobacteria (blue-green algae) form symbiotic associations with a wide variety of organisms in nature. For example, in lichens, cyanobacteria form symbiotic association with fungi. In bryophyte with *Anthoceros*, in pteridophytes with *Azolla* and in gymnosperms with coralloid roots of *cycas*. This has aroused some hope that such an association may be formed with the rice plant, leading to enhancement cyanobacterial nitrogen fixation in paddy fields.

Gamborg and Bottino (1981) were able to incorporate a non-nitrogen-fixing unicellular cyanobacterium into the protoplasts of a flowering plant. However, a similar attempt to incorporate the filamentous, nitrogen-fixer, *Anabaena variabilis* into isolated protoplasts of tobacco was largely unsuccessful (Meeks *et al.*, 1978). Since some nitrogen-fixing cyanobacteria can grow in the dark, a symbiotic system involving cyanobacteria and rice plant (even rice roots) might turn out to be more advantageous than free-living cyanobacteria, whose nitrogen-fixation declines with the shading effects of the growing rice canopy. In fact, Whitton *et al.* (1989) have reported endophytic growth of *Nostoc* and *Calothrix* with senescent leaf sheaths of cultivated deepwater rice plants. The contribution of fixed nitrogen by such cyanobacteria to the rice is, however, not known.

There is some scope for improving the symbiotic microflora. Although there exist several strains of *Rhizobium* for inoculating annual legumes *e.g.* soybean, cowpea and chickpea. There is a woeful paucity of strains that can nodulate leguminous trees. The fact that a few such strains have been found to nodulate *Leucaena leucocephala* raises the hope that similar strains may in future be found for other such trees. Another promising approach is to engineer novel strains of *Rhizobium* and *Frankia* to contain multiple copies of the major genes involved in the symbiosis, also genes for nitrogen fixation, nodulation, and those involved in interstrain competition. Certain nitrogen-fixing bacterial strains synthesize siderophores. Such strains can actively fix nitrogen in iron-deficient soils planted with groundnut and moong crops.

Acknowledgements

YS is very much thankful to Dr. H.D. Kumar, ex Professor and Head, CAS in Botany, BHU, Varanasi for providing information, literatures, and suggestions. Thanks are also due to A. Lembi and J.R. Waaland for their work on Algae and Human Affairs, and Pranjal, Pratyush and Anupama for their support in the preparation of manuscript, and in joy and despair.

References

Agrawal A (1979). Blue-green algae to fertilize Indian rice paddies. *Nature*, 279: 181.

AICPA [All–India Coordinated Project on Algae] (1979). *Algal Biofertilizers for Rice*. Indian Council of Agricultural Research, New Delhi.

Chen YX, He YF, Yang Y, Yu YI, Zheng SJ, Tian GM, Luo YM and Wong MH (2003). Effect of Cadmium on nodulation and nitrogen–fixation of soybean in contaminated soils. *Chemosphere*, 50: 781–787.

Coxson DS and Kershaw KA (1983). Rehydration response of nitrogenase activity and carbon-fixation in terrestrial *Nostoc commune* from *Stipa-Bouteloa* grassland. *Can J Bot*, 61: 2686–93.

Dao TT and Tran QT (1979). Use of *Azolla* in rice production in Vietnam. In: *Nitrogen and Rice*. International Rice Research Institute, Philippines, p 395–405.

Dreyfus BL, Diem HG and Dommergues YR (1988). *Future Directions for Biological Nitrogen Fixation Research*. Kluwer Academic Publishers, *p*. 191–99.

Fox JE, Gulledge J, Engelhaupt E, Burow ME and McLachlan JA (2007). Pesticides reduce symbiotic efficiency of nitrogen-fixing rhizobia and host plants. *Proc Natl Acad Sci, USA*, 104(24): 10282–10287.

Gage DJ (2004). Infection and invasion of roots by symbiotic, nitrogen-fixing rhizobia during nodulation of temperate legumes. *Microbiol Mol Biol Rev*, 68: 280–300.

Gamborg OL and Bottino PJ (1981). Protoplasts in genetic modifications of plants. In: *Advances in Biochem Engineering,* (Ed) Fiechter A . Springer Verlag, Berlin, p. 239–263.

Grant IF and Seegers R (1985). Movement of straw and algae facilitated by tubificids (Oligochaeta). in low land rice soil. *Soil Biol Bechem,* 5: 729–30.

Guillen–Navarro K, Encarnacion S and Dunn MF (2005). Biotin biosynthesis, transport and utilization in rhizobia. *FEMS Microbiol Lett*, 246: 159–65.

Haselkorn R (1986). Organization of the genes for nitrogen fixation in photosynthetic bacteria and cyanobacteria. *Ann Rev Microbiol,* 40: 525–47.

Henzell EF (1988). The role of biological nitrogen fixation research in solving problems in tropical agriculture. *Plant and Soil*, 108: 15–21.

Kaushik BD, Krishna Murti GS and Venkataraman GS (1981). Influence of blue-green algae on saline-alkaline soils. *Science and Culture*, 47: 169–170.

Krishnan HB, Kim WS and Sun-Hyung J (2007). Calcium regulates the production of nodulation outer proteins and precludes pili formation by *Sinorhizobium fredii* USDA257, a soybean symbiont. *FEMS Microbiol Lett*, 271: 59–64.

Krouma A, Drevon JJ and Abdelly C (2006). Genotypic variation of nitrogen-fixing bean (*Phaseolus vulgaris L.*). in response to iron deficiency. *J Plant Physiol*, 163: 1094–100.

Kumar A and Kumar HD (1988). Nitrogen fixation by blue-green algae. In: *Plant Physiological Research in India*, (Ed) Sen SP. Soc for Plant Physiol Biochem, New Delhi, p. 85–103.

Laguerre G, Depret G, Bourion V and Due G (2007). *Rhizobium leguminosarum* bv. *Viciae* genotypes interact with pea plants in developmental responses of nodules, roots and shoots. *New Phytol*,176: 680–90.

Liu ZZ (1979). Use of *Azolla* in rice production in China. In: *Nitrogen and Rice*. International Rice Res Instt, Philippines, p. 375–394.

Lumpkin TA (1987). Environmental requirements for successful *Azolla* growth. In: *Azolla Utilization Proc Workshop on Azolla Use*. Fuzhou, China. International Rice Res Instt, Manila, Philippines, p. 87–97.

Lumpkin TA and Plucknett DL (1982). *Azolla as a Green Manure: Use and Management in Crop Production*. Westview Tropical Agricultural Press, Boulder, Co.

Ma W, Penrose DM and Glick BR (2002). Strategies used by rhizobia to lower plant ethylene levels and increase nodulation. *Can J Microbiol*, 48: 947–54.

Meeks JC, Malmerg RL and Wolk CP (1978). Uptake of auxotrophic cells of a heterocyst-forming cyanobacterium by tobacco protoplasts and the fate of their association. *Planta*, 133: 56–60.

Metting B (1981). The systematics and ecology of soil algae. *Bot Rev*, 47: 195–312.

Metting B (1987). Dynamics of wet and dry aggregate stability from a three year microalgal soil-conditioning experiment in the field. *Soil Science*, 143: 139–43.

Metting B (1988). Microalgae and agriculture. In: *Microalgal Biotechnology*, (Eds) Borowitzka MA and Borowitzka LA. Cambridge University Press, Cambridge, p. 288–304.

Metting B and Pyne JW (1986). Biologically active compounds from microalgae. *Enzyme Microbiol Technol*, 8: 386–394.

Metting B, Rayburn WR and Reynaud PA (1988). Algae and agriculture. In: *Algae and Human Affairs*, (Eds) Lembi A and Waaland JR. Cambridge University Press, Cambridge, p. 335–370.

Mnasri B, Mrabet M, Laguerre G, Aouani ME and Mhamdi R (2007). Salt-tolerant rhizobia isolated from a Tunisian oasis that are highly effective for symbiotic nitrogen-fixation with *Phaseolus vulgaris* constitute a novel biovar of *Sinorhizobium meliloti*. *Arch Microbiol*, 187: 79–85.

Nicholas DJD (1986). Biochemistry of nitrogen fixation–achievements and challenges. In: *Proc 8th Australian Nitrogen Fixation Congress*, (Eds) Wallace W and Smith SE, p. 1–4.

Perret X, Staehelin C and Broughton WJ (2000). Molecular basis of symbiotic promiscuity. *Microbiol Mol Biol Rev*, 64: 180–201.

Pipe AE and Shubert LE (1984). The use of algae as indicators of soil fertility. In: *Algae as Ecological Indicators*, (Ed) Shubert LE. Academic Press, New York, p. 213–233.

Reddy PM and Roger PA (1988). Dynamics of algal populations and acetylene reducing activity in five rice soils inoculated with blue-green algae. *Biol Fertil Soils*, 6: 14–21.

Reynaud PA (1982). The use of *Azolla* in West Africa. In: *Biological Nitrogen Fixation Technology for Tropical Agriculture, (Ed)* Graham PH and Harris SC. Centro Internationale de Agric Tropical, Cali, Colombia, p. 365–366.

Reynaud PA and Laloe F (1985). La méthode des suspensions-dilutions adoptée al estimation des populations' algales dans une riziere. *Rev Ecol Biol Sol*, 22: 161–92.

Roger PA (1989). Blue-green algae (cyanobacteria). in agriculture. In: *Microorganisms that Promote Productivity*, (Eds) Dawson JO and Dart P. Martinus Nijhoff, Dordrecht.

Roger PA and Kulasooriya SA (1980). *Blue-green Algae and Rice*. International Rice Res Instt, Philippines.

Roger PA and Reynaud PA (1979). Ecology of blue-green algae in rice fields. In: *Nitrogen and Rice*. International Rice Res. Instt., Philippines, p. 289–309.

Roger PA and Reynaud PA (1982). Free living blue-green algae in tropical soils. In: *Microbiology of Tropical Soils and Plant Productivity*, (Eds) Dommergues YR and Diem HG. Martinus Nijhoff, Dordrecht, p. 147–68.

Roger PA and Watanabe I (1986). Technologies for utilizing biological nitrogen fixation in wetland rice: Potentialities, current usage and limiting factors. *Fertilizer Res*, 9: 39–77.

Singh RN (1961). *Role of Blue-green Algae in Nitrogen Economy of India*. Indian Council of Agricultural Research, New Delhi.

Singh VP and Trehan T (1973). Effects of extracellular products of *Aulosira fertilissima* on the growth of rice seedlings. *Plant and Soil*, 38: 457–64.

Singh Y (2002). Algal biofertilizer for paddy crop as a small rural industry. *Udyamita*, CEDMAP, Bhopal, p. 21–23.

Singh Y (2003). Biofertilizer potential of nitrogen-fixing cyanobacteria. *Science Tech Entrep*, 11: 24–26.

Soto MJ, Sanjuan J and Olivares J (2006). Rhizobia and plant pathogenic bacteria: common infection weapons. *Microbiology*, 152: 3167–3174 (Review).

Stanier RY and Cohen-Baziere G (1977). Phototrophic prokaryotes: the cyanobacteria. *Ann Rev Microbiol*, 31: 225–74.

Stewart WDP (1980). Systems involving blue-green algae. In: *Methods for Evaluating Biological Nitrogen Fixation*, (Ed) Bergersen FJ. Wiley Interscience, New York, p. 583–635.

Suzuki S, Aono T, Lee KB, Suzuki T, Liu CT, Miwa H, Wakao S, Iki T and Oyaizu H (2007). Rhizobial factors required for stem nodule maturation and maintenance in *Sesbania rostrata–Azorhizobium caulinodans* ORS571 symbiosis. *Appl Environ Microbiol*, 73: 6650–6659.

Thompson JA (1980). Production and quality control of legume inoculants. In: *Methods for Evaluating Biological Nitrogen Fixation*, (Ed)Bergersen FJ. Wiley, Chichester, p. 489–533.

Venkataraman GS (1972). *Algal Biofertilizers and Rice Cultivation*. Today and Tomorrow's Publishers, New Delhi.

Venkataraman GS (1975). The role of blue-green algae in tropical rice production. In: *Nitrogen Fixation by Free-Living Microorganisms*, (Ed) Stewart WDP. Cambridge University Press, Cambridge, p. 207–218.

Voisin AS, Salon C, Jeudy C and Waremboug FR (2003). Symbiotic N_2-fixation activity in relation to carbon economy of *Pisum sativum L.* as a function of plant phenology. *J Exp Bot*, 393: 2733–2744.

Watanabe I (1985). *Summerized Report on INSFFER Azolla Program*. International Rice Res Instt, Philippines.

Watanabe I and Roger PA (1984). Nitrogen fixation in wetland rice fields. In: *Current Development in Biological Nitrogen Fixation*, (Ed)Suba Rao NS. Oxford and IBH, New Delhi. 237–276.

Whitton BA, Aziz A, Kawecka B and Rother JA (1989). Ecology of deepwater rice fields in Bangladesh. *Hydrobiologia*, 169: 31–42.

Whitton BA and Roger PA (1989). Uses of blue-green algae and *Azolla* in rice culture. In: *Microbial Inoculation of Crop Plants*, (Eds) Campbell R and Mac Donald RM. Oxford University Press, Oxford. p. 89–100.

Soil Microflora, 2009 *Pages* **229–245**
Editor: **Rajan Kumar Gupta, Mukesh Kumar & Deepak Vyas**
Published by: **DAYA PUBLISHING HOUSE, NEW DELHI**

Chapter 19

Agrobacterium: A Natural Genetic Engineer

Ashutosh Bahuguna, Madhuri K. Lily and Koushalya Dangwal*
Department of Biotechnology, Modern Institute of Technology, Dhalwala, Rishikesh – 249 201, Uttarakhand

ABSTRACT

Agrobacterium is a genus of Gram-negative bacteria that uses horizontal gene transfer to cause tumors in plants. *Agrobacterium tumefaciens* is the most commonly studied species in this genus. *Agrobacterium* is well known for its ability to transfer DNA from its genome to plant genome, and for this reason it has become an important tool for plant improvement by genetic engineering. Overall species of *Agrobacterium* can transfer T-DNA to a broad group of plants (both angiosperm and gymnosperm) and other organism. Yet, individual *Agrobacterium* have a limited host range, the molecular basis for the limited host range is unknown.

Keywords: *T-DNA, Ti and Ri Plasmid,* vir *genes, Crown gall, Binary vector.*

Introduction

The genus *Agrobacterium* is a group of Gram negative soil bacteria found associated with plants (Figure 19.1). Many members of this group cause disease on plants. Infections at wound sites by *Agrobacterium tumefaciens* cause crown gall tumors on a wide range of plants including most dicots, some monocots, and some gymnosperms. Infections by *A. rhizogenes* cause hairy root disease. *A. vitis* causes tumors and necrotic lesions on grape vines and is commonly found in the xylem sap of infected

* E-mail: kdangwal1@yahoo.co.in, Phone: 0135-2435220

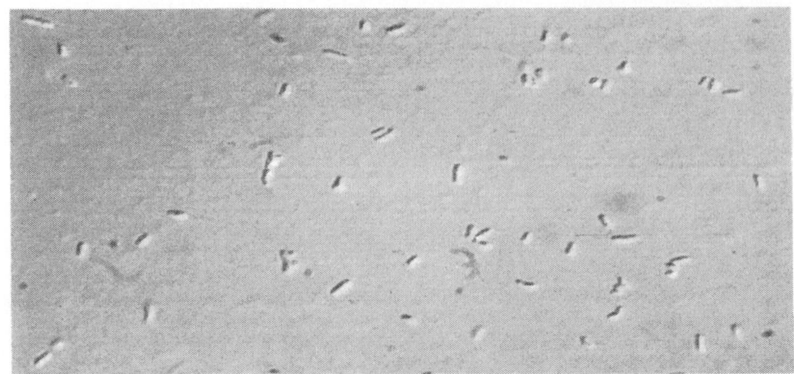

Figure 19.1: Agrobacterium Cells Observed in the Light Microscope

plants. Despite the general perception that most of the agrobacteria cause disease, *A. radiobacter*, is a non-virulent member of this group most often isolated from soil. (Matthysse, 2005)

Habitat

Agrobacteria are usually found in soil in association with roots, tubers, or underground stems. The bacteria also cause tumors from which they can be isolated. Tumors may be prevalent on grafted plants at the graft junction; examples include grapes, roses, poplars, and fruit trees. In some cases, the bacteria can be isolated from the xylem of infected plants such as *A. vitis which can be isolated* from the xylem of infected grapevines.

Phylogeny

On the basis of similar physiological characteristics, agrobacteria and rhizobia are placed in the same family *Rhizobiaceae*. More recent studies such as 16S rDNA and chromosomal gene sequence homology suggested that these two groups of bacteria are indeed closely related (Williams, 1993). Both physiological characteristics and molecular characteristics place these bacteria in the α-subgroup of the Proteobacteria.

Taxonomy

The genus is divided into species largely based on pathogenic properties, although other physiological characteristics correlate with pathogenic properties. The major species are *A. radiobacter* (nonpathogenic), *A. tumefaciens* (the causative agent of crown gall tumors), *A. rhizogenes* (the causative agent of hairy root disease), and *A. vitis* (the causative agent of tumors and necrotic disease on grapevines). There are also less well studied species such as *A. rubi* isolated from cane galls on *Rubrus* species. Agrobacteria also have been divided into biotypes (biovars) based on physiological properties. (Matthysse, 2005)

Biovar 1, which includes most strains of *A. tumefaciens*, has no growth factor requirements and will grow in the presence of 2 per cent NaCl. Most strains produce 3-ketolactose. Biovar1 bacteria also produce acid from dulcitol, melizitose, ethanol, and arabitol in addition to mannitol and adonitol. Some biovar 1 strains are able to grow at 37°C however; they may lose the Ti plasmid, which is required for virulence. Biovar 2 includes most strains of *A. rhizogenes*. These bacteria require biotin for growth. They fail to grow in the presence of 0.5 per cent NaCl or at 37°C. Some biovar 2 strains can grow on tartrate producing alkali. Biovar 3 strains include most *A. vitis* strains. Like biovar 1 strains, these bacteria will grow in the presence of 2 per cent NaCl but generally do not grow at 37°C. Both biovar 2 and 3 strains fail to produce 3-ketolactose. Biovar 3 strains can produce alkali from tartrate. Some biovar 3 strains require biotin for growth (Table 19.2).

Isolation

Agrobacteria can be isolated from soil obtained from the vicinity of infected plants, from galls formed by the bacteria, or, in the case of grapevines, from the xylem sap of infected plants. The bacteria are not numerous in older galls and may be easier to isolate from the surrounding soil than from the tumor tissue. Agrobacteria can grow readily in culture on complex or defined media (Table 19.1). Nutrient agar [with or without yeast extract (0.5 per cent)] or yeast mannitol agar supports the growth of most strains. Some strains require vitamin B-complex for growth, usually 0.2 mg/liter each of biotin, pantothenic acid and/or nicotinic acid. Many strains, including most *A. rhizogenes* isolates, are sensitive to salt and hence do not grow on media with high salt such as Luria-Bertani agar media. The colonies are generally white or slightly cream or pale pink in color in the absence of distinctive pigment. Large amounts of extracellular polysaccharide may be produced on some media giving the colonies a watery appearance. The agrobacteria grow at a moderate rate. *A. tumefaciens* usually require 2 to 4 days to form colonies on complex media. Some strains of *A. rhizogenes* are slow growing and may require as much as 1 week to form colonies on complex media. Optimal growth temperature for most strains is between 25°C and 28°C, although the optimal temperature for plant infection may be lower (22°C). (Matthysse, 2005)

Table 19.1: Media for Growth of Agrobacteria (Matthysse, 2005)

General Media		*General media*	
Luria Agar (for biovar1 and some biovar 3 strains)		**Mannitol glutamate agar (for all biovars)**	
Tryptone	10 g	Mannitol	10 g
Yeast extract	5 g	L-Glutamic acid	2 g
NaCl	5 g	KH_2PO_4	0.5 g
Water	1 liter	NaCl	0.2 g
Agar	14 g	$MgSO_4$ $7H_2O$	0.2 g
		Biotin	0.002 g
Yeast Mannitol Agar (for all biovars)		Water	1 liter
Mannitol	10 g	Agar	15 g
Yeast extract	1 g	Adjust pH to 7.0 before autoclaving.	
K_2HPO_4	0.5 g		
$CaCl_2$	0.2 g	**H4 Minimal Medium (for biovars 1 and 3, biovar 2 will grow very slowly on this medium)**	
NaCl	0.2 g	NH_4Cl	5 g
$MgSO_4$.7H2O	0.2 g	NH_4NO_3	1 g
$FeCl_3$	0.01 g	Na_2SO_4	2 g
Water	1 liter	K_2HPO_4	3 g
Agar	15 g	KH_2PO_4	1 g
Adjust to pH 7.0. For biovar 2 add biotin, calcium pantothenate, and nicotinic acid, all at 200 g/liter.			

Contd...

Table 19.1–Contd...

General Media	
$MgSO_4$ $7H_2O$	0.1 g
Water	1 liter

Dissolve salts in the order given; adjust pH to 7.2; add 10ml of sterile 20 per cent glucose after autoclaving

AB Minimal Medium

K_2HPO_4	3 g
NaH_2PO_4	1 g
NH_4Cl	1 g
$MgSO_4$ $7H_2O$	0.3 g
KCl	0.15 g
$CaCl_2$	0.005 g
$FeSO_4$ $7H_2O$	0.0025 g

Water 1 liter Dissolve salts in the order given; adjust pH to 7.2; add after autoclaving 10 ml of sterile 20 per cent glucose or sucrose.

Selective Media

Selective Medium of Biovar 1*

L (-) Arabitol	3.04 g
K_2HPO_4	1.04 g
KH_2PO_4	0.54 g
NH_4NO_3	0.16 g
$MgSO_4.7H_2O$	0.25 g
Sodium taurocholate	0.29 g
Water	1 liter
1 per cent Crystal violet	2 ml
Agar	15 g

Add after autoclaving 10 ml of 2 per cent cyclohexamide and 10 ml of 1 per cent Na_2SeO_3. $5H_2O$. On this medium colonies of agrobacteria are white, circular, raised, and glistening. They may become mucoid.

General media

Selective Medium for Biovar 2*

Erythritol	3.05 g
K_2HPO_4	1.04 g
KH_2PO_4	0.54 g
NH_4NO_3	0.16 g
$MgSO_4$ $7H_2O$	0.25 g
Sodium taurocholate	0.29 g
Yeast extract	0.01 g
Malachite green	0.005 g
Water	1 liter
Agar	15 g

Add after autoclaving 10ml of 2per cent cyclohexamide and 10ml of 1per cent $Na_2SeO_3.5H_2O$. On these medium colonies of agrobacteria are white, circular, raised, and glistening. They may turn brown as they age.

Selective Medium for Biovar 3*

Adonitol	4.0 g
K_2HPO_4	0.9 g
KH_2PO_4	0.7 g
NaCl	0.2 g
$MgSO_4$	0.2 g
Yeast extract	0.14 g
Boric acid	1.0 g
Water	1 liter
Agar	15 g

Adjust pH to 7.2 before autoclaving. After autoclaving add 10 ml of 2.5 per cent cyclohexamide, 1 ml of 8 per cent triphenyltetrazolium chloride, 1 ml of 2 per cent D-cycloserine, and 1 ml of 2 per cent trimethoprin. On these medium colonies of agrobacteria have dark red centers with white edges.

* Note that these media are only semi-selective. Other organisms may grow. Additional tests are necessary to positively identify an isolate as Agrobacterium.

Identification

Agrobacteria have been traditionally identified as Gram-negative bacteria which are non spore forming, showing dimension of 0.6-1.0 x 1.5 – 3.0 µm and G+C content (mol per cent) of 57-63 per cent (Prescott, 2003). Agrobacteria are motile rods with peritrichous flagella. Agrobacteria can also be identified on the basis of sugar fermentations and ketolactose production (Table 2.2). In recent years, lipid and fatty acid profiles have been used to identify both virulent and avirulent Agrobacteria

(Jarvis, 1996). Polymerase chain reaction (PCR) has also been used in identification of pathogenic and nonpathogenic strains. The PCR primers designed from *vir* genes such as *virD2* can be used to identify potentially pathogenic strains (Hass, 1995). Pathogenic strains have also been identified and grouped by their ability to grow on different opines and by their ability to produces particular types of opines. *Agrobacterium* genus is classified into five species on the basis of sugar fermentations, fatty acid profiles, PCR, opine production and utilization, and genome organization as mentioned below. (Matthysse, 2005)

Table 19.2: Traits Uused for Identification of Biovars of *Agrobacterium*

Characteristic	*Biovar 1*	*Biovar 2*	*Biovar 3*	A. rubi
Growth factor requirements	None Biotin	Biotin for some strains	Biotin acid, nicotinic acid	Pantothenic
3-Ketolactose production	Most strains	No	No	No
Growth on 2 per cent NaCl	Yes	No	Yes	Yes
Growth at 37°C	Yes	No	No	Yes
Acid production from mannitol	Yes	Yes	Yes	Yes
Adonitol	Yes	Yes	Yes	Yes
Erythritol	No	Yes	No	No
Dulcitol	Yes	Yes	No	No
Melizitose	Yes	No	No	No
Ethanol	Yes	No	No	No
Arabitol	Yes	No	No	No
Alkali production from tartrate	No	Yes	Yes	No

Source: Matthysse, 2005.

Systemic Classification of *Agrobacterium* (Fischer, 2006)

Kingdom: Bacteria

Phylum: Proteobacteria

Order: Rhizobiales

Subdivision	*Family*	*Genus*	*Species*
α-Proteobacteria	Rhizobiaceae	*Allorhizobium*	
		Mesorhizobiuum	
		Sinorhizobium	
		Rhizobium	
		Azorhizobium	
		Bradyrhizobium	
		Agrobacterium	*A. tumefaciens*
			A. rhizogenes
			A. rubi
			A. vitis
			A. radiobacter

Comprasion of *Rhizobium* and *Agrobacterium* (Fischer, 2006)

(Symbiotic vs. pathogenic plant microbe interaction)

Sl.No.	Properties	Rhizobium	Agrobacterium
1.	Family	*Rhizobiaceae*	*Rhizobiaceae*
2.	Habitat	Soil	Soil
3.	Gram reaction	Negative	Negative
4.	Shape	rod shape	rod shape
5.	Host Interaction	Yes, symbiotic	Yes, pathogenic
6.	Chemotaxis	yes	yes
7.	Host range	specific legumes	dicot plants and few monocots
9.	Signals	(iso-) flavonoids,	phenolic compounds, sugars, lipooligosaccarides pH, opines,
10.	Site of infection	root hairs	wounded tissues
11.	Infection	intracellular bacteroids	DNA transfer

Historical Landmarks *in Agrobacterium* Biology (Fischer, 2006)

Year	Discovery
1853	First written report of crown gall disease.
1897	*Agrobacterium vitis* identified as the causal agent of crown gall in grapes
1907	*Agrobacterium tumefaciens* identified as causal agent of crown gall in paris daisy (Margeritte*; Agyranthemum frutescens*)
1947	Report showing sterile plant tumor tissue indefinite proliferation in hormone free medium. Tumor cells are proposed to be "transformed" by an *Agrobacterium* derived tumor inducing principle (TIP).
1956	Identification of unusual low molecular weight nitrogenous compounds (opines) in tumor tissues.
1971	*Agrobacterium tumefaciens* loses virulence when grown at 37°C. The TIP can be transferred from virulent to non-virulent strain of *A. tumefaciens strain.*
1974	*A. tumefaciens* virulence depends on the presence of a large "tumor inducing" (Ti) plasmid. The TIP is probably a component of the Ti plasmid.
1977	The T-DNA region of Ti plasmid is present in the genome of crown gall tumor cells: the T-DNA is TIP.
1980	The opine concept: states that the synthesis of opine by transformed cells creates an ecological niche for the infecting strain of *Agrobacterium.*
1883	First report implicating use of *A. tumefaciens* as vector in the transformation of plant
1984	T-DNA oncogenes that mediate overproduction of auxin and cytokinin are identified
1985	The *vir A/vir G* two-component regulatory system is identified as a central component of signal perception and transduction in *Agrobacterium* transformation.
1986	Elucidation of the *vir*-gene encoded T-DNA transfer process; identification of plant genes involved in *A. tumefaciens* transformation; extension of *A. tumefaciens* host range for transformation of monocots.
2001	Publication of the complete genome sequence of two *A. tumefaciens* strains

Ti and Ri Plasmid

Agrobacterium tumefaciens possess a Ti plasmid (200Kb) (Figure 19.2) while *Agrobacterium rhizogenes* has the Ri plasmid which are responsible for crown gall disease and hairy root disease in dicots respectively (Brown, 2001). Ti and Ri plasmids share similarity in several general features, therefore these plasmids can be interchanged between the two species. Both plasmids carry three regions including T-DNA, *vir* and host specificity region (Brown, 2001; Fischer, 2006). The T-DNA can vary from approximately 12 to 24 kilobase pairs (kb) and harbors genes for opine metabolism and phytohormone production. T-DNA is transferred into the plant cell and is integrated into their genome during infection. *vir* region is located on the 35 kb region of these plasmids that lies outside the T-DNA region and carries at least seven and possibly eight different *vir* genes. The products of these *vir* genes are responsible for excision, transfer of T-region and integration into genome of plant cell (Glick and Pasternak, 2001). The *vir* and the host specificity region together constitute the 170 kb of the plasmid (Brown, 2001).

Two aspects which make the pTi and pRi as unique bacterial plasmids are:

1. The presence of the genes located within their T-DNA which have eukaryotic regulatory sequences solely recognized by plant cells, thereby conferring these genes with the ability to be expressed only in the plant cells rather than *Agrobacterium*. However, the remaining genes have prokaryotic regulatory sequences.

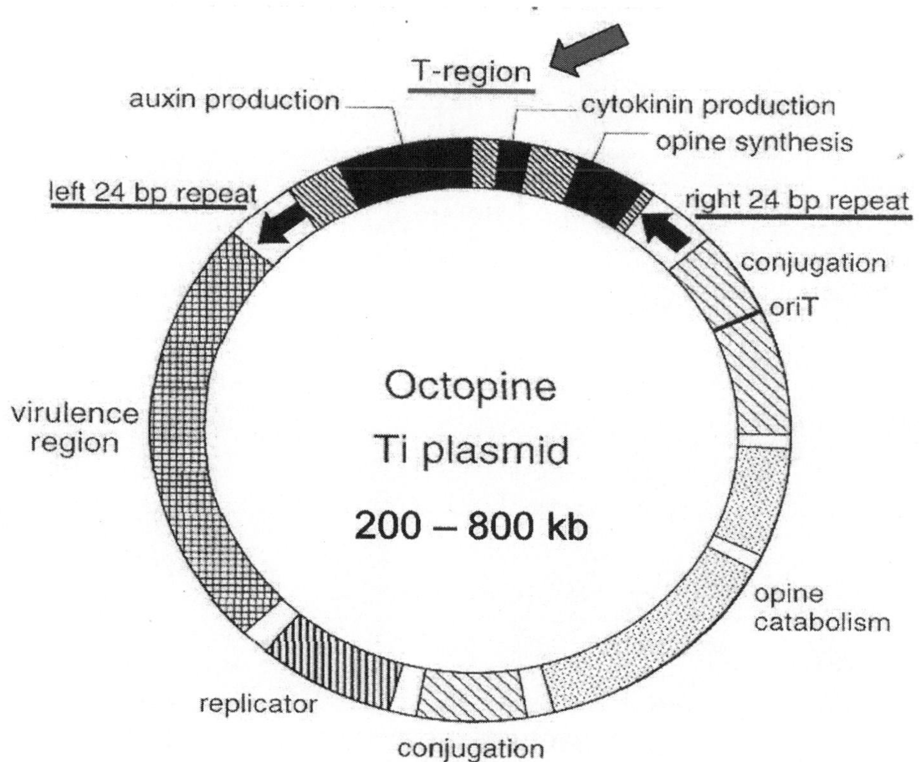

Figure 19.2: Structure of a Ti Plasmid (Hooykaas, 2000)

2. The capability of these plasmids to naturally transfer the T-DNA into the genome of host plant, which claims *Agrobacterium,* to be aptly called as natural genetic engineer (Glick and Pasternak, 2001).

The Ti plasmids are classified into different types based on the type of opine produced by their genes. The opines are unique and unusual condensation products of an amino acid and a keto-acid, or an amino acid and a sugar. The different opines specified by pTi are octopine, nopaline, agrocinopine and mannopine (Figure 19.3) (Singh, 2006; Fischer, 2006). These opines are used as carbon source and sometimes also as nitrogen source by any *Agrobacterium tumefaciens* or *Agrobacterium rhizogenes* that carries genes for the catabolism of that particular opine on their plasmid (Glick and Pasternak, 2001). The Ti plasmids found in various species of *Agobacterium tumefaciens* can fall into two general categories:

1. Octopine type.
2. Nopaline type.

Both octopine and nopaline type pTi contain the following important functional region: (Singh, 2006; Fischer, 2006).

(*a*) T-DNA which carries oncogenes (phytohormone production genes) and opines synthesis genes. This region is transferred into host plant genome.

(*b*) Vir region that regulates T-DNA transfer into plant cells.

(*c*) Opine catabolism regions producing enzymes necessary for the utilization of opines by *Agrobacterium*

(*d*) Conjugative transfer (ori T or tra) region responsible for the conjugative transfer of the plasmid, as well as for T-DNA transfer, when T-DNA borders are deleted.

(*e*) Origin of replication for propagation in *Agrobacterium.*

Molecular Insight of *Agrobacterium* Infection

The *Agrobacterium tumefaciens* infects dicot plants and results in the transfer of T-DNA into the plant genome. The infection process is governed by both chromosomal and the plasmid borne genes of

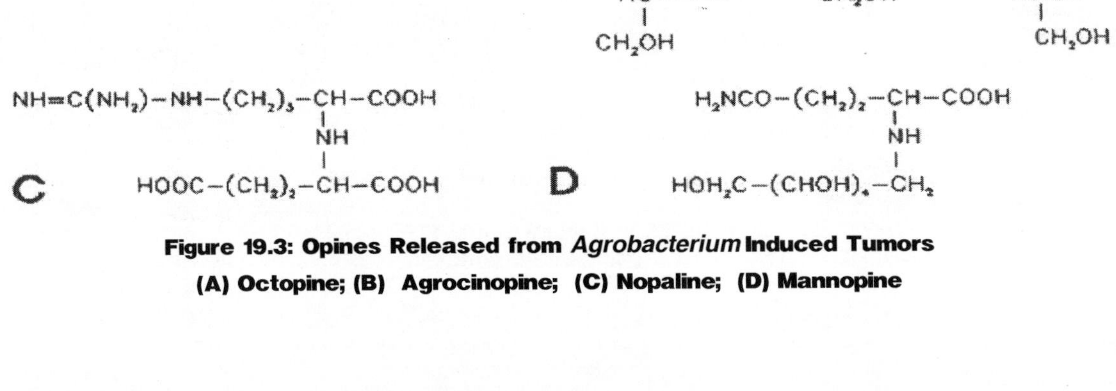

A: $NH{=}C(NH_2){-}NH{-}(CH_2)_3{-}CH{-}COOH$ | NH | $CH_3{-}CH{-}COOH$

B: HC–$HCOH$–$HOCH$–$HOCH$–HC–CH_2OH (ring O); $O{-}C(CH_2OH){-}O$; $HOCH$–HC–HC–CH_2OH; PO_4; CHO–CH–$HCOH$–$HCOH$–CH_2OH

C: $NH{=}C(NH_2){-}NH{-}(CH_2)_3{-}CH{-}COOH$ | NH | $HOOC{-}(CH_2)_2{-}CH{-}COOH$

D: $H_2NCO{-}(CH_2)_2{-}CH{-}COOH$ | NH | $HOH_2C{-}(CHOH)_4{-}CH_2$

Figure 19.3: Opines Released from *Agrobacterium* Induced Tumors
(A) Octopine; (B) Agrocinopine; (C) Nopaline; (D) Mannopine

Agrobacterium tumefaciens. Infection begins with the attachment of *Agrobacterium* cells to appropriate host plant cells (Figure 19.4); which is governed by bacterial chromosomal genes, generally the *chv* (Chromosomal virulence) genes (Table 19.3).

Table 19.3: Chromosomal Genes and their Function

Chromosomal Gene	*Function*
ChvA	Encodes an inner membrane protein essential for the transport of β-1, 2-glucan from cytoplasm to periplasm.
ChvB	Encode an inner membrane protein most likely involved in the synthesis of β-1, 2-glucan.
ChvD and *ChvE*	Needed for optimal expression of vir genes of pTi
Exo locus	genes Biosynthesis of attachment polysaccharide.
exoC	Encodes an enzyme directly in the biosynthesis of β-1,2-glucan
cel genes	Cellulose fibril synthesis especially during the early phases of infection that bacterial cells become firmly adhered to plant cells

Most of the genes such as *chv B*, *exo* genes, *cel* genes are concerned with the biosynthesis of cell attachment polysaccharides due to which the bacterial sells become firmly adhered to plant cells and two genes, viz, *chv D* and *chv E* are needed for an optimal expression of pTi *vir* genes. These chromosomal genes are expressed constitutively in agrobacterium (Singh, 2006; Fischer, 2006)

Infection by *A. tumefaciens* (Figure 19.5) produces tumor like growth from which roots and/or shoots may sometime be produced (Figure 19.6) while infection by *A. rhizogenes* gives rise to "hairy roots" which may often show negative geotropism, in some species shoots may regenerate from the

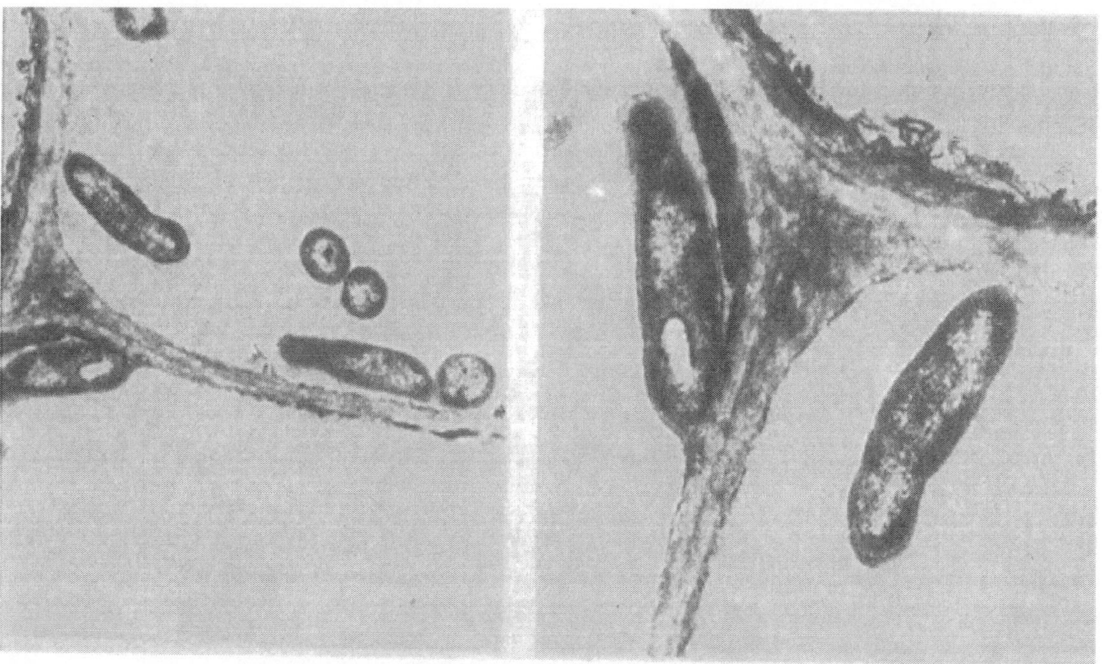

Figure 19.4: Attachement of *Agrobacterium* Cells to Plant Cell Walls

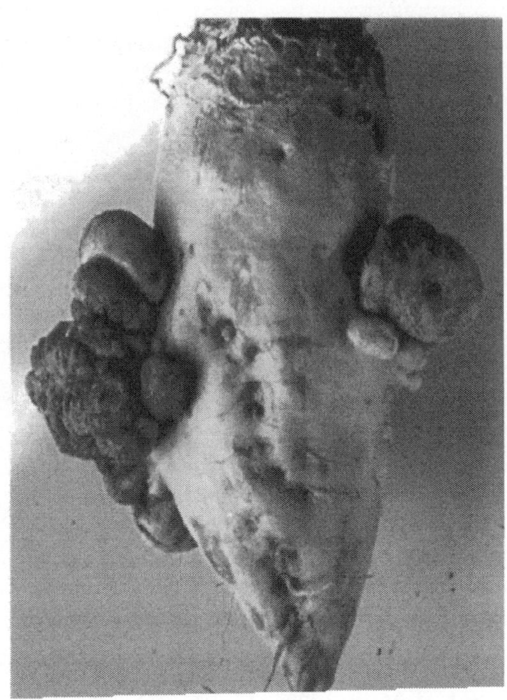

Figure 19.5: *Agrobacterium* Induced Tumor on a Turnip

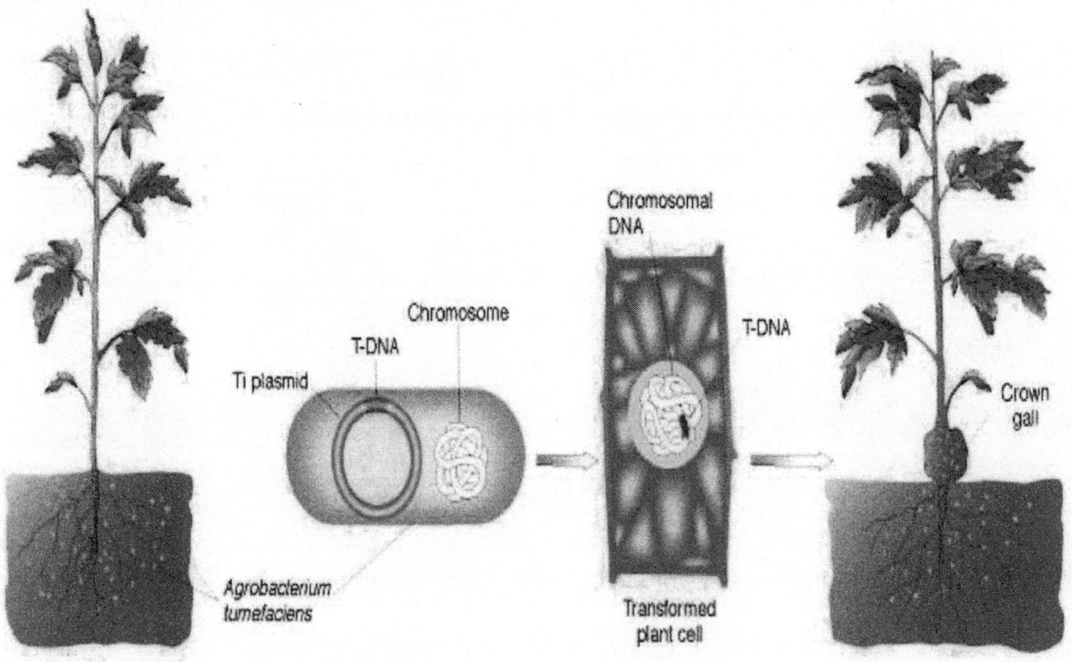

Figure 19.6: Key Events in the Formation of Crown Gall Tumors

roots giving rise to complete plants (Brown, 2001). Both hairy root and crown gall cells (free of *Agrobacterium* cells) show phytohormone independence while normal plant cells need exogenous auxin and or cytokinin in culture. The crown gall and hairy root cells also synthesize unique nitrogenous compounds called opine which are neither produced by normal plant cells nor utilized by them but only by *Agrobacterium* cells as their carbon and nitrogen source. *A. tumefaciens* usually produces octopine and nopaline, while *A. rhizogenes* produces either agropine or mannopine. The genes for opines production and utilization are present in respective pTi or pRi.

This plasmid also carries genes for indole acetic acid (IAA) and cytokinin production (Figure 19.7) which is responsible for indefinite growth of crown gall on a GR-free culture medium. When pTi is introduced into *Rhizobium trifoli*, it gains the ability to utilize opines and produce gall.

During *Agrobacterium* infection, the T-DNA (transferred DNA), 23 kb (Figure 19.8) possessed by Ti/Ri plasmid is transferred into the host plant genome. The T-DNA is flanked by 24 bp direct repeat border sequence on both the sides, and contains the genes for tumor /hairy root induction and those for opine biosynthesis. pTi possesses three genes responsible for crown gall formation (Table 19.4). Two of these genes (*iaaM* and *iaaH*) encode enzymes, which together convert tryptophan into IAA (indole, 3-acetic acid). The third gene, *ipt*, encodes an enzyme which produces the zeatin–type cytokinin isopentenyl adenine. In addition, genes involved in opine biosynthesis is located near the right border of T-DNA (Glick and Pasternak, 2001; Fischer, 2006). The T-DNA is organized in two distinct regions called TL (left T-DNA) and TR (right T-DNA).

Table 19.4: The Important Genes/Sequences of pTi (Nopaline Type) and their Functions

Gene/Operon	*Function*
T-DNA	
iaaM (auxl, tms l)	Auxin biosynthesis; encodes enzyme tryptophan-2-Mono-oxyganse, which converts tryptophan into indole- 3-acetamide (IAM).
iaaH (aux2, tms2)	Auxin biosynthesis; encodes enzyme indole -3-acetamide hydrolase, which converts IAM into IAA (indole -3-acetic acid)
ipt (tmr, cyt)	Cytokinin biosynthesis; encodes enzyme isopentenyl transferase, which catalyzes the formation of isopentenyl adenine, nos nopaline biosynthesis; encodes the enzyme nopaline synthase from arginine and pyruvic acid
24 bp left and right	Site of endonuclease action during T-DNA transfer, border sequences are essential for T-DNA transfer
Vir Region	
VirA (1)	Encodes a receptor for acetosyringone that functions as an autokinase; also phosphorylates VirG protein; constitutive expression
VirB (11)	Membrane proteins; possibly form a channel for T-DNA transport (conjugal tube formation); VirB11 has ATPase activity
VirC (2)	Helicase; binds to the overdrive region just outside the right border; involved in unwinding of T-DNA.
VirD (4)	VirD1 has topoisomerase activity; it binds to the right border of T-DNA; VirD2 is an endonuclease; it nicks the right border
Vir E (2)	Single –strand binding proteins (SSBP); bind to T-DNA during its transfer.
Vir F (1)	Not well understood
Vir G (1)	DNA binding protein; probably forms dimer after phosphorylation by VirA, and induces the expression of all vir operons; constitutive expression
VirH (2)	Not well known

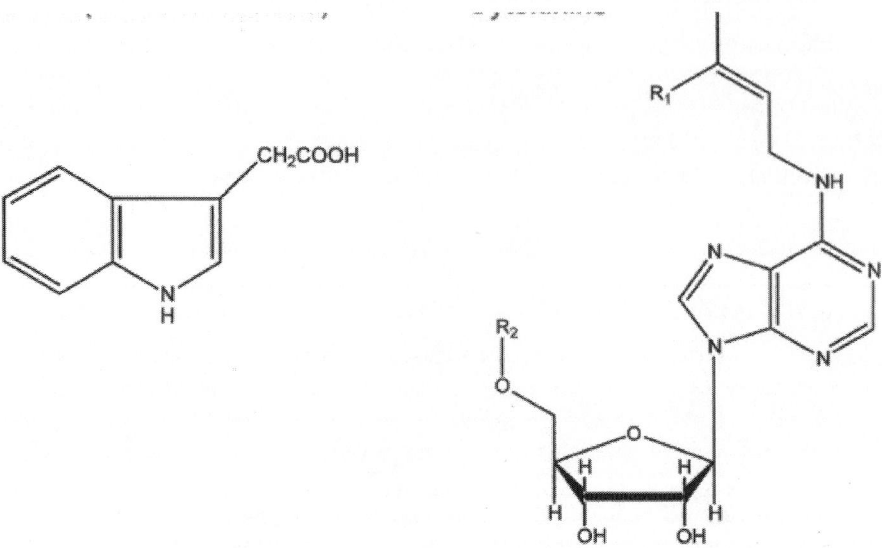

Figure 19.7: Plant Hormones Present in *Agrobacterium* Induced Tumors (A) Auxin (Indole acetic acid); (B) Cytokinins

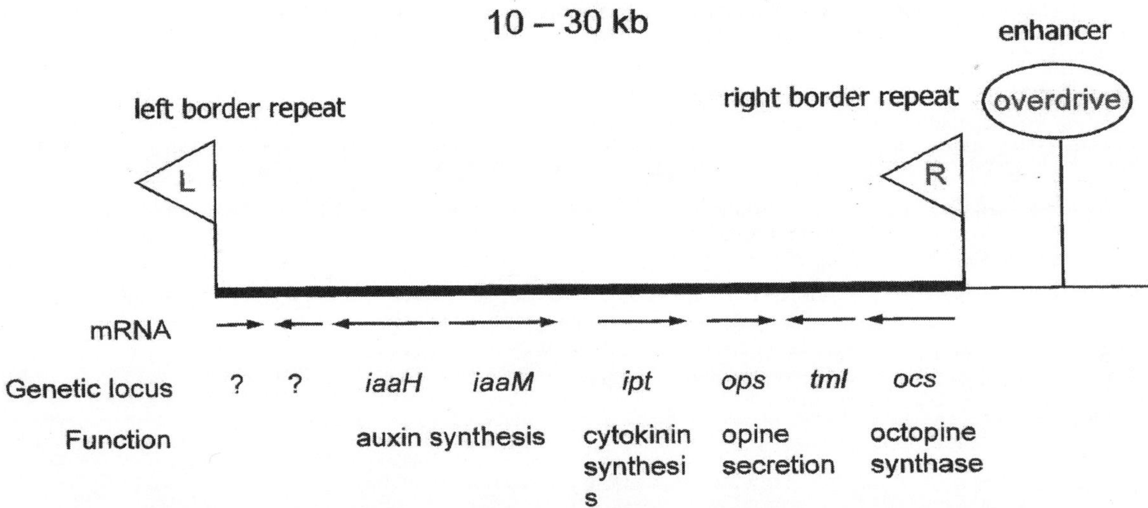

Figure 19.8: Structure of T-DNA (T-DNA encode genes have eukaryotic regulatory sequences. Thus they are not expressed in Agrobacterium but only in plants) (Ream, 1991)

The *vir* region (Figure 19.9) that spans about 40 kb of DNA has 8 operons (designated as *virA, virB, virC, virD, virE, virF, virG,* and *virH*) consisting of 25 genes. The transfer of T-DNA into plant genomes and the virulence or production of crown gall / hairy root disease is mediated by this region; therefore, it is called as the virulence region or *vir* region. Out of 8 *vir* operons, essential role for virulence is played by the 4 operons viz., *virA, virB, virD,* and *virG* whereas remaining 4 operons play

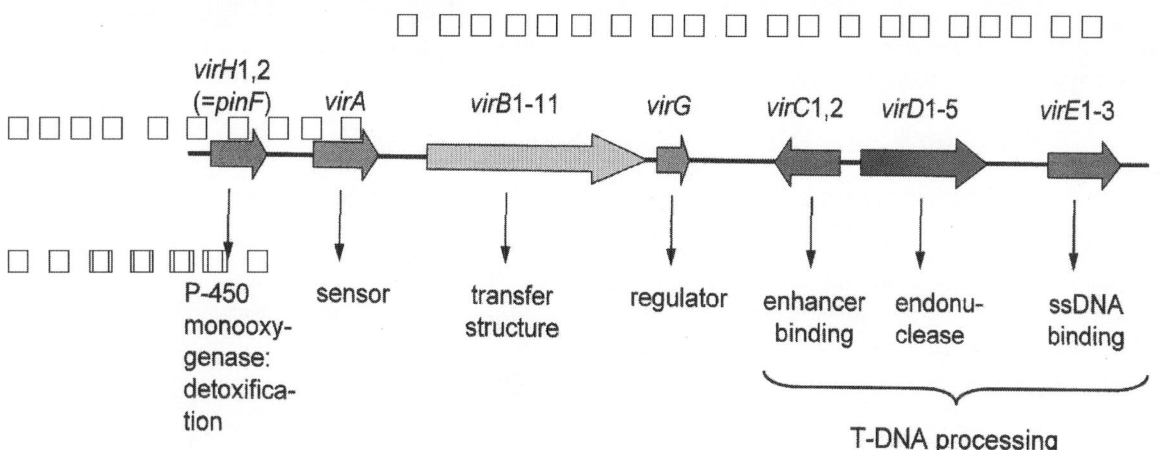

Figure 19.9: Organisation and Functions of Virgenes (Zhu *et al.*, 2000)

an accessory role (Glick and Pasternak, 2001; Fischer, 2006). The expression of operons *virA* and *virG* occurs constitutively and it regulates all the other vir operons. The other *vir* operons encode various proteins involved in T-DNA transfer. The genes possessed by the T-DNA are not required for its transfer, only the 24 bp direct repeat left and right border of T-DNA acting as recognition sequence are essential for the transfer.

The phenolic signal molecules namely acetosyringone, α-hydroxyacetosyringone and others, (Figure 19.10) produced by wounded tissues of virtually all dicot plant species, activates all the vir operons (Glick and Pasternak, 2001). These phenolics bind to the *virA* gene product that acts as

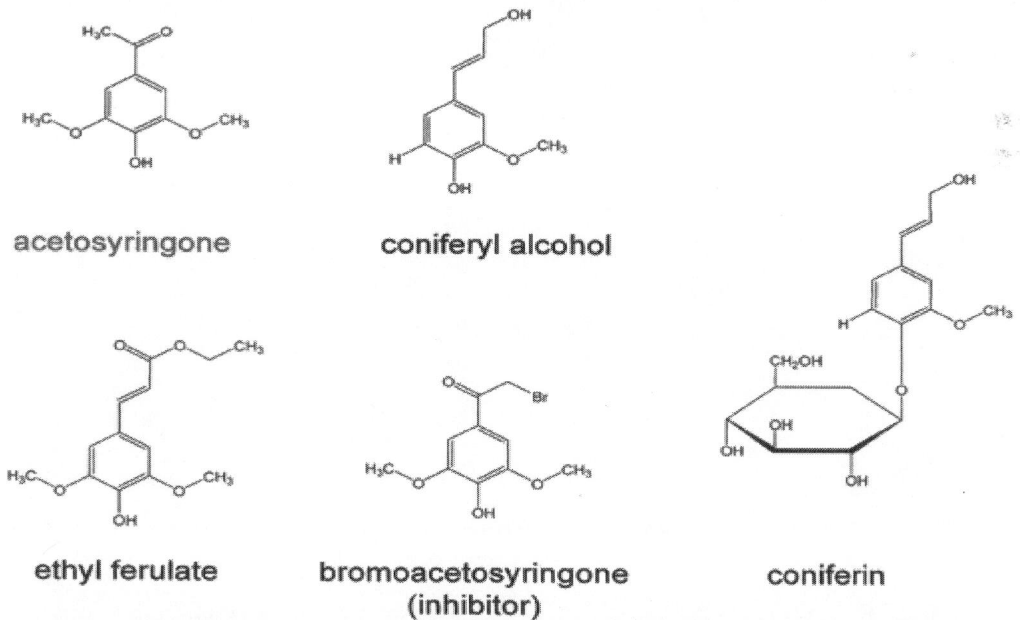

Figure 19.10: Inducer of the Vir Regulon Present in Wounded Plant Tissue (Winans, 1992)

receptor leading to activation of the VirA protein. VirA which is a potent kinase when activated undergoes autophosphorylation and then phosphorylates VirG protein. Phosphorylated VirG a DNA binding protein, probably undergoes dimerisation and induces transcription of all the other vir operons (Figure 19.11) (Singh, 2006; Fischer, 2006).

VirD1 protein, an topoisomerase (Figure 19.12) and binds to the right border sequence leading to relaxation of super coiling thereby facilitating the action of protein VirD2. VirD2, an endonuclease; it nicks at the right border and covalently binds (and remains bound during the T-DNA transfer) to the 5′-end so generated. At the site of nick, the 3′-end produced serves as a primer for DNA synthesis; as a result, the T-DNA strand is again nicked at the left border to generate a single-strand copy of T-DNA. About 600 copies of VirE2 protein which is single-strand binding protein bind to the single –strand T-DNA and protect it from nuclease action (Fischer, 2006).

virB operon has 11 genes, which encode mostly membrane bound proteins. Together with VirD4 protein, VirB proteins, participate in conjugal tube formation between the bacterial and plant cells, providing a channel for T-DNA transfer. An ATPase, VirB11 generates energy needed for the delivery

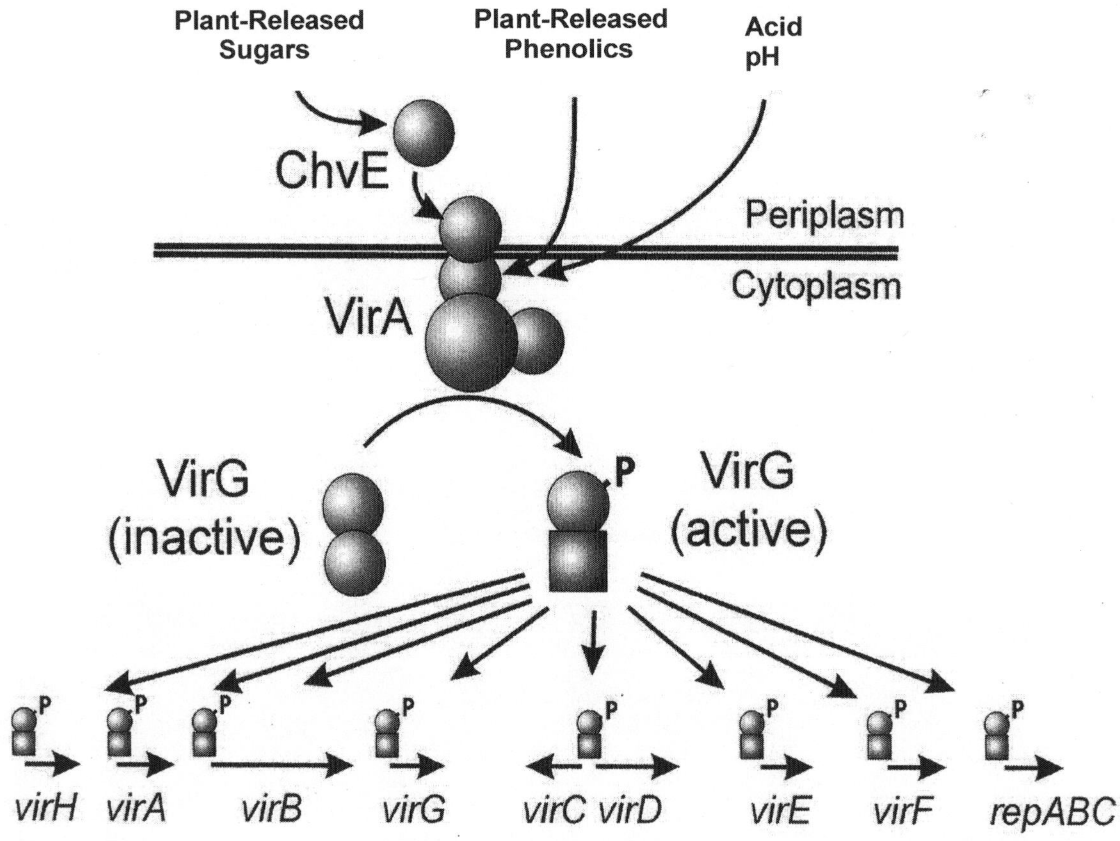

Figure 19.11: Vir Gene Regulation by the Vir AG Two Component Regulatory System (Brencic and Winans, 2005)

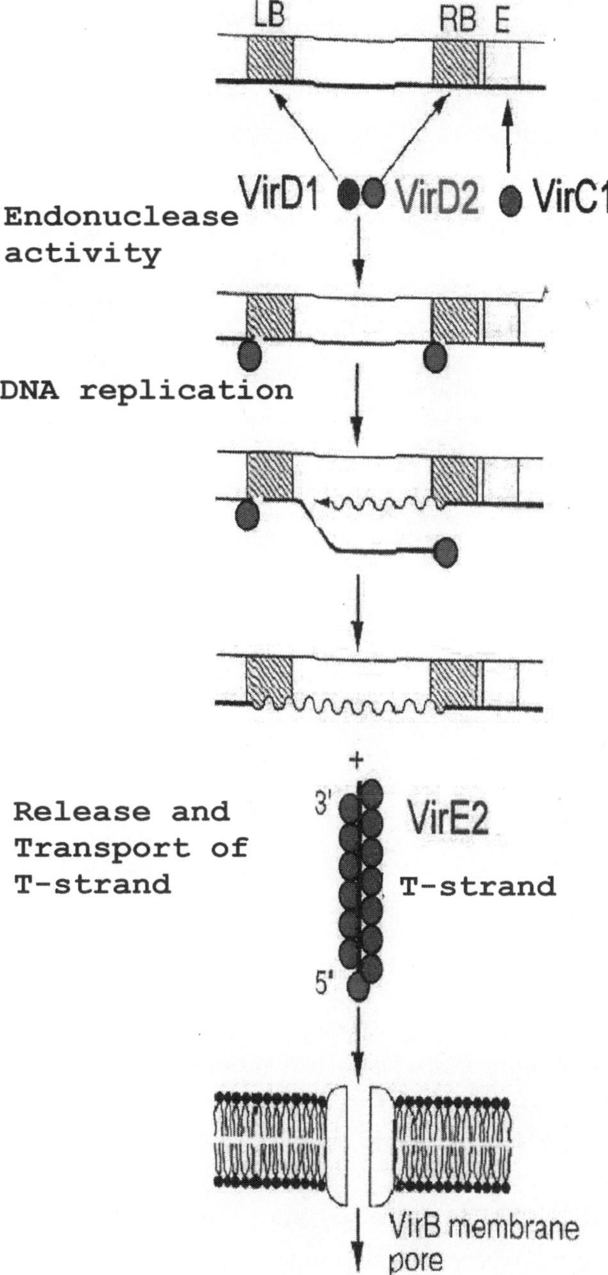

Figure 19.12: T-DNA Synthesis and Transport (Hooykass, 2000)

of T-DNA into the plant cells. The endonuclease VirD2, which nicks the right border and remains covalently bound to the 5′ end of the single-strand T-DNA copy, has a signal sequence, which drives it towards the nucleus of the transformed plant cell (after the delivery of T-DNA into the plant cell). The transport of T-DNA into the nucleus occurs through nuclear pore complex (Singh, 2006; Glick and Pasternak, 2001; Fischer, 2006).

The single stranded form of T-DNA enters plant cells and thereafter gets converted into a double stranded form (Glick and Pasternak, 2001). The double stranded T-DNA integrates in the host plant genome at random sites most likely by a process of illegitimate recombination due to a homology in short segments of the host DNA and its integration is generally accompanied by short deletions of 23-79 bp at the site of recombination (target site). The T-DNA is generally integrated in low copy number per cell but up to 1 dozen copies/cells have been recorded (Singh, 2006).

Application (Glick and Pasternak, 2001; Brown, 2001; Fischer, 2006)

1. Binary vector system: A two plasmid system in *Agrobacterium* for transferring a T-DNA region that carries cloned genes into plant cells. The virulence genes are on one plasmid, and the engineered T-DNA region is on the other plasmid
2. Transformation of *Agrobacterium* containing disarmed Ti plasmid
3. Infection of plant cells
4. Selection of transformed plant cells
5. Regeneration of transformed plant.

References

Brencic A and Winans SC (2005). Detection of and response to signals involved in host-microbe interaction by plant associated bacteria. *Microbiol Mol Biol Rev,* 69: 155–194.

Brown TA (2001). *Gene Cloning and DNA Analysis, 4th Edition: Cloning Vectors for Higher Plants*, p. 139–144

Glick BR and Pasternak JJ (2001). *Molecular Biotechnology: Principle and Application of Recombinant DNA, 2nd Edition: Genetic Engineering of PLants : Methodology.* ASM Press, Washington, p. 427–433.

Fischer HM (2006). *Molecular Microbiology* SS 2006, May 11, 2006.

Gelvin SB (2003). *Agrobacterium*-mediate plant transformation: The biology behind the "gene-jockeying" tool. *Microbiol Mol Bio Rev,* 67: 16–37.

Hass JH, Moore LW, Ream W and Manulis S (1995). Universal PCR primers for detection of phytopathogenic *agrobacterium* strain. *Applied and Environmental Microbiology,* 61: 2879–2884

Hooykaas PJJ (2000). *Agrobacterium.* In: *Encyclopedia of Microbiology,* Vol 1, 2nd edn. (Ed) J Lederberg. Academic Press, San Diego, USA, p. 78–85.

Jarvis BDW, Sivakumaran S, Tighe SW and Gillis M (1996). Identification of *Agrobacterium* and *Rhizobium* species based on cellular fatty acid composition. *Plant and Soil* 184: 143–158.

Matthysse Ann G (2005). *The Prokaryotes,* Part 1. Springer, New York. The Genus *Agrobacterium*, p. 91–95.

Prescott, Harley, Klein (2003). *Microbiology,* 5th edn. Mc Graw Hills Publication. Bacteria: The Proteobacteria, p. 488.

Ream W (1991). The essentials of *Agrobacterium* genetics. In: *Modern Microbial Genetics,* (Eds) UN Streips and RE Yasbin. Wiley-Liss, New York, p. 431– 453.

Singh BD (2006). *Biotechnology*, 2nd Edition. Kalyani Publishers, Ludhiana, Trasgenic plants: Gene Constructs, Vectors and Transformation Method, p. 339–349.

William A and Collins MD (1993). Phylo genetic analysis of rhizobia and agrobacteria based on 16s rDNA sequence. *International Journal of Systemic Bacteriology*, 43: 305–313.

Winans SC (1992). Two way chemical signaling in *Agrobacterium*-plant interactions. *Microbiol Rev*, 56: 12–31.

Zhu J, Oger PM, Schrammeijer B, Hooykaas PJJ, Farrand SK and Winans SC (2000). The bases of crown gall tumorigenesis. *J Bacteriol*, 182: 3885–3895.

Soil Microflora, 2009 *Pages* **246–250**
Editor: **Rajan Kumar Gupta, Mukesh Kumar & Deepak Vyas**
Published by: **DAYA PUBLISHING HOUSE, NEW DELHI**

Chapter 20

Association of Vesicular Arbuscular Mycorrhizas with Ornamental Plant *Petunia*

***Deepak Vyas*[*1], *Deepali Bilthare*[1], *Pramod Kumar Richhariya*[1] and *Rajan Kumar Gupta*[2]**

[1]*Lab of Microbial Technology and Plant Pathology, Department of Botany, Dr. H. S. Gour University, Sagar – 470 003, M.P.*
[2]*Department of Botany, Govt. P.G. College, Rishikesh – 249 201*

ABSTRACT

Petunia is a wildly cultivated genus of flowering plants of family Solanaceae. Present study deals with association of vesicular arbuscular mycorrhiza (VAM) with *Petunia*. The result obtained from our experiment suggest that VAM fungi vary at different stages of plant growth. It is clearly evident from the result that root colonization by VAM fungi occurs after 15 days of seedling transplantation. At this stage, spore population was sparsely scattered and only 3 VAM fungi were recorded and identified in the rhizosphere soil of *Petunia*. Increases in root colonization, spore population and VAMF species was recorded with the growth of the plant. However, arbuscules were observed between 15 to 30 days of plant growth. Maximum number of spores, higher percentage of root colonization and greater number of VAMF was observed in 60 days old plants. At later stage of plant growth arbuscules were disappeared and vesicles were appeared. When plants were uprooted after 90 days, spore population as well as VAM fungi were found reduced.

Keywords: *Petunia, Colonization, Rhizosphere, VAMF, Fungi.*

---

* E-mail: dvyas64@yahoo.co.in

Introduction

Vesicular Arbuscular Mycorrhizal (VAM) fungi are important components of natural eco-systems and strongly influence the plant community composition and eco-system function (Hart and Klironomos, 2002). VAM fungi are known to geographically ubiquitous and occur over a broad range of dissimilar environments (Gerdemann and Trappe 1974; Koske, 1987) from the arctic to the tropics and occupy a wide range of ecological niches (Shrivastava *et al.*, 1996). They are commonly associated with plants in agriculture, horticulture pastures and tropical forests (Rao *et al.*, 2000; Read *et al.*, 1976). More than 80 per cent of all plants form relationship with VAM fungi in their root system (Smith and Read, 1997; Harley and Smith, 1983). According to Gupta *et al.* (1994) VAM fungi have also been observed in association with ornamental and vegetables plants. Mycorrhizal relationship is complex, subtle and affect above ground community structure in a variety of ways.

Petunia belongs to family solanaceae. Its most of the varieties seen in gardens are hybrids. These plant grow in most soil types and do very well in poor soils and in pots. Nutritional value in the plant depends upon the availability of nutrients in soil where they are growing wild as well as cultivated in botanical garden. Keeping with view that mycorrhization may influence synthesis of essential nutrients in *Petunia* therefore, we have taken this study.

Material and Methods

This study was conducted at botanical garden of Dr. H. S. Gour University, Sagar (M.P.). Soil and Root samples were collected from the rhizosphere soil of *Petunia*. Physico-chemical properties of rhizospheric soil was determined using (Jackson, 1958). The sampling was done in different intervals. VAM spores were isolated by wet-sieving and decanting method (Gerdamann and Nicolson, 1963). VAM colonization in the roots determined by method of Phillips and Hayman (1970) and per cent root colonization was determined after staining with Lacto-phenol cotton blue. Mycorrhizal spores are identified using conventional taxonomic key of Schenck and Perez (1990), Giovannetti and Mosse (1980). Morton (1998), Morton and Redecker (2001).

Result

The soil of this region is greyish black, sandy clay having about 7.2 pH and annual temperature is 28°C, Table 20.1. The result mentioned in Table 20.2 suggests that 15 days of seedling, plants shows 52 per cent root colonization, spore population 104 spores, 100 gm^{-1} of soil and 3 VAM fungal species were recorded and identified in Table 20.3. 30 days old plants, showed 71 per cent root colonization, 120 spores 100 gm^{-1} of soil and 4 VAMF species respectively. Further, increase in root colonization, spore population and VAMF species was recorded with the increase in plant growth. Therefore 45 and 60 days old plant show 76 and 84 per cent, 216 and 320 spores 100 gm^{-1} of soil show 5 and 6 VAMF species respectively. However, when 90 days old plants were examined for mycorrhization, it was found that not only percent root colonization was reduced but also number of spores and VAM fungal species were found decreased in comparison to 60 days old plant. Moreover, arbuscules were seen in early stages of plants growth and at later stages of growth they disappeared, vesicles were found in 60 days old plants. Dominance of *Glomus* was observed among the VAMF obtained during the study period.

The VAM fungi species recorded during the present investigation at different intervals from the rhizosphere soil are *Acaulospora bireticulata* (ABRT), *Acaulospora dilata* (ADLT), *Acaulospora denticulata* (ADTC), *Acaulospora gardemmnii* (AGPM), *Acaulospora lacunose* (ALCN), *Acaulospora scrobiculata* (ASCB), *Glomus ambisporum* (LABS), *Glomus clarum* (LCLR), *Glomus clerodium* (LCRD), *Glomus citricolum* (LCTC),

Glomus fusiculatum (LFSC), *Glomus hoi* (LHOI), *Glomus heterosporum* (LHTS), *Glomus lacteum* (LLCT), *Glomus leptoticum* (LLPT), *Glomus mosseae* (LMSS), *Glomus pubescens* (LPBS) and *Sclerocystis puchycaulis* (SPCC) now grouped under *Glomus* sp. (www.amfphylogeny). It is clearly evident from the result that *Glomus spp.* dominated in the occurrence in association with *Petunia* at intervals of plant growth.

Table 20.1: Physico-chemical Properties of Rhizospheric Soil of *Petunia*

Properties		*Study Soil*		
Physical	Colour	Reddish brown	Brown	Brown to black
	Texture	Graval sand	Sandy clay	Clay
	Temperature	28°C	26°C	25°C
	pH	7.2	7.3	7.2
Chemical	Nitrogen (N) $Kgha^{-1}$	378	395.10	383.62
	Phosphorus (P) $Kgha^{-1}$	9.98	12.81	10.87
	Potassium (K) $Kgha^{-1}$	302.57	377.90	276.00
	% Organic Carbon	1.45	1.70	1.73

Table 20.2: VAM Association with *Petunia* in Different Growing Intervals

Sl.No.	*Growing Intervals*	*Root Colonization (%)*	*Spore Population 100 gm^{-1} of Rhizospheric Soil*	*Name of Dominant VAM Species*
1.	15	52	104	ADTC, LLCT, LHOI
2.	30	71	120	LCTC, LCLR, LHOI, LMSS
3.	45	76	216	ADLT, ASCB, LFSC, LPBS, LMSS
4.	60	84	320	ABRT, ALCN, LCTC, LHOI, LHTS, LLPT
5.	90	70	288	AGDM, LABS, LCRD, LFSC, LMSS

Species Code as per Perez and Schenck (1990)

ABRT: *Acaulospora bireticulata* (Rothwell and Trappe); ADTC: *Acaulospora denticulata* (Siverding and Taxo); ADLT: *Acaulospora dilata* (Morton); AGDM: *Acaulospora gerdemmnii* (Schenck and Nicolson); ALCN: *Acaulospora lacunosa* (Morton); ASCB: *Acaulospora scrobiculata* (Trappe); LABS: *Glomus ambisporum* (Smith and Schenck); LCLR: *Glomus clarum* (Nicolson and Schenck); LCTC: *Glomus citricolum* (Tang and Zong); LFSC: *Glomus fasiculatum* (Thaxter); LHOI: *Glomus hoi* (Brech and Trappe); LHIS: *Glomus heterosporum* (Smith and Schenck); LLCT: *Glomus lacteum* (Rose and Trappe); LLPT: *Glamus leptotichum* (Schenck and Smith); LMSS: *Glomus mosseae* (Gerdemann and Trappe); LPBS: *Glomus pubescens* (Trappe and Gerdemann).

Discussion

Petunia is a well-known ormamental plant, herbaceous generally cultivated in gardens particularly in summer season. The occurrence of mycorrhizal fungi at early stages of growth shows its mycorrhizal dependency. It shows good association of VAM fungi with its roots. *Petunia* shows greater colonization in the roots suggest that host plant develop strong mutualistic association with only few VAM fungi. The results obtained from the study, clearly indicates that in *Petunia* occurence of VAM fungi is affected by couple of possible reasons. During the early phase of plant growth, plants were irrigated within the short span of time. Therefore wet condition were sustained for larger time. It has been reported that wet condition are detrimental for mycorrhization. Since plants during early stage of their growth, photosynthesis was less resulting, lesser production of photosynthate and subsequently lesser root exudation due to this, competition of carbon in the rhizoshphere of *Petunia* restricted the growth of VAM fungi. Later stage of plant growth which required lesser amount of water, and leaves started photosynthesis, might be providing enough carbon resulting good growth of VAM fungi. 90 days old plant showed reduction in per cent colonization, spore population and VAMF species. It is deduced that at the dying stage in the plants neither required nutrient nor carbon. It has been reported by Sanderes and Koide (1994) that fully grown plants divert there photosynthate resulting lesser leaching of root exudation. Carbon scarisity in mycorrhizosphere zone ultimately responsible for the reducing trends of the VAM fungi in *Petunia*. Dominance of *Glomus* in soil of Botanical garden in Dr. H.S. Gour University is already been reported by Vyas and Soni, 2004)

Acknowledgements

Authors are thankful to Head Department of Botany, Dr. H.S.G. University, Sagar, for providing lab facilities. One of the author (RKG) is thankful to UGC, New Delhi.

References

Gerdemann JW and Nicolson TH (1963). Spores of mycorrhizal endogone species extracted from soil by wet sieving and decanting. *Trans Br Mycol Soc*, 46: 235–244.

Gerdemann JW and Trappe JM (1974). The endogonaceae in the Pacific Northwest. *Mycologia Mem*, 5: 1–76

Giovannetti M and Mosse B (1980). An evaluation of techniques for measuring vesicular-arbuscular mycorrhizal infections in roots. *New Phytol*, 84: 489–500.

Gupta ML, Mohankumar V and Janardhanan KK (1994). Distribution of vesicular-arbuscular mycorrhizae in some important medicinal and aromatic plants. *KAVAKA*, 22/23: 29–33.

Harley JL and Smith SE (1983). *Mycorrhizal Symbiosis*. Academic Press, London, pp. 483.

Hart M and Kirlonomos JN (2002). Diversity of arbuscular mycorhizal fungi and eco-system functioning. In: *Mycorrhizal Ecology*, (Eds) Van der Heijden MGA, Sander IR. Springer, Berlin Heidelberg, New York, pp. 225–242.

Jackson ML (1958). *Soil Chemical Analysis*. Prentice-Hall International, Inc, Eagliwood Cliffs, New Jersey.

Koske RE (1987). Distribution of VAM mycorrhizal fungi along a latitudinal temperature gradient, *Mycologia*, 79: 55–58.

Morton JB (1998). Taxonomy of vesicular-arbuscular mycorrhizal fungi; classification nomenclature and identification. *Mycotaxon*, 32: 267–324.

Morton JB and Redecker D (2001). Two new families of Glomales, Archaeosporaceae and Paraglomaceae, with two new genera *Archaeospora* and *Paraglomus* based on concordant molecular and morphological character. *Mycologia*, 93: 181–195.

Phillips JM and Haymann DS (1970). Improved procedures for clearing roots and staining parasitic and vesicular arbuscular mycorrhizal fungi for rapid assessment of infection. *Trans Br Mycol Soc*, 55: 158–163.

Rao GV, Manoharachary C, Kunwar IK and Rao BRR (2000). Arbuscular mycorrhizal fungi associated with some economically important spices and aromatic plants. *Philippine J Sci,* 129: 51–55.

Read DJ, Koucheki HK and Odgson J (1976). Vesicular-arbuscular mycorrhiza in natural vegetation systems I. The occurrence of infection. *New Phytol,* 77: 641–653.

Sanders IR and Koide RT (1994). Nutrient acquisition and community structure in co-occurring mycotrophic and two mycotrophic old-field annuals. *Functional Ecology,* 8: 77–84.

Schenck NC and Perez (1990). *A Manual for Identification of Vesicular Arbuscular Mycorrhizal Fungi.* INVAM University of Florida Gainesville.

Shrivastava D, Kapoor R, Shrivastava SK and Mukerji KG (1996). Vesicular arbuscular mycorrhizal: An overview. In *Concepts in Mycorrhizal Research,* (Ed) KG Mukerji. Kluwer Academic Publishers, Netherlands, pp. 1–39.

Smith SE and Read DJ (1977). *Mycorrhizal Symbiosis,* 2nd edn. Academic Press, San Diego.

Vyas D and Soni A (2004). Diversity and distribution of VAMF in the seminatural grassland. *Indian J Ecology,* 31: 170–171.

www.amfphylogeny.

Soil Microflora, 2009 *Pages* **251–259**
Editor: **Rajan Kumar Gupta, Mukesh Kumar & Deepak Vyas**
Published by: **DAYA PUBLISHING HOUSE, NEW DELHI**

Chapter 21

Cyanobacterial Biodiversity in the Soils of Kumaon Region

Anjali Khare[1], Mukesh Kumar[2] and Promod Kumar[3]
[1]*Department of Botany, Advance Institute of Science and Technology, Dehradun*
[2]*Department of Botany, Sahu Jain (P.G.) College, Najibabad*
[3]*Department of Botany, Hindu College, Moradabad*

ABSTRACT

Soil microflora consists of only about 01 per cent of algae, out of which Cyanobacteria forms a very small part yet they play an important role in soil conditioning and vitality of the soil. Different species of Cyanobacteria are known to suppress used growth, reduce the loss of applied chemical nitrogen fertilizer, can be used as an animal feed, human food, a medicine and water purifies.

Keeping this point in view, the present study has been made to explore the presence of Cyanobacteria in the soil of the sub-Himalayan Belt of Kumaon Region. The observations reveal that all in all 57 terrestrial species have been reported from the region. The information can be further utilized so as to utilize these forms of Cyanobacteria by culturing them and apply wherever they have the potential use.

Introduction

Biodiversity is the variety of the world's organisms. It is the scientific terminology for the natural biological wealth that influences human life and well-being. The breadth of the concepts of Biodiversity reflects on the interrelationship of genes, species and ecosystems (Singh, 2004). Biodiversity is no doubt, the very basis of man's being, but due to some environmental hazards and man's interference it is in serious risk of extinction. The current losses to biodiversity can be attributed to direct causes including habituate loss and fragmentation, invasion of introduced species, over exploitation of living resources, and modern agriculture and forestry practices (Miller *et al.*, 1992).

Algae are the heterogenous assemblage of plants that includes prokaryotic and eukaryotic organisms. They are the pioneer colonizers both in hydrosphere and xerosphere and occupy the base of the trophic pyramid. These organisms have been found to synthesize 0.8x10" tonnes of organic matter, constituting about 40 per cent of the total organic matter synthesized annually on this planet (Goyal, 2002)

Cyanobacteria constitute the most diversity group of plant kingdom. They occur in a variety of habitats including terrestrial, lithophytic, epiphyitc etc. These have the ability to survival in diverse ecological conditions. They are polymorhic, prokaryotic microorganisms with single cell colonial and trichome organization. They are named variously *i.e.* cyanophytes, cyanophyceae and most recently cyanoprokaryotes (Fabbro and Mc Gregor, 2003). Endowed with the remarkable capacities to adapt to varying environmental conditions they colonize almost all kinds of terrestrial ecosystems, often in some specific environments where no other vegetation can exist.

The basic significance of the ecological observations on the abundance of BGA in Indian rice field soils became apparent when it was recognized that heterocystous forms could fix atmospheric nitrogen that is made available to the plants during life cycle and after its death by decomposition of cells, which became available to the subsequent crops. The extensive research during last few decades has strengthened our knowledge towards economic use of this group of organisms. They are widely used as biofertilizers. Their growth not only adds nitrogen to the soil but also help reducing soil erosion, decreasing soil compaction, adding organic matter librating growth regulator (Venkataraman, 1986; Goyal, 193). Certain Cyanobacteria perform the process of O_2 sensitive N_2 fixation and O_2 evolving photosynthesis simultaneously (Stewart, 1969).

Cyanobacteria is an important component of soil microflora although it forms very small part, yet its vast use as biofertilizers has got the capability to fix atmospheric nitrogen. Keeping this point in view, the different approaches for measuring cyanobacterial biodiversity in the soil of the sub-Himalayan belt of Kumaon region have been considered.

Methodology

Sites of Study

The study area is situated in South-East to North of Kumaon Himalayas (between 20°00′ N to 78°80′ E), Uttarakhand, India. The present investigations are based on four research sites *i.e.* Kashipur, Rudrapur, Ramnagar and Haldwani, located in two districts, Udham Singh Nagar and Nainital of Kumaon Himalayas and exhibit a sub- tropical climate. The first research site, Kashipur is situated at an altitude of 235m amsl, and 20°13′N latitude and 28°59′ longitude. The second research site, Rudrapur is situated at an altitude of 244 m amsl, and a latitude of 29°03′N and longitude of 79°3′ N. The third research site, Ramanagar is situated at an altitude of 330m amsl, latitude of 79°05′N and longitude of 79°05′E where as the fourth research site, Haldwani is located at a latitude of 79°13′ N and longitude of 79°3′E at 432 m amsl.

Collection of Samples

The cyanobacterial samples for the present study were collected several times in different seasons *i.e.* winter, summer and rainy, of the year from various habitats. Cyanobacteria growing on soil, bark of trees, walls of buildings, ponds etc. were collected for taxonomic studies from different localities of all research sites of the Kumaon Himalayan range. The samples were stored into sterile plastic vials, then washed thoroughly with water and preserved in 4 per cent formalin solution.

Identification of Cyanobacteria

The microscopic slides were prepared from fixed as well as fresh samples. Preparations were observed under the trinocular research microscope. Camera-lucida drawings have been sketched and the drawings were analyzed on the basis of morphological observations consulting the pertinent literature in the field (Desikachary, 1959), the cyanobacterial specimens have been identified at the level of class, order, family, genus and species.

Pedological Studies

For pedological studies, the soil samples from different research sites, with or without visual plant community of blue green algae were collected from surface as well as from different depths between 10-15 cm and transported to the laboratory for physical as well as chemical analysis. The physical (texture, temperature, pH and electrical conductivity) as well as chemical contents (total nitrogen, available phosphorus and potassium) of the soil present at different sites have been studied.

Results and Discussion

The soil in loam clay and silt in Kashipur, sandy loam and silt sand in Rudrapur, silty loam in Ramnagar and Loamy sand and clay in Haldwani. The soil samples were collected from different localities of the research areas and have been subjected to physical as well as chemical analysis to bring out the complete picture of the soils of the area (Table 21.2 and Figures 21.1 and 21.2).

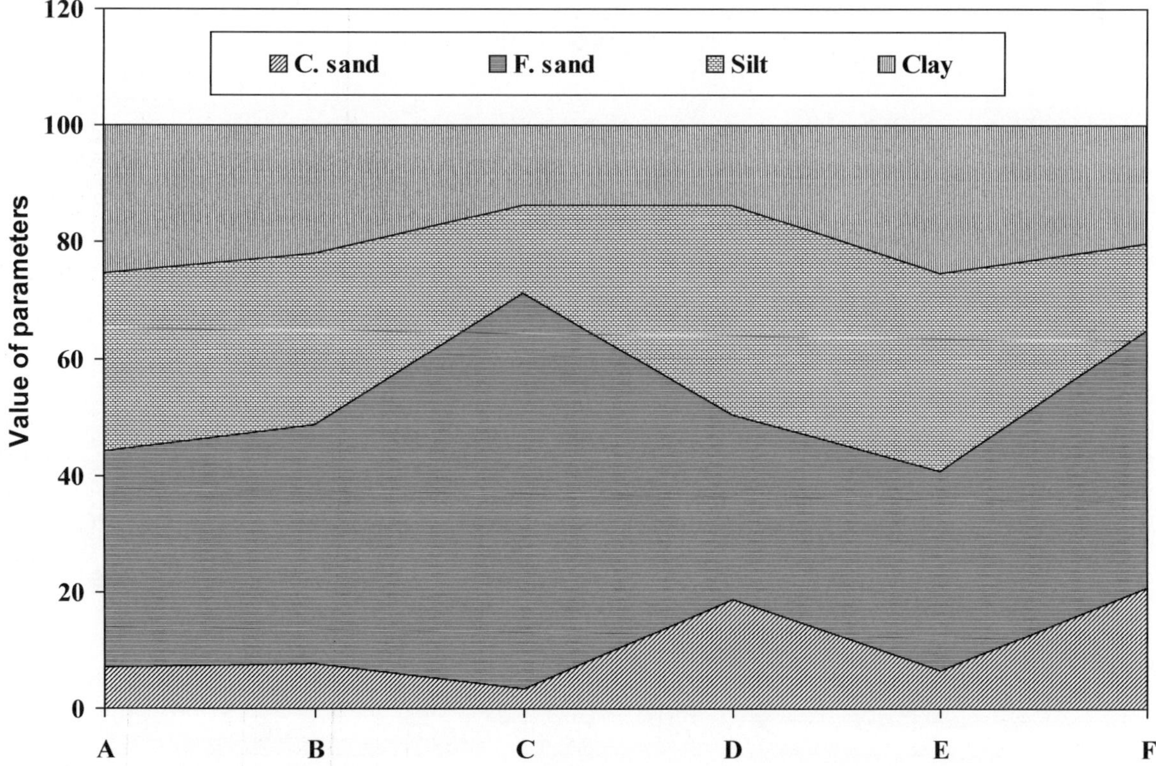

Figure 20.1: Mechanical Analysis Showing Texture of Soil at Different Sites

(A) Kashipur Awas Vikas Colony; (B) Kashipur Nagarpalika; (C) Rudrapur Pattharchata Nala; (D) Rudrapur Danpur Village; (E) Ramnagar; (F) Haldwani

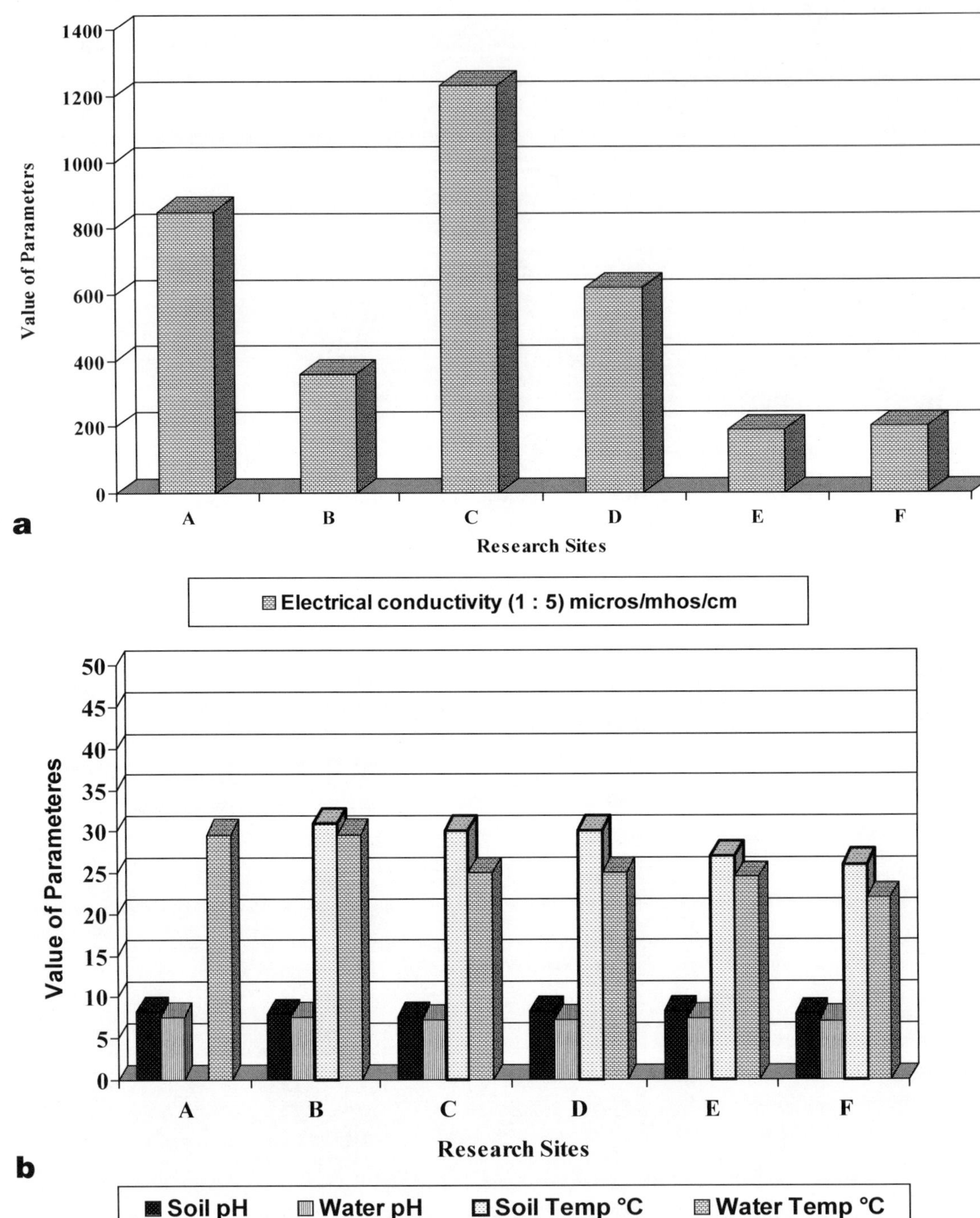

Figure 21.2a,b: Physical Parameters of the Soils at Different Research Sites

(A) Kashipur Awas Vikas Colony; (B) Kashipur Nagarpalika; (C) Rudrapur Pattharchata Nala; (D) Rudrapur Danpur Village; (E) Ramnagar; (F) Haldwani

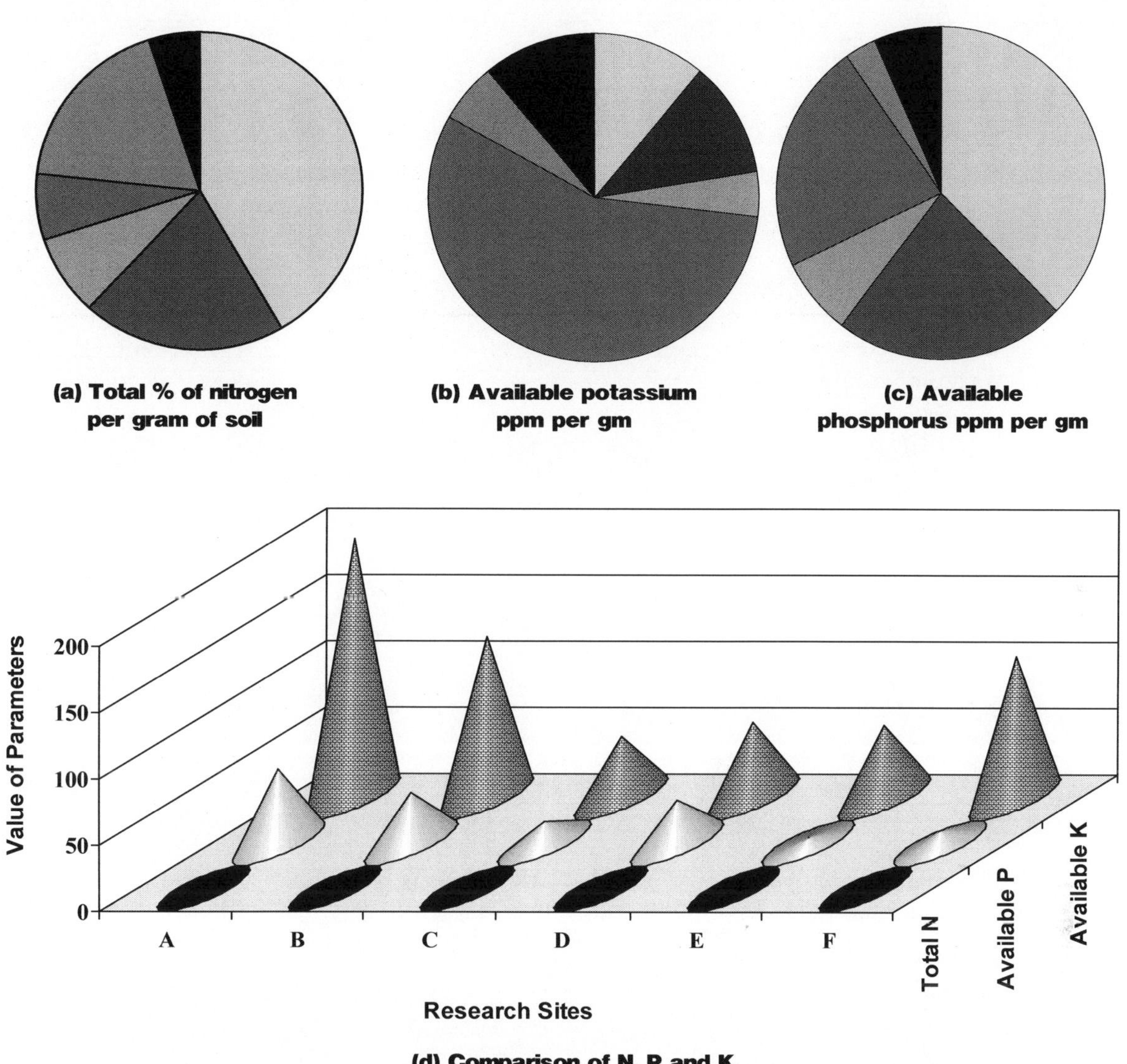

Figure 21.3: Chemical Parameters of Soils at Different Research Sites (A) Kashipur Awas Vikas Colony; (B) Kashipur Nagarpalika; (C) Rudrapur Pattharchata Nala; (D) Rudrapur Danpur Village; (E) Ramnagar; (F) Haldwani

A total of 170 species, 38 genera, 11 families and 04 orders have been recorded from all the research sites (Khare, 2007). Out of these a total of 57 species were terrestrial. Terrestrial BGA were formed sub-dominantly in all research sites. Rudrapur has highest terrestrial species (23) and Kashipur and Haldwani has same number (11) Ramnagar have 14 terrestrial forms, respectively (Table 20.1).

Table 21.1: Terrestrial Cyanobacterial Species in Kumaon Region

Sl.No.	Name of species	Research Sites			
		Kashipur	Rudrapur	Ramnagar	Haldwani
1.	*Microcystis orissica* v.nov	–	+	–	–
2.	*M. viridis*	–	–	+	–
3.	*Chroococcus pallidus*	–	–	+	–
4.	*C. gomontii*	–	+	–	–
5.	*C varius*	–	–	+	–
7.	*Synechocous cedrourum*	+	+	–	–
8.	*Synechocous pevalekii*	+	–	+	–
9.	*Dactylococus raphidioides*	–	–	–	+
10.	*Chlorogloea microcystoides*	–	–	+	–
11.	*C. fritschi*	–	+	–	–
12.	*Arthrosphira platensis var non constricta*	–	–	+	–
13	*A. tenuis*	–	–	+	–
14.	*A massartii*	–	–	+	–
15.	*A spriulinoides f. tenuis*	–	–	–	+
16.	*Spriulina gigantea*	+	–	–	–
17.	*S. labyrinthiformis*	–	+	–	–
18.	*S. meneghiniana*	–	–	–	+
19.	*Oscillatoria formosa*	+	–	–	–
20.	*O. boryana*	+	–	–	–
21.	*O. tenuis*	+	–	–	–
22.	*O princeps*	–	–	+	–
23.	*O. animalis*	+	–	–	–
24.	*O. animalis f. tenuior*	–	–	+	–
25.	*O. acuta*	–	+	–	–
26.	*O. brevis*	–	+	–	–
27.	*O. grunowiana*	+	–	–	–
28.	*O. obscura*	–	+	–	–
29.	*O. jasorvensis*	+	–	+	–
30.	*O. terebriformis*	–	–	–	+
31.	*O. agardhii*	–	+	–	–
32.	*O. splendida*	–	+	–	–
33.	*O. schultzii*	–	+	–	–
34.	*O. angusta*	–	+	–	–
35.	*O. subbrevis*	+	–	–	–
36.	*O. vizagapatensis*	–	+	–	–

Contd...

Table 21.1–Contd...

Sl.No.	Name of species	Research Sites			
		Kashipur	Rudrapur	Ramnagar	Haldwani
37.	*O. prolifica*	–	+	–	–
38.	*O. salina f. major*	–	+	–	–
39.	*O. anguina*	–	–	–	+
40.	*O. amphibia*	–	–	–	+
41.	*O. chilkensis*	–	–	–	+
42.	*O. curviceps*	–	–	–	+
43.	*Phormidium fragile*	–	+	–	–
44.	*P. fragile*	–	–	+	–
45.	*Lyngbya gracilis*	–	+	–	–
46.	*L. aerugineo– coerulea*	–	+	–	–
47.	*L. cryptovaginata*	–	–	–	+
48.	*L. corticola var. minor*	–	–	+	–
49.	*Cylindrospermum alatosporum*	–	+	–	–
50	*C. majus*	–	+		–
51.	*Nostoc muscorum*	–	+	+	–
52.	*N. punctiforme*	–	–	+	–
53.	*Pseudanabaena catenata*	–	–	–	+
54.	*Aulosira prolifica*	+	–	–	–
55.	*A. laxa*	–	–	+	–
56.	*Hormothamnion solutum*	–	–	+	–
57.	*Mirochaete aeruginea* var. *minor* (v. nov)	–	+	–	–
	Site Wise Number of Terrestrial species	**11**	**23**	**16**	**11**

Note: +: observed; –: Not observed.

The presence of highest number of terrestrial species in Rudrapur clearly indicates that sandy loam soil with silt is best suited for luxurious growth of Cyanobacteria, on the other hand; Kashipur has loam clay and sitly soil whereas Haldwani has loamy sand and clay and lowest number of terrestrial species has been observed from these both sites. This brings the author to the conclusion that the soil having clay as present in Kashipur and Haldwani is not good for the growth of Cyanobacteira.

Conclusion

Cyanobacteria can be considered as an important component of soil microflora as many of the BGA species make major contributions to the world food supply by naturally partilizing soil and rice paddies. It has also been reported by several workers that the introduction of blue-green algae to saline and alkaline soils in the state of Uttar Pradesh increases the soils content of nitrogen and organic matter and also their capacity for holding water. This treatment has enabled formerly barren soils to grow crops.

Table 21.2: Pedological Features of Various Research Sites

Sites	Physical Parameters								Chemical Parameters		
	Coarse Sand%	Fine Sand%	Silt%	Clay%	Soil Texture	pH Soil	Temp. Soil	E.c micros/ mhos/cm	Total N %	P (ppm)	K (ppm)
Kashipur (Awas-Vikas) 235m amsl	7.08	37.24	30.24	25.44	Loam clay and silt	8.2	31.00	850	0.25	0.0055	0.019
Kashipur (Nagar palika) 235m amsl	7.72	40.92	29.44	21.92	Loam clay and silt	7.9	31.00	360	0.12	0.0037	0.012
Rudrapur (Patthachala) 244m amsl	3.3	67.9	15.2	13.6	Loamy sand	7.6	30.00	1230	0.05	0.0016	0.0045
Rudrapur (Danpur village) 244m amsl	18.8	31.6	36.0	13.6	Sandy loam and silt	8.2	30.00	620	0.04	0.0033	0.056
Ram Nagar 330m amsl	6.68	33.96	34.08	25.28	Silty loam	8.2	27.00	190	0.11	0.0004	0.005
Haldwani 432m amsl	20.72	44.40	14.72	20.16	Loamy sand and silt	7.9	26.00	200	0.03	0.0009	0.01

Another report reveals that a coating of blue greens on prairie soil binds the particles of the soil, their mucilage coating, maintains a high water content and reduces erosion. The blue green algae work as biofertilizers also.

Thus, all the terrestrial species of Cyanobacteria observed form the research sites reveal the presence of this soil microflora of immense importance to a great extent. Further investigations are required to explore all possibility of being new taxonomic specimens with interesting features. The strains which are responsible for biological nitrogen fixation or act as biofertilizer can be cultured and applied to the soils to increase the yield of the crop without cussing any harm to the soil or the environment.

Hence, it can be concluded although Cyanobaceria forms a very small part of the soil microflora, yet it is of immense importance and its study is beneficial for upraising of the agricultural field or in other words to increase soil biodiversity.

Acknowledgements

The authors are thankful to the UGC, New Delhi for providing financial support in the form of a Major Research Project on Cyanobacteria.

References

Fabbro L and Mc Gregor G (2003). *Blue green Algae*. General Information NRM facts water seces W3, Queensld Govt. Natural Resources and Mines, Australia.

Goyal SK (2002). A profile on Algae Biofertilizer. In: *Biotechnology of Biofertilizers*, (Ed) S Kannaiyan. TNA, Coimbatore, Tamil Nadu, pp 250–258.

Goyal SK (1993). Algae Biofortilizers for vital soil and free nitrogen. *Proc Indian Natn Sci Acad,* 3: 295–302.

Khare A (2007). Cyanobacterial Biodiversity of the Sub-Himalyan Belt of Kumaon Reigon. *PhD Thesis,* MJP Rohilkhand University, Bareilly

Mitter KR, Raid WV and Barber V (1992). The global biodiversity strategy and its significance for sustainable agriculture. In: *Biodiversity: Implication for Global Food Security,* (Eds) MS Swaminathan and S Jena. Macmillan Publ., Madras, p 326.

Singh BK (2004). *Biodiversity: Conservations and Management.* Mangaldeep Publ, Jaipur, p. 586.

Stewart WDP (1969). Biological and ecological aspects of nitrogen fixation by free living microorganisms. *Proc R Soc London,* B112: 376–388.

Venkataraman LV (1980). Cyanobacteria as biofertilizers: In: *Handbook of Micrologal Mass Culture.* CRC Press, p. 455–471.

Soil Microflora, 2009 *Pages* **260–271**
Editor: **Rajan Kumar Gupta, Mukesh Kumar & Deepak Vyas**
Published by: **DAYA PUBLISHING HOUSE, NEW DELHI**

Chapter 22

Ecological Diversities in Soil Microorganisms

P.B. Tiwary

Department of Botany, S.M. P.G. College, Chandausi, Moradabad, Uttar Pradesh

ABSTRACT

Ecological biodiversity in soil is a way to study the presence and contributions of microorganisms in soil through their activities, to places where they occur. Knowledge on microbial occurrence and its contribution to soil and inter associations with plants and environment create an interesting theme of ecosystem. Microbial functions in a physical location that can be described as its microenvironment. The resources available in a microenvironment and their time of use by a microbe, describe the niche. One microorganism may grow on another microbe as an ectosymbiont or endosymbiont. Positive or negative interactions between microorganisms involve a competition for space or nutrient. Different ecological factors can change the frequency or quality of such interactions.

In usual cases of soil, organic matter accumulation occurs through the direct activities of primary producers or by the import of preformed organic materials. Soil can be formed in regions such as the antarctic area where there are no vascular plants. Many soil microorganisms play important role in the dynamics of greenhouse gases such as Carbon dioxide, nitrous oxide, nitric oxide and methane etc. These microorganisms can contribute to production and consumption of these gases.

Introduction

It is apparent that a great variety of plants cover the earth's surface. At first glance, one is impressed with the diversity of plants, which populate the earth. A problem of interest of layman and scientists alike, involves the grouping of the different types of microorganisms, covering the earth's surface, in such a manner as to make their study more convenient. Along with the progress of civilization, based on agricultural and pathological technologies, scientists recognized a related series from the simplest microorganisms to more complex form of life and that so many forms are distinguished from closer one by slight differences (Fish, S., 1970; McCarthy, 1978; Pritts, 2006).

Different microscopic organisms occur in/on soil are termed as soil microorganisms. Pathologists have accepted that a broader outlook is required to search the entire dynamics of microbial ecology. (Prescott *et al.*, 2003). Simply a study of life cycles is only a prerequisite to the proper understanding of soil related microorganisms in relation to its surrounding. An environment of a microorganism in the soil is biologically complex and the interactions between organisms themselves produce balance.

Soil as an Environment

In soil, so many substances useful as food for microorganisms, tend to be absorbed upon surfaces.

Actually a comparable relationship occurs in soil. For example, in top soil, hundred to billions of microbes have reportedly lived, multiplied, struggled for space, food and their survival. This figure of microorganisms includes variety of viruses, bacteria, fungi, algae, protozoans, nematodes etc. A range in variations of number is summarized in Table 22.1.

Table 22.1: Microbial Frequency in per Gram of Soil

Microorganism	*Lower Limit*	*Usual Range*	*Higher Limit*
True Bacteria	1,000–10,000	1,000,000–10,000,000	1,000,000,000–10,000,000,000
Actinomycetes	100–1,000	1,00,000–10,00,000	5,000,000–10,000,000
Molds	01–100	1,000–1,00,000	2,00,000–5,00,000
Algae	none–100	1,000–1,00,000	2,00,000–5,00,000
Protozea	none–100	10,000–1,00,000	5,00,000–10,00,000

Source: From Sarles *et al.*, 1985. *Microbiology: General and Applied,* 2nd edn. Harper and Row.

1. From disintegrating rocks, soil minerals contain different inorganic particles ranging in size and shape. Due to their irregularities in shape, interestices contain variable amount of water, air, carbon dioxide, hydrogen sulfide, ammonia and some other gases in trace form. Their ratio depends upon process of pedogenesis, rainfall, atmospheric pressure, flow of wind, temperature variations, relative humidity, microbial interactions etc.
2. In average agricultural soil, water may be considered as a dilute nutrient. Both contain different ions like CO_3^{2-}, PO_4^{3-}, Na^+, Ca^{++}, NO_3^-, SO_4^{2-}, Mg^{++}, Fe^{+++}. (Blakeslee and Broad, 1996; Jeanthon, 2000). Other available data shows variations in soil pH [6.0 to 8.0- best for most of the soil microorganisms]. Soil temp. (15 to 45°C), free Oxygen, availability of Hydrogen donor and acceptor compounds, physico-chemical properties. All these make water (in an average fertile soil) as an excellent culture medium for most of microorganisms. (Ford and Monroe, 1971)
3. Thus soil makes a variable and specific environment which may favour variety of microbial flora in universe. Different physico-chemical, biochemical and redox processes regulate the nature, frequencies, properties and density of microorganisms. (Maramorosch, 1973; Vern Grubinger *et al.*, 2007). Termendous growth of microbes occur after plowing under manures and green crops temporarily depletes the soil water of nutrient compounds. These are

removed from solution by combination as new microbial substance (Demain, 1976; Sermonti, 1979; Pierre and Brubaker, 2008).

Nutritional Relation in Soil

During development and establishment of ecological diversities, different microorganisms provide nourishment for each other. It helps in the development of communities of simpler form to complex form of microbial flora. Certain species of Pseudomonadaceae, Cytophagaceae, Actinomycetales, Eumycetes can hydrolyse cellulose (a complicated polymer of glucose) in to Cellobiose (a simpler polymer of glucose) [Tong Zhonghua *et al.*, 2007]. Again Cellobiose, Maltose, Dextrins like di or oligosaccharides are decomposed by other species of microbes to glucose. Glucose is utilized by its aerobic or anaerobic oxidations in different groups of microorganisms. Proteins are also hydrolyzable into peptones, peptoses and amino acid molecules. (Ruby, 1999; Barkay, 2000).

The least digestible part of plant tissues like lignin, resins and animal carcasses like waxes, hair, horn, bone etc. show slow decomposition. These decaying remains make up some residual complex that is humus. Humus serves as a reserve store of slowly released food for microorganisms and crop plants.

1. A region of increased microbial presence, its growth and activities in the soil around the root of plants termed as rhizosphere. Certain diseases are caused by soil borne microbes which enter the host plant through roots or other underground parts. These include some important and widespread disease like wilt of cotton, Rhizoctonia rot of cotton, food rot of wheat, wilt of pigeon pea, root rot of jute, flax wilt etc. (Mehrotra, 1999). The intensity of the disease may be increased, in some cases by collecting soil from other diseased fields and scattering it over the test plot or by inoculating the soil with cultures of the causal organism grown in sand oatmeal medium or other nutrient media. The same soil areas is then used again in succeeding years as a sick plot. Glasshouse tests may be made to determine resistance of growing varieties in containers filled with infected soil. The infected soil is obtained from diseased fields or by mixing cultures of the causal organism with sterilized soil. Glasshouse tests often differentiate varieties better than field tests, because temperature and moisture favourable for the growth and development of the microorganisms can be maintained (Fletcher, 1991; Sarbu *et al.*, 1996; Uttech Sara, 2008).
2. The rhizosphere may be mentioned as one of the most active factory for the transformation of essential elements into living constituents as well as for returning the essential elements to the soil upon the death and decomposition of different microorganisms (Baker and Snyder, 1970; Paul, Elder, 2006).

Action-Interactions in Microroganisms

This is an important part of microbial diversities where one organism may grow on the surface of another one as an ectosymbiont or inside other microbe as an endosymbiont. Sometimes microbes may have other organisms on their surface and inside them at the same time, such combination is known as ecto/endosymbiosis as in Thiothrix (a sulfer using bacterium). It is attached to the surface of a mayfly larva and which itself contains a parasitic bacterium.

There is some type of physical contact provides no information on the types of interactions that might be occuring. Such interactions can be positive (Mutualism, Protocooperation, Commensalism) or negative (Parasitism, Predation, Amensalism, Competition).

Positive Interaction

Mutualism

It is mutually beneficial and obligatory relationship between protozoan- termites, Phycobiont + Mycobiont= Lichens, Methane fixing microbes.

Protocooperation

Another mutually beneficial relationship but it is not obligatory as in between Desulfovibrio and Chromatium, Azotobacter and a Cellulose degrader-Cellulomonas.

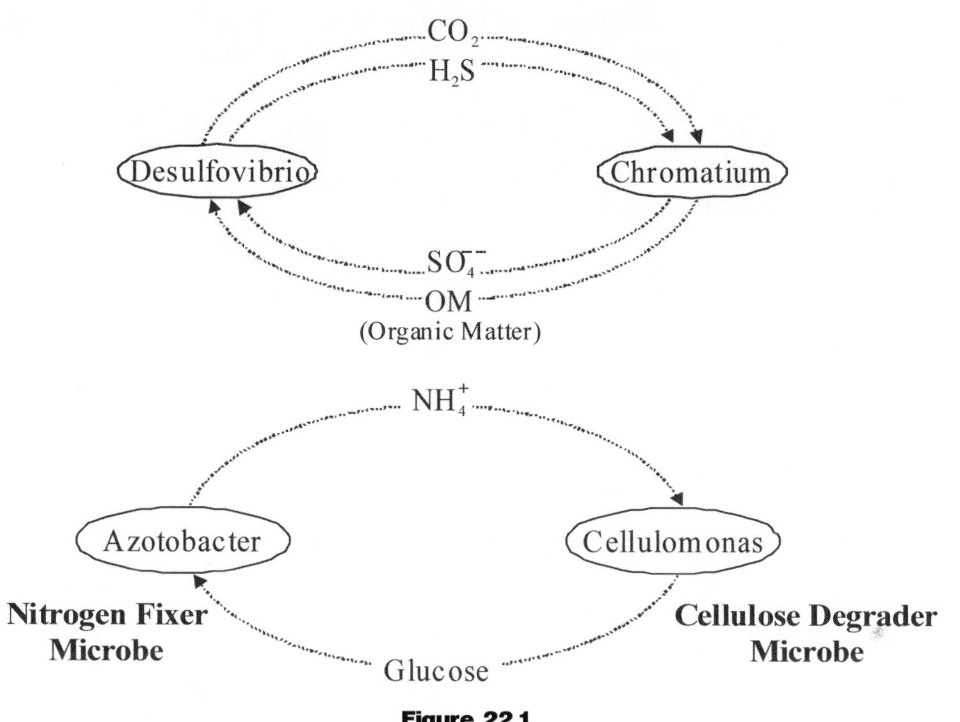

Figure 22.1

Commensalism

In such relationship, product of one organism can be used beneficially by another microorganism. It is unidirectional process. Bacteria like Nitrobacter benefits from its association with Nitrosomonas because it uses nitrite to obtain energy for growth.

Negative Interaction

Parasitism

This complex microbial relationship shows a longer term internal maintenance of another organism or a cellular infectious agents. Parasitic fungi include Rhizophydium sphaerocarpum with Spirogyra. Rhizoctonia solani is a parasite of Mucor and Pythium. Some bacteriophages can establish a lysogenic relationship with their hosts. Many members of protozoans, fungi, Bacteria, viruses, mycoplasma etc. show pathogenic nature under parasitism.

Predation

In this relationship, predator engulfs or attacks a larger or smaller prey. Bdellovibrio, Vampirococcus, Daptobacter are one of the best predatory bacteria which show specific mode of attack against a susceptible bacterium. Some excellent predators like Ciliates ingest Legionella to protect this microbe from chlorine. Actually Ciliates serve as a reservoir host.

Some fungal members like Arthrobotrys trap nematodes by using constricting rings.

Parasitism and predation are closely related. Predation has many positive effects on populations of predators and prey. These include the microbial loop (returning minerals immobilized in organic matter to mineral forms for reuse by Chemotrophic and photosynthetic primary producers), protection of prey from heat and damaging chemicals and possibly aiding pathogenicity, as with Legionella.

Amensalism

It is a unidirectional move based on the release of a specific compound by one organism which has a negative effect on another organism. The best example of amensalism is production of antibiotics that can inhibit or kill susceptible microbes.

Competition

Such interaction involves organisms competing for space or a limiting nutrient. In 1934, EF Gause had described the competitive exclusion principle and found that if the two competing ciliates overlapped so much in terms of their resource use, one of the protozoan populations was excluded.

The quality of all these actions and interactions can change, multiply or minimize depending on the environment and the characteristics of the particular organisms.

Cycling of Different Elements

During active actions and interactions with plants, animals and the environment, microorganisms act important roles in nutrient cycling. Assimilatory processes show incorporation of different elements

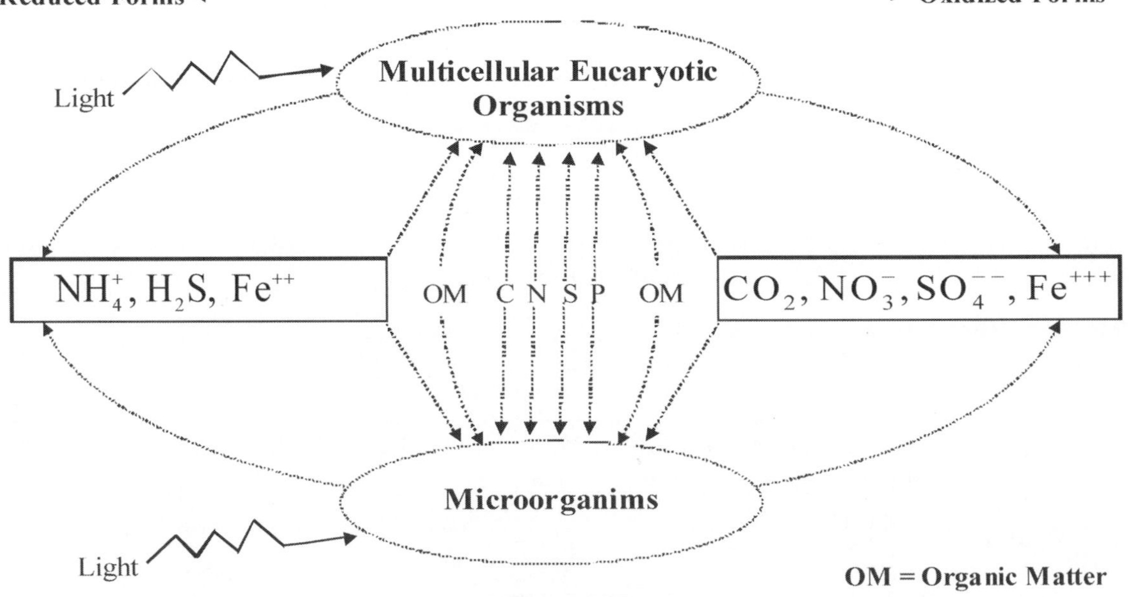

Figure 22.2

into organism's biomass during anabolism. On the other hand, dissimilatory processes regulate the release of different elements to the environment after catabolic steps.

Some elements are essential components of protoplasm, undergo cyclic alternations. Such repeated transformation of elements from living protoplasm to the free state in nature constituted the cycle of elements or nutrients. Some essential elements undergoing biological transformation are Carbon, Oxygen, Nitrogen, Sulfer, Phosphorus, Iron etc.

Carbon Cycle

Transformation of Carbon occurs constantly from its most oxidized state, CO_2 and is reduced primarily by photosynthesis.

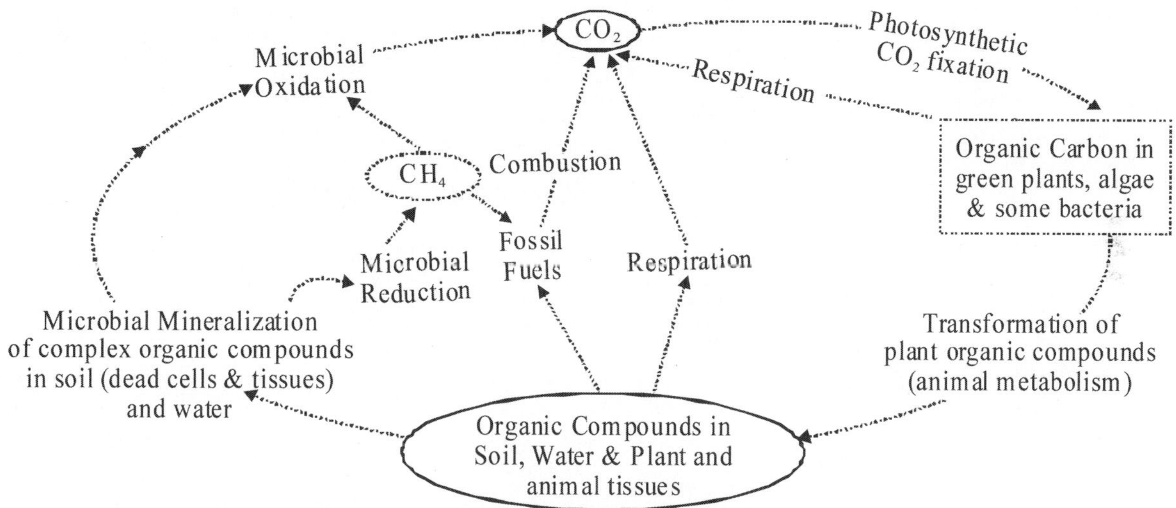

Figure 22.3: Carbon Cycle

The Carbon in carbonates may be returned to the atmosphere by acids (HNO_3) produced by microorganisms. Anaerobic decomposition of organic materials may yield and products such as CH_4, H_2 and CO_2 in addition to various organic acids and alcohols. Some of the organisms capable of producing CH_4 from oxidation of hydrogen and reduction of CO_2 are species of Methanobacterium, Methanococcus, Methanosarcina and some species of Clostridium. (Larsen, E.I. 1999; Yrukov, V.V. 1998; Madigan, M.T., 2003; Tomasz Alexander, 2006). Even Carbon monoxide is relished as a source of energy and carbon by Carboxydomonas oligocarbophilia.

Oxygen Cycle (Figure 22.4)

All aerobic organisms require Oxygen for their entire redox processes of metabolism. However anaerobes lack such requirements. Oxygen is necessary in the combustion of fossil fuels as well as alcohols, wood etc. and such oxygen is usually combined with carbon to form CO_2.

Nitrogen Cycle (Figure 22.5)

Nitrogen is the most important component for plant growth. It requires for the synthesis of amino acids, proteins, enzymes, chlorophylls, nucleic acids etc. Green plants obtain nitrogen from the soil solution in the form of ammonium, nitrate and nitrite ions.

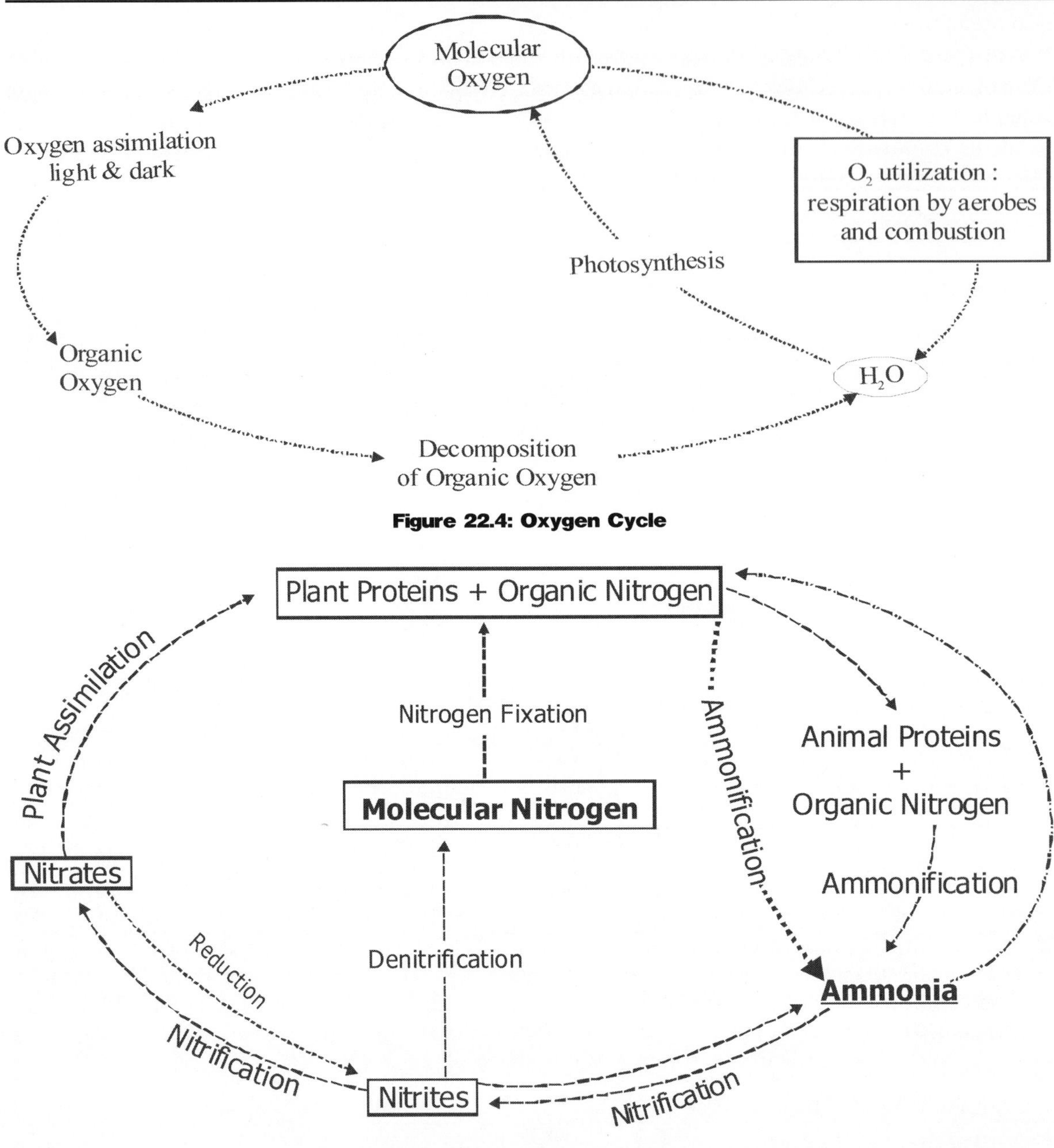

Figure 22.4: Oxygen Cycle

Figure 22.5: Nitrogen Cycle

Atmospheric Nitrogen is not directly available to the organisms with the exception of some prokaryotes like blue green algae and nitrogen fixing bacteria. Nitrogen cycle is briefly illustrated under following heads:

Nitrogen Fixation

Physical or biological conversion of free nitrogen of atmosphere into the biologically acceptable form of nitrogenous compounds is known as nitrogen fixation. Some Nitrogen fixing organisms are listed in Table 22.2.

Table 22.2

Symbiotic Organisms	*Free Living*		
	Bacteria	*Yeast*	*Blue Green Algae*
Rhizobium species with legume plants	Nonphotosynthetic sp. of	Species of *Pullularia*	Species of *Anabaena*
Klebsiella species	*Azotobacter*	*Rhodotorula*	*Calothrix*
with *Psychotria*	*Azotomonas*		*Chlorogloea*
Actinomycetes	*Bacillus*		*Cylindrospermum*
and/or fungi with	*Beijerinckia*		*Fischerella*
sp. of	*Chromobacterium*		*Mastigocladus*
Alnus	*Clostridium*		*Nostoc*
Casuarina	*Derzia*		*Scytonema*
Ceratozamia	*Desulfovibrio*		*Stigonema*
Ceanothus	*Enterobacter*		*Tolypothrix*
Cercocarpus	*Nocardia*		
Comptonia	*Pseudomonas*		
Coriaria	*Spirillum*		
Cycas			
Discaria	Photosythetic sp. of		
Dryas	*Chlorobium*		
Elaeagnus	*Chromatium*		
Encephalartos	*Methanobacterium*		
Hippophae	*Rhodomicrobium*		
Macrozamia	*Rhodopseudomonas*		
Myrica	*Rhodospirillum*		
Podocarpus			
Purschia			
Stangeria			
Shepherdia			

Source: *Microbiology*, 3rd edn. W.B. Saunders Co., 1972.

Nitrogen

Assimilation Inorganic nitrogen in the form of nitrates, nitrites and ammonia is absorbed by the green plants and converted into nitrogenous organic compounds. Animals derive their nitrogen requirement from the plant proteins.

Ammonification

Different microorganisms like *Bacillus ramosus, B. vulgaris, B. mesenterilus* and *Actinomycetes* utilize organic compounds in their metabolism and release ammonia.

Nitrification

Microorganisms like Nitrosomonas, Nitrococcus, Nitrosogloea, Nitrospira convert ammonia into nitrites.

$$2NH_4^+ + 2O_2 \rightarrow NO_2^- + 2H_2O + \text{energy}$$

Nitrites are converted into nitrates by Nitrobacter, Nitrocystis, Penicillium sp. etc.

$$2NO_2^- + O_2 \rightarrow 2NO_3^- + \text{energy}$$

Denitrification

It is a conversion of Ammonia and nitrates into free nitrogen by Thiobacillus denitrificans, Micrococcus denitrificans, Pseudomonas aeruginosa etc.

$$2NO_3^- + 2NO_2^- \rightarrow 2NO \rightarrow N_2O \rightarrow N_2$$

Table 22.3 illustrates genera of chemosynthetic microorganisms containing species reported to reduce nitrate dissimilatively and to denitrify.

Sulphur Cycle

Photosynthetic microorganisms transform sulfer by using sulfide as an electron source, allowing Thiobacillus and similar chemolitho autotrophic genera to act. Desulfovibrio use sulfate as an oxidant. Desulfuromonas, Thermophilic archaea, Cyanobacteria in hypersaline sediments have been found to carry out dissimilatory elemental sulfur reduction. Alteromonas, Clostridium, Desulfovibrio, Desulfotomaculum etc. can reduce sulfite to sulfide. (Lovely *et al.*, 1995; Lens and Pol, 2000; Coleman David *et al.*, 2006).

Phosphorus Cycle

Phosphorus is an important component of Nucleic Acids, ADP, ATP, NADP, Phospholipids etc. It occurs in the soil in five forms that is P_1 (stable organic), P_2 (labile organic), P_3 (labile inorganic), P_4 (soluble) and P_5 (mineral form).

The dissolved phosphorus is absorbed by plants and converted to organic form. When the plants and animals die, decomposer microorganisms attack them and liberate phosphorus to the environment. This process proceeds in cyclic way.

Iron Cycle

In nature, Iron exists in ferrous (Fe^{++}) and Ferric (Fe^{+++}) state. These two forms are readily convertible under the influence of pH and redox potential of the environment. Some organisms like Ferro bacillus ferroxidans and Thiobacillus ferroxidans can oxidize ferrous form to ferric hydroxide.

Most Iron reduction is carried out by specialized iron respiring microorganisms such as Geobacter metallireducens, G. sulfur reducens, Ferribacterium limneticum, Shewanella putrefaciens etc. (Ehrenreich and Widdel, 1994; Straub *et al.*, 1996; Cummings *et al.*, 1999; Sabev *et al.*, 2006; Ajcann, 2008)

Table 22.3

Nitrate Respiring Bacteria		Denitrifying Bacteria
NO_3^- →	NO_2^-	$NO_3^- \rightarrow NO_2^- \rightarrow NO \rightarrow N_2O \rightarrow N_2$
Achromobacter	*Halobacterium*	*Achromobacter*
Actinobacillus	*Leptothrix*	*Alcaligenesi*
Aeromonas	*Micrococcus*	*Bacillus*
Agarbacterium	*Micromonospora*	*Chromobacterium*
Agrobacterium	*Mycobacterium*	*Corynebacterium*
Alginomonas	*Nocardia*	*Halobacterium*
Arizona	*Pasteurella*	*Hyphomicrobium*
Arthrobacter	*Propionibacterium*	*Micrococcus*
Bacillus	*Proteus*	*Moraxella*
Beneckea	*Providencia*	*Nitrosomonas*
Brevibacterium	*Pseudomonas*	*Propionibacterium*
Cellulomonas	*Rettgerella*	*Pseudomanas*
Chromobacterium	*Rhizobium*	*Spirillum*
Citrobacter	*Salmonella*	*Thiobacillus*
Corynebacterium	*Sarcina*	*Xanthomonas*
Cytophage	*Selenomonas*	
Enterobacter	*Serratia*	
Erwinia	*Shigella*	
Escherichia	*Spirillum*	
Eubacterium	*Staphyloccus*	
Flavobacterium	*Streptomyces*	
Haemophilus	*Vibrio*	
	Xanthomonas	

Source: *Bact. Rev.*, 37:409, 1973.

Conclusion

Ecological diversities influence the seasonal development and geographical distribution of different microorganisms in/on soil. All the external conditions affecting actions, interactions, life and development of microorganisms, which also include temperature, light and moisture with some living factors too. Decreased species diversity usually occurs in extreme environments and many microorganisms that can function in such habitats, called extremophiles. These have specialized growth requirements. Major organic compounds used by microorganims differ in structure, linkage, elemental composition and susceptibility to degradation under aerobic and anaerobic conditions. Layers of microbes or biofilms are widespread and are formed on a wide variety of living and nonliving surfaces. These are important in disease occurrence and the survival of pathogens. Biofilms can develop to form complex layered ecosystem.

References

Ajcann (2008). Multiple hosts and Lyme disease. *Proc Biol Sci*, 275: 227–235.

Baker KF and Snyder WC (Eds.) (1970). *Ecology of Soil-borne Plant-Pathogens: Prelude to Biological Control.* University of California Press, Berkeley.

Barkay T (2000). *Encyclopedia of Microbiology*, 2nd edn, Vol 3, (Ed-in-Chief) J Lederberg. Academic Press, San Diego, p. 171–181.

Blackeslee S and Broad WJ (1996). Earth's dominant life form is also its smallest: The microbe. *The New York Times*, October 15, Science Times, Section p. B5.

Coleman David C *et al.* (2006). *Fundamental of Soil Ecology*, 2nd edn. Academic Press, USA.

Crosson PR and Brubaker Sterling (2008). *Resource and Environmental Effects of US Agriculture: Soil Conservation Programs*. Questia Media America, p.160–178.

Cummings DE, Caccavo F Jr, Spring S and Rosenzweig RF (1999). Ferribacterium limneticum, gen. nov. Sp nov., an Fe (III) reducing microorganism isolated from miningimpacted fresh water lake sediments. *Arch Microbiol,* 171: 183–188.

Demain AL (1976). Industrial aspects of maintaining germ plasm and genetic diversity. In: *The Role of Culture Collections in the Era of Molecular Biology*, (Ed) R R Colwell (Editor). *Am Soc Microbiol*, Washington DC.

Ehrenreich A and Widdel F (1994). Anaerobic oxidation of ferrous iron of purple bacteria, a new type of phototrophic metabolism. *Appl Environ Microbiol*, 60: 4517–4526.

Fletcher M (1991). The physiological activity of bacteria attached to solid surface. *Adv Microb Physiol,* 32: 53–85.

Ford JN and Monroe JE (1971). *Living Systems: Principles and Relationships*. Canfield Press, San Francisco.

Grubinger V, Magdoff F and Harold V (2007). *Soil Microbiology: A Primer.* Sustainable Agri Publications, Hills Building, Burlington.

Jeanthon C (2000). Molecular ecology of hydrothermal vent microbial communities. *Antonie van Leeuwenhoek*, 77: 117–133.

Larsen EI, Shy LI and McEwan AG (1999). Adsorption and oxidation by whole cells and a membrane fraction of Pedomicrobium sp. ACM 3067. *Arch Microbiol,* 171: 257–264.

Lens P and Pol LH (2000). Sulfer Cycle. In: *Encyclopedia of Microbiology*, 2nd edn, Vol 4, (Ed-in-Chief) J Lederberg. Academic Press, San Diego, p. 495–505.

Lovely DR, Phillips EJP, Lonergan DJ and Widman PK (1995). Fe (III) and S reduction by Pelobacter Carbinolicus. *Appl Environ Microbiol*, 61: 2132–2138.

Madigan MT, Martinko JM and Parker J (2003). *Brock's Biology of Microorganisms*, 10th edn. Prentice Hall.

Maramorosch K. (Ed) (1973). Mycoplasma and mycoplasma like agents of human, animal and plant diseases. *Ann NY Acad Sci*, p. 225.

Mehrotra RS (1999). *Plant Pathology*. Tata McGraw-Hill Publishing Company Limited, New Delhi.

Paul Eldor A (2006). *Soil Microbiology: Ecology and Biochemistry*, 3rd edn. Academic Press, USA.

Prescott, LM, Harley JP and Klein DA (2003). *Microbiology*. McGraw Hill Higher Education Publication.

Pritts MP (2006). Cover crop rotations alter soil microbiology and reduce plant disorders in strawberry. *Horti Science*, 41(5): 1303–1308.

Ruby EG (1999). Ecology of a benign "Infection": Colonization of the squid luminous organ by vibrio fischeri. In: *Microbial Ecology and Infectious Disease*, (Ed) E Rosenberg. Washington DC, p. 217–231.

Sabev HA, Handley PS and Robson GD (2006). Fungal colonization of soil buried plasticized polyvinyl chloride and the impact of incorporated biocides. *Microbiology*, 152: 1731–1739.

Sarbu SM, Kane TC and Kinkle BK (1996). A chemoautotrophically based cave ecosystem. *Science*, 272: 1953–1955.

Sermonti G (1979). Mutation and microbial breeding. In: *Genetics of Industrial Microorganisms*, (Eds) OK Sebek and AI Laskin. Am Soc Microbiol, Washington DC.

Straub KL, Benz M, Schink B and Widdle F (1996). Anaerobic, nitrate-dependent microbial oxidation of ferrous ion. *Appl Environ Microbiol*, 62: 1458–1460.

Tomasz Alexander (2006). Microbilogy: Drug resistance and soil microbes. *Science*, 311: 342.

Tong Zhonghua *et al.* (2007). Impact of fullerene (C_{60}) on a soil microbial community. *Environ Sci Technol*, 41(8): 2985–2991.

Uttech Sara (2008). *The Sweet World of Soil Microbiology*. Soil Sci Society of America, Washington DC.

Yrukov VV and Beatly JT (1998). Aerobic anoxygenic phototrophic bacteria. *Microbiol Mol Biol Rev*, 62: 695–724.

Soil Microflora, 2009 *Pages* **272–278**
Editor: **Rajan Kumar Gupta, Mukesh Kumar & Deepak Vyas**
Published by: **DAYA PUBLISHING HOUSE, NEW DELHI**

Chapter 23

Biocontrol of Leaf Rot of Pan Caused by *Phytophthora parasitica* var. *Piperina* by Native Fungal Species

Deepak Vyas[1], Rajesh Yadav[1] and Rajan Kumar Gupta[2]
[1]Lab of Microbial Technology and Plant Pathology, Department of Botany, Dr. H.S. Gour University, Sagar – 470 003, M.P.
[2]Department of Botany, Govt. P.G. College, Rishikesh – 249 201, Uttarakhand

ABSTRACT

Three native fungi *viz. Aspergillus, penicillium* and *Trichoderma* sp. were isolated from native field and are tested against *Phytophthora parasitica var. piperina* causal organism of root rot of pan. The results suggest that all test bioagents show potential to control the growth of pathogen. Among the three fungal species, *Trichoderma* sp. produces the best results and able to protect the test organism with the reduction of 66.7 per cent in the growth of pathogen under *in vitro* condition, where as *Penicillium* reduces 46.7 per cent and *Aspergillus* reduces 33.4 per cent. It was also observed that *Trichoderma* not only completely check the sporulation of pathogen but also reduces about 70 per cent of leaf rot under *in vivo* condition. In comparison to *Trichoderma, penicillium* and *Aspergillus* are less effective against the test pathogen.

Introduction

Pan is one of the cash crop of our country. Pan is vernacular hindi name used for betlevine plants. It is a perennial, evergreen creeper and climber belonging to the family piperaceae; probably a native of Malaya and Indonesia but originally comes from Java now cultivated in India. Betelvine is cultivated in warm and humid regions of country *viz.*, Assam, West Bengal, Kerela, Madhya Pradesh, Uttar Pradesh, Bihar Andhra Pradesh, Gujarat, Karnataka, Orissa.

Pan is chewed almost in every part of the country, beside chewing, it has many uses. In our country in every religious ceremony pan is essential thing, no rituals is completed without pan. Above all, it has got good medical properties and leaves are useful for carminative, antiseptic astringent mild stimulant as well as tonic for brain, heart and lever.

The crop faces various climatic changes which renders the crop to a number of diseases. Out of many diseases root and leaf rot of pan caused by *Phytophthora parasitica*

Modern agricultural plant diseases are controlled by the chemical substances. It is well known fact that these chemical substances have detrimental effect on targeted as well as nontargeted organism living in agroecosystem. The danger inherent in chemical substances has brought forth an awareness to find out other alternative like bio-control agents. The microorganisms used in biological control of plant diseases are termed as 'Antagonists'. An 'antagonist' is a microorganism that adversely affects another microorganism growing in association with it. Antagonism is the balance wheel of nature. It operates through competition, parasitism and antibiosis. Biological control is a natural phenomenon, nature's own way of keeping diseases from getting catastrophic

In order to minimize or check the chemical application biological management of plant pathogen is advocated now and then. The idea came through the first report by Millard and Taylor (1927). Weidling published the first of brilliant series of papers on the parasitism by *Trichoderma viridae* on soil (Weidling, 1932). Wright (1956) showed that it produced significant amount of antibiotic gliotoxin. Nearly 32 crop diseases have been controlled by different species of *Trichoderma* from 1990-2003 Gangawane (2008). Considering all such facts present piece of work was undertaken.

Materials and Methods

Effect of Biocontrol Agents on the Mycelial Growth and Sporulation of *P. parasitica var. piperina*

For the present study three biocontrol agents *viz. Trichoderma* sp., *Penicillium* sp. and *Aspergillus* sp. were selected to test against the growth and sporulation of *P. parasitica var. piperina.*

The antagonistic potential of biocontrol agents were tested under *in vitro* condition using oat meal agar medium, by dual culture technique. Circular mycelial discs of test pathogen was inoculated in the centre of each petriplate. Then mycelial discs of each biocontrol agents were inoculated near the periphery of petriplate. These discs were inoculated after 2-days inoculation of test pathogen, then incubated at 27±1°C. Without having any biocontrol agent served as control. Each experiment was designed in triplicate. Data were recorded after every 24 hours up to a period of 144 hours. On the basis of this mycelial growth, percent inhibition and sporulation was recorded.

The Effect of Biocontrol Agents on the Development of Leaf Rot

In order to study the effect of bio-control agents on the development of leaf rot, healthy leaves of pan were collected from natural habitat (bareja) and washed with running tap water and followed by sterilized distilled water. Then leaves were dipped in each solution of biocontrol agents for an hour, these were inoculated with leaves which were injured with sterilized needle at centre of leaves. Each experiment was performed in triplicates with suitable control. Plates were incubated at 27±1°C. Leaf rot development was recorded after every 24 hours upto a period of 144 hours. On the basis of this, percent reduction of disease was recorded.

Results

Data regarding the effect of biocontrol agents on the mycelial growth of test pathogen are presented in Tables 23.1–23.4, Figure 23.1 and Figure 23.2. As evident from these results, *Aspergillus* sp. and *Penicillium* sp. showed 33.4 per cent and 46.7 per cent reduction in mycelial growth respectively, Table 23.4, whereas *Trichoderma* causes 66.7 per cent reduction in growth of test pathogen. The sporulation of pathogen was also affected adversely in the presence of all the three biocontrol agents. *Aspergillus* sp. restricted sporulation of *Phytophthora parasitica* moderately while *Penicillium* sp. caused poor sporulation. However, *Trichoderma* sp. caused complete suppression of sporulation of test pathogen. Under *in vivo* condition, *Aspergillus* sp., *Penicillium* sp. and *Trichoderma* sp. showed 34.6 per cent, 46.1 per cent and 69.2 per cent reduction of leaf rot respectively (Table 23.5 and Figure 23.2). Effectiveness of biocontrol agents against test pathogen presented in Figure 23.2.

Table 23.1: Effect of *Trichoderma* on the Growth of *P. parasitica var. piperina*

Incubation Period (Hours)	*In Vitro*		*In Vivo*	
	Growth (mm)		*Leaf Rotten (mm)*	
	Control	*Trichoderma*	*Control*	*Trichoderma*
24	08	00	06	00
48	20	08	12	06
72	38	10	22	08
96	54	14	34	10
120	72	22	42	12
144	90	30	52	16

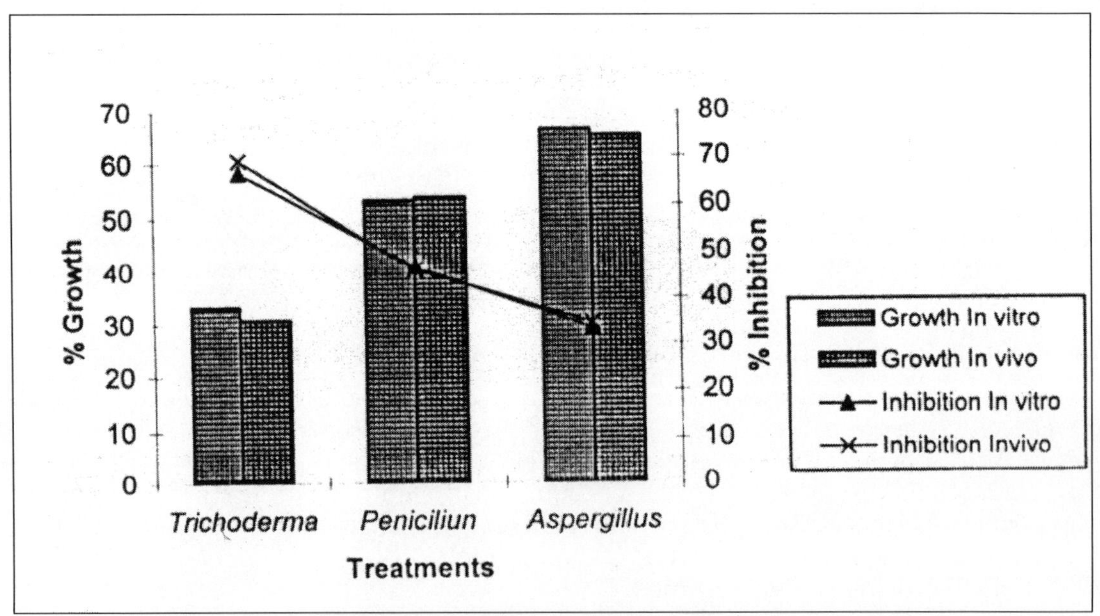

Figure 23.1: Effect of Biocontrol Agents on the Per cent Growth and Per cent Inhibition of *P. parasitica var. piperina*

Table 23.2: Effect of *Penicillium* on the Growth of *P. parasitica var. piperina*

Incubation Period (Hours)	*In Vitro*		*In Vivo*	
	Growth (mm)		*Leaf Rotten (mm)*	
	Control	*Trichoderma*	*Control*	*Trichoderma*
24	08	08	06	06
48	20	15	12	08
72	38	24	22	10
96	54	32	34	14
120	72	40	42	20
144	90	48	52	28

Table 23.3: Effect of *Aspergillus* on the Growth of *P. parasitica var. piperina*

Incubation Period (Hours)	*In Vitro*		*In Vivo*	
	Growth (mm)		*Leaf Rotten (mm)*	
	Control	*Trichoderma*	*Control*	*Trichoderma*
24	08	08	06	06
48	20	20	12	10
72	38	32	22	15
96	54	40	34	22
120	72	50	42	28
144	90	60	52	34

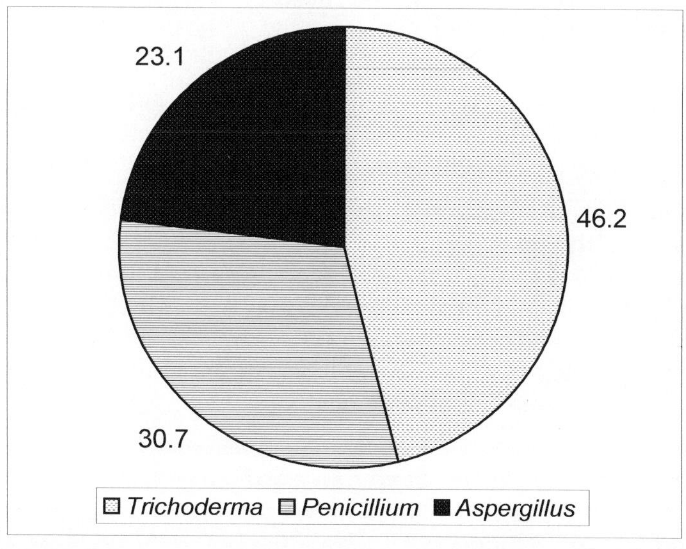

Figure 23.2: Effectiveness of Different Biocontrol Agents under *In vivo* Condition

Table 23.4: Effect of Biological Agents on the Mycelial Growth, Per cent Inhibition and Sporulation of *P. parasitica var. piperina*

Biocontrol Agents	Mycelial Growth (mm)	Inhibition (Per cent)	Sporulation
Control	90	00	+++
Trichoderma	30	66.7	-
Penicillium	48	46.7	+
Aspergillus	60	33.4	++
CD (P = 0.05)	2.77	2.83	ND

ND: Not detected; –: No sporulation; +: Poor sporulation; ++: Moderate sporulation; +++: Good sporulation.

Table 23.5: Effect of Biocontrol Agents on the Development and Per cent Reduction of Leaf Rot of Pan

Biocontrol Agents	Leaf Rooten (mm)	Reduction (per cent)
Control	52	00
Trichoderma	16	69.2
Penicillium	28	46.1
Aspergillus	34	34.6
CD (P = 0.05)	2.64	1.75

Discussion

The reduction of inoculum density or disease producing activities of a pathogen in its active or dormant state by one or more organisms was defined as Biological control by Baker and Cook (1974). This can be accomplished naturally by manipulating the environment, host or antagonist, or by mass introduction of one or more antagonists. Use of biological control rarely eliminates a pathogen from the site, but may reduce propagules or ability to produce disease instead. In these situations control may be achieved with little or no reduction in population of the pathogen.

The antagonism of various fungi to *Phytophthora* species has been reported by Utkhede (1992). A limited numbers of observations of antibiotic production by soil fungi such as *Penicillium* and *Trichodermal* species have been observed. *Trichoderma* species however has the ability to induce development of sex organs in some normally sterile isolates of *Phytophthora* species (Brasier, 1975).

As evident from the present results that when both *Phytophthora* and *Trichoderma* species were grown on oatmeal medium the *Trichoderma* over grew the *Phytophthora* culture and parasitized the hyphae. Earlier, Dennis and Webster (1971a, b) have made similar observation. According to Malajczuk (1983) hyphal lysis of *Phytophthora* trigered by *Trichoderma* species involved in the contact and coiling of the parasite around the hyphae before penetration. However, attraction of antagonistic fungi to *Phytophthora* hyphae is mediated by exudation of hyphae. The study also suggests that *Trichoderma* species is more effective then *Penicillium* species.

Inhibitory substances produced by *Aspergillus* spp. are known to play a role in soil fungistasis (Johri and Singh, 1975; Lee and Wu, 1979). Such substances have been reported from culture filtrates of *Aspergillus niger, A. flavus, A. candidus* (Shukla and Dwivedi, 1979). *A niger* exhibited greatest antagonism both in culture and on host during investigation on damping-off disease of egg plant caused by *F. solani, A. alternata* and *R. solani* (Bora, 1977). *A. terreus* was reported to block the capacity of production of toxin by *F. oxysporum*. The decline in wilt was attributed to antibiotic effect upon pathogen (Chadova *et al.*, 1980). *Aspergillus* spp. are well known for producing various kinds of active compounds including antifungal and antibacterial agents (Buch *et al.*, 1983; Fujimoto *et al.*, 1993).

Suppression of foot rot and root rot of *Piper betle* caused by *P. parasitica var. piperina*, by *T. viride* was always found particularly in the soil after fumigation (Tiwari and Mehrotra, 1973), *T. viride* is also known to control the root knot nematode in betelvine (Bhatt *et al.*, 2002), also foot rot disease of black pepper with many species of *Trichoderma* (Rajathilagam and Kannabiran, 2001).

Penicillium sp. the most frequent isolate from the soils has been found to be an active antagonist against many species of *Fusarium* (Kamyshiko *et al.*, 1976; Bora *et al.*, 1982). Culture filtrates of *P. citrinum* was found most inhibitory to the linear growth of *F. oxysporum* (Ashour *et al.*, 1980). However, mechanism by which *Penicillium* spp. serve as biocontrol agent is not known with certainity, although mycoparasitism has been reported against *Rhizoctonia solani* (Boosalis, 1956). It is because they are efficient colonisers. Substrate occupation is most likely to be a key mode of biocontrol (Naik and Sen, 1992). The efficacy of *P. parasitica* can be attributed to either its slower growth on oat meal agar medium or metabolites secreted by *Aspergillus* sp. were less toxic in nature.

Considering the above facts it may be concluded here that *Trichoderma* species was found to be most effective against leaf rot of Pan and followed by *Penicillium* and species of *Aspergillus*.

Acknowledgement

Authors are thankful to Head, Department of Botany, Dr. H.S. Gaur University, Sagar, M.P., R.K.G. thankfullly acknowledge, U.G.C. for financial support.

References

Ashour WA, Aly AA and Elewa IS (1980). Interaction of soil microorganisms and rhizosphere of different onion cultivars with *Fusarium oxysporum f.* sp. cepae, the cause of basal root rot disease of onion. *Agr Res Rev*, 2: 129–142.

Baker KF and Cook RJ (1974). *Biological Control of Plant Pathogens*. WH Freeman and Company, San Francisco, p. 433.

Bhatt J, Sengupta SK and Chaurasia RK (2002). Management of *Meloidogyne incognita* by *Trichoderma viride* in betelvine. *Indian Phytopath*, 55: 97–98.

Boosalis MG (1956). Effect of soil temperature and green manure amendments of unsterilized soil on parasitism of *Rhizoctonia solani* by *Penicillium vesmiculatum* and *Trichoderma* sp. *Phytopathol*, 40: 473–478.

Bora T (1977). *In vitro* and *in vivo* investigations on the effect of some antagonistic fungi against the damping off diseases of egg plant. *J Turk Phytopath*, 6: 17–18.

Bora T, Yildiz M, Akinci C and Nemli T (1982). Investigation on fungistasis with respect to wilt diseases in important cultivated soils of the Western Aegean region. *J Turk Phytopath*, 11: 1–13.

Brasier CM (1975). Stimulation of sex organ formation of *Phytophthora* by antagonistic species of *Trichoderma* I. The effect *in vitro*. *New Phytol*, 74: 183–194.

Buch G, Francisco MA and Muray WW (1983). Aspersitin: A new metabolite of *Aspergillus parasiticus*. *Tetrahedron Lett*, 24: 2527–2530.

Chadova ZS, Sedova SA and Kurakova TI (1980). The effect of plants in the interrelation between the pathogen of *Fusarium* wilt and its antagonist. *Izvestiya Akademi Nauk Turkmenskoi SSR. Biologicheskikh Nauk*, 4: 75–77.

Chengappa R (1989). Poison in your food. *India Today*, p. 82–83.

Dennis C and Webster J (1971a). Antagonistic properties of species-group of *Trichoderma* I. Production of non-volatile antibiotics. *Trans Brit Mycol Soc*, 57: 25–39.

Dennis C and Webster J (1971b). Antagonistic properties of species-group of *Trichoderma* II. Production of non-volatile antibiotics. *Trans Brit Mycol Soc*, 57: 41–48.

Fujimoto Y, Miyagawa H, Tsurushima T, Trie H, Okamura K and Ueno T (1993). Structures of antafumicins AaA and B, novel antifungal substances produced by the fungus *Aspergillus niger* NH-401. *Biosci Biotech Biochem*, 57: 1222–1224.

Gangawane LV (2008).Glimpses of phyopathology for sustainable agriculture. *Indian Phytophathol*, 61: 2–8.

Johri BN and Singh SC (1975). Volatile sporostatic factors of *Aspergilli* and their role in soil fungistasis. *Curr Sci*, 44: 59–61.

Kamyshiko OP, Tupenevich SM, Chumakov AE and Shekunova EG (1976). Mikroflora v eksperimental nomsevooborote i ee antagonistich eskaya activnost k. poschverrnym fitopatogenam. *Mikol Fitoputol*, 10: 32–36.

Lee YA and Wu WS (1979). Management of the sclerotinia disease with biological and chemical methods. *Memoirs Coll of Agri Nat Taiwan Univ*, 19: 96–107.

Malajczuk N (1983). Microbial antagonism to *Phytophthora*. In: *Phytophthora: Its Biology, Taxonomy, Ecology and Pathology*, (Eds) Erwin DC, Bartnicki-Garcia S and Tsao PH. The American Phytopathological Society, St Paul MN, p. 197–218.

Millard WA and Taylor CB (1927). Antagonism of microorganism as the controlling factor in the inhibition of scab by green manuring. *Ann Appl Biol*, 14: 202–215.

Naik MK and Sen B (1992). Biocontrol of plant disease caused by *Fusarium* species. In: *Recent Developments in Biocontrol of Plant Diseases*, (Eds) Mukerji KG, Tewari JP, Arora DK and Saxena G. Aditya Books Pvt Ltd, New Delhi, p. 36–51.

Rajathilagam R and Kannabiran B (2001). Antagonistic effect of *Trichoderma viride* against anthracnose fungus *Colletotrichum capsici*. *Indian Phytopath*, 54: 135–136.

Shukla AN and Dwivedi RS (1979). Survival of *Rhizoctonia solani* Kuhn under the influence of stalling growth products of some *Aspergilli* and its growth response to some phenolic substances. *Proc Indian Nat Sci Acad*, 45: 269–272.

Tiwari DP and Mehrotra RS (1973). Survival and control of *P. parasitica var. piperina* in fumigated soils. *J Indian Bot Soc*, 52: 138–146.

Utkhede RS (1992). Biological control of *Phytophthora* on fruit trees. In: *Recent Development in Biocontrol of Plant Disease*, (Eds) Mukerji KG, Tiwari JP, Arora DK and Saxena G. Aditya Books Pvt Ltd, New Delhi, p. 1–16.

Weidling R (1932). *Trichoderma linnorum* as parasite of other soil fungi. *Phytopatholgy*, 22: 837–845.

Wright JM (1956). The production of anibiorics in soil III. Production of gliotoxin in wheat straw buried in soil. *Appl Biol*, 44: 461–466.

Soil Microflora, 2009 *Pages* **279–314**
Editor: **Rajan Kumar Gupta, Mukesh Kumar & Deepak Vyas**
Published by: **DAYA PUBLISHING HOUSE, NEW DELHI**

Chapter 24

Azotobacter: Recent Advances

***Sangeeta Paul*[1] *and Bishwajeet Paul*[2]**
[1]Division of Microbiology, [2]Division of Entomology,
Indian Agricultural Research Institute, New Delhi – 110 012

Soil microorganisms are a very important component of the soil habitat. They play a pivotal role in soil ecosystems by controlling nutrient recycling which is very essential for maintaining soil fertility and in turn, plant health and by contributing to the genesis and maintenance of the soil structure. The root-soil interface, or rhizosphere, is the site of greatest microbial activity within the soil matrix. Microorganisms colonize the rhizosphere of the plant and remain in close association with roots. There, the microbial community influences growth of the plant by exerting beneficial or deleterious effects.

Azotobacter is one of the most extensively studied plant growth promoting microorganism because its inoculation benefits a wide variety of crops. The first species of the genus *Azotobacter*, named *Azotobacter chroococcum*, was isolated from the soil in Holland in 1901 by Beijerinck. Since then, these bacteria have been studied by numerous authors, who have made a number of significant discoveries about this genus. Azotobacteria represent the main group of heterotrophic free-living nitrogen-fixing bacteria. *A. chroococcum* was studied by Beijerinck (1901). In subsequent years, several other types of azotobacteria group have been found in the soil and rhizosphere such as *A. vinelandii*, Lipman (1903); *A. nigricans*, Krassilnikov (1949); *A. paspali*, Döbereiner (1966); *A. armenicus*, Thompson and Skerman (1981); *A. salinestris*, Page and Shivprasad (1991). These nitrogen-fixing bacteria are important for ecology and agriculture.

Along with nodular bacteria, Azotobacter was considered to be the most extensively studied genus among the saprophytes (Beijerinck, 1908; Winogradsky, 1938; Horner *et al.*, 1942). Winogradsky (1932) discovered that azotobacteria release ammonia into the soil. The first period of research on the genus was marked by studies on its morphological, cytological, and biochemical characteristics (Allison and Gaddy, 1940; Lipman and Mac Lees, 1940; Lee and Burris, 1943). Other areas of research included nutrient media used to grow azotobacteria, their reproduction methods, etc. (Jensen, 1955; Johnson and Magee, 1956; Prssa, 1963).

Azotobacter: The Microbe

Azotobacter belongs to the family Azotobacteriaceae. These are Gram-negative, non symbiotic diazotrophic, aerobic, heterotrophic, catalase positive bacteria. These bacteria proliferate in the rhizosphere of most of the plants from where they can be easily isolated.

Morphological Characteristics

Azotobacter cells are pleomorphic, the size of young rod shaped cells varies from (2.0-7.0 x 1.0-2.5 μ) and occasionally an adult cell may increase up to 10-12 μ (Figure 24.1). The most predominant cell-forms are oval, spherical or rod shaped cells, frequently distorted shapes like peanut and spindle. Breed *et al.* (1957) suggested that *Azotobacter* cells may occur in various forms such as in pairs, packets and even in chains surrounded by slimy membrane of variable thickness. Mobility is seen in *A. chroococcum*, *A. vinelandii*, *A. armeniacus*, and *A. paspali* by means of peritrichous flagella. *A. beijerickii* and *A. nigricans* are non-motile. Young cells have peritrichous flagella which serve as locomotory organs (Mishustin and Shilnikova, 1969) while the older cells are shorter, nonmotile, and coccoid in shape and surrounded by a thick mucoid capsule (Martin *et al.*, 1965). Very small cellular forms called germinal, L-forms or microcondidia were also isolated from soil (Gonzalez-Lopez and Vela, 1981). These were incapable of dinitrogen fixation but could generate large nitrogen fixing forms, which in turn produced the small forms.

Physiological Characteristics

Azotobacter can grow well on simple nitrogen free nutrient medium containing phosphate, magnesium, calcium, iron, molybedenum and carbon source. Though it can metabolize wide range of carbon compounds from organic acids to simple alcohols, the preferred carbon sources are glucose, sucrose or mannitol. Its catabolic versatility in utilizing several aromatic compounds like benzoate, p-hydroxy benzoate, protocatechuic acid, 2,4-D, chlorophenols like 2-chlorophenol, 4-chlorophenol, 2,6-dichlorophenol and 2,4,6-trichlorophenol, lindane, aniline, benzene, toluene etc. has been well noted (Hardisson *et al.*, 1969; Balajee and Mahadevan, 1990; Li *et al.*, 1991; Gahlot and Narula, 1996; Moreno *et al.*, 1999; Revillas, 2000; Gaofeng *et al.*, 2004; Paul and Anupama, 2008; Thakur, 2007). Abd-Alla (1994) reported that *A. chroococcum* strain MH-1 was able to utilize 4-hydroxy benzoic acid, resorcinol, pyrocatechol and vanillic acid as sole carbon source. Nitrogenase activity was detected with the first three phenolic compounds but medium with vanillic acid showed no nitrogenase activity. It can also utilize various nitrogen sources like urea, amino acids, nitrates, nitrites etc. in addition to the ability to use atmospheric nitrogen. The energy requirement for the process of nitrogen fixation is met by a very high rate of aerobic metabolism. High oxygen demand is believed to contribute for the maintenance of a minimal intracellular oxygen tension, a requirement of the oxygen sensitive nitrogenase to accomplish N_2-fixation (Robson and Postgate, 1980).

On minimal media, optimum growth occurs at temperature between 25°C and 30°C (Thompson and Skerman, 1979). Growth can occur under micro-aerobic conditions but is improved with aeration. The respiratory system of *Azotobacter* is known for its adaptability to a wide range of dissolved oxygen concentrations. Under optimum conditions, N_2 fixing growth proceeds with a generation time as low as two hours. The growth of *A. chroococcum* is inhibited by nitrification inhibitors *viz.* dicyanidiamide (10 ppm) and nitrapyrin (100 ppm), whereas 100 ppm of thiourea brought about only a small reduction in growth (Zacheri and Amberger, 1990). Fekete *et al.* (1983) found iron to be growth limiting for nitrogen fixing *Azotobacter* cultures at a concentration as high as 12 mM and iron was sufficient for growth at 25 mM concentration. Sub-optimal concentration of iron increased the mean generation time of *Azotobacter* resulting in poor growth (Sevinc and Page, 1992). In the presence of low level of iron under aerobic condition, *A. chroococcum* produced a catechol melanin as aero adaptive mechanism (Shivaprasad and Page, 1989).

Azotobacter is recognized as a non-symbiotic nitrogen-fixing organism. This microbe possesses three genetically distinct nitrogenase complexes and the expression of these nitrogenases varies with molybdenum, vanadium and ammonium in the culture medium (Bishop *et al.*, 1980). One of the enzymes nitrogenase I, which is expressed only when molybdenum is present in the medium. Nitrogenase II is expressed only when vanadium is present in the medium, while nitrogenase III is expressed when both molybdenum and vanadium are absent (Chisnelle *et al.*, 1988; Prema Kumar *et al.*, 1988; Bishop and Joerger, 1990; Harvey *et al.*, 1990; Fallik *et al.*, 1991; Joerger *et al.*, 1991). Ammonia is responsible for the repression of synthesis of all the three nitrogenases. This enzyme is very sensitive to oxygen. Along with a very high metabolic rate, there also exist other mechanisms for protection of the nitrogenase enzyme complex. These include respiratory protection (Philips and Johnson, 1961; Haddock and Jones, 1977), membrane protection (Haaker and Veegar, 1977) and conformational protection (Shethna *et al.*, 1968; Veegar *et al.*, 1980). Energy requirement for nitrogen fixation is obtained from EMP and TCA cycle (Jackson and Dawes, 1976). Acetate is utilized via glyoxalate pathway. Kleiner and Kleinschmidt (1976) demonstrated the presence of both GS and GOGAT for NH_4^+ assimilation. Incubation of *A. chroococcum* in the presence of micromolar concentration of $MnCl_2$ prevents nitrogenase activity (Ruiz *et al.*, 1990).

Polysaccharide or gum production is one of the characteristic features of *Azotobacter* (Mulder and Brontonegoro, 1974). Some species produce polysaccharides in copious quantities forming a capsule around the cell. The extra-cellular polysaccharide of *A. vinelandii* has a composition similar to alginic acid, an industrially important polymer (Horan *et al.*, 1983). Cote and Krull (1988) found two distinct polysaccharides by ion exchange chromatography in *A. chroococcum* NRLLB-14341. Exopolysaccharide I consisted of rhamnose, mannose and galactose in molar ratio 1:2:2 along with trace amounts of glucose. Exopolysaccharide II, however, consisted of mannuronic acid with small concentrations of galacturonic acid. The role of exopolysaccharides in nature has not been clearly established and is probably diverse and complex. It has been suggested that they may protect against desiccation, mechanical stress, phagocytosis and phage attack, participate in uptake of metal ions, as adhesive agents or ATP sinks or to be involved in interactions between plants and bacteria (Fyfe and Govan, 1983; Hammad, 1998). Cysts have been reported to survive in dry soil for several years because of presence of this polysaccharide coating. Under favourable conditions, including the presence of water, the alginate coating swells and the cyst germinates.

Pigments are produced by all species of *Azotobacter*. Their colour, property and chemical composition differ widely. *A. chroococcum* produces characteristic black water- insoluble melanin like pigment in old cultures. There is variation in the brownish-black pigments produced by the older culture of *A. chroococcum* (Zinovyeva, 1962 and James, 1970) (Figure 24.2). This pigmentation is due to the oxidation of tyrosine by a copper containing tyrosinase enzyme (Mulder and Brontoneyro, 1974 and Mishustin and Shilnikova, 1971). A water-soluble, yellow-green, fluorescent pigment is excreted by *A. vinelandii* and *A. paspali*; a red-violet or brownish-black pigment is seen in *A. nigricans*, *A. armeniacus*, and under certain conditions in *A. paspali*.

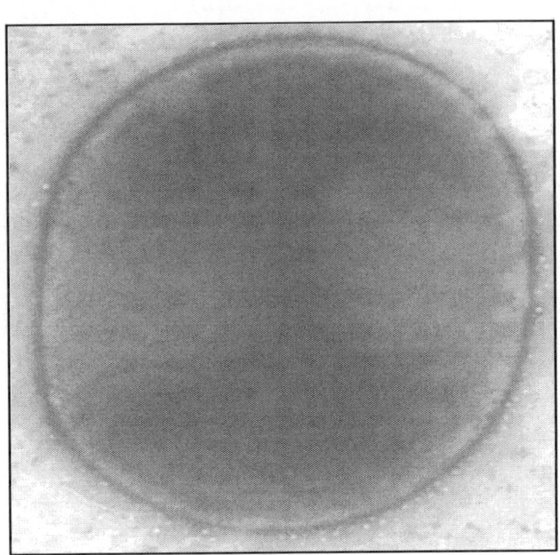

Figure 24.1: Electron Micrograph of *Azotobacter* Cell

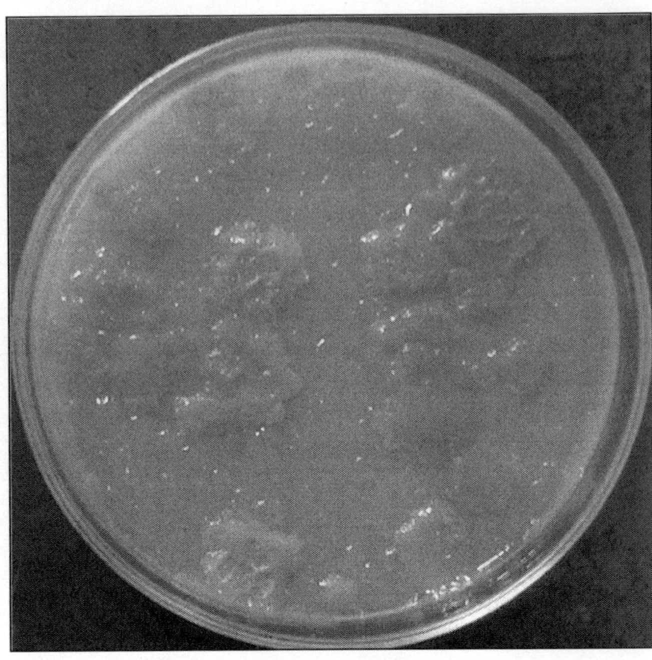

Figure 24.2: *Azotobacter chroococcum* colonies on Jensen's Medium

Azotobacter undergoes cyst formation during adverse growth conditions. Cyst is a living dormant cell with two coats namely exocystorium and two layers of exine. The cyst is found to be rich in poly β-hydroxy butyric acid (PHB). With the onset of favourable conditions, the cyst gives rise to vegetative cells. Calcium is essential for cyst formation (Page and Sadoff, 1975).

Classification of *Azotobacter*

Azotobacter belongs to the family *Azotobacteraceae*. Traditionally, this family includes various gram negative, aerobic, heterotrophic bacteria, which are able to fix dinitrogen nonsymbiotically under normal atmospheric partial pressure of oxygen. In *Azotobacteraceae*, three genera- *Azotobacter, Beijerinckia* and *Derxia* were recognized earlier. However, De Ley (1968) did DNA analysis and DNA hybridization tests of different species of *Azotobacter, Beijerinckia* and *Derxia*. Based on the differences in per cent (G+C) composition between species, they concluded that the genera *Beijerinckia* and *Derxia* were genetically distinct from the genus *Azotobacter* and now based on the dissimilarity of rRNA cistrons, Bergey's Manual limits this family to two genera *Azotobacter* and *Azomonas* excluding the previous members *Beijerinckia* and *Derxia* (Tchan, 1984). Both family members can be differentiated from each other by cyst formation and DNA base composition. Differential characteristics of *Azotobacter* and other morphologically or physiologically similar genera as given in Bergey's Manual of Systematic bacteriology (Tchan, 1984) are summarized in Table 24.1.

According to Bergey's Manual of Systematic Bacteriology, Volume I (1984), there are 6 species coming under genus *Azotobacter*. These include, *A. chroococcum, A. vinelandii, A. beijerinckii, A. nigricans, A. armenicus* and *A. paspali*. In addition to these 6 species, a new species *A. salinestris* for the Na^+

Table 24.1: Differential Characteristics of *Azotobacter* and Other Morphologically or Physiologically Similar Genera[a]

Characteristics	*Azotobacter*	*Azomonas*	*Beijerinckia*	*Derxia*	*Azospirillum*	*Rhizobium/ Bradyrhizobium*	*Klebsiella*	*Pseudomonas*	*Azotomonas*
Cell morphology:									
Pleomorphic:ovoid to rod shaped	+	+	–	–	–	–	–	–	–
Dumbellshaped with granules positioned at poles	–	–	+	–	–	–	–	–	–
Rod containing granules which give beaded appearance to the cell	–	–	–	+	–	–	–	–	–
Cell ≥ 2mm diameter x ≥ 3 mm length	+	+	–	–	–	–	–	–	–
Motility	D	+	+	+	+	+	–	+	+
Flagellar arrangement:									
Monotrichous	–	D	–	+	+[b]	D	–	D	–
Lophotrichus	–	D	–	–	–	–	–	D	–
Peritrichus	D	D	+	–	+[b]	D	–	–	+
Cysts (*sensu stricto*) produced	+	–	–	–	–	–	–	–	–
Nitrogen fixation under atmospheric partial pressure of O_2	+	+	+	+	–	–	–	–	–
Nitrogen fixation only under anaerobic or micro-aerophilic conditions	–	–	–	–	+	+	+[c]	–	–
Autotrophic use of H_2 to fix N_2	–	–	–	+	D[d]	D	–	–	–
Root-associated nitrogen fixation:									
Not producing root hypertrophy	D	–	–	–	+	–	–	–	–
Producing root hypertrophy	–	–	–	–	–	+	–	–	–
Mol % G + C of DNA	63.2–67.5	52–58.6	55–61	69–73	69–71	59–65	53–58	58–70	58–60.5

[a]: Symbols: see standard definitions.

[b]: When cultured in liquid media, *Azospirillum* strains have a single polar flagellum, but when cultured on solid media at 30°C they also form numerous lateral flagella having a shorter length.

[c]: Not all strains fix nitrogen.

[d]: Some strains of *A. lipoferum* appear to be capable of hydrogen autotrophy.

dependent strains isolated from soil in Alberta (Canada) was discovered by Page and Sivaprasad (1991). In fact, the main criteria used for the taxonomy of *Azotobacter* are flagellation, pigmentation and cyst formation. Recently DNA analysis has also been taken into consideration. Differential characteristics of various species of *Azotobacter* as given in Bergey's Manual of Sytematic bacteriology (Tchan, 1984) are summarized in Table 24.2.

Ecological Distribution of *Azotobacter*

Azotobacter is wide spread in nature. The different habitats of *Azotobacter* include soils, leaves, roots and marine and fresh water. This bacterium can occur at any temperature, ranging from tropical to temperate whereas at a pH ranging from 3.0 to 9.0 Becking (1981) and Gordon (1981) reported that its species were very specific in distribution. The predominant species in nature is *A. chroococcum*.

In soil

Azotobacter is found worldwide, in climates ranging from extremely northern Siberia to Egypt and India. *Azotobacter* sp. is not found in all types of soil. Moreover, its abundance varies as per the depth of the soil profile (Vojinoviv, 1961; Sariv, 1969a; 1969b). Thompson (1989a; 1989b; 1989c) determined azotobacteria abundance in vertisols, and Reddy and Reddy (1989) studied the competitive saprophytic survival of azotobacteria in black cotton soil. Mrkovaki *et al.* (1997a; 1998a; 1998b) determined the number of azotobacteria in chernozem soils of Vojvodina province.

Table 24.2: Differential Characteristics of the Species of the Genus *Azotobacter*[a]

Characteristics	*A. chroococcum*	*A. vinelandii*	*A. beijerinckii*	*A. nigricans*	*A. armeniacus*	*A. paspali*
Motility	+	+	–	–	+	+
Long filaments in young cultures	–[b]	–[b]	–[b]	–[b]	–[b]	+
Water soluble pigments:						
Yellow-green fluorescent[c]	–	+	–	–	–	+
Green	–	d	–	–	–	–
Brown-black	–	–	–	d	–	–
Brown-black to red-violet	–	–	–	+	+	–
Red-violet	–	d	–	d	+	+
Utilization as carbon source:						
Rhamnose	–	+	–	–	–	–
Caproate	+	+	–	–	–	–
Caprylate	–	+	–	–	+	–
meso–inositol	–	+	d	–	d	–
Mannitol	+	+	d	d	+	–
Malonate	d	+	+	d	–	–

[a]: Symbols: see standard definitions

[b]: These species may sporadically produce filamentous forms of different lengths.

[c]: On iron deficient medium.

Azotobacter sp. are widely distributed in non-acidic soils of India and other parts of the world excepting in the arid regions of the middle east where they are found to be poor in numbers. Their population may range from nil to several thousands per gram and rarely exceed 10^4/g (Alexander, 1961; Brown *et al.*, 1962). Studies on the occurrence of *Azotobacter* in some soil types of India have been done by Rangaswami and Sadasivam (1964). The maximum number per gram reported are 1.1x10^4 in Haryana soils (Sindhu and Lakshminarayana, 1986) and 8x10^4 in the forest soils of Karnataka (Channal *et al.*, 1989) and 10^5-10^6 in forest soils of Maharashtra (Nagraj, 1989). Kole *et al.* (1988) reported the occurrence of *A. chroococcum* in Eastern Canadian soils ranging from 1-2.5 x10^4 cells/g of soil. Micev (1971), Döbereiner (1974) and Barea *et al.* (1978) reported the abundance of this bacterium in chernozem to be 9-37 × 10^2 cfu/g.

Azotobacter requires neutral to slightly alkaline pH for growth. The low population of *Azotobacter* is generally attributed to low pH in acidic soils. However, Dobereiner (1953) observed a very frequent occurrence of *A. chroococcum* in 22 out of 27 acid soil samples collected in Brazil. In other soils, low numbers may be due to inhibition or destruction by associated microorganisms (Rubenchik, 1960). It was found that non-availability of the organic matter in soil is one of the main limiting factors in the proliferation of *Azotobacter* in soil rather these species are restricted to environment rich in organic matter. The beneficial effects of small amounts of humus on growth of *Azotobacter* and its nitrogen fixation are well known (Jensen, 1951; Iswaran, 1958; Gaur and Mathur, 1966; Bhardwaj and Gaur, 1970).

Inorganic fertilization of soil influences *Azotobacter* numbers. While addition of nitrogenous fertilizers to soil inhibit the growth of *Azotobacter*, addition of phosphate fertilizers improve bacterial growth and proliferation. Formation of resting structures–the cyst, enables *Azotobacter* to survive under deleterious conditions and also help in long term survival in dry soils. *A. chroococcum* and other species of *Azotobacter* survived in dry soils, stored in laboratory for periods of time as long as 24 years (Moreno *et al.*, 1986).

In Rhizosphere and Rhizoplane

Studies on azotobacterial abundance in the rhizosphere of certain crops, which began almost simultaneously with research on the soil as the habitat of these bacteria, revealed that azotobacteria are much more abundant in the rhizosphere of plants than in the surrounding soil and that this abundance depends on the crop species (Sariv and Rassoviv, 1963a; 1963b). However, their abundance in the soil depends on the plant species grown and it varies during the growing season (Sariv, 1978).

The rhizospheres of soybean and other legumes contained considerably larger number of these bacteria than the nearby soil, and the abundance also varied from one rhizosphere zone to the other. In the rhizoplanes of maize, soybean, sunflower and sugar beet azotobacteria were observed. No azotobacteria were found in the rhizoplane of wheat during the entire growing season, however, their number was higher in the rhizosphere and the adjoining zone than in the nearby soil (Sariv and Rassoviv, 1963b). Eweda and Vlassak (1988) found that *A. chroococcum* was able to colonize better in maize rhizosphere. Rao (1991) reported the presence of greater number of *A. chroococcum* in the rhizosphere of sugarcane as compared to that in root free soils collected from 30 cm depth. Mrkovaki (1997) and Mrkovaki *et al.* (1997a; 1998a) determined the number of azotobacteria in the rhizosphere of sugar beet to be 3-12 × 10^3 cfu/g. Rhizosphere soils of chickpea and wheat were analysed for the diversity of diazotrophic microorganisms by sequencing the *nifH* gene (Sarita *et al.*, 2008). *Azotobacter* sp. was observed to be one of the predominant diazotrophs present in both rhizosphere soils.

Azotobacter established well on the roots of cotton. The population was in the range from 10^2 to 10^4 cells g^{-1} soil which increased only 2-3 folds with cotton seed bacterization with *Azotobacter*. The initial growth period of cotton after germination *i.e.* till 30 days seemed critical for the establishment of *Azotobacter* in cotton rhizosphere. Some varieties exerted repressive effect, where the proliferation of *Azotobacter* at later period of growth was hampered. The inoculation successfully overcame the repressive rhizosphere effect and helped in the establishment of *Azotobacter* population in all varieties of cotton (Pandey *et al.*, 1989).

In the rhizosphere the root exudates, which contain amino acids, sugars, vitamins and organic acids together with the decaying portions of root system serve as energy sources for *Azotobacter* multiplication (Vancura and Macura, 1961). Federov (1944) showed that *Azotobacter* sp. established in the maize rhizosphere using root exudates as energy and carbon source. The ability to efficiently utilize the root exudates of a given crop confers competitive advantage to the microbial culture in its preferential establishment in the rhizosphere of the crop. Higher organic matter content and better humification seem to be important factors for predicting *A. chroococcum* behaviour in the rhizosphere (Requene *et al.*, 1997). The population of *Azotobacter* sp. in organic amended soils was found to be high and on inoculation their proliferation was more pronounced in barley rhizosphere (Negi *et al.*, 1987).

Rubia *et al.* (1989) observed that rhizosphere of sorghum showed the presence of *A. chroococcum* possessing high rate of growth and nitrogenase activity. Troitskii *et al.* (1989) obtained two strains of *A. chroococcum* comprising of new plasmids which excreted new proteins developing affinity to the surface of tomato and barley roots which resulted in improved adhesion to rhizosphere and movement with growing roots, thereby improving growth and yield of these crops.

Azotobacter cells are not only present in the rhizosphere but are also abundant on the rhizoplane. The ability of the various *Azotobacter* cultures to flourish on the root surface of these crops varies and such preferential establishment of the inoculated cultures may be ascribed to the competitive advantage conferred by the genotype of the culture to utilize root exudates of a given crop efficiently. However, toxic effects of root excretions of certain crops on *Azotobacter* have also been reported (Polumuri, 1989).

In Spermosphere

Ota *et al.* (1991) observed the establishment of *A. chroococcum* in spermosphere *i.e.* around soaked seed prior to germination and that promoted the subsequent development in rhizosphere.

In Phyllosphere

Presence of *A. chroococcum* have also been reported in the phyllosphere of *Brassica* sps. (Agarwal and Shende, 1987).

In Endorhizosphere

Though is *Azotobacter* considered a non-symbiotic nitrogen fixer in soil and rhizosphere, its presence in the swollen roots of *Cyperus rotundus* (L.) (Iswaran and Sen, 1959) and aerial roots of banyan tree (Iswaran and Subba Rao, 1967) was noted. Colonization of the root interior in maize and grasses by a diazatroph *Azospirillum* was observed by Patriquin and Doberiener (1978) who examined root sections under microscope using tetrazolium dye reduction technique to stain the root cells harbouring dinitrogen fixers. Similar technique was employed by Bhide and Purandare (1979) to report the presence of *Azotobacter* within the root cells of *Cyanodon dactylon* (L.), while Tikhe *et al.* (1980) observed the ability of *A. chroococcum* to establish within the cortical cells of 30 species of monocots.

Under the light microscope, the sections of *Pathos scandens* (L.) showed that the aerial clinging roots were relatively high in *A. chroococcum* population (30-60x10^3) cells mg^{-1} fresh weight and they possessed a high nitrogenase activity (Sharma *et al.*, 1985). Polumuri (1989) demonstrated *Azotobacter* cells in the cortical cells of roots of cotton, maize and wheat. These observations were supported by work carried out by Kumar (1992), Malik (1992) and Jose and Paul (2004). Agarwal and Shende (1987) and Jose and Paul (2004) using tetrazolium reduction technique reported the establishment of *A. chroococcum* inside the root tissue of *Brassica* sp. and wheat respectively.

Azotobacter chroococcum cells have also been found in maize tissue (Hallberg, 1995; Li *et al.*, 1995; Raieviv *et al.*, 1995a; 1995b). The penetration of azotobacteria into plant tissue was studied *in vitro* using the tissue culture procedure. Callus mass increased in association with azotobacteria (Mezei *et al.*, 1997/98). Mrkovaki *et al.* (1995a) confirmed the presence of these bacteria in sugar beet calluses in vertical as well as in horizontal direction

Plant Growth Promoting Activities

Rhizosphere-colonizing bacteria that possess the ability to enhance plant growth when applied as an inoculant to seeds, roots or tubers are called plant growth-promoting rhizobacteria (PGPR) (Kukreja *et al.*, 2004). The term PGPR was for the first time used by Kloepper and Schroth (1978). *Azotobacter* is a plant growth promoting *rhizobacterium*. Its use as a biofertilizer was first advocated by Gerlach and Vogel (1902), with the purpose of supplementing soil N with biologically fixed N due to the activity of this microbe. Since then, it has been observed to play a multifaceted role in plant growth promotion (Table 24.3). It has been observed to not only fix atmospheric dinitrogen under free-living conditions but also to possess other plant growth promoting activities like P solubilization, production of plant growth hormones like auxins, gibberellins, cytokinins, vitamins and amino acids.

Table 24.3: Plant Growth-promoting Activities of Different Strains of *Azotobacter chroococcum* (Apte and Shende, 1981 a)

Strain	*Nitrogen Fixation Efficiency (mg N/g sucrose consumed)*	*Phytohormones*	
		Excretion of IAA (mg/ml of culture filtrate)	*Production of GLS (grades of fluorescence)*
A-41	7.9	3.0	+++
B-1	5.3	1.8	++
B-2	6.6	1.1	++
C-1	9.5	1.7	+++
C-2	7.9	2.7	+++
M-2	10.0	2.8	+++
M-4	ND[a]	0.8	+
M-6	1.8	ND	+
P-1	6.2	1.5	++
P-2	6.2	2.5	++
P-4	6.5	3.0	+++
W-2	1.5	1.6	+
W-3	9.2	1.6	+
W-5	7.5	2.5	++

+: Yellowish green fluorescence; ++: Greenish yellow fluorescence; +++: Greenish fluorescence; [a]: Not detectable.

N_2 Fixation

Fixation of nitrogen is a prerequisite for the synthesis of cell proteins, the basic structural and functional unit of both plant and animal systems. Certain microbes are selectively endowed with a unique synthesis mechanism called "Biological nitrogen fixation" by which molecular nitrogen is reduced to ammonia (Roberts and Brill, 1981). Microorganisms that fix nitrogen are called diazotrophs. The ability to fix elemental nitrogen is a vital physiological characteristic of *Azotobacter* sp. It fixes N_2 independently in soil as well as in water.

Azotobacter sp. is capable of converting nitrogen to ammonia (Newton *et al.*, 1953; Bishop *et al.*, 1982), which in turn is taken up by the plant. Nitrogen fixing ability of a microbe can be estimated by acetylene reduction activity (ARA). *Azotobacter* has been observed to possess a very high ARA activity. The range of fixation is observed to be between 2-15 mg N fixed/g of glucose consumed (Apte and Shende, 1981a), although higher values have often been reported. However, most of this fixed nitrogen is bound to the microbe only and is released only after the death and lysis of the cell. Very low amounts 10-15 kg/ha of the fixed nitrogen is released into the soil and thus is available for uptake by the plant (Jensen, 1951).

Ammonia is the only demonstrated intermediate in the process of N_2 fixation (Burris, 1972). Most of the fixed ammonia is taken up by the organism itself. However, there are a number of strains of *Azotobacter* which have the ability to secrete ammonia into the medium. Narula *et al.* (1980) reported varying amounts of ammonia excretion by *A. chroococcum* strains when these were grown in nitrogen free medium under stationery conditions. *A. chroococcum* could excrete as much as 45 mg ammonia ml^{-1} of the culture broth in a sucrose supplemented synthetic medium (Narula *et al.*, 1981). Ammonia excreting strains showed the presence of GOGAT pathway while the ammonia non-excreting strains showed the presence of GDH pathway. Bali *et al.* (1992) have reported generating *nif L* mutant of *A. vinelandii* which excreted significant quantities of ammonium during diazotrophic growth.

Nitrogen fixation activity (nitrogenase) is repressed by ammonium in *Azotobacter* sp. Therefore, the presence of high amounts of nitrogenous fertilizer in the field leads to fixation of very low amounts of nitrogen under such conditions thereby reducing their effectiveness as biofertilizers. Mutants have been isolated that can fix nitrogen in the presence of ammonium and excrete ammonia (Gordon and Brill, 1972; Gordon and Jacobson, 1983; Bela *et al.*, 1986; Lakshminarayana *et al.*, 2000). Bela *et al.* (1986) and Lakshminarayana *et al.* (2000) isolated *Azotobacter* mutants resistant to analogues of ammonia *viz.*, methyl alanine (mal), methionine sulfoximine (Msx) and methyl ammonium chloride (Mac). Derepressed nitrogenase activity and ammonia excretion was studied in all the mutants. There was derepression of nitrogenase activity in case of Msx and Mal mutants however Mac mutants showed higher nitrogenase but no derepressed nitrogenase activity. All these classes of mutants showed early ammonia excretion.

Phosphate Solubilization

Phosphorus is an important plant nutrient, next only to nitrogen and grouped along with nitrogen and potassium as a major plant nutrient. The soluble forms of phosphorus, when applied to soil as phosphate fertilizers to supplement soil phosphorus are rendered insoluble within a short period and thus are unavailable to growing plants. Certain groups of soil microorganisms have the ability to solubilize this fixed phosphorus thereby making it available to the crop plants. *Azotobacter* sp. has been observed to possess the ability to solubilize phosphates (Shende *et al.*, 1975). The phosphate solubilization ability ranged from 8-16 per cent.

Kumar *et al.* (1999) generated mutants of *Azotobacter* capable of better P solubilizing potential. The Phosphorus solubilized was in the range of 1.5-1.7 ìg ml^{-1} of TCP and 0.19-0.22 ìg ml^{-1} of MRP. These were evaluated for their ability to improve crop yields of wheat and mustard and were observed to be very efficient. Similar results were obtained by Narula *et al.* (2002). Kumar *et al.* (2000) reported phsophate solubilizing analogue mutants of *A. chroococcum* capable of showing positive interaction on sunflower. Deubel and Merbach (2005) also demonstrated the contribution of *A. chroococcum* on the solubilization of calcium phosphates.

Azotobacter strains may solubilize insoluble form of phosphorus by production of organic acids or chelating substances, in microenvironments in the vicinity of rock phosphate or in the rhizosphere. Under these conditions P is converted to an available form. Proton release is thought to be the main mechanism that increases phosphorus availability (Illmer and Schinner, 1995; Villegas and Fortin, 2002). However, other phosphate solubilizing substances should also be included for consideration (Bajpai and Sundara Rao, 1971; Banic and Dey, 1981; Whitelaw, 2000; Staunton and Leprince, 1996).

Production of Plant Growth Regulating Substances

Favourable action of *Azotobacter* on growth of plants, inspite of maintaining a low population in soil is now attributed to production of plant growth regulating substances. These substances include vitamins of B group, hormones like indole acetic acid, gibberellin-like substances and cytokinins.

Growth substances, or plant hormones, are natural substances that are produced by microorganisms and plants alike. They have stimulatory or inhibitory effects on certain physiological-biochemical processes in plants and microorganisms. Bacteria of the genus *Azotobacter* synthesize auxins, cytokinins, and gibberellin-like substances under *in vitro* conditions (Brown *et al.*, 1968; Martinez-Toledo *et al.*, 1989; Nieto and Frankenberger, 1991; Reliv *et al.*, 1987; Salmeron *et al.*, 1990; Gonzales Lopez *et al.*, 1991).These growth materials are the primary substance controlling the enhanced growth of plant (Jackson *et al.*, 1964; Barea and Brown, 1974; Azcorn and Barea, 1975). These hormonal substances, which originate from the rhizosphere or root surface, affect the growth of the closely associated higher plants.

Apte and Shende (1981a) observed the ability to synthesize GLS and IAA was very wide spread in this genus. Brakel and Hilger (1965) showed that azotobacteria produced indol-3-acetic acid (IAA) when tryptophan was added to the medium. Vancura and Macura (1960), Burlingham (1964) and Hennequin and Blachere (1966), on the other hand, found only small amounts of IAA in old cultures of azotobacteria to which no tryptophan was added. Lee *et al.* (1970) reported production of maximum concentration of about 10^{-7} M IAA, while higher IAA production in the presence of 0.02 per cent tryptophan was reported by El-Essawy *et al.* (1984). Tryptophan is the precursor of IAA and is converted to IAA through a primary Trp-aminotransferase reaction. Inoculation with these *Azotobacter* strains improved the germination rate of seeds and accelerated the growth of the inoculated plants (Apte and Shende, 1981b).

Three gibberellin-like substances were detected by Brown and Burlingham (1968) in an *Azotobacter chroococcum* strain. The amounts found in the 14-day old cultures ranged between 0.01 and 0.1 µg GA3 equivalent^{-ml}. Reliv (1989) obtained 9.6-19.8 µg GA equivalent^{-l} medium of substance with gibberellic activity. Five cytokinins were identified in an *A. chroococcum* culture filtrate (Nieto and Frankenberger, 1989). A study by Govedarica *et al.* (1993) on the production of growth substances by nine *Azotobacter chroococcum* strains isolated from a chernozem soil has show that these strains have the ability to produce auxins, gibberellins and phenols. Miliv and Mrkovaki (1995) reported production of gibberellins by *Azotobacter* sp. equivalent to a GA3 concentration of 0.003-0.1 µg/cm^3 culture.

In addition to growth promoting substances like GLS and IAA, *Azotobacter* cultures were found to produce various kinds of vitamins especially B group of vitamins (Yatskuenena, 1962; Rubenchick *et al.*, 1965 and Rodelas *et al.*, 1997), nicotinic acid and pantothenic acids (Bershova and Kozlova, 1965), biotin and cyanocobalamine (El-Essawy *et al.*, 1984); Revillas *et al.* (2000) found that *A. chroococcum* strain 423 was able to produce B group vitamins with phenolic compounds as sole carbon source under diazotrophic and adiazotrophic conditions. Gonzalez-Lopez *et al.* (1999) reported *A. chroococcum* was capable of producing amino acid lysine.

Seed Germination

Improvement in the seed germination of some agricultural crops through seed bacterization with *Azotobacter* have been observed (Shende *et al.*, 1987; Harper and Lynch, 1979; Stoklasa, 1908) (Figure 24.3). *A. chroococcum* strains isolated by Shende and Apte (1982) from water growing *Pathos* sp. increased germination of rice and cotton seeds. *Azotobacter* inoculation resulted in enhancing the germination of rice seeds by 24 hours and different strains of *A. chroococcum* brought improvement in the germination of maize, rice and wheat (Table 24.4) (Shende *et al.*, 1977; Apte and Shende, 1981b). Paul *et al.* (2002a) reported improved seed germination and seedling health in onion due to *Azotobacter* inoculation. Inhibitory effects of *Azotobacter* inoculation on germination have also been reported. Shende *et al.* (1977) observed decrease in the percentage seed germination over uninoculated control.

Table 24.4: Effect of *Azotobacter* Inoculation on Seed Germination

Crop	*% Germination Over Control*	*Remarks*
Cotton	1.7-33.3	Some strains (B1, B2, M6 and W3) had depressive effect
Rice	1.6-16.8	In control no germination up to 48 hr but 45-61 per cent germination due to inoculation
Maize	3-27	All the strains tested had beneficial effects
Wheat	4.6-24	B1 and M 6 strains had depressive effect

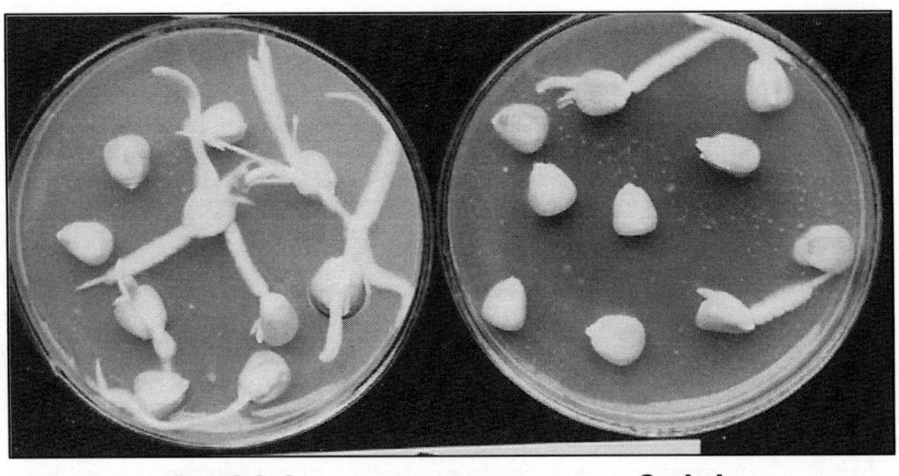

Inoculated **Control**

Figure 24.3: Effect of *Azotobacter* Inoculation on Germination of Maize Seed

Biocontrol

Azotobacter is also known to improve plant growth indirectly through suppression of phytopathogens or by reducing their deleterious effects (Table 24.5). Inoculation of seeds of crop plants with *Azotobacter* was found to reduce the incidence of fungal, bacterial and viral diseases in the crops (Sidorov, 1954; Samitsevich, 1962; Singh, 1977; Meshram, 1984). Antagonistic action of *A. chroococcum* on phytopathogens has been studied by various researchers (Schroth and Hancock, 1982; Meshram and Jager, 1983; Weller, 1988; Verma *et al.*, 2001), but whether this inhibition is due to siderophores or antifungal properties was not clear. Beetroot seeds inoculated with *A. chroococcum* were observed to have enhanced peroxidase activity, which is one of the defense enzymes induced by the plant when it is invaded by pathogens (Stajner *et al.*, 1997).

Beniwal *et al.* (1996) conducted extensive field experiments to evaluate the effect of *A. chroococcum* strains/mutants on the incidence of flag smut. The results revealed that flag smut incidence in wheat was significantly less under *Azotobacter* inoculation compared to the control. Bansal *et al.* (2000) determined the effects of rhizospheric bacteria on plant growth of wheat infected with *Heterodera avenae* (Wollenweber) in potted plants. Out of four rhizospheric diazotrophs tested, *A. chroococcum* (HT 54) showed maximum reduction in nematode infection (48 per cent) followed by *Pseudomonas* (11 per cent) and *Azospirillum* (4 per cent). In contrast, *Rhizobium ciceri* was totally ineffective. Chahal and Chahal (1988) carried out *in vivo* and *in vitro* tests and reported that *A. chroococcum* inhibited the hatching of egg masses of *Meloidogyne incognita* (Kofoid and White) and did not allow the larvae to penetrate into the roots of brinjal to form crown galls.

Table 24.5: Inhibitory Effects of *Azotobacter chroococcum* Strains on Growth of Various Fungi (Pandey and Kumar, 1990)

Fungal Strain	*Zone of Inhibition (mm) Produced by Strains of A. chroococcum*			
	A41	*C2*	*W5*	*M4*
Sclerotiorum rolfsii	22	16	14	6
S. sclerotiorum	20	17	8	11
Fusarium moniliforme	18	14	17	5
F. solani	18	12	15	6
F. oxysporum	18	11	15	10
Cephalosporium maydis	2	2	2	2
Alternaria brassicola	17	17	15	6
Collectotrichum falcatum	23	16	14	5
Exserohilum turcicum	19	2	11	2
Chaetomium globosum	11	6	8	10
Penicillium chrysogenum	17	13	11	6
Trichoderma viride	31	15	12	12
Drechslera tetramera	17	12	16	9
Cladosporium herbanum	32	23	18	17

Mechanism of biocontrol of plant pathogens by *Azotobacter* has been elucidated by many workers. The main biocontrol activities observed in *Azotobacter* are siderophore production and production of inhibitory substances.

Siderophore Production

Competition for iron is one of the well known mechanisms of biocontrol. Under iron limiting conditions, bacteria produce a range of iron chelating compounds or siderophores which have a high affinity for ferric ion. These siderophores bind most of the Iron (Fe^{3+}) available in the rhizosphere thereby making it unavailable to phytopathogens. Since, the plant pathogens may not have the cognate ferrisiderophore receptor for the uptake of iron-siderophore complex, they are prevented from proliferating in the immediate vicinity because of lack of iron (O'Sullivan and O'Gara, 1992).

A very high incidence of siderophore producing ability has been observed in *Azotobacter* sp. (Sevinc and Page 1992; Kumar, 1997; Suneja *et al.*, 1996 and Suneja and Lakshminarayana, 2001). Suneja and Lakshminarayana (1993) have reported siderophore synthesis by 79 per cent of screened *A. chroococcum* strains. A variety of siderophores, based on their chelating groups, are reportedly produced in *Azotobacter* sp. namely catechol type, hydroxamates and pyoveridins (Table 24.6).

Table 24.6: List of Siderophores Produced by *Azotobacter* sp. According to their Chelating Groups

Name of the Siderophore	*Chelating Group (s)*	*Organism*	*Reference*
		Catechol Type	
Aminochelin	Polyamine of 2,3-dihydroxy benzoyl putrescine	*A. vinelandii*	Page and Von Tigerstrom (1988)
Azotochelin	Polyamine of 2,3-dihydroxy lysine benzoyl	*A. vinelandii*	Corbin and Bulen (1969)
		Hydroxamates	
Unidentified hydroxamates	—	*A. chroococcum*	Page (1987)
A. chroococcum	Fekete *et al.* (1989)		
		Pyoveridins	
Azotobactins	Catechol, hydroxyl and α-hydroxyl acid	*A. vinelandii*	Page *et al.* (1991)

A. vinelandii seems to regulate low and high affinity iron uptake pathway depending on the level of iron stringency. There is constitutive synthesis of 2,3-dihydroxy benzoic acid, which appears to function as siderophores in *A. vinelandii* at a basal level. When the iron stringency increases, the catechol siderophores are produced coordinately. If iron continues to be limiting, then azotobactin, a high affinity siderophore is excreted (Page and Von Tigerstrom, 1988; Page and Huyer, 1984). Azotobactin can mobilize iron from a variety of insoluble iron minerals (Page and Huyer, 1984). Iron regulated outer membrane protein, implicated in the transport of their cognate ferrisiderophores, has been detected in *Azotobacter* sp. (Fekete, *et al.*, 1989; Page, 1987).

A. chroococcum strain W-5 capable of producing hydroxamate siderophores (Kumar, 1997) was observed to inhibit hatching of egg masses of *Spodoptera litura* (Fab.), *Spilarctia obliqua* (Walker) and *Corcyra cephalonica* (Stainton) (Paul *et al.*, 2002c). There was also drastic reduction in number of eggs laid/female, per cent pupation and the emergence of adults from pupae (Paul *et al.*, 2002c). Culture filtrate of *Azotobacter* RRLJ 203 containing hydroxamate siderophore inhibited fungal pathogen *Fusarium* sp. (Saikia and Bezbharuah, 1995). Siderophores produced by *A. chroococcum* decreased nematode infection by decreasing the cyst formation by 6-60 per cent Bansal *et al.* (1999).

Siderophores were also observed to help in increasing competitive ability in the rhizosphere of the producer strain thereby helping in better establishment of the producer strain in wheat rhizosphere as compared to non-producing strains (Kumar, 1997). However, the rhizosphere population was high in younger plants which on aging harboured less population.

Production of Inhibitory Substances

Azotobacter can influence plant growth indirectly by maintaining a balance between harmful and beneficial organisms in soil, thereby improving plant health. There was reduction of loose smut of wheat (*Ustilago tritici*), yellow leaf spot of wheat (*Helminthosporium tritici-vulgaris*), powdery mildew (*Erysiphe* sp.) and bacterial blight of bean (*Xanthomonas phaseoli*) by application of *Azotobacter* (Beltyukova, 1953). Sidorov (1954) observed a reduction in the attack of *Phytophthora infestans* and *Streptomyces scabies* on potato crop due to *Azotobacter* application. Chakrabarti and Yadav (1991) reported that *Azotobacter* treatment resulted in lowering of the disease incidence of downy mildew in opium poppy crop.

Work carried out by Pandey and Kumar (1990) demonstrated that *A. chroococcum* strains have a wide range of fungistatic activity against a number of phytopathogenic fungi (*Sclerotium sp., Fusarium sp, Cephalosporium maydis, Alternaria brassicola, Colletotrichum falcatum* etc.). These strains were assayed *in vitro* for their inhibitory effect on growth of these fungi. *Azotobacter* sp. was reported to produce a thermolabile ether-soluble fungistatic substance which inhibited the growth of phytopathogenic fungus *Fusarium monoliformae* (Lakshmi Kumari *et al.*, 1972), *Alternaria* and *Helminthosporium* (Singh 1977) under *in vitro* conditions. These bacteria reduced the germination of conidia to a remarkable extent which was followed by vacuolation, curling, distortion, breakage and lysis of hyphae (Lakshmi Kumari *et al.*, 1972). Meshram and Jager (1983) demonstrated antagonism of isolates of *A. chroococcum* to *Rhizoctonia solani* on agar plates and these isolates were able to control *R. solani* infection of potato sprouts.

Mishustin and Shilnikova (1969) showed that antibiotic substance produced by *A. chroococcum* belonged to conactin group which suppressed the growth of *Alternaria, Candida, Monilia, Aspergillus* and *Penicillium* sp. Sardaryan (1972) had reported the production of physiologically active compounds by *A. chroococcum* which were analogous to phenolic inhibitors.

Interaction of *Azotobacter* with Other Soil Microorganisms

Rhizosphere seems to be a very complex ecosystem where various organisms co-exist and interact with each other exerting competition for nutrients, inhibition or acceleration of growth, utilization or exchange of metabolites and sometimes follow symbiotic existence. *Azotobacter* can influence plant growth directly by providing plant growth promoting substances and indirectly by changing the microflora of rhizosphere and maintaining the balance between harmful and beneficial organisms in soil.

Azotobacter inoculation of seeds has been found to increase the microbial activity in the rhizosphere involved in ammonification, aerobic and anaerobic nitrogen fixation, phosphate mineralization, nitrification and cellulose degradation (Berezova, 1939; Beltyukova, 1953; Naumova, 1958; Shende, 1965; Patel, 1969 and Zinovyeva, 1959). Khariton (1953) and Zaremba and Sinyevskaya (1954) reported that *Azotobacter* markedly stepped up the total bacterial population in the rhizosphere. Population of N_2 fixing green and blue green algae, such as *Chlorella, Anabaena, Tolypothrix, Calothrix, Amorphonostoc* and *Stratonostoc* was found to increase due to *Azotobacter* applications and there was increase in nitrogen and carbon content of soil (Dhar and Arora, 1968; Koptyeva *et al.*, 1970). Lynch (1990) reported beneficial interactions between microorganisms and roots. Burns *et al.* (1981) found enhanced nodulation of leguminous plant roots by mixed culture of *A. vinelandii* and *Rhizobium*.

Leuck and Rice (1976) reported suppression of biological nitrogen fixation by a variety of plants including *Aristida*. Rhizosphere bacteria from the roots of *A. oligantha* were isolated and tested for inhibitory or stimulatory effects on *Azotobacter*. Marked inhibition of *Azotobacter* strains by bacteria isolated from the rhizosphere namely *Xanthomonas axonopodis, Arthrobacter citreus, Arthrobacter simplex, Enterobacter aerogenes, and Micrococcus luteus* was observed, although some stimulatory bacteria namely *Bacillus cereus* and *B. megaterium* were also reported. Inhibition of *A. chroococcum* by *Pseudomonas* sp. was reported by Chan *et al.* (1963) and the cause of inhibition was found to be production of acidic end products which lower the pH excessively.

Plant-*Azotobacter* Interactions

A. chroococcum is a free living nitrogen fixing bacterium. But, of late, due to its superb colonization on and inside the roots and extraordinary plant-microbe relationship, the doubts are expressed about its free living status. Moreover, information has also been piling up which reveal that *A. chroococcum* functions in a close association with plant (Zinovyeva, 1962 and Malik, 1992).

Role of Root Exudates

Root exudates play an important role in the plant microbe interaction. The relatively high populations of microorganisms in the rhizosphere as compared to the non rhizosphere region is by virtue of the chemotactic activity of the microorganisms towards root exudates and metabolism of some of the root exudate components by them. In the absence of root exudates the concept of rhizosphere would not have itself existed.

Bacteria occur in large number in rhizospheres of various plants and the nature of plant exudates may selectively promote the growth of certain bacteria. The major sources of substrates for microbial activity in the rhizosphere and on the rhizoplane have been identified as rhizodeposition products like exudates, lysates, mucilage secretions, dead cell material and also the gases including carbon dioxide. Rhizodeposition may constitute even upto 40 per cent of dry matter produced by plants (Lynch and Whipps, 1990). Microbial growth in the rhizosphere is thus, stimulated with the continued input of readily assimilable organic substances in the exudates by roots leading to them attaining a higher population than in soil. About 1-10 per cent of the rhizospheric population may be diazotrophs.

Zinovyeva (1962) reported that strains which had been trained or acclimatized with a particular crop performed better with the crop as compared to the untrained strains. According to her the root exudate was one of the factors governing the populations of *Azotobacter* in the root zone. *Azotobacter* sp. can utilize maize root exudates as energy and carbon sources, thus enabling them to become well established in maize rhizosphere (Federov, 1944; Martinez-Toledo *et al.*, 1988).

Polumuri (1989) observed that seed diffusates of maize, cotton and wheat had varying effects on growth pattern of *A. chroococcum* strains. Seed diffusates of wheat and cotton had stimulatory effect and were efficiently metabolized by *A. chroococcum* while maize seed diffusates had bacteriostatic effect. Malik (1992) showed that the seed diffusates of Egyptian cotton were efficiently metabolized by *A. chroococcum* strain C-2 which was isolated from American cotton. Martinez-Toledo *et al.* (1990) detected nitrogenase activity on barely roots, which harboured *A. chroococcum*.

Vancura and Macura (1961) reported that growth of *Azotobacter* was promoted in the presence root exudates of wheat. Plant root exudates of pea stimulated the growth of *Azotobacter* sp. when introduced into the root zones of plants grown in a root exudation apparatus (Marathe and Rangaswmi, 1973). Abdel Nasser and Moawad (1975) reported that soil sample amended with synthetic root exudates of wheat in different concentrations stimulated *Azotobacter.* Their number was higher in the presence of high concentration of root exudates than in the low concentration. Martinez-Toledo *et al.* (1988) showed that the growth of *A. chroococcum* strain H23 in the N free medium was stimulated in the presence of Maize hybrid AE 703 root exudates.

Specificity of *Azotobacter* to plants

A large number of studies carried out have confirmed that a particular *Azotobacter* strain may perform exceptionally better with one crop but it may not prove so successful with other crops (Bukatsch and Heitzer, 1952; Petrenko, 1961; Wani *et al.*, 1976). Apte and Shende, 1981c) reported that strains of *A. chroococcum* which established better on roots of crop brought greater increase in yield of the same crop. Generally strains isolated from a particular crop itself found a congenial habitat in their respective rhizospheres and thus performed better. They described these as homologous strains for compatibility. Martinez–Toledo *et al.* (1990) observed similar results in barley. Agarwal (1985) reported that W-5 and M-4 strains of *A. chroococcum* showed perfect specificity to *Brassica juncea* (L.) and *B. napus* (L.) respectively and either strain was ineffective for other genome (Table 24.7).

Table 24.7: Grain Yield of Irrigated Mustard Crop (q ha^{-1}) during *Rabi* 1985 and 1986

Seasons	*B. napus*			*Brassica juncea*		
	Control	*M-4*	*W-5*	*Control*	*M-4*	*W-5*
1985	13.0	16.4 (25)	13.7 (5)	13.0	12.4	18.6 (42)
1986	9.7	11.0 (11.3)	10.4 (10.5)	7.8	8.0 (2.3)	9.8 (2.5)

Figures in parenthesis are percentage increase over control.

Dobereiner (1966) described the close association between Brazilian C-4 grass, *Paspalum notatum* (L.) and *Azotobacter* sp., which appeared to be very specific although non-symbiotic. The species was later on named as *Azotobacter paspali.* The association was observed to be so specific that *Azotobacter paspali* has not been reported in the absence of this grass and flourishes only in its rhizosphere. Microscopic examination of roots of *P. notatum* showed bacteria like objects associated with outer mucilage of the root (Dobereiner *et al.*, 1972).

Jose (2003) reported induction of cell wall degrading enzymes namely cellulase and pectinase in *A. chroococcum* strains showing good endorhizosphere occupancy (compatible strains). In contrast to the compatible strains, higher induction of plant defense enzymes *viz.* phenylalanine lyase and peroxidase was observed in case of the strains unable to invade plant roots. Qualitative as well as quantitative changes could be observed in the profile of flavonoid-like substances from roots of the inoculated wheat plants.

Chemotaxis

This is an important feature of motile microorganisms that allows navigation through various environments and towards their optimum environment. It enables them to detect nutrients and to avoid unfavorable conditions. Root exudates are an important source of nutrients for microorganisms present in the rhizosphere and participate in the colonization process of by functioning as chemoattractants for soil microorganisms. Chemotaxis towards the root exudates is considered to be the initial step in plant microbe interaction and it helps microbes to efficiently colonize the rhizosphere of a plant. This property is made use of by both plant growth promoting as well as pathogenic bacteria for the effective colonization of the plant root surface. It provides a competitive and selective advantage to the microbe, whether a plant growth promoter or a pathogen in interaction process.

Azotobacter has been reported to exhibit chemotaxis towards simple molecules like sugars, amino acids, organic acids etc. (Haneline *et al.*, 1991; Sood, 2003; Surya Kalyani, 2005). These molecules have been reported to be present in wheat root exudates (Vancura, 1964) towards which also *Azotobacter* was observed to possess chemotaxis (Surya Kalyani, 2005). Sood (2003) also reported chemotactic response of *A. chroococcum* towards vesicular arbuscular mycorrhizal roots of tomato. The chemotactic behaviour of *Azotobacter* strains was also studied using cotton (Kumar *et al.*, 2007). Analysis of the root exudates revealed the presence of sugars and simple polysaccharides, amino acids and organic acids. Differences between cotton cultivars in root exudates composition were observed which influenced chemotactic response in *Azotobacter*. Unlike the simple molecules present in the root exudates, instantaneous response of *A. chroococcum* towards flavoniod like substances present in wheat root exudates was observed indicating that communication exists between *Azotobacter* and wheat plant facilitated through specific compounds in root exudates (Surya Kalyani, 2005).

Role of *Azotobacter* in Crop Productivity

A. chroococcum and *A. vinelandii* have long been used as soil inoculants. Yield of a number of crops have been improved by inoculation with *Azotobacter* (Table 24.8). Gerlach and Vogel (1902) started the artificial inoculation of seeds of crop plants with *A. chroococcum*. They conducted pot culture experiment and observed an increase of dry matter in buck wheat by 42 per cent due to *A. chroococcum* inoculation. Kostychev *et al.* (1926) recommended the use of *Azotobacte*r to improve growth of agricultural plants and soil properties.

Table 24.8: Improvement in Crop Production Due to *Azotobacter* Inoculation

Sl.No.	Crop	Per cent Increase in Grain Yield
1.	Cotton	15-23
2.	Wheat	6-17
3.	Maize	15-20
4.	Sorghum	8-35
5.	Potato	6-14
6.	Pea	60
7.	Cabbage	33.5
8.	Rice	17.7
9.	Onion	10-17
10.	Pea	36
11.	Chickpea	19-42
12.	Finger millet	37-39
13.	Pearl millet	10-12

Beneficial effects of *Azotobacter* inoculation on cereals have been well documented. The effect of *Azotobacter chroococcum* on vegetative growth and yields of maize has been studied by numerous authors (Hussain *et al.*, 1987; Sariv *et al.*, 1987; Martinez-Toledo *et al.*, 1988; Miliv and Sariv, 1988; Nieto and Frankenberger, 1991; Mishra *et al.*, 1995; Pandey *et al.*, 1998; Radwan, 1998), as well as the effect of inoculation with this bacterium on wheat (Emam *et al.*, 1986; Rai and Gaur, 1988; Gasiv *et al.*, 1990; Sariv *et al.*, 1990a; Badiyala and Verma, 1991; Tippanavar and Reddy, 1993; Elshanshoury, 1995; Pati *et al.*, 1995; Fares, 1997).

Sheloumova (1935) reported higher yield of bean, corn and potato ranging from 10-18 percent by *Azotobacter* inoculation. Krasilnikov (1945) reported an increase in yield of wheat by 10-23 per cent, 13-19 per cent in oat and 14-27 per cent in clover under field conditions. It was found that wheat seeds inoculated with *Azotobacter* increased the number of tillers significantly (Barea and Brown, 1974). They attributed this to the growth promoting substances produced by *Azotobacter* which led to better plant differentiation and development. *Azotobacter* inoculation also significantly increased the weight of plant, grain yield and nitrogen content of plant (Konde and Desai, 1976). Apte and Shende (1981d) carried out work on wheat, maize and cotton crops inoculating them with different strains of *A. chroococcum*. Homologous selection of a pigmented strain C-2 for cotton, M-4 for maize and W-5 for wheat gave higher yield increase than non-pigmented strains. The foliar spray of *Azotobacter* significantly increased the grain and straw yield of rice crop (Kannaiyan *et al.*, 1980). Significant increases in wheat grain yield, number of tillers and dry matter accumulation and NPK uptake has been reported (Narula *et al.*, 2000; Kumar, *et al.* 2001b and Malik *et al.*, 2005). Synergistic effects of *A. chroococcum* and AMF interaction have also been observed (Narula *et al.*, 1981; Kumar *et al.*, 2001). Wheat's response to inoculation with *Azotobacter* varied with varieties and genotypes (Manske *et al.*, 1998; Manske *et al.*, 2000; Singh *et al.*, 2002; Behl *et al.*, 2003; Singh *et al.*, 2004). The genes/QTLs for response to inoculation of *A. chroococcum* for micronutrient uptake are present on 6B, 1D, 7B and 5D disomic chromosomes (Vasudeva *et al.*, 2002; Singh *et al.*, 2005).

The use of *Azotobacter* inoculation has a great potential in oilseeds. There was an increase in yield of mustard due to *Azotobacter* inoculation (Gerlach and Vogel, 1902; Schmidt, 1960). *Azotobacter* inoculation in sunflower was found to increase the yield (Badve *et al.*, 1977). Yadav *et al.* (1996) reported that germination of sunflower seeds inoculated with *A. chroococcum* strain Mac 27 was in the range of 76-99 per cent. In Brassica crops, under field conditions, yield increases upto 35 per cent have been reported (Agarwal, 1985). The maximum benefits to the plant, however, were observed in the absence of nitrogen fertilizer (Singh and Bhargava, 1994).

Various vegetable crops like tomato, brinjal, cabbage, onion, potato, radish, chillies and sweet potato responded positively to azotobacterisation (Joi and Shinde, 1976; Imam and Badaway, 1978; Khuller *et al.*, 1978; Konde *et al.*, 1980; Banerjee and Singh, 1986; Paul *et al.*, 2002a). Kumaraswamy and Madalagiri (1990) reported that tomato seedlings inoculated with *A. chroococcum* along with 20 kg P_2O_5 and 50 kg K_2O/ha gave the best quality of fruits. Casual trials conducted on sugar cane (Agarwal *et al.*, 1987; Durai and Mohan, 1991), fruit trees (Kerni and Gupta, 1986) and forest trees (Pandey *et al.*, 1986) have also shown some beneficial effects of *Azotobacter* inoculation.

Wani *et al.* (1988) observed 13.6 per cent increase in pearl millet yield over the uninoculated control with *A. chroococcum*. Seed bacterization of rainfed pearl millet with *A. chroococcum* increased grain yield by 22 per cent (Jadhav *et al.*, 1991). Goudreddy *et al.* (1989) found increased yield of sorghum by seed inoculation with *Azotobacter*.

Azotobacter inoculation response has also been observed in fibre crops like jute (Poi and Kabi, 1974) and cotton (Apte and Shende, 1981d; Pandey, 1985; Paul *et al.*, 2002b). There was significant increase in the yield of seed cotton and dry matter production. The response has been observed to be much higher in cotton as compared to cereals specifically at lower fertilizer application where fertilizer nitrogen was not added at all (Srivastava *et al.*, 1975; Pothiraj, 1979).

Co-inoculation studies with *Rhizobium* on various leguminous crops have also yielded interesting results. Synergistic effect of co-inoculation of *Azotobacter* with *Rhizobium* in pea (Paul and Verma, 1999), chickpea (Verma *et al.*, 2000) and groundnut (Rashid *et al.*, 1999) was observed. Higher pod yield, plant biomass and nodules mass were obtained. Similar studies carried out on chickpea did not show any synergistic effect of co-inoculation, however, the increase in pod yield due to azotobacterization was at par with that obtained due to rhizobial inoculation.

Azotobacter is a broad spectrum biofertilizer and can be used as inoculant for most of agricultural crops. Earlier, its utility as a biofertilizer was relegated to back ground due to its relatively low populations in the rhizosphere of the plant. But at the same time, the fact could not be overlooked that seed inoculation in several crops brought about phenomenal increases in the yields (Apte, 1978 and Meshram, 1981). Besides, its well known nitrogen nutritional function, it is now recognized to play a multiple role in helping the plant to improve its growth potential and yield and this has revived interest in this rhizobacterium.

Genetics of *Azotobacter*

Azotobacter contains more DNA than most other bacteria. The size of its genome is typical of prokaryotes. Renaturation rate of denatured DNA of *A. vinelandii* strain ATCC 12837 (strain M.S) was similar to that of denatured DNA of *E. coli* (Sadoff *et al.*, 1979). Restriction pattern of genomic DNA from *A. chroococcum* by two dimensional gel-electrophoresis summed up to 2000 Kb which is approximately half the size of *E. coli* genome (Robson *et al.*, 1984). Contrary to this report, Pulse Field Gel Electrophoretic (PFGE) analysis of *Ase I* and Dm 1 digests of genomic DNA from *A. chroococcum* M-4 showed its size was around 5300 Kb which is more than *E. coli* chromosome (4700 Kb) in size (Smith *et al.*, 1987; Manna and Das, 1994). The comparative values of chromosome size of *A. vinelandii* varied from 4500 Kb to 4700 Kb by PAGE analysis as reported respectively by Manna and Das (1993) and Maldonado *et al.* (1994).

The number of chromosome equivalents per cell has been found to be 40-80 in *A. vinelandii* UW (Sadoff *et al.*, 1979 and Nagpal *et al.*, 1989) and 20-25 in *A. chroococcum* (Sadoff *et al.*, 1979). Maldonado *et al.* (1994) found that the DNA content in *A. vinelandii* changes during its growth in rich medium. In early exponential phase, it shows low ploidy but there is an increase in DNA content during exponential phase of growth cycle. However, in stationary and late stationary phase it was found to have equivalents of 80 to 100 chromosomes per cell. No change in ploidy levels is observed in cells grown on minimal medium. The reason for such change in ploidy level in *Azotobacter* is not known, but it may be related to the fact that *Azotobacter* cell is approximately 10 fold larger than other bacteria (Sadoff *et al.*, 1979). The G + C content of *Azotobacter* DNA is 65-68 per cent (Becking, 1981).

The laboratory strains of A. *chroococcum* do not contain native plasmids. On the other hand, of eight A. *chroococcum* strains studied, including six new soil isolates, had two to six plasmids of size ranging from 5.5-200 Md (Robson *et al.*, 1984). However, the plasmids have not been encoded for specific properties.

References

Abd Alla MH (1994). Utilization of some phenolic compounds by *A. chroococcum* and their effect on growth and nitrogenous activity. *Folia Microbiol*, 39: 57–60.

Abdel-Nasser M and Moawad H (1974). Occurrence of certain autotrophic and heterotrophic soil microorganisms in the rhizosphere and rhizoplane of different plants. *Zentralbl Bacteriol Parasitenkd Infectionskr Hyg II*, 128: 397–404.

Abdel-Nasser M and Moawad H (1975). Changes in numbers of microorganisms during decomposition of root exudates in soil. *Zentral Bakt II*, 130: 738–744.

Agarwal S and Shende ST (1987). Tetrazolium reducing microorganisms inside the root of *Brassica* sp. *Curr Sci*, 56: 187–188.

Agarwal S (1985). Interaction of strains of *Azotobacter chroococcum* with cultivars of mustard (*B. juncea* and *B. napus*). *MSc Thesis*, PG School, IARI, New Delhi.

Aggarwal ML, Shishodia OPS, Khan ZA and Dayal R (1977). Response of *Azotobacter* inoculation on sugarcane yield. *Indian Sugar Crops J*, 4: 66.

Alexander M (1961). *Introduction of Soil Microbiology*. John Wiley and Sons, Inc, New York.

Allison FE and Gaddy VL (1940). Synthesis of coenzyme R by certain rhizobia and by *Azotobacter chroococcum*. *J Bacteriol*, 39: 273.

Apte R and Shende ST (1981a). Studies on *Azotobacter chroococcum* I. Morphological, biochemical and physiological characteristics of *A. chroococcum*. *Zbl Bakt II Abt*, 136: 548–554.

Apte R and Shende ST (1981b). II. Effect of *Azotobacter chroococcum* on germination of seeds of agricultural crops. *Zbl Bakt II Abt*, 136: 555–559.

Apte R and Shende ST (1981c). Establishment of *A. chroococcum* on roots of crop plants. *Zbl Bakt II*, 136: 560–562.

Apte R and Shende ST (1981d). Seed bacterization with strains of *A. chroococcum* and their effect on crop yields. *Zbl Bakt II*, 136: 637–640.

Apte R (1978). Screening of *Azotobacter* for bacterisation. *PhD Thesis*, PG School, IARI, New Delhi.

Azcorn R and Barea JM (1975). Synthesis of auxins, gibberellins and cytokinins by *Azotobacter vinelandi* and *Azotobacter beijerinckii* related to effects produced on tomato plants. *Plant Soil*, 43: 609–619.

Badve DA, Konde BK and More BB (1977). Effect of azotobacterization in combination of different levels of nitrogen on yield of sunflower (*Helianthus annuus*) Laboratory studies. *Food Farming Agric*, 8: 23.

Bajpai PD and Sundara Rao WVB (1971). Phosphate solubilizing bacteria. *Soil Sci Plant Nutri*, 17: 41–53.

Balajee S and Mahadevan A (1990). Utilization of chloroaromatic substances by *Azotobacter chroococcum*. *Syst App Microbiol*, 13: 194–198.

Bali A, Blanco G, Hill S and Kennedy C (1992). Excretion of ammonium by *nif L* mutant of *Azotobacter vinelandii* fixing nitrogen. *Appl Environ Microbiol*, 58: 1711–1718.

Banerjee VS and Singh MDK (1986). Effect of different organic manures and biofertilizers on growth and yield of potatoes. *Indian Agriculturist*, 30: 117.

Banic S and Dey BK (1981). Phosphate-solubilizing microorganisms of a lateritic soil. 1. Solubilization of inorganic phosphates and production of organic acids by microorganisms, isolated in sucrose calcium phosphate agar plates. *Zentralbl Bakt Abt II*, 136: 478–486.

Bansal RK, Dahiya RS, Lakshminarayana K, Suneja S, Anand RC and Narula N (1999). Effect of rhizospheric bacteria on plant growth of wheat infected with *Heterodera avenae*. *Nematol Med*, 27: 311–314.

Barea JM and Brown ME (1974). Effects on plant growth produced by *Azotobacter paspali* relating substances. *J Appl Bacteriol*, 37: 583–593.

Barea JM, Ocampo JA, Azcon R, Olivares J, Montoya (1978). Effects of ecological factors on the establishment of *Azotobacter* in the rhizosphere. *Ecol Bull*, 26: 325–330.

Becking JH (1981). The family *Azotobacteraceae*. In: *The Prokaryotes*, (Eds) MP Starr, H Stolp, HG Truper, A Balous, HG Schlegel. Springer Verlag, Berlin, Heidelberg, New York, pp 794–817.

Behl RK, Sharma H, Kumar V and Narula N (2003). Interaction amongst mycorrhiza, *Azotobacter chroococcum* and root characteristics of wheat varieties. *J Agron Crop Sci*, 189: 151–153.

Beijerinck MW (1901). Über ologonitrophile mikroben. *Zentralbl Bakteriol Parasitenkd Infektionskr II Abt*, 7: 561–582.

Beijerinck MW (1908). Fixation of free atmospheric nitrogen by *Azotobacter* in pure culture. *Koninel Ned Acad Weteucchap Prac*, p. 11.

Bela S, Dahiya P and Laksminaryana K (1986). Isolation and characterization of mutants of *Azotobacter chroococcum* derepressed for nitrogenase. In: *Current Status of Biological Nitrogen Fixation*, (Eds) R. Singh, HS Nainawatee and SK Sawhney. Bombay-Food and Agri Commun, DAE, Government of India and Hisar-Haryana Agri University Publishers, pp. 31.

Beltyukova (1953). Effects of Azotobacterin on the vulnerability of farm plants to bacterial diseases. *Izad V Akad, Naukukr SSR Kie V*. pp. 123–134.

Beniwal MS, Karwasara SS, Lakshminarayana K and Narula, N (1996). Integrated management of flag smut of wheat. In: *Resource Management in Agriculture*, (Eds) RC Dogra, RK Behl and AL Khurana. CCS HAU, Hisar and MMB, New Delhi, pp. 151–157.

Berezova EF (1939). Bacteria effective against fungal diseases of agricultural plants. *Mikrobiologiya (USSR)*, 8: 186–197.

Bershova OI and Kozlova IA (1965). Action of various combination of vitamins by soil microorganisms. *Kiev Nauk Dumkq*, pp. 87–90.

Bhardwaj KKR and Gaur AC (1970). The effect of humic and fulvic acids on the growth and efficiency of nitrogen fixation of *A. chroococcum*. *Folia Microbiol*, 15: 364–367.

Bhide VP and Purandare AG (1979). Occurrence of *Azotobacter* within root cells of *Cyanodon dactylon*. *Curr Sci*, 48: 913–914.

Bishop PE and Joerger RD (1990). Genetics and molecular biology of alternative nitrogen fixation system. *Ann Rev Plant Physiol Plant Mol Biol*, 41: 109–125.

Bishop PE, Jarlenski DM and Heterington DR (1982). Expression of an alternative nitrogen fixation system in *Azotobacter vinelandii*. *J Bacteriol*, 150: 1244–1251.

Bishop PE, Jarlenski DML and Hetheringtion DR (1980). Evidence for an alternative nitrogen fixation system in *Azotobacter vinelandii. Proc Natl Acad Sci, USA*, 77: 7342–7346.

Brakel J and Hilger F (1965). Etude qualitative et quantitative de la synthese de substances de nature auxinique par *Azotobacter chroococcum in vitro. Bull Inst Agron Stns Rech Gembloux*, 33: 469–487.

Breed RS, Murray EGD and Smith NR (1957). *Bergey's Manual of Determinative Bacteriology*. Will and Wilk, Baltimore.

Brown ME, Burlingham SK and Jackson RM (1962). Studies on *Azotobacter* species in soil. II. Populations of *Azotobacter* in the rhizosphere and effects of artificial inoculation. *Plant Soil*, 17: 320.

Brown ME, Burlingham SK (1968). Production of plant growth substances by *Azotobacter chroococcum. J Gen Microbiol*, 53: 135–144.

Brown ME, Jackson RM and Burlingham SK (1968). Growth and effects of bacteria introduced into the soil. In: *Ecology of Soil Bacteria*, (Ed) TRG Cray and Parkinson, Liverpool, p. 531–551.

Bukatsch F and Heitzer J (1952). Contributions to the knowledge of physiology of *Azotobacter. Arch Mikrobiol*, 17: 79.

Burlingham SK (1964). Growth regulators produced by *Azotobacter* in culture media. *Ann Rep Rothampstead Exp Stat*, pp. 92.

Burns TA, Bishop PE and Israel DW (1981). Enhanced nodulation of leguminous plant roots by mixed culture of *Azotobacter vinelandii* and *Rhizobium*. *Pl Soil*, 62: 399–412.

Burris RH (1972). Nitrogen fixation: Assay methods and techniques. In: *Methods in Enzymology*, (Eds) SP Colowick and NO Kaplan. Academic Press, New York, 24: 415–423.

Chahal, PPK and Chahal VPS (1988). Biological control of root-knot nematode of brinjal (*Solanum melongena* L.) with *Azotobacter chroococcum*. In: *Advances in Plant Nematology*, (Eds) MA Maqbool, AM Golden, A Gbaffilr, LR Krusberg. In: *Proceedings of the US–Pakistan International Workshop on Plant Nematology*, April 6–8, 1986, Karachi, Pakistan, pp. 257–263.

Chakrabarti DK and Yadav AL (1991). Effect of *Azotobacter* species on incidence of downy mildew (*Poronospora arborescens*) and growth and yield of opium poppy (*Papavar somniferum*). *Indian J Agric Sci*, 61: 287.

Chan ECS, Katzhelson H and Rovatt JW (1963). The influence of soil and root extracts on the associative growth of selected soil bacteria. *Can J Microbiol*, 9: 187–197.

Channal HT, Algawadi AR, Bharama Gowder TD, Udupa SG, Patil PL and Mannikeri M (1989). *Azotobacter* population as influenced by soil properties in some soils of Northern Karnataka. *Curr Sci*, 58: 70–71.

Chisnelle JR, Premkumar R and Bishop PE (1988). Purification of a second alternative nitrogenase from a nif HDK deletion strain of *A. vinelandii. J Bacteriol*, 170: 27–33.

Corbin JL and Bulen WA (1969). The isolation and identification of 2,3-dihydrobenzoic acid and 2-N, 6-N-di(2,3-dihydroxy benzoyl)-L-Lysine formed by iron-deficient *Azotobacter vinelandii. Biochem*, 8: 757–762.

Cote GL and Krull LH (1988). Characterization of extracellular polysaccharides from *Azotobacter chroococcum. Carbohyd. Res.*18: 143–152.

De Ley J (1968). DNA base composition and classification of some more free-living nitrogen-fixing bacteria. *Antonie Von Leewenhoek*, 32: 6–16.

Deubel A and Merbach W (2005). Influence of microorganisms on phosphorus bioavailability in soils. In: *Soil Biology, Vol 3: Microorganisms in Soils + Roles in Genesis and Functions*, (Eds) F Buscot and A Varma. Springer-Verlag, Berlin, Heidelberg, pp. 177–191.

Dhar NK and Arora SK (1968). Effect of algae and *Azotobacter* on carbon nitrogen transformation in Gangapur soil. *Proc Natl Acad Sci, India*, 38: 439–455.

Döberainer J (1974). Nitrogen fixing bacteria in the rhizosphere. *Biological Nitrogen Fixation*, 33: 86–120.

Dobereiner J, Day JM and Dart PJ (1972). Nitrogenase activity and oxygen sensitivity of the *Paspalum notatum– Azotobacter paspalum* association. *J Gen Microbiol*, 71: 103.

Dobereiner J (1953). *Azotobacter* em solos acidos. *Bol Inst Ecol Exp Agr*, 11: 1–36.

Dobereiner J (1966). *A. paspali* sp. n. Uma bacteria fixadora, de nitrogenio na rizosfera de *Paspalum*. *Presq Agropec Bras*, 1: 357–365.

Durai R and Mohan JR (1991). Study on *Azotobacter* in economizing fertilizer requirement in sugarcane. *Co-operative Sugar*, 22: 599–600.

El-Essawy AA, El-Sayed MA and Mohamed YAH (1984). Production of cyanocobalamine by *A. chroococcum*. *Zentrabl fur Mikrobiol*, 139: 335–342.

Eweda E and Vlassak K (1988). Seed inoculation with *Azospirillum brasilense* and *A. chroococcum* on wheat and maize growth (Egypt). *Annals Agri Sci Ainshams University, Egypt*, 32: 833–856.

Fallik E, Chan YK and Robson RL (1991). Detection of alternative nitrogenase in aerobic gram-negative nitrogen fixing bacteria. *J Bacteriol*, 173: 365–371.

Federov MV (1944). The effect of corn root excretions on atmospheric nitrogen fixation by *A. chroococcum* grown in pure culture. In: *Azotobacter and its Use in Agriculture*, (Ed) LI Rubenchick (1963). National Science Foundation, Washington DC, pp. 165.

Fekete FA, Lanzi RA, Beaulien JB, Longcope DC, Sulay AW, Hayer RN and Mabott GA (1989). Isolation and preliminary characterization of hydroxamic acids formed by nitrogen-fixing *Azotobacter chroococcum* B-8. *Appl Environ Microbiol*, 55: 298–305.

Fekete FA, Spence JT and Emery T (1983). Siderophore produced by nitrogen–fixing *Azotobacter vinelandii* in iron–limited continuous culture. *Appl. Environ. Microbiol*. 46: 1297–1300.

Fyfe JAM and Govan JRW (1983). Synthesis, regulation and biological function of bacterial alginate. *Prog Industr Microbiol*, 18: 45–83.

Gahlot R and Narula N (1996). Degradation of 2,4-Dichlorophenoxy acetic acid by resistant strains of *Azotobacter chroococcum*. *Indian J Microbiol*, 36: 141–143.

Gaofeng W, Hong X and Mei J (2004). Biodegradation of chlorophenols, a review. *Chemical Journal on Internet*, 10: 67. www.chemistrymag.on/cji/2004/.

Gaur AC and Mathur RS (1966). Stimulating influence of humic substances on nitrogen fixation by *Azotobacter*. *Sci Cult*, 32: 319.

Gerlach M and Vogel J (1902). Nitrogen fixing bacteria. *Zentrabl Bakt Abt*, 2: 817.

Gonzales-Lopez J, Martinez-Toledo MV, Reina S, Salmeron V (1991). Root exudates of maize on production of auxins, gibberellins, cytokinins, amino acids and vitamins by *Azotobacter chroococcum* chemically defined media and dialysed soil media. *Toxicol Environ Chem*, 33: 69–78.

Gonzalez-Lopez J and Vela GR (1981). True morphology of the *Azotobacteraceae* filterable bacteria. *Nature*, 289: 588–590.

Gonzalez-Lopez, J, Martinez-Toledo MV, Rodelas B and Salmeron V (1999). Effect of some herbicides on the production of lysine by *Azotobacter chroococcum*. *Aminoacids*, 17: 165–173.

Gordon JK and Brill WJ (1972). Mutants that produce nitrogenase in the presence of excess ammonia. *Proc Natl Acad Sci*, USA, 69: 3501–3503.

Gordon JK and Jackson MR (1999). Isolation and characterization of *Azotobacter vinelandii* mutant strains with potential as bacterial fertilizer. *Can J Microbiol*, 29: 973–978.

Gordon JK (1981). In: *The Prokaryotes: A Handbook on Habitats, Isolation and Identification of Bacteria*. (Eds) MP Starr, H Stolp, HG Truper, A Balous, HG Schlegel. SpringerVerlag, Berlin/Heidelberg/New York, pp. 781–794.

Goudreddy BS, Patil VS, Radder GD and Chittapur BM (1989). Response of *rabi* sorghum to nitrogen FYM and *Azotobacter* under irrigated condition. *J Maharasthra Agric Univ*, 14: 266–268.

Govedarica M, Miliv V, Gvozdenoviv DJ (1993). Efficiency of the association between *Azotobacter chroococcum* and some tomato varieties. *Plant Soil*, 42: 113–120.

Haaker H and Veeger C (1977). Involvement of the cytoplasmic membrane in nitrogen fixation by *A. vinelandii*. *Eur J Biochem*, 77: 1–11.

Haddock BA and Jones CW (1977). Involvement of the cytoplasmic membrane in nitrogen fixation by *Azotobacter vinelandii*. *Bacteriol Rev*, 431: 47–99.

Hallberg B (1995). Nitrogen fixing bacteria associated with maizes native of Oaxaca. In: *Proceedings of the 10th International Congress on Nitrogen Fixation*, St Petersburg, Russia, No. 638.

Hammad AMM (1998). Evaluation of alginate encapsulated *Azotobacter chroococcum* as a phage–resistant and an effective inoculum. *J Basic Microbiol*, 1: 9–16.

Haneline S, Conelly CJ and Melton T (1991). Chemotactic behavior of *Azotobacter vinelandii*. *Appl Environ Microbiol*, 57: 825–859.

Hardisson C, Sala-Trepat JM and Stanier RY (1969). Pathways for the oxidation of aromatic compounds by *Azotobacter*. *J Gen Microbiol*, 59: 1–11.

Harper SHT and Lynch JM (1980). Microbial effects on the germination and seedling growth of barley. *New Phytol*, 84: 473–481.

Harvey I, Arber JM, Eady RR, Smith BE, Garner CD and Hasnain SS (1990). Iron K-edge X-ray absorption spectroscopy of the iron-vanadium cofactor of the vanadium nitrogenase from *A. chroococcum*. *Biochem J (London)*, 266: 929–931.

Hennequin JR, Blachere H (1966). Recherches sur la synthese de phytohormones et de composes phenoliques par *Azotobacter* et des bacteries de la rhizosphere. *Ann Inst Pasteur*, 3: 89–102.

Horan NJ, Jarman TR and Dawes EA (1983). Studies on some enzymes of alginic acid biosynthesis in *Azotobacter vinelandii* grown in continuous culture. *J Gen Microbiol*, 129: 2985–2990.

Horner CK, Burk D, Allison FE, Sherman MS (1942). Nitrogen fixation by *Azotobacter* as influenced by molybdenum and vanadium. *J Agric Res*, 65: 173–193.

Illmer P and Schinner F (1995). Solubilization of inorganic calcium phosphates solubilization mechanisms. *Soil Biol Biochem*, 27: 257–263.

Imam MK and Badaway FH (1978). Response of three potato cultivars to inoculation with *Azotobacter. Potato Res*, 21: 1–6.

Iswaran V and Subba Rao NS (1967). *Azotobacter chroococcum* in the aerial roots of the Banyan tree. *Nature*, 214: 814.

Iswaran V (1958). Effect of humus of legume and non-legume origin on the nitrogen fixation in *Azotobacter chroococcum*. *Curr Sci*, 27: 489–490.

Iswaran, V and Sen, A (1959). *Azotobacter chroococcum* in the roots of Cyperus grass. *Curr Sci*, 28: 449.

Jackson FA and Dawes EA (1976). Regulation of tricarboxylic acid and poly-β-hydroxy butyrate metabolism in *Azotobacter beijerenckii* grown under nitrogen or oxygen metabolism. *J Gen Microbiol*, 97: 303–312.

Jackson RM, Brown ME, Burlingham SK (1964). Similar effects on tomato plants of *Azotobacter* inoculation and application of gibberellins. *Nature*, 203: 851–852.

Jadhav AS, Shaikh AA and Harinaryana G (1991). Response of rainfed pearlmillet (*Pennisetum glaucum*) to inoculation with nitrogen fixing bacteria. *Ind J Agric Sci*, 61: 268–271.

James E (1970). Strain variation in *A. chroococcum*. *MSc Thesis*, PG School, IARI, New Delhi.

Jensen H (1955). The *Azotobacter*–flora of some Danish water courses. *Dansk Bot Tidsskr*, 52: 143–157.

Jensen HL (1951). Notes on the biology of *Azotobacter*. *Proc Soc Appl Bacteriol*, 14: 89–94.

Joerger RD, Elizabeth DW and Bishop PE (1991). The gene encoding dinitrogenase reductase 2 is required for expression of the second alternative, nitrogenase from *Azotobacter vinelandii*. *J Bacteriol*, 173: 4440–4446.

Johnson MK, Magee LA (1956). Some factors affecting the respiratory response of *Azotobacter* to 24–D, and related compounds. *Appl Microbiol*, 4: 169–178.

Joi MB and Shinde PA (1976). Response of onion crops to Azotobacterization. *J Maharasthra Agric Univ*, 1: 161.

Jose TC and Paul S (2004). Biochemical studies on endorhizospheric establishment of *Azotobacter chroococcum* in wheat. In: *Proc. of National Symposium on Geoinformatics Applications for Sustainable Development*, IARI, New Delhi on 17–19 February.

Jose TC (2003). Biochemical studies on endorhizopheric establishment of *Azotobacter chroococcum* in wheat. *MSc Thesis*, PG School, IARI, New Delhi.

Kannaiyan S, Govindarajan K and Lewin HD (1980). Effect of foliar spray of *Azotobacter chroococcum* on rice crop. *Plant Soil*, 56: 487–490.

Kerni PN and Gupta A (1986). Growth parameters affected by Azotobacterization of mango seedlings in comparison to different nitrogen doses. *Res Develop Factor*, 3: 77.

Khariton EG (1953). *Symposium on Problems of the Bacterial Fertilizers.* Tzv. USSR No. 5.

Khonde BK and Desai JN (1976). Influence of doses of *Azotobacter* on growth and yield of wheat. *MSc Thesis*, MPKV, Ahmednagar, Maharashtra.

Khuller S, Chahal VPS and Kaur PP (1978). Effect of *Azotobacter* inoculation on chlorophyll content and other characters of carrot, radish, brinjal and chillies. *Indian J Microbiol,* 18: 138.

Kleiner D and Kleinschmidt JA (1976). Selective inactivation of nitrogenase in *Azotobacter vinelandii* batch cultures. *J Bacteriol,* 128: 117–122.

Kloepper JW and Schroth MN (1978). Plant growth-promoting rhizobacteria on radishes. In: *Proc of the 4th Internat Conf on Plant Pathogenic Bacter.* 2, Station de Pathologie Vegetale et Phytobacteriologie, INRA, Angers, France. pp. 879–882.

Kole MM, Altasaar J, Page WJ (1988). Distribution of *Azotobacter* in eastern Canadian soils and in association with plant rhizospheres. *Can J Microbiol,* 34: 815–817.

Konde BK, Desai JN and More BB (1980). Studies on associative action of *Azotobacter* and seed borne fungi on onion (*Allium cepa*). *Food Farmg Agric,* 17: 170.

Koptyeva JP, Tanstyurenko OV and Smyrmova MM (1970). Nitrogen fixing ability of blue green algae of paddy fields in the south of USSR. *Microbiol Zh,* 32: 707–714.

Kostychev SP, Sheloumova AM and Shulgina OG (1926). Investigation of the biodynamics of soils. *Microbiologia,* 1: 5–46.

Krasilnikov NA (1945). The microbiological principles of bacterial fertilizers. *Izad. AN USSR* (Published by the Academy of Sciences of the USSR, Moscow).

Krasilnikov NA (1949). Opredeliteli Bakterii i Aktinomicetov. Izd. AN sssr.M.

Kukreja K, Suneja S, Goyal S and Narula N (2004). Phytohormone production by *Azotobacter*: A review. *Agricul Rev,* 25: 70–75.

Kumar L (1992). Interaction between *A. chroococcum* and cultivars of wheat (*Triticum aestivum* L.). *MSc Thesis*, PG School, IARI, New Delhi.

Kumar L (1997). Siderophore producing *Azotobacter chroococcum* strains and their rhizospheric colonization. *PhD Thesis*, PG School, IARI, New Delhi.

Kumar RK, Bhatia R, Kukreja K, Behl RK, Dudeja SS and Narula N (2007). Establishment of *Azotobacter* on plant roots: chemotactic response, development and analysis of root exudates of cotton (*Gossypium hirsutum* L.) and wheat (*Triticum aestivum* L.). *J Basic Microbiol,* 47: 436–439.

Kumar V and Narula B (1999). Solubilization of inorganic phosphates and growth emergence of wheat as affected by *A. chroococcum* mutants. *Biol Fert Soils,* 28: 301–305.

Kumar V, Aggarwal NK and Singh BP (2000). Influence of analogue resistant mutants of *A. chroococcum* solubilizing phosphate on yield and quality of sunflower. *Folia Microbiol,* 45: 343–347.

Kumar V, Behl RK and Narula N (2001a). Establishment of phosphate solubilizing strains of *Azotobacter chroococcum* in rhizosphere and their effect on wheat under green house conditions. *Microbial Res,* 156: 87–93.

Kumar V, Behl RK and Narula N (2001b). Effect of phosphate solubilizing strains of *Azotobacter chroococcum* on yield traits and their survival in the rhizosphere of wheat genotypes under field conditions. *Acta Agron Hungarica,* 49: 141–149.

Kumaraswamy D and Madalagiri BB (1990). Effect of *Azotobacter* inoculation on tomato. *South Indian Hort*, 38: 345–346.

Lakshmi Kumari M, Vijayalakshmi K and Subba Rao NS (1972). Interaction between *Azotobacter* sp. and fungi. *In vitro* studies with *Fusarium monoliformae* Sheld. *Phytopath Z*, 75: 27–30.

Lakshminaryana K, Bela S, Sandhu SS, Kumari P, Narula N and Sheoran RK (2000). Analogue resistant mutants of *Azotobacter chroococcum* derepressed for nitrogenase activity and early ammonia excretion having potential as inoculants for cereal crops. *Indian J Exptl Biol*, 38: 373–378.

Lee M, Breckenridge C and Knowles R (1970). Effect of some culture conditions on the production of indole-3-acetic acid and gibberellin-like substance by *Azotobacter vinelandii*. *Can J Microbiol*, 16: 1325–1330.

Lee SB, Burris RH (1943). Large scale production of *Azotobacter*. *Industr Ind Eng Chem*, 35: 112–121.

Leuck EE and Rice EL (1976). Effect of rhizosphere bacteria of *Aristida oligantha* on *Rhizobium* and *Azotobacter*. *Bot Gaz*, 137: 160–164.

Li F, Liu R, Wang CC, Wand YZ (1995). Research on distribution of ^{32}P labeled nitrogen fixing bacteria in plant. In: *Proceedings of the 10th International Congress on Nitrogen Fixation*, St Petersburg, Russia, No. 623.

Li YD, Eberspacher J, Wagner B, Kuntzer J and Lingens F (1991). Degradation of 2,4,6-Trichlorophenol by *Azotobacter* sp. strain GP1. *Appl Environ Microbiol*, 57: 1920–1921.

Lipman CB, Mac Lees E (1940). Dissociation of *Azotobacter chroococcum* (Beijerinck). *Soil Sci*, 50: 75–82.

Lipman JG (1903). *Report on the New Jersey Agricultural Experiment Station*, 24: 217–285.

Lynch JM and Whipps JM (1990). Substrate flow in the rhizosphere. *Plant Soil*, 129: 1–10.

Lynch JM (1990). Beneficial interactions between microorganisms and roots. *Biotechnol Adv*, 8: 335–346.

Maldonado R, Jimenez J and Casadesus J (1994). Changes of ploidy during the *Azotobacter vinelandii* growth cycle. *J Bacteriol*, 176: 3911–3919.

Malik BS, Paul S, Sharma RK, Sethi AP and Verma OP (2005). Effect of *Azotobacter chroococcum* on wheat (*Triticum aestivum*) yield and its attributing components. *Indian J Agric Sci*, 75: 600–602.

Malik RK (1992). Studies on interaction between strains of *A. chroococcum* and genotypes of Egyptian cotton (*Gossypium barbadense* L.). *PhD Thesis*, PG School, IARI, New Delhi.

Manna AC and Das HK (1993). Determination of the size of *Azotobacter chroococcum* chromosome. *Mol Gen Genet*, 241: 719–722.

Manna AC and Das HK (1994). The size of the chromosome of *Azotobacter chroococcum*. *Microbiol*, 140: 1237–1239.

Manske GGB, Behl RK, Luttger AB and Vlek PLG (1998). Enhancement of mycorrhizal (AMF) infection, nutrient efficiency and plant growth by *Azotobacter chroococcum* in wheat: Evidence for varietal effects. In: *Azotobacter in Sustainable Agriculture*, (Ed) N Narula. CBS Publisher and Distributors, New Delhi, pp. 136–147.

Manske GGB, Qritz-Monasterio JI, Van Ginklel M, Gozzalez RM, Rajaram S, Molina E and Vlek PLG (2000). Traits associated with improved P-uptake efficiency in CIMMYTs semi dwarf spring bread wheat grown on an acid Andisol in Mexico. *Plant Soil*, 221: 189–204.

Marathe GV and Rangaswmi G (1973). Influence of foliar chemical sprays on microorganisms in rhizosphere of three crop plants. *Indian J Exptl Biol,* 11: 468–469.

Martin JP Ervin JO and Shepherd RA (1965). Decomposition and binding action of polysaccharides from *Azotobacter indicus* and other bacteria in soil. *Soil Sci Soc Amer Proc,* 29: 397–399.

Martinez-Toledo MV, Gozalez-Lopez J, De la Rubia T, Moreno J, Ramos Cormenzana A (1988). Effect of inoculation with *Azotobacter chroococcum* on nitrogenase activity of *Zea mays* roots grown in agricultural soils under aseptic and non-sterile conditions. *Biol Fert Soils,* 6: 170–173.

Martinez-Toledo MV, Salmeron V and Gonzalez-Lopez J (1990). Effect of *Azotobacter* inoculation on nitrogenase activity of *H. vulgare. Chemosphere,* 21: 243–250.

Meshram SU and Jager G (1983). Antagonism of *Azotobacter chroococcum* isolates to *Rhizoctonia solani. Neth J Pl Path,* 89: 191–197.

Meshram SU (1981). Effect of *Azotobacter chroococcum* on maize crop. *PhD Thesis,* PG School, IARI, New Delhi.

Meshram SU (1984). Suppressive effect of *Azotobacter chroococcum* on *Rhizoctonia solani* infestation of potatoes. *Neth J Pl Path,* 90: 127–132.

Mezei S, Popoviv M, Kovaev L, Mrkovaki N, Nagl N, Malenov D (1997/98). Effect of *Azotobacter* strains on sugar beet callus proliferation and nitrogen metabolism enzymes. *Biol Plantarum,* 40: 277–283.

Micev N (1971). Results of investigation of free nitrogen fixation (*Azotobacterium*) in the rhizosphere of some plants and soil types of Macedonia. *Agrohemija,* 7–8: 309–317.

Miliv V, Mrkovaki N (1995). Production of growth substances gibberellin type in strains of *Azotobacter chroococcum.* In: *Proceedings of VII Congres Microbiologists of Yugoslavia,* Herceg Novi, p. 28.

Mishustin EN and Shilnikova VK (1969). Free living nitrogen fixing bacteria of the genus *Azotobacter* In: *Soil Biology, Reviews of Research,* UNESCO Publication, pp. 72–124..

Mishustin EN and Shilnikova VK (1971). *Biological Fixation of Atmospheric Nitrogen.* MacMillan, London.

Moreno J, Gonzalez-Lopez J and Vela GR (1986). Survival of *Azotobacter* sp. in dry soil. *Appl Environ Microbiol,* 51: 123–125.

Moreno J, Vargas-Garcia C, Lopez, MJ and Sanchez-Serrano (1999). Growth and exopolysaccharide production by *Azotobacter vinelandii* in soil phenolic compounds. *J Appl Microbiol,* 86: 439–445.

Mrkovaki N, Mezei S, Kovaev L, Sklenar P, Miliv V (1997a). Number of *Azotobacter chroococcum* in soil and rhizophere of inoculated sugar beet plants. In: *Papers of the IX Congress of the Yugoslav Society of Soil Science,* Novi Sad, pp. 443–448.

Mrkovaki N, Mezei S, Kovaev L, Sklenar P (1998a). The effect of inoculation of sugar beet with *Azotobacter chroococcum* on the bacteria's number on the root and in the rhizosphere. *Arch Biol Sci,* 50: 189–193.

Mrkovaki N, Mezei S, Kovaev L (1995a). Specific relationship between *Azotobacter* strains and sugar beet plants. *Soil Plant,* 44: 9–17.

Mrkovaki N, Ov N, Mezei S, Miliv V (1998b). Effect of inoculation on the number of *Azotobacters* in soil and rhizosphere during sugarbeet growing season. *Acta Agricult Serbica,* 5: 53–59.

Mulder EG and Brontonegoro (1974). Free-living heterotropic nitrogen-fixing bacteria, pp 37–85. In: *Biology of Nitrogen Fixation*, (Ed) A Quispel. North Holland, Amsterdam.

Nagpal D. Jafri S, Reddy MA and Das HK (1989). Multiple chromosomes of *A. vinelandii*. *J Bacteriol*, 171: 3133–3138.

Nagraj R (1989). Occurrence of *Azotobacter* and *Beijerinckia* in forest soils of Maharashtra (India). *Indian J Forestry*, 12: 112–116.

Narula N, Kukreja K, Kumar V and Lakshmi Narayana K (2002). Phosphate solubilization by soil isolates of *Azotobacter chroococcum* and their survival at different temperatures. *J Agril Trop Subtropics*, 103: 81–87.

Narula N, Kumar V, Behl RK, Deubel A, Gransee A. and Merbach W (2000). Effect of P-solubilizing *A. chroococcum* on N, P, K uptake in P responsive wheat genotypes under green house conditions. *J Plant Nutr Soil Sci*, (Germany), 163: 393–398.

Narula N, Lakshminarayana K and Tauro P (1980). Field evaluation of *Rhizobium japonicum* and soybean varieties by acetylene reduction method. *Indian J Microbiol*, 20: 298–301.

Narula N, Lakshminarayana KL and Tauro P (1981). Ammonia excretion by *Azotobacter chroococcum*. *Biotech Bioeng*, 23: 467–470.

Naumova AN (1958). Survival of *Azotobacter* in podsol soils. Kiev. *Akad Nauk Ukr, SSR.*

Negi M, Sadasivam KV and Tilak KVBR (1987). Establishment of *Azotobacter* and *Azospirillum* in the rhizosphere of barley (*Hordeum vulgare* L.) in organic amended soils. *Zentrabl Mikrobiol*, 142: 149–154.

Newton JW, Wilson PW and Burris RH (1953). Direct demonstration of ammonia as an intermediate in nitrogen fixation by *Azotobacter*. *J Biol Chem*, 204: 445–451.

Nieto KF and WT, Frankenberger Jr (1991). Influence of adenine, isopentyl alcohol and *A. chroococcum* on the vegetative growth of *Zea mays*. *Plant Soil*, 135: 213–222.

Nieto KF, and Frankenberger WT (1989). Biosynthesis of cytokinins by *Azotobacter chroococcum*. *Soil Biol Biochem*, 21: 967–972.

Nieto KF, Frankenberger WT (1991). Influence of adenine, isopentyl alcohol and *Azotobacter chroococcum* on the vegetative growth of Zea mays. *Plant Soil*, 135: 213–221.

Ota H, Kurihara Y, Satoh S and Eashi Y (1991). Development of acetylene reduction (N_2 fixation) activity on and around imbibed plant seeds. *Soil Biol Biochem*, 23: 9–14.

Page WJ and Sadoff HL (1975). Relationship between calcium and uronic acids in the encystment of *Azotobacter vinelandii*. *J Bacteriol*, 122: 145–151.

Page WJ and Shivprasad S (1991). *Azotobacter salinestris* sp. nov a sodium dependent, microaerophilic and aeroadaptive nitrogen fixing bacterium. *Int J Syst Bacteriol*, 41: 369–376.

Page WJ and Von Tigerstrom M (1988). Aminochelin a catecholamine siderophore produced by *Azotobacter vinelandii*. *J Gen Microbiol*, 134: 453–502.

Page WJ, Collinson, SK, Demange P, Dell A and Abdullah MA (1991). *Azotobacter vinelandii* strains of disparate origin produce azotobactin siderophores with identical structures. *Biol Metals*, 4: 217–222.

Page WJ (1987). Iron dependent production of hydrxamate by sodium dependent *Azotobacter chroococcum*. *Appl Environ Microbiol*, 53: 1418–1424.

Page WJ. and Huyer M (1984). Derepression of *Azotobacter vinelandii* siderophore system using iron containing minerals to limit iron repletion. *J Bacteriol*, 158: 496–502.

Pandey A and Kumar S (1990). Inhibitory effect of *Azotobacter chroococcum* and *Azospirillum brasilense* on a range of rhizosphere fungi. *Indian J Exp Biol*, 28: 52–54.

Pandey A, Shende ST and Apte RG (1989). Effect of *Azotobacter chroococcum* seed inoculation on its establishment in rhizosphere, on growth and yield attributing parameters of cotton (*G. hirsutum*). *Zentrabl Microbiol*, 114: 596–604.

Pandey A (1985). Interaction between strain of *Azotobacter chroococcum* and varieties of cotton (*Gossypium hirsutum*) and wheat (*Triticum aestivum*). *PhD Thesis*, PG School, IARI, New Delhi.

Pandey A and Kumar, S (1990). Inhibitory effects of *Azotobacter chroococcum* and *Azospirillum brasilense* on a range of rhizospheric fungi. *Indian J Exptl Biol*, 28: 52–54.

Pandey R.K, Bahl RK and Rao PRT (1986). Growth stimulating effects of nitrogen fixing bacteria on oak seedling. *India Forester*, 112: 75.

Patriquin DG and Dobereiner J (1978). Light microscopy observations of tetrazolium reducing bacteria in the endorhizosphere of maize and other grasses in Brazil. *J Microbiol*, 24: 7343.

Paul S and Verma OP (1999). Influence of combined inoculation of *Azotobacter* and *Rhizobium* on the yield of chickpea (*Cicer arietinum* L.). *Indian J Microbiol*, 39: 249–251.

Paul S, Paul B and Verma OP (2002c). Effect of *Azotobacter chroococcum* on lepidopteran insects. *New Bot*, 29: 163–168.

Paul S, Verma OP and Rathi MS (2002b). Potential of homologous and heterologous *Azotobacter chroococcum* strains as bioinoculant for cotton. *New Bot*, 29: 169–174.

Paul S, Verma OP, Rathi MS and Tyagi SP (2002a). Effect of *Azotobacter* inoculation on seed germination and yield of onion (*Allium cepa*). *Annals Agril Res*, 23: 297–299.

Paul S. and Anupama KS (2008). Potential of *Azotobacter chroococcum* for lindane degradation. In: *Proc. of First International Conference on Agrochemicals Protecting Crop, Health and Natural Environment*, IARI, New Delhi on 8–11 Janruary.

Petrenko GY (1961). Factors influencing the symbiotic relationships of *Azotobacter* and higher plants. *Microorganism and Effective Soil Fertility*, 11: 11.

Phillips DH and Johnson J (1961). *J Biochem Microbiol Technol Engg*, 3: 277–309.

Poi SC and Kabi MC (1979). Effect of *Azotobacter chroococcum* inoculation on the growth and yield of jute and wheat. *Indian J Agric Sci*, 49: 478.

Polumuri SK (1989). Studies on initial stages of interaction between plant and *A. chroococcum*. *MSc Thesis*, PG School, IARI, New Delhi.

Polumuri SK (1995). Interaction of strains of *Azotobacter chroococcum* with cultivars of maize. *PhD Thesis*, PG School, IARI, New Delhi.

Pothiraj P (1979). Effect of *Azotobacter* on the yield of rainfed cotton. *Madras Agric J*, 66: 70.

Premakumar R, Chisnell JR and Bishop PE (1988). A comparison of the three dinitrogenase reductases expressed by *A. vinelandii*. *Can J Microbiol*, 35: 344–348.

Prsa M (1963). The cyclic development of *Azotobacters* isolated from terra rossa of Instria. *Bull Sci Council Acad*, RSFY, 8: 89.

Raieviv V, Sariv M, Sariv Z, Bogdanoviv V (1995a). *Azotobacter* movement in maize root. In: *Proceedings of the 10th International Congress of Nitrogen Fixation*, St. Petersburg, Russia, No. 607.

Raieviv V, Sariv M, Sariv Z, Bogdanoviv V (1995b). Colonization and adsorption of some *Azotobacter* strains in maize root. In: *Proceedings of the 10th International Congress of Nitrogen Fixation*, St. Petersburg, Russia, No. 608.

Rangaswami G and Sadasivam KV (1964). Studies on the occurrence of *Azotobacter* in some soil types. *J Indian Soc Soil Sci*, 12: 43–49.

Rao BV (1991). Bacterial flora in the rhizosphere of sugarcane plant (*Saccharum officinarum*). *World J Microbiol Biotechnol*, 7: 431–432.

Rashid *A*, Musa M, Aadal NK, Yaqub M and Chaudhary GA (1999). Response of groundnut to *Bradyrhizobium* and a diazotrophic bacteria inoculum under different levels of nitrogen. *Pak J Soil Sci*, 16: 89–98.

Reddy MVR, Reddy TKR (1989). Competititive saprophytic survival of *Azotobacter chroococcum* in black cotton soil. *Curr Sci*, 58: 139–140.

Reliv B, Govedarica M, Nes¢kovis¢ M (1987). Plant hormone activity in *Azotobacter* culture. In: *Book of Abstracts of VIII Simposum of Yugoslav Society of Plant Phisiology*.

Reliv B (1989). Plant horomone activity in *Azotobacter* culture and effect on wheat. *Master Thesis*, Faculty of Natural Sciences, University Novi Sad.

Requena N, Baca TM and Azcon R (1997). Evolution of humic substances from unripe compost during incubation with lignolytic/cellulolytic microorganisms: effects on the lettuce growth promotion mediated by *Azotobacter chroococcum*. *Biol Fert Soils*, 24: 59–65.

Revillas JJ, Rodelas B, Pozo C, Martinez-Toledo MV, Gonzalez Lopez J (2000). Production of B group vitamins by two *Azotobacter* strains with phenolic compounds as sole carbon source under diazotrophic and adiazotrophic conditions. *J Appl Microbiol*, 89: 486–493.

Roberts GP and Brill WJ (1981). Genetics and regulation of nitrogen fixation. *Ann Rev Microbiol*, 35: 207–235.

Robson RL and Postgate JR (1980). O_2 and hydrogen in biological nitrogen fixation. *Ann Rev Microbiol*, 34: 183–207.

Robson RL., Cheshyre JA, Wheeler C, Jowes R, Woodley PR and Postgate JR (1984). Genome size and complexity in *A. chroococcum*. *J Gen Microbiol*, 130: 1603–1612.

Rodelas B, Gonzalez-Lopez J and Martinez-Toledo MV (1997). Production of vitamins by soil diazotrophic microorganisms– Recent Research Developments in soil. *Biol Biochem*, 1: 39–45.

Rubenchik LI, Smaly VT, Zinovyeva KG and Bershnova OI (1965). Formation of vitamins by microorganisms of the rhizosphere of agricultural plants. In: *Role of Microorganisms in Plant Feeding and Increasing the Efficiency of Fertilizers*. Leningrad, pp. 14–21.

Rubenchik LI (1960). *Azotobacter and its Use in Agriculture*. Kier, AA Ukr, SSR.

Rubia de la T, Gonzalez-Lopez J, Moreno J, Martinez-Toledo MV and Ramos-Cormenzana A (1989). Isolation and characterization of *Azotobacter* species from roots of *Sorghum bicolor*. *Microbias,* 57: 113–119.

Ruiz MT, Cejudo FJ and Paneque A (1990). Effect of divalent cations on the short term NH_4^+ inhibition of nitrogen fixation in *Azotobacter chroococcum*. *Arch Microbiol,* 154: 313–316.

Sadoff HL, Shieurei B and Elisa S (1979). Characterization of *A. vinelandii* deoxyribonucleic acid and folded chromosomes. *J Bacteriol,* 138: 871–877.

Saikia N and Bezbaruah B (1995). Iron dependent plant pathogen inhibition through *Azotobacter* RRLJ 203 isolated from iron rich acid soil. *Indian J Exptl Biol,* 33: 571–575.

Salmeron V, Martinez-Toledo MV, Gonzalez-Lopez J (1990). Nitrogen fixation and production of auxins, gibberellins and cytokinin by an *Azotobacter chroococcum* strain isolated from root of *Zea mays* in presence of insoluble phosphate. *Chemosphere,* 20: 417–422.

Samitsevich SA (1962). Preparation, use and effectiveness of bacterial fertilizers in the ukrainian SSR. *Mikrobiologiya,* 31: 923–933.

Sardaryan EO (1972). Phenol compound produced by *Azotobacter chroococcum*. *Biol Zh Arm,* 25: 31–35.

Sarita S, Priefer UB, Prell J and Sharma PK (2008). Diversity of *nifH* gene amplified from rhizosphere soil DNA. *Curr Sci,* 94: 109–115.

Sariv Z, Rassoviv B (1963a). The influence of the maize on the dynamic of *Azotobacter* in the soil. *Soil Plant,* 13: 273–277.

Sariv Z, Rassoviv B (1963b). The effect of some plants on the dynamics of *Azotobacter* in the soil. *Annals Scient Work Faculty Agric,* Novi Sad, 7: 1–11.

Sariv Z (1969a). Biogenic levels of the horizons of calcerous chernozem in Vojvodina. *Contemp Agric,* 17: 819–825.

Sariv Z (1969b). Biogenity of limeless chernozem inVojvodina. *Annals Scient Work the Instt Res Agric,* Novi Sad, 7: 145–151.

Sariv Z (1978). The influence of mineral fertilizers on the population of *Azotobacter* and oligonitrophilic bacteria in chernozem. *Mikrobiology,* 15: 153–166.

Schmidt OC (1960). *Azotobacter* inoculation. *Soil Fertilizers,* 12: 1368.

Schroth MN and Hancock JG (1982). Disease suppressive soil and root colonizing bacteria. *Science,* 216: 1376–1381.

Sevinc MS and Page WJ (1992). Generation of *A. vinelandii* strains defective in siderophore production and characterization of a strain unable to produce known siderophores. *J Gen Microbiol,* 138: 587–596.

Sharma S, Jain N and Shah AK (1985). Occurrence of *Azotobacter chroococcum* in *Pathos scandens*. *Curr Sci,* 54: 142–143.

Sheloumova AM (1935) The use of *Azotobacter* as a bacterial manure for non-leguminous plants. *Bull State Inst Agril Microbiol* (USSR), 6: 48.

Shende ST and Apte R (1982). *Azotobacter* inoculation: A highly remunerative input for agriculture. In: *Biological Nitrogen Fixation, Proceedings of the National Symposium* IARI, New Delhi, pp. 532–543.

Shende ST, Apte RG and Singh T (1975). Multiple action of *Azotobacter*. *Indian J Gen Plant Breeding*, 35: 314.

Shende ST, Apte RG and Singh T (1977). Influence of *Azotobacter* on germination of rice and cotton seeds. *Curr Sci*, 49: 675–676.

Shende ST, Singh M and Singh VP (1987). Effect of seed bacterization with *Azotobacter chroococcum* on germination and seedling height of upland cotton. *Indian J Agric Sciences*, 58: 206–209.

Shende ST (1965). Role of Azotobacteria in producton of rice. Work of the People's friendship Univ. Moscow Ser. Agr. Sci. No. I, M.

Shethna TJ, Dervatanian DV and Beinert H (1968). Non-heme (Iron sulphur) proteins of *Azotobacter vinelandii*. *Biochem Biophys Res Commun*, 31: 862–868.

Shivaprasad and Page WJ (1989). Catechol formation and melanization by Na^+ dependant *Azotobacter chroococcum*: a protective mechanism for aeroadaptation. *Appl Environ Microbiol*, 55: 1811–1817.

Sidorov EF (1954). The effect of Azotobacterin on yield of potatoes. *Dokl Akad S Kh Nauk*, (Cited *Soils Fert*, 17, 1952).

Sindhu SS and Lakshminarayana K (1986). Distribution of *Azotobacter* in Haryana soils and effect of bacteriostasis on *Azotobacter* survival. *Environ Ecol*, 4: 536–540.

Singh KP, Behl RK, Bansal R and Sharma SK (2002). Genotypes for warmer areas by exploiting supernumerary spikelet genes. In: *Proc. 2nd International Group Meeting on Wheat Technologies for Warmer Areas*, Agharkar Research Institute, Pune, India, September 23–26, pp 15.

Singh P and Bhargava SC (1994). Changes in growth and yield components of *Brassica napus* in response of *Azotobacter* inoculation at different rates of nitrogen application. *J Agric Sci*, 122: 241–247.

Singh R, Behl RK, Singh KP, Jain P and Narula N (2004). Performance and gene effects for wheat yield under inoculation of arbuscular mycorrhiza fungi and *Azotobacter chroococcum*. *Pl Soil Environ*, 50: 409–415.

Singh SK, Behl RK, Lokesh, Singh A and Narula N (2007). Variability among Disomic Chromosome Substitution Lines of Wheat for content and uptake of important micronutrients. In: *Crop Production in Stress Environments: Genetic and Management Options*, (Eds) DP Singh, VS Tomar, RK Behl, SD Upadhyaya, MS Bhale and D Khare. Agrobios (International), Jodhpur.

Singh T (1977). Studies on interaction between *Azotobacter chroococcum* and some plant pathogens. *PhD Thesis*, IARI, New Delhi.

Smith CL, Econome JG, Schutt A, Kleos and Cantor CR (1987). A physical map of the *E. coli* K_{12} genome. *Science*, 236: 1448–1453.

Sood SG (2003). Chemotactic response of plant-growth-promoting bacteria towards roots of vesicular-arbuscular mycorrhizal tomato plants. *FEMS Microbiol Ecol*, 45: 219–227.

Srivastava SP, Singh AP and Dave PV (1975). Note on bacterial fertilization in dry land agriculture. *Sci Cult*, 41: 344.

Stajner D, Kevresan S, Gasic O, Mimica-Dukic N and Zongli H (1997). Nitrogen and *Azotobacter chroococcum* enhance oxidative stress tolerance in sugarbeet. *Biologia Plantarum*, 39: 441–445.

Staunton S and Leprince F (1996). Effect of pH and some organic anions on the solubility of soil phosphate+ implications for P bioavailability. *Eur J Soil Sci*, 47: 231–239.

Stoklasa J (1908). The chemical changes involved in the assimilation of free nitrogen by *Azotobacter* and *Radiobacter*. *Centr Bakt Par*, 21: 484–509.

Suneja S and Lakshminarayana A (2001). Isolation of siderophore negative mutants of *Azotobacter chroococcum* and studies on the role of siderophores in mustard yield. *Indian J Plant Physiol*, 6: 190–193.

Suneja S and Lakshminarayana K (1993). Production of hydroxamate and catechol siderophores by *Azotobacter chroococcum*. *Indian J Expt Biol*, 31: 878–881.

Suneja S, Narula N, Anand RC and Lakshminarayana K (1996). Relation of *Azotobacter chroococcum* siderophores with nitrogen fixation. *Folia Microbiol*, 4: 154–158.

Surya Kalyani S (2005). Chemotaxis of *Azotobacter* sp. towards wheat root exudates. *MSc Thesis*, PG School, IARI, New Delhi.

Tchan YT (1984). *Azotobacteraceae*. In: *Bergey's Manual of Systematic Bacteriology*, Vol. 1, (Eds) Krieg JG, Holt. Williams and Wilkins, Baltimore, London, pp. 219–225.

Thakur JK (2007). Utilization of aromatic hydrocarbons by *Azotobacter chroococcum*. *MSc Thesis*, PG School, IARI, New Delhi.

Thompson JP and Skerman VBD (1979). *Azotobactericeae: The Taxonomy and Ecology of the Aerobic Nitrogen-fixing Bacteria*. Academic Press, London.

Thompson JP, Skerman VBD (1981).Validation list No 6. *Int J Syst Bacteriol*, 31: 215–218.

Thompson JP (1989a). Counting viable *Azotobacter chroococcum* in vertisoils 1: Comparison of media. *Plant Soil*, 117: 9–16.

Thompson JP (1989b). Counting viable *Azotobacter chroococcum* in vertisoils 2: The non-proportionality phenomenon. *Plant Soil*, 117: 17–29.

Thompson JP (1989c). Counting viable *Azotobacter chroococcum* in vertisoils 3: Methods for preparation of soil suspensions. *Plant Soil*, 117: 31–40.

Tikhe PR, Purandare AG and Kaephanis RN (1980). Occurrence of intracortical *Azotobacter chroococcum* in some monocots. *Curr Sci*, 49: 794.

Troitskii NA, Novitskaya MA and Troitskaya TM (1989). Associative *Azotobacter chroococcum* strains in rhizophil bacteria fertilizers. *Soviet Biotechnol*, 5: 79–83.

Vancura V and Macura J (1961). The effect of root excretions on *Azotobacter*. *Folia Microbiol*, 6: 250.

Vancura V.1964. Root exudates of plants: Analysis of root exudates of barley and wheat in their initial stages of growth. *Plant Soil*, 21: 231–248.

Vasudeva M, Behl RK, Karb P, Yadava RK, Narula N, Merbach W and Vashisht R K (2002). Response of disomic lines of wheat with *A. chroococcum* grown under soil types. In: *Proceedings 2nd International Conference on Sustainable Agriculture for Food, Energy and Industry*, (Ed) Li Dajue, Beijing, China

Veeger D, Laane C, Shering G, Matz L, Haaker H and Vanjeeland Wolbers L (1980). In: *Nitrogen Fixation*, (Eds) Newton, WE and Ornae Johnson, WH, 7: 208–274.

Verma OP, Paul S and Rathi MS (2000). Synergistic effect of co-inoculation of *A. chroococcum* and *Rhizobium* on pea (*Pisum sativum*). *Ann Agril Res*, 21: 418–420.

Verma S, Kumar V, Narula N and Merbach W (2001). Studies on *in vitro* production of antimicrobial substances by *Azotobacter chroococcum* isolates/mutants. *J Pl Diseases Protect*, 108: 152–165.

Villegas J and Fortin JA (2002). Phosphorus solubilization and pH changes as a result of the interactions between soil bacteria and arbuscular mycorrhizal fungi on a medium containing NOÀÛ-Ü as nitrogen source. *Can J Bot*, 80: 571–576.

Vojinoviv Z (1961). Mikrobiological properties of main types soils in Serbia for nitrogen cycling. *J Scient Agril Res*, 43: 3–25.

Wani SP, Chandrapalaih S, Zambre MA and Lee KK (1988). Association between nitrogen fixing bacteria and pearl millet. *Plant Soil*, 110: 289–302.

Wani SP, Shinde PA and Konde BK (1976). Response of rice (*Oryza sativa* L.) to *Azotobacter* inoculation. *Curr Res*, 5: 209.

Weller DM (1988). Biological control of soil borne plant pathogens in the rhizosphere with bacteria. *Ann Rev Phytopath*, 26: 379–407.

Whitelaw MA (2000). Growth promotion of plants inoculated with phosphate-solubilizing fungi. *Adv Agron*, 69: 100–153.

Winogradsky S (1938). Etudes sur la microbiologie du sol et des eaux. sur la morphologie et l'oecologie des *Azotobacter*. *Ann Inst Pasteur*, 60: 351–400.

Winogradsky SN (1932). Sur la synthese de l' ammoniaq par les *Azotobacters* du sol. *Ann Inst Pasteur*, 48: 269–300.

Yadav KS, Suneja S and Sharma HR (1996). Seed bacterization studies with *Azotobacter chroococcum* in sunflower (*Helianthus annuus*). *Crop Res, Hisar*, 11: 239–243.

Yatskuneue AB (1962). Increasing the vitamin content of carrots by the action of *Azotobacter*. *Bull nt. Inf. Lit. N. i. in t Zivotnovodstva*, 3: 46–50.

Zacheri B and Amberger A (1990). Effect of nitrification inhibitors on nitrogen–fixing bacteria *Rhizobium leguminosarum* and *Azotobacter chroococcum*. *Fert Res*, 22: 137–139.

Zaremba VP and Sinyevskaya IA (1954). *Nauch. Tr. Ukr. –1 inst. Sots. Zemledeliya*, 7.

Zinovyeva KG (1959). Interrelationships between *Azotobacter* and higher plants. *Thesis Kiev*, 1959.

Zinovyeva KG (1962). *Azotobacter* and agricultural plants. Kiev. Sel' hozgiz, Ukr. SSR.

Zuanazzi JS, Clergeot PH, Quirion JC, Husson HP, Kondorosi A and Ratet P (1998). Production of *Sinorhizobium meliloti* nod gene activator and repressor flavonoids from *Medicago sativa* roots. *Mol Plant Microbe Interact*, 11: 784–794.

Zucker M (1965). Induction of phenylalanine deaminase by light and its reaction to chlorogenic acid synthesis in potato tuber tissue. *Plant Physiol*, 40: 779–784.

Soil Microflora, 2009 *Pages* ***315–323***
Editor: **Rajan Kumar Gupta, Mukesh Kumar & Deepak Vyas**
Published by: **DAYA PUBLISHING HOUSE, NEW DELHI**

Chapter 25

Prospects and Potential of *Azolla-Anabaena* System

G. Abraham, Sudheer Saxena and Dolly Wattal Dhar*
Centre for Conservation and Utilization of Cyanobacteria,
Indian Agricultural Research Institute, New Delhi – 110 012

ABSTRACT

Nitrogen fixing systems are ecologically sound and considerably reduce external inputs by improving the internal resources. It is in this context that the biological nitrogen fixation assumes great significance in the low land rice ecosystem that provides about 80 per cent of the World's rice. An *Azolla-Anabaena* system plays a very crucial role in the nitrogen balance in low land rice cultivation. In addition to nitrogen fixation the system adds organic matter to the soil and improves properties of the soil and offer tremendous potential for the future due to their eco-friendly nature and economic feasibility. In addition to the biofertilizer potential *Azolla* has many other uses that could be exploited successfully. The present article briefly highlights the importance of *Azolla* as biofertilizer and strategies for successful exploitation of this organism as an efficient biofertilizer.

Keywords: *Azolla-Anabaena, Biofertilizer, Nitrogen fixation, Rice-paddy fields*

Introduction

Rice is staple food for several millions of people in India. Availability of water and nitrogen are two key factors that lead to higher yields in rice crop. Since the soil nitrogen status is poor more and

* Corresponding Author: E-mail: gabraham1@rediffmail.com

more nitrogen in the form of chemical fertilizers need to be added to the soil to increase yield. According to an estimate the rice plant absorbs about 20 Kg N/t of paddy produced. The ever increasing population, demand for more food grains, dwindling prices of the crude in the International market and the depleting soil health has compelled Agricultural Scientists to think about other viable alternatives such as biofertilizers as nutrient inputs. *Azolla* is an important biofertilizer for flooded rice and it can grow along with rice (Figures 25.1 and 25.2). The aquatic pteridophyte *Azolla* is able to fix atmospheric nitrogen due to the presence of heterocystous cyanobacterium *Anabaena azollae* (Moore, 1969). This system is agronomically important because of the ability to fix atmospheric nitrogen at a rate of 1.1 Kg N ha^{-1} per day (Watanabe *et al.*, 1977). The cyanobacterium is confined to the dorsal leaf cavity of the fern and the cavity is formed by enfolding of the adaxial epidermis in the dorsal leaf lobe. *Azolla* has a global distribution and occurs in fresh water habitats of tropical, sub-tropical and warm temperate regions. Rice paddy fields are an ideal habitat for the growth and multiplication of *Azolla*. According to Lumpkin and Plucknett (1980) rapid multiplication and quick decomposition coupled with wide adaptability makes this fern superior to other green manures. According to Watanabe *et al* (1977) *Azolla pinnata* double its biomass in a span of 3-5 days under optimum laboratory conditions while the doubling time taken in the field are generally more. Wagner (1997) has listed many uses of this symbiotic system such as animal feed, human food, medicine, hydrogen fuel, production of biogas, water purifier, weed control and reduction of ammonia volatilization after application of chemical nitrogen and called it as "green gold mine". *Azolla* can even absorb P from the soil and accumulate it and after the death of the plant this P gets released in to soil. Therefore in addition to the property of supplying N to the rice crop it can increase the soil organic matter and fertility also. The present article describes the biology, taxonomy and production of *Azolla* for biofertilizer purposes.

Distribution

Svenson (1944) has reviewed the distribution of *Azolla* species and it has been reported to occur throughout the World inhabiting a range of ecosystems such as freshwater, temperate and tropical regions. Luxuriant growth was reported in the rice paddy fields. *Azolla pinnata* is the most widely distributed species in India. Several strains of *Azolla* have been maintained at Centre for conservation and utilization of blue green algae, Indian Agricultural Research institute, New Delhi and Central Rice Research Institute, Cuttack.

Taxonomy

The genus *Azolla* belongs to the monotypic family Azollaceae and many extant and fossil species have been reported (Hills and Gopal, 1967; Konar and Kapoor, 1972). Similarity in the vegetative characters makes the identification process complex. Svenson (1944) categorized *Azolla* into two subgenera *viz. EuAzolla* and *Rhizosperma* on the basis of the number of megaspore floats. The subgenus *EuAzolla* is characterized by the presence of three megaspore floats and consists of species such as *A. caroliniana, A. filiculoides, A. mexicana, A. rubra and A. microphylla*. In contrast the subgenus Rhizosperma consists of nine megaspore floats. *A. pinnata* and *A. nilotica* belong to this subgenus. Another approach to classify *Azolla* was on the basis of the morphological variabilities (Lumpkin and Plucknett, 1982). Variation in the rhizome surface has also been used as a taxonomic criterion in delineating the subgenus (Nayak and Singh, 1988). Somatic chromosome number of the species of *Azolla* was employed by Nayak and Singh (1989) as a valid taxonomic tool. Basic chromosome number of species belonging to all sections of *Azolla* and *Rhizosperma* has been employed by Sterigianou and Fowler (1990) to resolve taxonomic problems related to *Azolla*. The basic chromosome number in all species of the section as well as the section *Rhisosperma* is n =22 except in *A. nilotica* where n =26. The taxonomic assignment of *Azolla* is difficult because many accessions do not form sporocarps under culture conditions. Eskeu *et al* (1993) used DNA Amplification Finger printing (DAF) to distinguish between closely related accessions of *Azolla*.

Algal Symbiont

The nitrogen fixing symbiotic cyanobacterium is associated with the dorsal lobes of *Azolla* from the time of development and the symbiotic association is maintained throughout the life cycle of *Azolla*. Strasburger (1873) named the symbiotic algae as *Anabaena azollae*. It belongs to the family Nostocaceae and the order Nostocales (Shen, 1960). Electron microscopic studies by Grilli (1964) showed the development of heterocysts from the vegetative cells. Nierzwicki and Kannaiyan (1990) conducted molecular and ultrastructural studies on *Azolla-Anabaena* system and observed that the symbiont actually enters into the leaf cavity of the young sporophyte that emerges from the megaspore apparatus. The development of symbiosis by the partners has been found to be synchronous (Calvert and Peters, 1981). The symbiont contains chlorophyll a, phycobilins and carotenoids. However, the contribution of the total CO_2 fixed is not more than 5 per cent of the intact association. Ray *et al*. (1978) observed all enzymes associated with assimilation of ammonia in the crude extracts of *Azolla* grown with and without *A. azollae*.

Nitrogen Fixing Potential of *Azolla*

Nitrogen fixing potential of *Azolla* strains has been exploited successfully to use it as an efficient biofertilizer for rice paddy fields. The nitrogen fixing potential of the system has been reported to be 1.1 Kg N ha^{-1} per day (Watanabe *et al*., 1977). Singh (1988) reported that *A. pinnata* fixed 75 mg N g^{-1} dry weight day^{-1} and produced a biomass of 347 t fresh weight ha^{-1} in a year. This had contained 868 Kg N which was present in 1900 Kg urea. One crop of *Azolla* provides 20-40 Kg N/ha to the rice in about 20-25 days. A study conducted at Central rice Research Institute, Cuttack showed wide variability regarding growth and nitrogen fixation among different strains of *Azolla*. According to Kannaiyan (1993) and Singh and Singh (1995) the growth and nitrogen fixing potential of *Azolla* was determined by nutrient availability, rate of application and time of inoculation etc.

Production Technology

It is quite easy to multiply *Azolla* on commercial production by setting up small or large production units as per the requirement. Generally the production is carried out in nursery ponds, ponds/ditches/canals and concrete tanks/polythene lined ditches (Figure 25.3). The selected field for *Azolla* cultivation needs to be thoroughly prepared and leveled uniformly. 20 m × 2 m size plots can be made from the field with suitable bunds and irrigation channels. At least 10 cm depth of water is to be maintained. Fresh cow dung (10 Kg) mixed with water (20 liters) must be added in each plot and *Azolla* (8-10 Kg) is inoculated in each plot. Single super phosphate (100 g) is applied in 2-3 split doses at an interval of 4 days. Furadon or carbofuran (3 per cent active granules) can be applied in the plots (100 g/plot) after a week of inoculation. It is possible to harvest 100-150 Kg fresh *Azolla* from each plot after 15 days from each plot. The same methodology can be extended to production in bigger plots or and the quantity of the in puts have to be varied accordingly. It can also be maintained in nursery in trays or earthen or cemented pots of any dimension depending on the availability and need of the inoculum. There is no need to add any inputs like fertilizer or insecticide if the production is carried out in a pond or canal.

Figure 25.1: ***Azolla*** **Biomass**

Figure 25.2: ***Azolla*** **in Paddy Fields**

Source**: International** ***Azolla*** **News, 2008**

Figure 25.3: ***Azolla*** **Cultivation in Ponds Near Rice Fields**
Source**: International** ***Azolla*** **News, 2008**

Method of Application

Generally the mode of application in the field is as green manure or as dual crop along with rice. In case of application as green manure *Azolla* collected directly from ponds/ditches is applied in the field. It may be grown in nurseries as specified earlier and can also be applied in the field. After application a thick mat of *Azolla* will be formed in about 2-3 weeks time which can be incorporated in the soil. Subsequently rice can also be transplanted in the field. Single super phosphate (25-50 Kg/ha) is applied in split doses and the dosage of the same can be reduced after analyzing the soil P-status. Instead of single super phosphate cattle dung or slurry may also be used. Pest control must be done in case of pest infestation or attack. It was observed that this type of application leads *Azolla* contributing 20-40 Kg N/ha.

Dual cropping involves application and growth of *Azolla* along with rice (Figure 25.2). Each crop of *Azolla* in dual cropping contributes on an average 30 Kg N/ha. Fresh inoculum of azolla is applied in the field after 7-10 days of transplantation at the rate of 0.50-1.0 ton/ha. Single super phosphate is applied at the rate of 20 Kg/ha in split doses. A thick mat of *Azolla* will be formed in about 15-20 days time. Incorporated *Azolla* decompose in about 8-10 days and is able to release the fixed nitrogen. Similarly another crop of *Azolla* can be raised in the similar way during the crop cycle of rice.

The production and maintenance of *Azolla* is simple and not very expensive. This technology is very efficient in terms biomass accumulation and nitrogen fixation. The rice growing season is also conducive for the growth of *Azolla* plants and the dual application does not have any negative influence on the rice crops. *Azolla* is also used an excellent poultry, fish and cattle feed (Singh and Subudhi, 1978; Chu, 1987; Parthasarathy *et al.*, 2002; Kamalasanan Pillai *et al.*, 2002). Therefore once the infrastructure is being created for the multiplication the unit it will become self sustainable by providing the biomass for agriculture as well as feeding the poultry, fish and cattle. The poultry and cow dung are valuable in puts in the farming system and the farmers can benefit from this cheap and inexpensive technology that may go a long way in the coming years in view of the depleting soil health and high cost of fertilizers.

Factors Affecting Growth and Nitrogen Fixation

Water

Since *Azolla* is a free floating plant it requires water for its successful establishment and growth. A water depth of 5-10 cm is recommended for good growth but depths up to 30 cm does not adversely affect the growth (Singh, 1989). However, lack of water will lead to complete drying and death of the plant. Arora and Singh (2003) compared growth and nitrogen fixing capacity of six different species of *Azolla viz. A. pinnata, A. microphylla, A. filiculoides, A. caroliniana, A. rubra* and *A. mexicana* to ascertain their growth potential and nitrogen fixation. It was observed that *A. microphylla* performed better in terms of growth and nitrogen fixation.

Mineral Nutrients

It requires all the essential plant nutrients for growth and the deficiency of any of these elements will lead to reduction in growth and nitrogen fixation. It is an efficient scavenger of potassium and can serve as source of K for rice in K-deficient soils. When the plant decomposes it acts indirectly as a K-fertilizer (Van Hove, 1989). Phosphurus is a key element involved in the growth (Watanabe *et al.*, 1980) and its absence may lead to poor growth, reduced N content, pink or red coloration and curling of the roots. According to Vaishampayan (1992) use of *Azolla* as a source of nitrogen is severely limited by its high P requirement for replication and nitrogen fixation. Cobalt and molybdenum are necessary for efficient functioning of the N-fixing system. Dawar and Singh (2002) observed comparable results when they compared nutrient and soil based cultures of *Azolla*.

Light

Growth and nitrogen fixation of *Azolla* is influenced by both quality and intensity of light. Photosynthesis has been carried out by both the symbiont and the host to maintain a high degree of nitrogen fixation. The length of the day also affects the growth. During summers it prefers to grow under certain degree of shade.

Temperature

The response to the temperature varies with different species of *Azolla*. It is an important environmental factor that limits the growth of *Azolla*. Optimum temperature for growth is 25-30°C but it grows successfully up to a temperature of 40°C. Better growth is generally observed during July to December. A strain of *Azolla viz. A. filiculoides* can tolerate low temperature up to 5°C. Extreme temperature will result in intense reddish brown coloration.

pH and Salinity

Soil as well as media based cultures have been proposed for the growth of Azolla (Nickel, 1961; Malavolta *et al.*, 1981). Although it can survive in the pH range of 3.5-10.0, the optimum pH for growth of *Azolla* is 5-8. pH influences the availability of nutrients and thus influences the growth. Soil generally acts as a good buffering agent whereas pH of the medium rises later due to accumulated debris of decayed plant material and roots ((Dawar and Singh, 2001). Salinity also affects the growth adversely. Varietal differences in response to salt have been observed by many workers (Levan and Sobochkin, 1963; Ge Shi-an *et al.*, 1980; Rai and Rai., 1999; Masood and Abraham, 2006; Masood *et al.*, 2006).

Insects and Pests

Successful growth and productivity of *Azolla* is also influenced by efficient management of pests and insects. Singh (1977) identified the pests to lepidopterous and dipterous orders and they attack the plant in summer as well as winter seasons. Well developed *Azolla* mat may be completely affected and destroyed by pests in 3-5 days (Kannaiyan, 1982). They can be easily controlled by the application of common pesticides like carbofuran. However, good cultural practices such as always using healthy inoculum and avoiding overcrowding of *Azolla* mat can help to minimize the infection by pests.

Future Perspectives

Azolla is able to adapt and grow well in wet land rice fields and significantly contribute in the nitrogen economy by fixing nitrogen. Biological nitrogen fixation by *Azolla* is an important process as it provides nitrogen to increase the productivity of crop plants. In addition to this the plants decompose rapidly in the soil and supplies fixed nitrogen to the crop. Mineral ions such as phosphorus, iron, potassium, zinc, molybdenum and other micronutrients are also released to the soil on decomposition. *Azolla* is eco-friendly and organic acids released during the mineralization process may enhance the availability of phosphorous in the soil. The contamination of the ecosystem due to heavy metals and salinity is a major concern. Recently considerable interest has been generated on phyoremediation using *Azolla*. Arora and Saxena (2005) successfully cultivated *A.microphylla* biomass in secondary treated Municipal waters of Delhi. Tolerance and phytoremediation potential of three different species of *Azolla* to chromium has also been carried out by Arora *et al* (2006). This is a new area of applied research where *Azolla* biomass could be employed successfully to clean the polluted environment of the heavy metals. Further, the stratospheric levels of UV-B are also increasing due to the depletion of ozone layer. This has resulted in decreased growth and productivity of many aquatic plants including *Azolla* (Masood *et al.*, 2008). Scientists working on climate change have predicted gross changes in the climatic pattern and possible changes in the cropping pattern. Therefore research should be focused on screening and selection of *Azolla* strains with high biomass production and nitrogen fixing potential and tolerant to abiotic stresses. It is very difficult to observe the sporulation in *Azolla* strains. Control of the induction of sporulation is also a difficult task. This problem is a real constraint in the development of hybrid strains of *Azolla*. Recently in Tamilnadu Agricultural University, Coimbatore, Prof. Kannaiyan and his team successfully developed a hybrid *Azolla* strain which is highly adaptive and has high nitrogen fixing potential. The hybrid was developed by sexual hybridization. Similarly they have screened a strain of *A. microphylla* which is able to produce high biomass and tolerate high temperatures and salinity. According to Roger (1989) the area under *Azolla* cultivation is only 2 per cent of the total rice area of the world. Extension machinery must be strong enough to disseminate the benefits of the technology and the Government must come forward to provide incentives to farmers utilizing bio-manure like *Azolla*. Regular training programs and front line demonstrations need to be conducted for farmers for successful adoption of this technology. Innovations in molecular biology may be also be used for strain improvement. Precise knowledge of the sporulation and symbiosis employing the most modern genetic tools must be aimed at improving the performance of the symbiosis. *Azolla*, is an organism with tremendous potential and it is able to enrich soil organic matter, soil enzymes and the soil microbial population. However, it's potential has not yet been fully exploited due to lack of awareness of the benefits of the system.

Acknowledgement

Financial Assistance from Indian Council for Agricultural Research is gratefully acknowledged. We also thank the Director and Joint Director (Research) for encouragement.

References

Arora A and Singh PK (2003). Comparison of biomass productivity and nitrogen fixing piotential of *Azolla* spp. *Biomass and Bioenergy*, 24: 175–178.

Arora A and Saxena S (2005). Cultivation of *Azolla microphylla* biomass on secondary treated Delhi Municipal effluents. *Biomass and Bioenergy*, 29(1): 60–64.

Arora A, Saxena S and Sharma DK (2006). Tolerance and phytoaccumulation of chromium by three *Azolla* spp. *World J Microbiol Biotechnol*, 22(2): 97–100.

Calvert HE and Peters GA (1981). The *Azolla–Anabaena* relationships. ix. Morphological analysis of leaf cavity hair population. *New Phytol*, 89: 327–355.

Chu LC (1987). Re-evaluation of *Azolla* utilization in agricultural production. In: *Proceedings of the Workshop on Azolla Use*, Fuzhou, Fujian, China, International Rice Research Institute, Philippines, p. 67–76.

Dawar S and Singh PK (2001). Growth, nitrogen fixation and occurance of epiphytic algae at different pH in the cultutres of two species of *Azolla*. *Biol Fert Soils*, 34: 210–214.

Dawar S and Singh PK (2002). Comparison of soil and nutrient based medium for maiuntenance of *Azolla* cultures. *J Plant Nutr*, 25(12): 2719–2729.

Eskew, DL, Caetano-Anolles G, Bassain BJ and Gresshoff PM (1993). DNA amplification finger printing of the *Azolla-Anabena* symbiosis. *Plant and Soil*, 21: 363–373

Ge Shi-an S, Dai-Xing X and Zhi-hao S (1980). Salt tolerance of *Azolla filiculoides* and its effects on the growth of paddy in Xinwa Haitu. *Zhejiang Nongye Kexue*, 1: 17–20.

Grilli M (1964). Infrastructure di *Anabaena azollae* vivente nelle foglioline de *Azolla caroliniana*. *Ann Microbiol Enzymol*, 14: 69–90.

Hills LV and Gopal B (1967). *Azolla primaeva* and its phylogenetic significance. *Can J Bot,* 45: 1179–1191.

Kamalasanan Pillai P, Prema Latha S and Rajamony S (2002). *Azolla*: A sustainable feed substitute for livestock. *LEISA,* 4(1): 15–17.

Kannaiyan S (1982). *Azolla* and rice. In: *Multiplication and Use of Azolla Biofertilizer for Rice Production.* Tamil Nadu Agricultural University, Coimbatore, Tamil Nadu, India.

Kannaiyan S (1993). Biofertilizers for Rice. Tamil Nadu Agricultural University, Coimbatore.

Konar RN and Kapoor RK (1972). Anatomical studies on *Azolla pinnata. Phytomorphology,* 22: 211–223.

Levan K and Sobochkin AA (1963). The problems on the utilization of *Azolla* as a green manure in Democratic republic of Vietnam. *Proc Timuya Zev Moscow Agric Acad* Tamil Nadu 94: 93–97.

Lumpkin TA and Plucknett DL (1980). *Azolla*: Botany, physiology and uses as a green manure. *Econ Bot,* 34: 111–153.

Lumpkin TA and Plucknett DL (1982). *Azolla* as a green manure. In: *Use and Management in Crop Production.* Westview Press, Boulder, Colarado, p. 230.

Malavolta E, Acorsi WR, Ruschel AP, Krug LJ, Nakayama LI and Fiomi I (1981). Mineral nutrition and nitrogen fixation in *Azolla.* In: *Associative Nitrogen Fixation,* (Eds) Vose PB and Ruschell AP. CRC Press, Boca Raton, FL, p. 229–231.

Masood A and Abraham G (2006). Physiological response of *Azolla pinnata* plants to salinity stress. *Roum Biotech Lett,* 11(4): 2841–2844.

Masood A, Shah NA, Zeeshan M and Abraham G (2006). Differential response of antioxidant enzymes to salinity stress in two varietries of *Azolla* (*Azolla pinnata* and *Azolla filiculoides*). *Environ Exp Bot,* 58: 216–222.

Masood A, Zeeshan M and Abraham G (2008). Response of growth and antioxidant enzymes in *Azolla* plants (*Azolla pinnata* and *Azolla filiculoides*). exposed to UV–B. *Acta Biol Hung,* 59 (2): 247–257.

Moore AW (1969). *Azolla*: Biology and Agronomic significance. *Bot Rev,* 35: 17–35.

Nayak SK and Singh PK (1988). Some variation in dermal appendages of *Azolla. Ind Fern J,* 5: 170–175.

Nayak SK and Singh PK (1989). Cytological studies in the genus *Azolla. Cytologia,* 54: 275–286.

Nickell LG (1961). Physiological studies with *Azolla* under aseptic conditions. II. Nutritional studies and the effects of chemicals on growth. *Phyton,* 17: 49–54.

Nierzwicki–Bauer SA and Kannaiyan S (1990). Molecular and ultrastructural studies of the *Azolla–Anabaena* association. In: *Biotechnology of Biofertilizers,* (Ed) S Kannaiyan. Tamil Nadu Agricultural University, Coimbatore, Tamil Nadu, India, pp. 44–58.

Parthasarathy R, Kadirvel and Kathaperumal V (2002). *Azolla* as a particle replacement for fish meal in broiler rations. *Ind Vet J,* 79(2): 144–146.

Rai V and Rai AK (1999). Growth behaviour of *Azolla pinnata* at various salinity levels and induction of high salt tolerance. *Plant and Soil,* 206: 79–84.

Ray TB, Peters GA, Toia RE and mayne BC (1978). *Azolla-Anabaena* relationship. VII. Distribution of ammonia assimilating enzymes, protein and chlorophyll between host and symbiont. *Plant Physiol*, 62: 463–467.

Roger (1989). Blue-green algae (Cyanobacteria) in agriculture. In: (Eds) Dawson JO and Dart P. Martinus Nijhoff, Dordrecht.

Shen EYF (1960). *Anabaena azollae* and its host *Azolla pinnata*. *Taiwania*, 7: 1–7.

Singh PK (1977). *Azolla* plants as fertilizer and feed. *Indian Farming*, 27: 19–22.

Singh PK (1988). Biofertilization of rice crop. In: *Biofertilizers: Potentials and Problems, Plant Physiology Forum*, (Eds) Sen SP and Palit P. Calcutta, India, pp. 109–114.

Singh PK (1989). Use of *Azolla* in Asian agricuklture. *Appl Agric Res*, 4: 149–161.

Singh PK and Subudhi BPR (1978). Utilize *Azolla* in poultry feed. *Indian Farming*, 27(10): 37–38.

Singh DP and Singh PK (1995). Influence of rate and time of *A. caroliniana* inoculation and its growth and N_2 fixation and yield of rice (*Oryza sativa*). *Ind J Agric Sci*, 65(1): 10–16.

Sterigianou KK and Fowler K (1990). Chromosome numbers and taxonomic implications in the fern genus *Azolla* (Azollaceaea). *Pl Syst Evol*, 173: 223–229.

Strasburger E (1873). Ueber *Azolla* verlag Von Ambr. *Abel Jena*, Leipzing.

Svenson I IK (1944). The new World species of *Azolla*. *American Fern Journal*, 34: 69–84.

Vaishampayan A (1992). Recent advances in the molecular biology of *Azolla–Anabaena* symbiotic N_2 fixing complex and its use in agriculture. In: *Microbes and Environment*, (Ed) Prasad AB. Narendra Publications, India, p. 1–41.

Van Hove C (1989). *Azolla and it's Multiple Uses with Emphasis on Africa*. FAO, Rome, Italy.

Wagner GM (1997). *Azolla*: A review of its biology and and utilization. *Bot Rev*, 63: 1–21.

Watanabe I, Espinas CR, Berja NS and Alimango BV (1977). Utilization of the *Azolla–Anabaena* complex as a nitrogen fertilizer for rice. *IRRI Research Paper Series*, 11: 1–5.

Watanabe I, Berja NS and Del Rosario DC (1980). Growth of *Azolla* in paddy field soils as affected by phosphorus fertilizer. *Soil Sci Plant Nutr*, 26: 301–307.

Soil Microflora, 2009 *Pages* ***324–339***
Editor: **Rajan Kumar Gupta, Mukesh Kumar & Deepak Vyas**
Published by: **DAYA PUBLISHING HOUSE, NEW DELHI**

Chapter 26

Ecological and Biotechnological Relevance of Cyanobacteria

Kaushal Kishore Choudhary
Department of Botany, Banaras Hindu University, Varanasi–221 005
E-mail: kkc1970@gmail.com, k_k_c1970@yahoo.co.in

ABSTRACT

Cyanobacteria are photosynthetic organisms that inhabit all kinds of possible biomes. They show diversity in their cellular organization as well as in their habitats. Some of them are endowed with a specialized structure "heterocyst" contained with enzyme "nitrogenase complex" that converts atmospheric nitrogen into combined nitrogen. The ability to utilize elementary nitrogen and their conversion into soluble forms of nitrogen elevated them as an experimental organism. Cyanobacteria are an important component of many soils particularly rice fields where they play a vital role in soil conditioning by maintaining the soil fertility particularly nitrogen and phosphorus. They have also been reported to enhance the soil aggregation property and water holding capacity by building up extracellular polysachharides in the soil. Besides their role in agriculture, cyanobacteria have been reported to produce various kinds of primary and secondary metabolites of biotechnological interest. These include pigments, vitamins, toxins, bioameliorating agents and bioactive compounds of pharmacological interest. Recent advances in cyanobacterial exploitation for bioactive compounds have elevated them as promising group of organisms of present day biotechnological programme. The present article has been aimed to discuss the cyanobacterial distribution and their abundance in rice field in relation to light, temperature, nitrogen and phosphorus and their role in enhancement of soil fertility particularly in terms of nitrogen and phosphorus with little emphasis on growth-promoting harmones. The different metabolites including antimicrobial, antiviral and toxins will be discussed in brief.

Introduction

Blue-green algae commonly called cyanobacteria (Haselkorn, 1978) are unique creatures of the nature which shares metabolic activity of the eukaryotes and cellular structure of prokaryotes. They are photosynthetic oxygen evolving organisms and uses water as an electron donor during photosynthesis. Some of the cyanobacteria have been endowed with the special cell structure called "heterocyst' that enables them to fix and convert atmospheric nitrogen into ammonia with the help of enzyme "nitrogenase complex" contained in the heterocyst by deriving energy from the sun (Adams and Duggan, 1999). Such tropic independence of cyanobacteria for carbon and nitrogen together with a great adaptability to variations of environmental factors enables BGA to be ubiquitous. They are characterized by the presence of pigments like c-phycocyanin and c-phycoerythrin along with chlorophyll a, carotenes and xanthophylls. Schopf and Walter (1982) reported the origin of cyanobacteria in the "proterozoic era" and called it as "Age of Cyanobacteria". They have been reported to tolerate high metal content, low oxygen content and high level of free sulphide (Padan and Cohen, 1982), UV-B and UV-C and may utilize H_2S as hydrogen donor in addition to H_2O during photosynthesis (Cohen *et al.*, 1975).

Cyanobacteria show plasticity in their morphology as well as in their genomic organization. Their vegetative forms extend from simple unicells to multiseriate, colonial to branched, nearby colorless to intensely pigmented. They show diversity in habitats ranging from psychrophilic to thermophilic, fresh water to marine, free living to endosymbionts including epiphytes. The multiplication in filamentous forms of cyanobacteria is commonly facilitated by chains of uniform cells called hormogonia (Desikachary, 1959) and is associated with dispersal of the organisms. The growth rate of cyanobacteria ranges between thousand years for isolates of Antarctica (Nienow and Friedlmann, 1993) to 2.1 h for *Anacystis nidulans* (*Synechococcus* PCC 6803–Kratz and Myers, 1955). Hormogonia are possibly always rich in stores of nitrogen, phosphorus and perhaps other nutrients (Rippka *et al.*, 1979) and play crucial roles in a number of important physiological processes in cyanobacteria (Tandeau de Marsac, 1994) including formation of colonies of some cyanobacteria (*e.g. Rivularia*) by the aggression of a number of hormogonia. Additionally, cyanobacteria develop special structure called akinetes under adverse conditions which are usually much larger than vegetative cells and are mostly formed in heterocystous forms (Whitton, 1992). Vegetative cells of unicellular cyanobacteria range in diameter from about 0.4 µm to over 40 µm (*Chroococcus turgidis* UTEX 123) whereas filamentous forms of Oscillatoriaceae have been reported to be of 100 µm. Demoulin and Janssen (1981) have attributed this difference in cell-size to short-lived m-RNA occurring in prokaryotes. Cyanobacteria contain multiple copies of the genome. Even the small-celled and morphologically simple forms, *Synechococcus* and *Synechocystis*, have been reported to contain six identical "chromosomes" (Mann and Carr, 1974; Binder and Chisholm, 1990; Castenholtz, 1992).

With establishment of cyanobacterial potential in agriculture, their role in maintaining the fertility of soil has been extensively studied. Gupta and Shukla (1969) have reported the secretion of some aminoacids like aspartic acids and glutamic acids along with alanine, vitamins-like B_{12} and auxin-like substances into the soil. Not only this, many cyanobacteria mobilize the insoluble phosphate in the soil accumulated due to excessive use of the chemical fertilizers ultimately making the soil enriched with available phosphate to crop plants (Roger *et al.*, 1987). Cyanobacteria have been also found to play a crucial role in maintaining the stability of the surface crusts of semi-deserts and fertility of soils used for farming in arid regions. Many cyanobacteria like *Microcystis*, *Lyngbya* and *Oscillatoria* act as bioindicators of degree and type of evolution (Choudhary *et al.*, 2007).

Additionally, cyanobacteria have been considered as a most promising group of organisms for their ability to produce bioactive molecules. Their ability to produce pigments, vitamins, toxins, fine chemicals, pharmaceuticals probes (Thajuddin and Subramanian, 2005) are already in literature. Recent advances in cyanobacterial products for bioactive compounds have attracted attention of worldwide cyanobacteriologists. They have been reported to produce phenolic compounds by immobilized cells of cyanobacterium, *Anabaena doliolum* (Choudhary, 1999). Moreover, they produce antialgal, antiviral or antimicrobial or anti AIDS compounds. The present development in polysachharidic nature of outermost mucilaginous envelope has paid more attention. This mucilaginous sheath has been reported to act as emulsifying agents. Additionally, they play an important role in maintaining the soil structure and increase the water holding capacity of the soil. The pigments like carotenes and PBPs have been reported to serve as a second line defense against modification of low density proteins and fluorescence immunoassays and fluorescence microscopy for diagnostic and biomedical research. This article has been aimed to discuss the ecological significance of cyanobacteria in terms of distribution, abundance and their role in enhancement of productivity in rice soils with a short emphasis on biotechnological significance in relation to antimicrobial, antiviral and toxins production by cyanobacteria.

Ecology of Soil Cyanobacteria

The progress encountered so far in the field of cyanobacteria is centered on cyanobacteria of rice fields. This section of the chapter will deal with distribution and abundance of cyanobacteria in particular reference with the cultivated rice fields and the progress in their potential for enrichment and maintenance of soil productivity for crop plants in terms of nitrogen and phosphorus.

Distribution of Cyanobacteria

With the establishment of the agronomic potential of the cyanobacteria as nitrogen fixer, their enumeration and evaluation was taken into consideration worldwide. Cyanobacterial distribution and their abundance in either kind of environment are determined by the moisture, pH, mineral nutrients and combined nitrogen source available in the environment (Granhall, 1975). The literature cited so far suggest that the cyanobacteria is predominant in rice field as it provides favorable growth condition of light, water, high temperature and nutrient availability required for the optimal multiplication of cyanobacteria. This favours the higher abundance of cyaobacteria in paddy fields than in other cultivated soils (Watanable and Yamamoto, 1971). The cyanobacterial composition under widely different climatic conditions of India (Mitra, 1951), Japan (Okuda and Yamaguchi, 1956) and the Ukraine (Prikhod'kova, 1971) has been reported. Watanable (1959) in his study of soil samples collected from different countries in the South-East Asian region suggested that the nitrogen-fixing cyanobacteria are more dominant in tropical and sub-tropical regions than in temperate and sub-temperate regions. Cyanobacterial diversity in the rice field of 24 Parganas district of West Bengal showed the dominance of heterocystous forms of cyanobacteria (Singh *et al.*, 2001). Despite the favorable growth condition in rice fields, the divergence in cyanobacterial diversity in rice fields of different climatic conditions has been noticed. The rice fields cyanobacteria are represented by both unicellular and filamentous forms but the biomass of filamentous forms are greater. The study made so far suggested that the cyanobacteria constitute 86 per cent of total algal flora in South-East Iraq (Al-Kaisi, 1976). According to Mitra (1951), cyanobacteria constitute more than half the total number of algal species in north and south India whereas 0-76 per cent of the total algae has been reported in acidic soils of Kerala state of India (Aiyer, 1965). In contrary to above assumption of abundance of BGA in paddy fields, Venkataraman (1975) have observed the presence of only 33 per cent nitrogen-fixing cyanobacteria.

In spite of abundance of cyanobacteria in rice fields, their range of distribution and diversity depends on the time and space of the cultivation cycle of rice. In some cases, the cyanobacterial abundance may be observed just after the plantation of rice with high availability of light and fertilizers, whereas in other cases they may be abundantly present after two weeks (Kurusawa, 1956) or one month (Ichimura, 1954) of plantation. The subsequent decrease in algal biomass was observed after attaining the maximal population in given space and time of rice field and it was attributed to the deficient light and limitation in nutrient availability. In dry land rice fields in India, maximal biomass developed between tillering and panicle initiation as observed in rice field of Senegal whereas in wetland fields of India, the density of biomass was maximal a little later (Gupta, 1966). The gradual growth in rice plants results into decrease in the light intensity reaching to the soil surface and ultimately affects the diversity and distribution of cyanobacteria.

Effect of Physical Factors on Distribution of Cyanobacteria

Cyanobacterial distribution in the various ecosystems is governed by different physical and nutrient status of the environment. The paddy field soil at the early stage of plantation is characterized by low pH, an absence of plant cover and high level of Co_2 caused by soil moistening. These all conditions provide favorable condition for the development of members of Chlorophyceae and not the cyanobacteria. The role of physical factors including sunlight, temperature and nutrient status in terms of nitrogen and phosphorus has been described below.

Effect of Light on Composition of Cyanobacteria

Light

Light is the most important determining factor to affect the seasonal variation in cyanobacterial populations in different ecosystems (Ichimura, 1954). Cyanobacterial occurrence is restricted to the photic zone and is usually located in the upper 0.5 cm of the soil. The paddy field witnesses the cyanobacterial populations as surface scum or as crust-forming aggregates at the soil water interface. The abundance of cyanobacteria at different stages of rice cultivation is determined by the extent of light reaching to the surface of the soil. In the submerged soils, the light availability to the crop plants depends upon the season and latitude, the plant canopy and the turbidity of the water. In contrast, the light availability to rice crop in terrestrial habitats is dependent on the rice canopy. It has been demonstrated that light availability is decreased to 50 per cent after 15 days, 85 per cent after 30 days and 95 per cent after 60 days of rice plantation (Kurusawa, 1956). In a submerged soil, the light intensity may vary from deficiency to inhibitory levels. The variation in light intensity reaching to the surface of the soil has witnessed a shift in algal flora particularly cyanobacterial flora in accordance with the decreasing intensity of the light. The growth in rice plants have negative effect on the intensity of light reaching to the soil surface. The further observation of tolerance of light intensity by cyanobacteria suggested that the cyanobacteria are generally sensitive to high light intensities and may be regarded as low light species (Roger and Reynaud, 1979). This is the reason why the green algae and diatoms show their first appearance in rice fields during early stages of rice cultivation. However, certain cyanobacteria like *Cylindrospermum* (paddy fields in Mali) and *Aulosira fertilissima*, have been reported to grow under full sunlight of more than 100 klux intensity (Singh, 1976; Traore *et al.*, 1978). Thus, algal diversity is directly correlated with the light intensities. Cyanobacteria particularly nitrogen-fixing cyanobacteria are dominant only after the rice plant achieves a full growth with dense cover protecting the rice surface with high light intensities. It has been further demonstrated that the acetylene reduction activity (ARA) of the soil was highest in the most heavily shaded soil. On the other hand in a monsoon zone where light intensities are not high, cyanobacteria appear from the beginning of the cultivation cycle (Gupta, 1966). Simultaneously, the deficiency in light intensity may also affect the abundance of cyanobacterial diversity. In Japan, the productivity of phytoplankton increased in early summer but decreased in late July when the rice canopy decreased the light intensity below the compensation point of the phytoplankton (Ichimura, 1954). Not only that, the coloration in cyanobacteria is also controlled by light intensity. Cyanobacteria growing on soil surface in direct contact of light exhibited dark coloration (blue-dark, brown, red-brown or red) in comparison to those growing below the soil surface or in submerged soils. The dark coloration in cyanobacteria is the resultant of accumulation of UV-absorbing pigments, scytonemin, in the brown mucilaginous sheaths surrounding the photosynthetic trichomes of cyanobacteria (Garcia-Pichel and Castenholz, 1991). Dark coloration is more prominent in cyanobacteria growing in open than shaded condition.

Effect of Temperature on Composition of Cyanobacteria

Out of several factors that affect the cyanobacterial abundance in the rice fields, temperature is one among them. The optimal temperature for the proliferation of cyanobacteria in field condition is 30-35 ^{0}C. The requirement of optimal temperature by cyanobacteria is higher than eukaryotic algae which are competitor of cyanobacteria in rice field. The temperature during rice cultivation cycle in most part of the world remains in/below the range of required temperature. Thus, the temperature is not commonly inhibiting factor for growth of cyanobacteria in rice fields (Roger and Reynaud, 1979). The further observations with cyanobacterial abundance suggested that the species appearing in the early cultivation cycle is replaced by other species during the late cultivation cycle. In Japan, the paddy fields witnessing the proliferation of *Spirogyra setiformis* and *Anabaena oscillarioides* in summer were replaced by *Tetraspora gelatinosa* and related species in winter. During the dry season of temperate zones, a lower temperature at the beginning of the cultivation cycle may favour eukaryotic algae and inhibit cyanobacterial proliferation (Roger and Reynaud, 1977). In India experiments carried out in field and pot conditions showed rapid decrease in cyanobacterial abundance (Subrahmanyan *et al.*, 1965) and productivity in the cold season. The rise in temperatures above 35°C results into the decline in CO_2 assimilation (Kurusawa, 1956). Simultaneously, the high temperature (34-39°C) in Indian paddy water was found to be favourable for the proliferation of *Aulosira fertilissima*. Thus, light and temperature together are crucial factor in determining the cyanobacterial diversity, abundance and productivity in rice fields. Among the different inorganic fertilizer used and required for the growth and production of crop plants, nitrogen along with phosphorus and potassium are most common limiting nutrient for plant productivity.

Effect of Nitrogen on Composition of Cyanobacteria

Nitrogen is the most important inorganic fertilizer required for the proliferation of plants and plays an important role in determining the cyanobacterial diversity and their abundance in the soil. The proliferation of nitrogen-fixing cyanobacteria is greatly favored under nitrogen starved condition by a lack of competitiveness of the other algae if the other environmental factors are not limiting. In the nitrogenous fertilizer supplemented soil, the abundance and diversity of the nitrogen-fixing cyanobacteria is greatly reduced or their nitrogen-fixing activity is inhibited. In presence of sufficient nitrogen sources in the soil, the nitrogen-fixing cyanobacteria use mineral nitrogen for their growth and other metabolic activity. Under such conditions, they don't use their nitrogen-fixing metabolic machinery as this process is ATP dependent. Additionally, they compete with non nitrogen-fixing cyanobacteria and other eukaryotic algae for nitrogen sources. The inhibitory effect of nitrogenous fertilizers on nitrogen-fixing cyanobacteria has been reported under field and laboratory conditions (Rinaudo, 1974).

The pot experiments carried out by Yoshida *et al.* (1973) under controlled conditions demonstrated that the addition of nitrogenous substances in the growth media resulted into increase in total biomass but decrease in number of species *i.e.* more cyanobacteria in pots without nitrogen fertilizer. Than Tun (1969) reported that soil treated with ammonium sulphate or calcium cyanamide witnessed the dominance of green algae. The further observation made in terms of acetylene reduction activity (indices of nitrogen fixation) of the soil treated with nitrogen fertilizer confirmed the inhibitory affect of nitrogen on nitrogen-fixing activity of the cyanobacteria. The observation suugested that the nitrogen-fixing activity of the organisms is maintained under nitrogen starved condition. Simultaneously, the inhibitory affect of the nitrogen is not permanent and inhibitory effect is relaxed with decrease in nitrogen content of the soil with their regular uptake by cyanobacteria and other living systems of the soil. Thus, competition for mineral nitrogen among cyanobacteria and non-nitrogen-fixing cyanobacteria together with eukaryotic algae in the soil may lead to the development and establishment of non-nitrogen fixing systems. The addition of nitrogen fertilizer favors the growth of non-heterocystous forms along with other forms of algae whereas only nitrogen starved condition is favorable for the dominance of the nitrogen-fixing forms.

Effect of Phosphorus on Composition of Cyanobacteria

Phosphorus is the key component of the various cellular structure including nucleic acids and occupies second position among the inorganic fertilizers. Cyanobacteria assimilate more phosphorus than they require and store the excess as polyphosphate, which can be used under phosphorus starved conditions. Arora (1972) have observed that addition of phosphates in soluble (KH_2PO_4) or insoluble [$Ca_3 (PO_4)_2$] form stimulates algal growth and nitrogen fixation under laboratory conditions. *Anabaena* and *Tolypothrix* showed enhanced nitrogen fixation in phosphated soils than in unphosphated ones (Arora, 1972). Under field conditions, the addition of phosphate has stimulatory effect on algal growth and nitrogen fixing activity (De and Mandal, 1958; Srinivasan, 1978). In this way, the phosphorus is the key constituent of the soil that regulates the cyanobacterial diversity, dominance and nitrogen-fixing activity in the field conditions.

Cyanobacteria as Biofertilzer

The above discussion is all about the cyanobacteria, their abundance in time and space and the role of different physical factors (light and temperature) and inorganic fertilizers (nitrogen and phosphorus) on distribution, diversity and dominance of the cyanobacteria in rice fields. The preceding part of this chapter will focus on the ecological significance in terms of biofertilizers (nitrogen and phosphorus) and biotechnological significance in terms of their role in human life. The ever increasing population is posing a tremendous pressure on limited natural resources of cultivable land to fulfill the requirement of food grains. It is estimated that for feeding a population of 1.4 billion by 2025, our country will need to produce 3.1 mt food grain and to achieve this, India will need at least 45 mt plant nutrients, out of which at least 35 mt should be chemical fertilizer sources and rest 10 mt nutrient should come from other sources like biofertilizers or manure. The present day chemical fertilizers based agriculture is unsustainable and slowly leading to ecological crisis in terms of decreased soil fertility. The adverse effect of chemical fertilizers acts as stimuli to develop alternate system *i.e.* biological system to cope with the increasing demand of fertilizers to obtain the food grains by maintaining the ecosystem sustainable. In this way, the concept of 'Organic Agriculture' originated advocating minimum use of the chemical fertilizer and increasing dependence on biological inputs like compost, farm yard manure, green manures and biofertilizers. Amongst the array of biofertilizers developed for different crops, cyanobacteria constitute the most potential inputs in rice fields.

Cyanobacteria as a Source of Nitrogen in Rice Fields

With the establishment of agronomic potential of cyanobacteria by De (1939) and Singh (1942), these organisms were variously studied for their potential application in the soil in terms of nitrogen, phosphorus and other growth promoting substances around the world. Now, this is the established fact that cyanobacterial nitrogen-fixation play an important role in enhancement of soil fertility by maintaining the ecosystem eco-friendly as against the chemical fertilizer. Its importance was understood only after receiving some deleterious effect of chemical fertilizer on soil fertility. Additionally, the chemical fertilizer has chances of contaminating ground water and atmosphere ultimately posing risk on persisting living system on earth particularly human health. As mentioned above, some cyanobacteria are endowed with a highly specific cellular structure 'heterocyst' contained with oxygen-sensitive enzyme ' nitrogenase complex' that enable them to convert atmospheric elementary nitrogen to soluble form of ammonia. These specialized structures are formed under nitrogen starved condition (Fog, 1944; Castenholtz and Waterbury, 1989). The cyanobacterial diversity in the paddy soil has been surveyed and found to be most appropriate niche for their dominance and diversity. Singh *et al.* (2001) studied the cyanobacterial flora of the paddy fields of 24 Parganas of West Bengal and found the dominance of nitrogen-fixing diazotrophic heterocystous cyanobacteria in the paddy soil. In Mali also, heterocystous and nitrogen-fixing cyanobacteria dominated the paddy soil. In Japanese rice soils, the dominant forms of the cyanobacteria belonged to *Nostoc, Anabaena* and *Tolypothrix*. In India, Singh (1961) documented the cyanobacterial flora of rice fields of Uttar Pradesh and Bihar and reported the dominance of *Aulosira fertilissima* intermingled with *Anabaena* and *Cylindrospermum*. Mitra (1951) screened soil samples of north and south India and observed the presence of nitrogen-fixing species of *Nostoc, Anabaena* and *Calothrix*. In addition to heterocystous forms, the nitrogen fixation has also been observed in few unicellular and non-heterocystous forms belonging to *Aphanothece, Gloeocapsa* (*Gloeothece*), *Plectonema, Trichodesmium* (*Oscillatoria*) and others (Singh, 1973; Gallon *et al.*, 1975). The best documented nitrogen-fixing heterocystous forms commonly found in rice-field are *Anabaena, Nostoc, Cylindrospermum, Gloeotrichia, Scytonema, Tolypothrix, Calothrix* and *Aulosira ferttilissima*. The application of cyanobacteria as biofertilizer in paddy cultivation has been subsequently studied by many researchers in India and abroad. Notable among them are Venkataraman and Goyal (1968), Roger and Kulasooriya (1980), Watanable (1973), Bent (1961), Stewart *et al.*, (1979), and Singh and Bisoyi (1989).

Nitrogen fixed by cyanobacteria is released in the soil environment by autolysis, extracellular transport or through microbial decomposition after the cells death. It is in the form of small peptides as well as free amino acids (Misra and Kaushik, 1989). The nitrogen contribution by different members of the cyanobacteria has been estimated variously in terms of acetylene reduction activity. The estimated acetylene reduction activity ranged varied from 189 to 259 for *Aphanothece* and from 249 to 357 nmol ethylene mg^{-1} chl h^{-1} for *Aulosira*. The further study on nitrogen-fixation potential of cyanobacteria, the nitrogen-fixing capacity (mg N 50 ml^{-1} 28 $days^{-1}$) range from 0.84–2.45 in *Anabaena*; 0.77–5.83 in *Calothrix*; 0.31 -1.15 in *Hapalosiphon*; 0.44 4.04 in *Nostoc*; 0.30–3.41 in *Scytonema*; 2.55–5.67 in *Tolypothrix* and 0.60–2.34 in *Westiellopsis* (Kolte and Goyal, 1989). On an average cyanobacterial mat provides 15–25 kg biologically fixed N ha^{-1} per season in linear fashion with increase in productivity of rice grains up to 10-15 per cent. Additionally, the application of cyanobacteria as biofertilizer also increases the nitrogen content of the grain and straw along with increase in plant height, number of spikelet per panicle and amount of dry matter (Singh, 1961; Rao *et al.*, 1977). Pankratova and Vakrushev (1971) have reported the contribution of 3.3 kg N ha^{-1} by *Nostoc commune* in and on a loamy soil of Russia. The contribution of 10-30 kg N ha^{-1} has been reported by Roger and Kulasooriya (1980) and Kaushik (1998) whereas Traore *et al.* (1978) have observed the nitrogen contribution up to 80 kg N ha^{-1}. The algalization of rice field has showed enhanced growth of rice plants resulting into increased grain yield up to 2.2 to 28.9 per cent at an average of 7.2 per cent (Venkataraman, 1961) and straw yield. Besides an increase in grain and straw yield, the increase in nitrogen content of the yield and straw has been also reported (Sundara *et al.*, 1963). In this way, nitrogen is contributed differently to the soil by cyanobacteria with increased productivity of grains and straw along with maintaining the soil fertility and ecosystem ecofriendly.

Cyanobacteria as a Source of Phosphorus in Rice Fields

Phosphorus is the second main constituent of the living system with being a constituent of nucleic acids and source of energy. Most of the phototrophs including cyanobacteria utilize inorganic phosphate from their surrounding to driven out their metabolic activity for growth and development. But capability of living system to assimilate organic phosphate accumulating in various ecosystems is more important to maintain the terrestrial and freshwater ecosystem sound and sustainable. The organic phosphates are accumulating regularly in terrestrial ecosystems as agricultural, municipal and industrial effluents. In terrestrial ecosystem, organic phosphates are leading to the development of nutrient imbalances directly effecting the plant growth and development. However in freshwater, their accumulation is resulting into eutrophication leading to the appearance and overgrowth of particular group of organisms able to utilize organic phosphates efficiently and suppression of growth of other living systems present in that system. The organic phosphates are complex organic molecules able to supply all the phosphorus requirements to the cyanobacteria need to be dissociated before utilization by living system. The recycling of phosphates from complex forms is increasing concern of present day environmental programme. In the field condition, the recycling of phosphorus is restricted to certain group of organisms like cyanobacteria, algae and mosses.

The study reflects that organisms can utilize organic phosphates as a substrate for source of inorganic and soluble form of phosphates (Cembella *et al.*, 1984). However, some cyanobacteria can utilize insoluble organic phosphates (Whitton *et al.*, 1991). The ability of organisms to utilize the organic phosphates is linked with phosphatase activity of the organisms. Boavida (1990) has described phosphatase as 'whole bunch of enzymes' having different half saturation constants, temperature, pH optima and substrate specificity (Hoppe, 2003). The stimulation of phosphatases enzymes in the soil after algalization has been extensively studied in the field (Mishra *et al.*, 2005) and laboratory conditions (Whitton *et al.*, 1991). The experimental studied was performed by growing the axenic culture of the cyanobacterial strains in nutrient media supplemented with organic (monoesters β-glycerophosphate or *para*-nitrophenyl phosphate) but no inorganic phosphates for prolonged duration (Whitton *et al.*, 1991). They found vigorous growth suggesting possible utilization and degradation of organic phosphates. In the field conditions, the experiments were performed by inoculating the soil with or without cyanobacterial inoculum and increase in phosphatase activity of the soil inoculated with cyanobacterial inoculum were observed over untreated soils (Mishra *et al.*, 2005). These results favoured the possible application of cyanobacteria in hydrolysis of organic phosphates. Donald *et al.* (1997) also performed their experiment with *Synechococcus* WH 7803 and found that this organism can grow using the *para*-nitrophenyl phosphates, glucose 6-phosphates or glycerol phosphates.

Phosphatase enzymes are basically of two type *viz.* alkaline and acid phosphatase based on substrate types. However, depending on the presence of ester group, phosphatases are grouped into phosphomonoesterases and phosphodiesterases. Phosphatases have been variously defined in terms of their location inside or outside the cell. Phosphatses are said to be intracellular if their activity is centered inside plasmamembrane, surface bound if their activity is external to the membrane and extracellular if they are released outside the cell into surroundings. P-limitation condition in environment induces the synthesis and activity of alkaline phosphatase in cyanobacteria. Alkaline phosphatases are the first enzyme that remains active in free dissolve state in natural water (Berman 1969; Kobayashi *et al.*, 1984). In this way, the cyanoacteria has been endowed with different kinds of phosphatase enzymes that can regulate the hydrolysis of organic phosphates extracellularly or intracellularly, acidic or alkaline and release the inorganic phosphates to drive out the metabolic activity of organisms. Thus, cyanobacteria regulate the phosphorus status of the soil and water and maintain the soil and water system sound and healthy for growth of organisms.

Cyanobacteria as a Source of Growth Promoting Substance

Besides increasing soil fertility and sustaining rice yields, cyanobacteria have been reported to benefit crop plants by producing growth-promoting substances. Earlier confirmation of hormonal effect of cyanobacteria was noticed by treating the rice seedlings with cyanobacterial cultures or their extracts. The stimulatory effect of cyanobacterial exudates has been observed in enhancement of germination and seedling growth of rice. Increased length and number of ears, and number of grains per ear have been induced in rice by treating them with nitrogen-fixing cyanobacteria. The extracts from *Calothrix* spp. and *Anabaena* sp. have a rhizogenous effect and stimulate plant organs. Gupta and Lata (1964) have observed that the presoaking of rice seedlings in extracts of non-nitrogen-fixing cyanobacterium, *Phormidium* sp., resulted into enhanced germination. The further study with presoaking of rice seedlings in extracts of *Phormidium* sp. promoted the growth of roots and shoots (Gupta and Shukla, 1969), stimulated vegetative growth of plants and increased the protein content of the grains (Gupta and Shukla, 1967). The probable nature of these substances resemble to the gibberelin. In this way, the cyanobacteria not only contribute soluble carbon (being oxygenic photosynthetic orgnisms), nitrogen and phosphorus, they also enhance the growth promoting substances in the soil.

Biotechnological Significance

Cyanobacteria have been now considered as a promising group of organisms for their ability to produce various kinds of bioactive compounds. Some of them have been mentioned below:

Antimicrobial Compounds from Cyanobacteria

A large number of cyanobacterial extracts and/or extracellular products have been screened for the bioactive compounds and some of them have antimicrobial (antifungal, antibacterial, antialgal and antiprotozoal) activity. Antimicrobial substances produced by cyanobacteria include phenolics (De Cano *et al.*, 1990; Choudhary and Bimal, 2008), bromophenols (Pederson and DaSilva, 1973), isonitrile-containing indole alkaloids such as haploindole A and various toxins (Carmichael, 1992). The most of the studies on activity of cyanobacterial antimicrobial substances have been carried out *in vitro* and they have been found to be little or no application in medicine as they are too toxic or inactive *in vivo*. Bonjouklian *et al.* (1991) isolated an antimicrobial compound, tjipanazoles, from cyanobacterium *Tolypothrix tjipanensis* which show a little toxicity *in vitro* and no *in vivo* against *Candida albicans*. However, tjipanazole A1 and A2 show appreciable fungicidal activity against rice blast and leaf rust infection. Additionally, the antimicrobial substances with algicidal properties produced from cyanobacteria are of special importance in environmental management programme. The γ-lactone and cyanobacterin produced by *Scytonema hofmanni* (Gleason *et al.*, 1986), fischerellin from *Fischerella muscicola* (Gross *et al.*, 1991) and product from *Oscillatoria* sp. (Bagchi *et al.*, 1990) are important among them. The cyanobacterin produced from cyanobacteria has their role as herbicide (Gleason *et al.*, 1986). The literature on production of bioactive molecules by cyanobacteria is accumulating regularly. Volk (2005) has reported that the indole alkaloid norharmane [9H-Pyrido (3,4-b) indole] produced by cyanobacterium *Nodularia harveyana* and the phenolic compound 4,4′-dihydroxybiphenyl by cyanobacterium *Nostoc insulare* exhibited algicidal activity. Recently, Volk (2006) has reported the another exometabolites, β-carboline [norharmalane=3,4-dihydro-9H-Pyrido (3,4-b) indole] from the culture media of cyanobacterium *N harveyana* of algicidal activity. The cytotoxic compounds particularly algicidal molecules produced by cyanobacteria may be helpful in combating the worldwide problem of bloom formation in different water bodies.

Antiviral Substances from Cyanobacteria

The cyanobacteria have been also reported to accumulate or produce antiviral substances in their masses or in the nutrient medium (Patterson *et al.*, 1993a). Reinhart *et al.* (1981) have reported the antiviral activity of cyanobacterial extracts against *Herpes simplex* and respiratory syncytial virus. Cyanobacterial extracts have also been reported to have inhibitory effect on reverse transcriptases of avian myeloblastosis virus and human immunodeficiency virus. Cyanobacteria with anti-AIDS property are also in literature (Gustafson *et al.*, 1989).

Toxins and Pharmacological Compounds from Cyanobacteria

The toxins produced from cyanobacteria is of special interest in pharmacology. The anatoxin produced by *Anabaena flos-aquae* are potent post synaptic neuromuscular blocking agents (Carmichael, 1992). The other kind of toxins produced by cyanobacteria is hepatotoxins, microcystins and nodularin. These toxins are found to have inhibitory effect on protein phosphatase (Chen *et al.*, 1993) and are helpful in studies of cellular regulation. The cytotoxins produced from cyanobacteria have been reported to act as anti cancer drugs (Patterson *et al.*, 1991). Besides these toxins, the cyanobacteria produce many more toxins like scytophycins from *Scytonema pseudohofmanni* and *Tolypothrix conglutinate* var. *colorata* that show toxicity against the KB (a human nasopharyngeal carcinoma) cell line and Lewis lung carcinoma (Ishibashi *et al.*, 1986). The acutiphycins from cyanobacterium, *Oscillatoria acutissima* showed resemblance in activity with scytophycins (Barchi *et al.*, 1984). Helms *et al.* (1988) have reported calcium agonist compounds, scytonemin A from *Scytonema* sp.. Another interesting group of bioactive compounds isolated from cyanobacteria are the brominated bi-indoles from *Rivularia firma* that show anti-inflammatory and anti-amphitamine activity. Additionally, cyanobacteria produce some pharmacologically active compounds with inhibitory effect on angiotension-converting enzymes (Borowitzka, 1995). These are some bioactive compounds of human interest produced by cyanobacteria that have elevated them as biotechnological organisms.

Besides the above mentioned bioactive compounds, cyanobacteria produce many more compounds that has not been characterized or exploited by cyanobacteriologists. The cyanobacteria produce the pigment, scytonemin in their sheath which have potential to absorb the ultraviolet radiation (see Borowitzka 1995; Proteau *et al.*, 1993). Cyanobacteria are utilized by ancient people for feed and their role in agriculture in maintaining the fertility by enhancing the nitrogen status and mobilizing the phosphate are in literature and have been also discussed above. The production of vitamins like pro-vitamin A (β-carotene with anti-tumour and cancer preventing activity (Davison *et al.*, 1993), vitamin B_6, B_{12}, biotin etc. are already in literature (Borowitzka, 1988a). Cyanobacteria have been also reported to be important sources of poluunsaturated fatty acids like γ-linolenic acid, arachidonic acid and docosahexaenoic acid. Besides, Cyanobacteria also produce and accumulate different kinds of polysachharides of pharmacological interest in their mucilaginous sheath. Cyanobacteria with ability to grow under different extreme conditions with ease and ability to produce different bioactive compounds of human interest have made them an excellent and promising group of organisms for present day biotechnological programme worldwide.

Future Prospects

The future of cyanobacterial exploitation for application in soil conditioning and maintenance of system ecofriendly is promising. Although, cyanobacteria play a crucial role in building up the nitrogen and phosphorus status of the soil, it is needed to develop optimal condition particularly nitrogen-starved and maintenance of moisture in the rice field for the proliferation of nitrogen-fixing cyanobacteria. Furthermore, the establishment of cyanobacteria inside higher plants to get the maximum output is the urgent need of present day organic farming programme. And it may be achieved either by establishing the symbiotic association of cyanobacteria with desired organisms or by inserting the nucleotide sequences responsible for nitrogen-fixation into the organisms of interest. The present shift in application of cyanobacteria from biofertilizer to bioactive compounds has paid more attention in the recent past. Their ability to produce a large number of biomolecules of different nature may be helpful in combating the problems of different origin including ecological, environmental and pharmaceutical problems. In spite of their successful study in the laboratory, they have not been elevated at industrial scale. Therefore, the cyanobacteria require much more attention to develop some effective strategies for successful exploitation of cyanobacteria for different purposes of human welfare. Lastly, it may be concluded that the information gathered so far on cyanobacterial potential is in infancy stage and requires proper investigation to achieve the maximum efficient production from these tiny organisms.

References

Adams DG and Duggan PS (1999). Heterocyst and akinete differentiation in cyanobacteria. *New Phytol*, 144: 3–33.

Aiyer RS (1965). Comparative algological studies in rice fields in Kerala state. *Agric Res J, Kerala*, 3: 100–104.

Alkaisi KA (1976). Contributions to the algal flora of the rice fields of Southeastern Iraq. *Nova Hedwigia*, 27: 813–827.

Arora SK (1972). Effect of basic slags and blue-green algae on nitrogen mobilization in paddy fields. *Riso*, 21: 233–238.

Bagchi SN, Palod A and Chauhan VS (1990). Algicidal property of a bloom-forming blue-green alga, *Oscillatoria* sp. *J Bas Microbiol*, 30: 21–29.

Barchi JJ, Moore RE, Furusawa E and Patterson GML (1984). Acutiphycin and 20,21-didehydroacutiphycin, new antineoplastic agents from the cyanophyte *Oscillatoria acutissima*. *J Am Chem Soc*, 106: 8193–8197.

Berman T (1969). Phosphate release of inorganic phosphorus in lake Kinneret. *Nature*, 224: 1231–1232.

Binder BJ and Chisholm SW (1990). Relationship between DNA cycle and growth rate in *Synechococcus* sp. strain 6301. *J Bacteriol*, 172: 235–241.

Boavida MJ (1990). Natural plankton phosphatases and the recycling of phosphorus. *Verhandlung Internationale Vereinigung Limnologie*, 24: 258–259.

Bonjouklian R, Smitka TA, Doolin LE, Molloy RM, Debono M, Shaffer SA, Moore RE, Stewart JB and Patterson GML (1991). Tjipanajoles, new antifungal agents from the blue-green alga *Tolypothrix tjipanasensis*. *Tetrahedron*, 47: 7739–7750.

Borowitzka MA (1988a). Vitamins and fine chemicals. In: *Micro-algal Biotechnology*, (Ed) Borowitzka MA and Borowitzka LJ. Cambridge University Press, Cambridge, p. 153–196.

Borowitzka MA (1995). Microalgae as sources of pharmaceutical and other biologically active compounds. *J Appl Phycol*, 7: 3–15.

Bunt JS 1961. Nitrogen-fixing blue-green algae in Australian rice soils. *Nature*, 192: 479–480.

Carmichael WW (1992). A Review: Cyanobacteria secondary metabolites-the cyanotoxin. *J Appl Bact*, 72: 445–459.

Castenholtz RW (1992). Species usage, concept and evolution in the cyanobacteria (blue-green algae). *J Phycol*, 28: 737–745.

Castenholtz RW and Waterbury JB (1989). Group I. Cyanobacteria. Preface. In: *Bergey's Manual of Systematic Bacteriology*, (Eds) Staley JT, Bryant MP, Pfenning N and Holt Jg, 3: 1710–1727.

Cembella AD, Antia NJ and Harrison PJ (1984). The utilization of inorganic and organic phosphorus compounds as nutrients by eukaryotic microalgae: A multidisciplinary perspective. Part 1. *CRC Critical Rev Microbiol*, 10: 317–391.

Chen DZX, Boland MP, Smillie MA, Klix H, Ptak C, Andersen RJ and Holmes CFB (1993). Identification of protein phosphatase inhibitors of the microcystin class in the marine environment. *Toxicon*, 31: 1407–1414.

Choudhary, K.K. (1999). *Ex situ* conservation of cyanobacterial germplasm of north Bihar, India. *PhD Thesis*, BRA Bihar University, Muzaffarpur, Bihar, India.

Choudhary KK and Bimal R (2008). Phenolics production by cells of the cyanobacterium *Anabaena doliolum* Bharadwaza immobilized in calcium alginate. *Seaweed Res Util.*

Choudhary KK, Singh SS and Mishra AK (2007). Nitrogen fixing cyanobacteria and their potential applications. In: *Advances in Applied Phycology*, (Eds) Gupta RK and Pandey VD. Daya Publishing House, New Delhi, p. 142–154.

Cohen YJ, Padan E and Shilo M (1975). Facultative anoxygenic photosynthesis in the cyanobacterium *Oscillatoria limnetica*. *J Bacterol*, 123: 855–863.

Davison A, Rousseau E and Dun B (1993). Putative anticarcinogenic actions of carotenoids-nutritional implications. *Can J Physiol Pharmacol*, 71: 732–745.

De PK (1939). The role of blue-green algae in nitrogen fixation in rice fields. *Proc R Soc Lond*, B: 121–139.

De Cano MMS, De Mule MCZ, De Caire GZ and de Halperin DR (1990). Inhibition of *Candida albicans* and *Staphylococcus aureus* by phenolic compounds from the terrestrial cyanobacterium *Nostoc muscorum*. *J Appl Phycol*, 2: 79–81.

De PK and Mandal LN (1958). Fixation of nitrogen by algae in rice soils. *Soil Sci*, 81: 453–458.

Demoulin V and Jansen MP (1981). Relationship between diameter of the filament and cell shape in blue-green algae. *Br Phycol J*, 16: 55–58.

Desikachary TV (1959). *Cyanophyta*. Indian Council of Agricultural Research, New Delhi.

Donald KM, Scanlan DJ, Carr NG, Mann NH and Joint I (1997). Comparative phosphorus nutrition of the marine cyanobacterium *Synechococcus* (WH6803) and the marine diatom *Thalassiosira weissflogii*. *J Plankton Res*, 19: 1793–1813.

Fogg GE (1944). Growth and heterocyst production in *Anabaena cylindrica* Lemm. *New Phytologist*, 43: 164–175.

Gallon JR, Kurz WGW and LaRue TA (1975). The physiology of nitrogen-fixation by a *Gloeocapsa* sp.. In: *Nitrogen Fixation by Free-living Microorganisms*, (Ed) Stewart WDP. Cambridge University Press, Cambridge, p. 159–173.

Garcia-Pichel F and Castenholz RW (1991). Characterization and biological implications of scytonemin, a cyanobacterial sheath pigment. *J Phycol*, 27: 395–409.

Gleason FK, Case DE, Siprell KD and Magnuson TS (1986). Effect of the natural algicide cyanobacterin, on a herbicide resistant mutant of *Anacystis nidulans* R2. *Plant Sci*, 46: 5–10.

Granhall U (1975). Nitrogen fixation by blue-green algae in temperate soils. In: *Nitrogen Fixation by Free-living Microorganisms*, (Ed) Stewart WDP. Cambridge University Press, Cambridge, p. 189–198.

Gross EM, Wolk CP and Juttner F (1991). Fischerellin, a new allelochemical from the freshwater cyanobacterium *Fischerella muscicola*. *J Phycol*, 27: 686–692.

Gupta AB (1966). Algal flora and its importance in the economy of rice fields. *Hydrobiologia*, 28: 213–222.

Gupta AB and Lata KJ (1964). Effect of algal growth harmones on the germination of paddy seeds. *Hydrobiologia*, 24: 430–434.

Gupta AB and Shukla AC (1967). Studies on the nature of algal growth promoting substances and their influence on growth, yield and protein content of rice plants. *Labdev J Sci Technol, Kanpur*, 5: 162–163.

Gupta AB and Shukla AC (1969). Effects of algal extracts of *phormidium* species on growth and development of rice seedlings. *Hydrobiologia*, 34: 77–84.

Gustafson KR, Cardellina JH, Fuller RW, Wieslow OS, Kiser Rf, Snader KM, Patterson KML and Boyd ME (1989). AIDS: Antiviral sulfolipids from cyanobacteria (blue-green algae). *J Nat Cancer Ins*, 81: 1254–1258.

Haselkorn R (1978). Heterocysts. *Annual Rev Plant Physiol*, 29: 319–344.

Helms GL, Moore, RE, Niemczura WP, Patterson GML, Tomer KB and Gross ML (1988). Scytonemin A, a novel zalcium antagonist from a blue-green alga. *J Org Chem*, 53: 1298–1307

Hoppe HG (2003). Phosphatase activity in the sea. *Hydrobiologia*, 493: 187–200.

Ichimura S (1954). Ecological studies on the phytoplankton in paddy fields. I. Seasonal fluctuations in the standing crop and productivity of plankton. *Jpn J Bot*, 14: 269–279.

Ishibashi M, Moore RE, Patterson GML, Xu C and Clardy J (1986). Scytophycins, cytotoxins and antimycotic agents from the cyanophyte *Scytonema pseudohofmanni*. *J Org Chem*, 51: 5300–5306.

Kaushik BD (1998). Use of cyanobacterial biofertilizers in rice cultivation: A technology improvement. In: *Cyanobacterial Biotechnology*, (Eds) Subramanian G, Kaushik BD and Venkataraman GS. Science Publishers, Inc., Enfield USA, p. 211–222.

Kobayashi M, Takahashi E and Kawaguchi K (1967). Distribution of N_2-fixing microorganisms in paddy soils of Southeast Asia. *Soil Sci*, 104: 113–118.

Kolte SO and Goyal SK (1989). Natural variation in nitrogen potential of cyanobacteria. *Phykos*, 25: 166–170.

Kratz WA and Myers J (1955). Nutrition and growth of several blue-green algae. *Am J Bot*, 42: 275–280.

Kurusawa H (1956). The weekly succession in the standing crop of plankton and zoobenthos in the paddy field. Part I and Part II. *Bull Res Sci Japan*, 41–42: 86–98.

Mann NH and Carr NG (1974). Control of macromolecular composition and cell division in the blue–green alga, *Anacystis nidulans*. *J Gen Microbiol*, 83: 399–405.

Mishra U, Choudhary KK, Pabbi S, Dhar DW and Singh PK (2005). Influence of Blue Green Algae and Azolla inoculation on specific soil enzymes under paddy cultivation. *Asian J Microbiol Biotechnol Env Sc*, 1: 9–12.

Misra S and Kaushik BD (1989). Growth promoting substances of cyanoacteria II. Detection of aminoacids, sugars and auxins. *Proc Indian Natl Sci Acad*, B55: 499–504.

Mitra AK (1951). The algal flora of certain Indian soils. *Indian J Agric Sci*, 2: 357.

Nienow JA and Friedmann EI (1993). Terrestrial lithophytic (rock) communities. In: *Antarctic Microbiology*, (Ed) Friedmann EI. Willey-Liss, New York, p. 353–412.

Okuda A and Yamaguchi M (1952). Distribution of nitrogen-fixing microorganisms in paddy soils in Japan. In: *VI Cong Int Sci Sol Rap* C: 521–526.

Padan E and Cohen Y (1982). Anoxygenic photosynthesis. In: *The Biology of Cyanobacteria*, (Eds) Carr NG and Whitton BA. Blackwell, Oxford and University of California Press, Berkeley, p. 215–235.

Pankratova YM and Vakrushev AS (1971). Field determination of the fixation of the atmospheric nitrogen by blue-green algae using ^{15}N. *Soviet Soil Sci*, 3: 726–733.

Patterson GML, Baldwin CL, Bolis CM, Caplan FR, Karuso H, Larsen LK, Levine IA, Moore RE, Nelson CS, Tschappat KD, Tuang GD, Furusawa E, Furusawa S, Norton TR and Raybourne RB (1991). Antineoplastic activity of cultured blue-green algae (Cyanophyta). *J Phycol*, 27: 530–536.

Patterson GML, Baker KK, Baldwin CL, Bolis CM, Caplan FR, Larsen LK, Levine IA, Moore RE, Nelson CS, Tschappat KD, Tuang GD, Boyd MR, Cardellina JH, Collins RP, Gustafson KR, Snader KM, Weislow OS and Lewin RA (1993). Antiviral activity of cultured blue-green algae (Cyanophyta). *J Phycol*, 29: 125–130.

Pederson M and DaSilva EJ (1973). Simple brominated phenols in the blue-green alga *Calothrix brevissima* West. *Planta*, 115: 83–96.

Prikhod'kova LP (1971). Nitrogen-fixing blue-green algae of soils, rice fields and ephimeric basins of south of Ukraine. *Ukr Bot Zh*, 28: 753–758.

Proteau PJ, Gerwick WH, Garciapichel F and Castenholz RW (1993). The structure of scytonemin, an ultraviolet sunscreen pigment from the sheaths of cyanobacteria. *Experientia*, 49: 825–829.

Rao JL, Venkatachari A, Rao SWVB and Reddy RK (1977). Individual and combined effects of bacterial and algal inoculation on the yield of rice. *Curr Sci*, 46: 50–51.

Reinhert KL, Shaw PD, Shield LS, Gloer JB, Harbour GC, Koker MES, Samin D, Schwartz RE, Tymiak AA, Weller DL, Carter GT, Munro MHG, Hughes RG, Renis HE, Swynenberg EB, Stringfellow DA, Vavra JJ, Coats JH, Zurenko GE, Kuentzel SL, Li LH, Bakus GJ, Brusca RC, Craft LL, Young DN, Connor JL (1981). Marine natural products as sources of antiviral, antimicrobial and antineoplastic agents. *Pure Appl Chem*, 53: 785–817.

Rinaudo G (1974). Biological nitrogen-fixation in three types of rice soils in Ivory Coast. *Rev Ecol Biol Sol*, 11: 149–168.

Rippka R, Deruelles JB, Waterbury JB, Herdman M and Stanier RY (1979). Generic assignments, strain histories and properties of pure cultures of cyanobacteria. *J Gen Microbiol*, 111: 1–61.

Roger PA and Kulasooriya SA (1980). *Blue-green Algae and Rice*, IRRI, Los Banos.

Roger PA and Reynaud PA (1977). Algal biomass in rice fields of Senegal: Relative importance of cyanophyceae that fix nitrogen. *Rev Ecol Biol Sol*, 14: 519–530.

Roger PA and Reynaud PA (1979). *Ecology of Blue-green Algae in Paddy Fields*. International Rice Research Institute, Los Baṅos, Philippines, p. 289–309.

Roger PA, Santiagi-Ardales S, Reddy PM and Watanable I (1987). The abundance of heterocystous blue–green algae in rice soils and inocula used for application in rice fields. *Biol Fert Soils*, 5: 98–105.

Schoff JW and Walter MR (1982). Origin and early evolution of cyanobacteria. The geological evidence. In: *The Biology of Cyanobacteria*, (Eds) Carr NG and Whitton BA. Blackwell, Oxford and University of California Press, Berkeley, p. 543–564.

Singh BV, Choudhary KK, Dhar DW and Singh PK 2001. Occurrence of some *Nostocales* from 24 Parganas of West Bengal. *Phykos*, 40: 83–87.

Singh PK (1973). Nitrogen-fixation by the unicellular blue-green alga *Aphanothece*. *Arch Microbiol*, 103: 297–302.

Singh PK (1976). Algal inoculation and its growth in waterlogged rice fields. *Phykos*, 15: 5–10.

Singh PK and Bisoyi RN (1989). Blue-green algae in rice fields. *Phykos*, 28: 181–195.

Singh RN (1942). The fixation of elementary nitrogen by some of the commonest blue-green algae from paddy field soils of United Provinces and Bihar. *Indian J Agric Sci*, 2: 743–746.

Singh RN (1961). *The Role of Blue-green Algae in Nitrogen Economy of Indian Agriculture*. ICAR Publication, New Delhi, p. 175.

Srinivasan S (1978). Use of chemical fertilizers along with blue-green algae. *Aduthurai Rep*, 2: 145–146.

Stewart WDP, Rowell P, Ladha JK and Sampio MJA (1973). Blue-green algae (Cyanobacteria): Some aspects related to their role as sources of fixed nitrogen in paddy soils. In: *Nitrogen and Rice*, IRRI, Los Banos, p. 263.

Subrahmanyan R, Relwani LL and Manna GB (1965). Fertility build-up of rice field soils by blue-green algae. *Proc Indian Acad Sci*, 62B: 252–277.

Sundara Rao WVB, Goyal SK and Venkataraman GS (1963). Effect of inoculation of *Aulosira fertilissima* on rice plants. *Curr Sci*, 32: 366–367.

Tandeau De Marsac N (1994). Differentiation of harmogonia and relationships with other biological processes. In: *The Molecular Biology of Cyanobacteria*, (Ed) Bryant DA. Kluwer Academic Publishers, Dordrecht, The Netherlands, p. 825–842.

Thajuddin N and Subramanian G (2005). Cyanobacterial biodiversity and potential applications in biotechnology. *Curr Sci*, 89: 47–57.

Than Tun (1969). Effect of fertilizers on the blue-green algae of the soils of the paddy fields of Mandalay Agricultural Station. *Union Burma J Life Sci*, 2: 257–258.

Traore TM, Roger PA, Reynaud PA and Sasson A (1978). N_2-fixation by blue-green algae in a paddy field in Malia. *Cah. ORSTOM Ser Biol*, 13: 181–185.

Venkataraman GS (1961). The role of blue-green algae in agriculture. *Sci Cult*, 27: 9–13.

Venkataraman GS (1975). The role of blue-green algae in tropical rice cultivation. In: *Nitrogen Fixation by Free-living Microorganisms*, (Ed) Stewart WDP. Cambridge University Press, Cambridge, p. 207–218.

Venkataraman GS and Goyal SK (1968). Influence of blue-green algae inoculation on the crop yield of rice plants. *Soil Sci Plant Nutr*, 14: 249.

Volk RB (2005). Screening of microalgal culture media for the presence of algicidal compounds and isolation and identification of two bioactive metabolites, excreted by the cyanobacteria *Nostoc* insulare and *Nodularia* harveyana. *J Appl Phycol*, 18: 145–151.

Volk RB (2006). Antialgal activity of several cyanobacterial exometabolites. *J Appl Phycol*, 18: 145–151.

Watanable A (1959). Distribution of nitrogen-fixing blue-green algae in various areas of south and east Asia. *J Gen Appl Microbiol*, 5: 21–29.

Watanable A (1973). On the inoculation of paddy fields in the pacific area with nitrogen–fixing blue–green algae. *Soil Biol Biochem*, 5: 161.

Watanable A and Yamamoto Y (1971). Algal nitrogen fixation in the tropics. *Plant Soil*, Spl Vol: 403–413.

Whitton BA (1992). Diversity, ecology and taxonomy of the cyanobacteria. In: *Photosynthetic Prokaryotes*, (Eds) Mann NH and Carr NG. Plenum, New York, 1–51.

Whitton BA, Grainger SLJ, Hawley GRW and Simon JW (1991). Cell-bound and extracellular phosphatase activities of the cyanobacterial isolates. *Microbial Ecol*, 21: 85–98.

Yoshida T, Roncal RA and Bautista EM (1973). Atmospheric nitrogen-fixation by photosynthetic microorganisms in a submerged Philippine soil. *Soil Sci Plant Nutr*, 19: 117–123.

Soil Microflora, 2009 *Pages* **340–345**
Editor: **Rajan Kumar Gupta, Mukesh Kumar & Deepak Vyas**
Published by: **DAYA PUBLISHING HOUSE, NEW DELHI**

Chapter 27

Use of *Azolla* as Cheap and Sustainable Source of Feed and Other Utilities for Future

G. Abraham, Raghubir Shah and Dolly Wattal Dhar*
Centre for Conservation and Utilization of Cyanobacteria, Indian Agricultural Research Institute, New Delhi – 110 012

ABSTRACT

The free floating aquatic fern *Azolla* forms symbiotic association with the nitrogen fixing cyanobacterium *Anabaena azollae* and is being used as a biofertilizer. This fern is widely distributed in ponds, ditches and canals abundantly and conducts both photosynthesis and nitrogen fixation. However, very few information is available on the importance of *Azolla* as food or feed and its potential has not been tapped fully in this regard. This article highlights on the uses of *Azolla* as food or feed other than its conventional use as biofertilizer and green manure.

Keywords: *Azolla, Anabaena azollae, Biofertilizer, Feed.*

Introduction

Azolla, the aquatic pteridophyte has a global distribution and occurs in freshwater habitats of tropical, sub-tropical and warm temperate regions. It is important as a biofertilizer because of its association with the symbiotic cyanobacterium *Anabaena azollae*. According to an estimate the potential, capacity of *Azolla* to fix dinitrogen in the field has been established as 1.1 Kg N ha^{-1} (Watanabe *et al.*, 1977). Because of the remarkable ability to fix dinitrogen at high rates *Azolla* has long been used as a

* Corresponding Author: E-mail: gabraham1@rediffmail.com

green manure in rice cultivation (Tuan and Thuyet, 1979; Liu 1979). Although, rice paddy fields are ideal habitat for *Azolla* it has also been found in ponds, ditches and canals. According to Wagner (1997) *Azolla* has many uses such as use as human food, animal feed, medicine, production of biogas, hydrogen fuel, water purifier, weed control, reduction of ammonia volatilization and aptly referred it as "green gold mine". *Azolla* harvested in large quantities from water bodies in parts of tropical Africa, India and South west Asia was used as fodder for cattle and pigs (Sculthorpe, 1967). Pirie (1976) have attempted the use of weeds and planktonic algae as potential sources of food. However, no serious attempt was done on *Azolla* on its exploitation as a source of food or feed.

The nutritional potential of *Azolla* with respect to food and feed still remains as a least researched subject. In India Singh and Subudhi (1978) and Subudhi and Singh (1978) attempted utilization of *Azolla* as poultry feed. In the past however, few attempts have been made to explore the possibility of using weeds and aquatic algae from marine as well as fresh water habitats (Patterson *et al.*, 1967; Blum and Calet, 1975; Muztar *et al.*, 1976). People in China, Vietnam and Philippines are taking up *Azolla* cultivation in a big way but unfortunately it is not that popular in our country despite its wide range of availability. It can easily be adopted by farmers in India who do not have large areas of cultivable land for the production of fodder. The present article highlights the importance of *Azolla* as food and feed.

Poultry Feed

Azolla can be used for feeding poultry and both dried and fresh biomass has been found useful in this regard. *Azolla* is rich in proteins, essential amino acids, vitamins (vitamin A, vitamin B_{12}, β-carotene), growth promoting intermediates and minerals such as calcium, phosphorous, potassium, iron, copper and magnesium etc (Table 27.1). Nutritional value of *A. pinnata* has been analyzed by Subudhi and Singh (1978) and they observed that inclusion of fresh *Azolla* in the diet of chicken can replace about 20 per cent of commercial feed. Ali and Leeson (1995) observed increase in growth and body weight values in broilers fed with *Azolla*. This was similar to results obtained from using maize-soybean meal. Feeding up to 8 per cent fresh *Azolla* has resulted in the carcass quality of broiler chicken without any adverse impact (Ardakani *et al.*, 1996). Parthasarathy *et al.* (2002) also highlighted the importance of *Azolla* as feed. According to Becerra *et al.* (1995) fresh *Azolla* can partially replace whole soybeans up to a level of 20 per cent of the total crude protein in diet fattening duck.

Cattle Feed

Azolla is widely used along with commercial cattle feed in varying ratios. This has led to improvement in the milk production as well as quality of milk. (Kamalasanan Pillai *et al.*, 2002). According to them the overall increase in milk yield was about 15 per cent when 1.5-2.0 Kg of *Azolla* was combined with regular feed. A comparative study conducted by Kamalasanan Pillai *et al.* (2005) showed the superiority of *Azolla* with respect to other fodder plants in relation to the protein content (Table 27.2).

Fish Feed

Azolla has been thoroughly exploited as a fish feed in China and high value fishes such as *Tilapia rendalli* and *Cteerophanyngodon idellus* have been cultured in *Azolla* pond. Dried powder of *Azolla* was also used for feeding fish. Chu (1987) observed increase in the body weight *Tilapia nilotica* due to *Azolla* feeding. The economics of *Azolla* production has also been worked out by them and it was found out to be quite economically feasible (Table 27.3). According to Das *et al.* (1994) digested *A. pinnata* slurry is an ideal fish pond fertilizer and it has significantly increased the phytoplankton population. In China rice-Azolla-fish culture system is quite successful in Fujihan, China (Watanabe and Liu, 1992).

Table 27.1: Chemical Composition of *Azolla*

Constituents	*Percentage on Dry Matter Basis*
Ash	10.5
Crude fat	3-3.36
Crude protein	24-30
Nitrogen	4-5
Phospjorious	0.5-0.9
Calcium	0.4-1.0
Potassium	2-4.5
Magnesium	0.5-0.65
Manganese	0.11-0.16
Iron	0.06-0.26
Soluble sugars	3.5
Crude fibre	9.1
Starch	6.54
Chlorophyll	0.34-0.55

Source: Singh and Subudhi (1978).

Table 27.2: Comparison of Biomass and Protein Content of *Azolla* with Different Fodder Species (t/ha)

Species	*Annual Production of Biomass*	*Dry Matter Content*	*Protein Content*
Hybrid napier	250	50	4
Leucen	80	16	3.2
Cow pea	35	7	1.4
Sorghu	40	3.2	0.6
Azolla	730	56	20

Source: Kamalasanan Pillai *et al.* (2002).

Human Feed

Information is available on the use of *Azolla* in soup or "*Azolla-meat*" balls as a food for man (Van Hove, 1989). Although it is fit for human consumption as human food but the recipes are either not available or not published which further restricts its use in this direction.

Other Uses

The medicinal properties of *Azolla* were published in a book by Li-Shi-Zhen in China in the 16th Century (Shi and Hall, 1988). According to Wagner (1992) it has been used in Tanzania to treat people with cough. It can also be exploited as a source of hydrogen fuel. According to Newton (1976) the system is capable of evolving hydrogen at a rate of 760 nmol H_2 g^{-1} fresh weight $hour^{-1}$. Park *et al.* (1991) obtained hydrogen production of 83 ml H_2 g^{-1} Chlorophyll day^{-1} by further modifying the growth conditions. Since hydrogen is a non polluting fuel work in this direction may help us to tap the hydrogen energy in view of the current grim energy scenario. Another aspect of *Azolla* use is in the production of biogas. Van Hove (1989) reported that anaerobic fermentation of *Azolla* resulted in the production of methane. Similarly Das *et al.* (1994) obtained increase in gas production with a mixture of *Azolla* and cow dung as compared to *Azolla* alone.

Table 27.3: Economics of *Azolla* Production with NAEDEP Method

Sl.No.	*Cost of Production in 4 units in an year*	*Amount (Rupees)*
1.	Cost of 120 guages lipauline 2.8 x 1.6 x 4 m	400.00
2.	Labour charges for bed preparation	100.00
3.	Cow dung	146.00
4.	Superphosphate	7.50
5.	Magnesium sulphate	4.0
6.	Micronutrients	15.0
7.	New *Azolla* 200 g x 8 = 1600 g	5.0
	Total cost of production	677.50
	Total production in 4 units (1 Kg/unit/day for 350 days)	1050.00
	Cost/Kg	0.65 Rs/Kg

Source: Kamalasanan Pillai *et al.* (2002).

Azolla has the weed control property and according to Krock *et al.* (1991) there is relation between *Azolla* cover and weed population. Another important aspect of *Azolla* is the property to control mosquitoes. It was observed that breeding by *Anopheles spp* was almost completely suppressed in water bodies due to a thick cover of *Azolla* (Ansari and Sharma, 1991). *Azolla* is gaining importance in the phytoremediation of polluted environment. Arora and Saxena (2005) successfully cultivated *A.microphylla* biomass in secondary treated Municipal waters of Delhi. Tolerance and phytoremediation potential of three different species of *Azolla* to chromium has also been carried out by Arora *et al.* (2006). Therefore, it appears that it may be successfully used to purify the polluted water bodies also.

Future Thrusts

The ever increasing population, demand for more food, escalating prices of crude in the International market and inflationary pressures etc. has put enormous burden on the economic policies of the Government. It is in this context that alternate sources of energy and food sources become important. More over they are non-polluting as compared to the chemical nitrogen fertilizers. Therefore it is high time that research priorities must be reoriented to exploit alternate sources of energy and food to save foreign exchange. Strain improvement could be attempted and biotechnological tools must be used to enhance the nitrogen fixing potential of *Azolla*. The exact nature of the symbiosis must be identified. Other efficient cyanobacterial strains may be used to achieve symbiosis and efficient nitrogen fixation. Hybrids resistant to abiotic stresses such as salinity, temperature, heavy metal pollution, UV-B etc. must be developed. Nutritional composition of different *Azolla* strains must be screened and attempts made to enhance the nutritional value of promising strains. Proper extension activities must be conducted to popularize the use of *Azolla* biomass as a human feed as well as poultry, cattle and fish feed. There should be concerted efforts to bring together Scientists from allied disciplines to carry out collaborative research programs to bring out the best from this wonderful organism. India has an agricultural based economy where agricultural farming, dairy farming, live stock sector etc. play a major role. The ever increasing population has resulted in need for more food in terms of milk and meat products. Fodder production has shown considerable decline owing to decrease in forest cover as well as grass lands. Therefore a new emphasis has to be laid on the identification of alternate and better fodder stuff which can be used in livestock production. It is in this context *Azolla* finds importance as a potential candidate as human feed, animal feed, fish feed or poultry feed.

Acknowledgement

Financial Assistance from Indian Council for Agricultural Research is gratefully acknowledged. We also thank the Director and Joint Director (Research) for encouragement.

References

Ali MA and Leeson S (1995). The nutritive value of some indigenous Asian poultry feed ingredients. *Anim Feed Sci Technol*, 55: 227–237.

Ansari MA and Sharma VP (1991). Role of *Azolla* in controlling mosquito breeding in Ghaziabad District Villages (U.P). *Ind J Malarial*, 28: 51–54.

Ardakani M, Shivazad H, Mehdizadeh SMS and Novrozian H (1996). Utilization of *Azolla filiculoides* in broiler nutrition. *J Agric Sci*, 2(5–6): 35–46.

Arora A and Saxena S (2005). Cultivation of *Azolla microphylla* biomass on secondary treated Delhi Municipal effluents. *Biomass and Bioenergy*, 29(1): 60–64.

Arora A, Saxena S and Sharma DK (2006). Tolerance and phytoaccumulation of chromium by three *Azolla* spp. *World J Microbiol Biotechnol*, 22(2): 97–100.

Becerra M, Preston TR and Ogle B (1995). Effect of replacing whole boiled soybeans with *Azolla* in the diets of growing ducks (http://www.fao.org)

Blum JC and Calet C (1975). Nutritive value of Spirulines algae for broiler chicks (in French). *Ann Nutr Alimentation*, 29: 651–674.

Chu LC (1987). Re-evaluation of *Azolla* utilization in agricultural production. In: *Proceedings of the Workshop on Azolla Use*, Fuzhou, Fujian, China, International Rice Research Institute, Philippines, p. 67–76.

Das D, Sikdar K and Chatterjee AK (1994). Potential of *Azolla pinnata* as biogas generator and as a fish feed. *Ind J Environ Health*, 36: 186–191.

Kamalasanan Pillai P, Prema Latha S and Rajamony S (2002). *Azolla*: A sustainable feed substitute for livestock. *LEISA*, 4(1): 15–17.

Kamalasanan Pillai P, Prema Latha S and Rajamony S (2005). *www.leisa.info*.

Krock T, Alkamper J and Watanabe I (1991). *Azolla's* contribution to weed control in rice cultivation. *Plant Res Develop*, 34: 117–125.

Liu CC (1979). Use of *Azolla* in rice production in China. In: *Nitrogen and Rice*. International Rice Research institute, Los Banos, Philippines, pp. 375–394.

Muztar AJ, Slinger SJ and Burton JH (1976). Nutritive value of aquatic plants for chicks. *Poultry Science,* 55: 1917–1922.

Newton JW (1976). Photoproduction of molecular hydrogen by a plant-algal symbiotic system. *Science,* 191: 559–561.

Park IH, Rao KK and Hall DO (1991). Photoproduction of hydrogen, hydrogen peroxide and ammonia using immobilized cyanobacteria. *Int J Hydrogen Energy,* 16: 313–318.

Parthasarathy R, Kadirvel and Kathaperumal V (2002). *Azolla* as a particle replacement for fish meal in broiler rations. *Ind Vet J,* 79(2):144–146.

Patterson SPT, Chandler E,B, Kalan AR, Leoblich III, Fuller G and Benson AA (1967). Food value of red tide (*Gonyaulax polyedra*). *Science,* 158: 789–790.

Pirie NW (1976). *Food Resources: Conventional and Novel*. Penguin Books, Harmondsworth.

Sculthorpe CD (1967). *The Biology of Aquatic Vascular Plants*. Edward Arnold, London.

Shi DJ and Hall DO (1988). The *Azolla–Anabaena* association: Historical perspectives, symbiosis and energy metabolism. *Bot Rev (Lancaster),* 54: 353–386.

Singh PK and Subudhi BPR (1978). Utilize *Azolla* in poultry feed. *Indian Farming,* 27(10): 37–38.

Subudhi BPR and Singh PK (1978). Nutritive value of the water fern *Azolla pinnata* for chicks. *Poultry Science,* 57(2): 378–380.

Tuan DT and Thuyet TQ (1979). Use of azolla in rice production in Vietnam. In: *Nitrogen and Rice,* International Rice Research institute, Los Banos, Philippines, pp. 395–405.

Van Hove C (1989). *Azolla and its Multiple Uses with Emphasis on Africa.* Food and Agricultural Organization, Rome.

Wagner GM (1992). Algae in agriculture with special emphasis on their application in rice production. In: *Proceedings of the First International Workshop on Sustainable Sea weed Resources Development in Sub-Saharan-Africa,* (Eds) Mihigeni KE *et al.* Windhoek, Namibia, 22–29th March.

Wagner GM (1997). *Azolla*: a review of its biology and utilization. *Bot Rev,* 63: 1–21.

Watanabe I and Liu CC (1992). Improving nitrogen fixing systems and integrating them into sustainable rice farming. *Plant and Soil,* 141: 57–67.

Watanabe I, Espinas CR, Berja NS and Alimango BV (1977). Utilization of the *Azolla–Anabaena* complex as a nitrogen fertilizer for rice. *IRRI Research Paper* Series, 11: 1–5.

Soil Microflora, 2009 *Pages* **346–352**
Editor: **Rajan Kumar Gupta, Mukesh Kumar & Deepak Vyas**
Published by: **DAYA PUBLISHING HOUSE, NEW DELHI**

Chapter 28

Soil Micro-diversity and its Importance in Agriculture

G.K. Sharma
P.G. Department of Botany, Hindu College, Moradabad

ABSTRACT

Soil biodiversity refers to the variability among the living organisms present in it. In brief, it refers to variability with in species or microorganism or some other living organism and the ecological complex on the earth. Soil biodiversity is the assemblage of different life forms. It can also be defined as the number of different microorganism and their relative frequency in the soil system. Soil biodiversity includes both flora and fauna. Soil flora includes bacteria, soil fungi, soil actinomycetes, algae, rhizoid, rhizome, root of higher plants. Soil fauna are protozoa, nematodes, insects, mites and earthworms.

In most terrestrial ecosystems the majority of the biodiversity present occurs below the soil surface, not above it. This is especially the case for agricultural systems in which emphasis is usually placed on maintaining a low diversity of plant species as well as small population of those above ground consumer organisms associated with them. Organisms present in below-ground soil perform various important process as decomposition, nutrient mineralization, energy flow and various transformation of the main nutrient cycles. These processes determine plant growth and this effects the yield of the crop. The soils also contain a broad range of herbivorous invertebrates, microbial pathogen and rhizosphere organism which can exert more direct effects on plant growth. There is linkage between the spectrum of soil microorganism and plant growth and organic matter return. (Setala and huhta,1991; Warde and Lavelle, 1971). Many agricultural practices such as tillage, use of fertilizers and water alter the soil biodiversity/organism. Agricultural yield depend upon the ability of soil organism to provide nutrients to the plant. The principal aim of agriculture is to provide food for people of the country. Soil biodiversity has direct consumptive value in agriculture. Russian scientist N.I. Vavilov estimated that about 80,000 edible plants have been used at one time or the other in human history, of which about

150 have even been cultivated on a large scale.

Today merely 10 to 20 plant species provide 80-90 per cent food requirements of the world. In India, rural communities, particularly the tribal obtain a considerable part of their daily food from wild plants. At one time, nearly all medicines were derived from biological resources. Around 20,000 plants species are believed to be used for medicine in the developing world. In India the knowledge about medicinal value of plants has evolved in the form of traditional systems of medicinal sciences like Unani. Ayurveda and Siddha. More than 8000 species are used in some 10,000 drug formulations. In agricultural soil systems, the main purpose of management is to produce harvestable producers, primarily through increased production of specific crop plants. The production of crop plants is effected by a range of disturbance factors (*i.e.* those involving rapid changes in environmental condition) and stress factors (*i.e.* those involving non varying harsh condition), which emerge from such practices as preparation of seed beds, controlling organism which compete with or consume crop plants, ensuring adequate plant nutrition and moisture, management of plant residues and harvesting of crop plants. These all have the potential to influence soil organism. Thus it is evident that soil biodiversity is determined by both disturbances and stress (Grime, 1979; Huston, 1994). If these disturbances and stress are not found in a given soil habitat, definitely soil biodiversity is characteristically poor because a small number of very successful species monopolize most of resources and effectively exclude others. Disturbances and stresses are necessary to suppress dominant organisms and to survive other organism. Thus both of them can coexist to each other and thus these are enhancing soil biodiversity. This is also possible that stress and disturbance cause a adverse condition to the soil organism resulting the reduced diversity. In this more species are lost and few of them can survive and colonize.

Except stress and disturbance, some other factors are also important in regulating soil biodiversity. The agricultural practices is also important to determine the number of soil organism. These practices exert a longer term influence on the soil organism through determining the composition of the plant community. These influences are very important and are manifested through adding the quantity and quality of organic matter in the soil. Thus these influences can be a powerful determinants of the soil microorganisms and some adequate resource input is required to maintain soil organic matter (the basic substrate for soil organisms) in the long term.(Parton *et al.*, 1987).

This chapter discuss some important aspects of soil which are relevant to agricultural practices and soil microorganism.

Introduction

There are three functionally different components of the soil biota: Soil associated food webs which regulate decomposition processes, microorganism involved in nutrient transformation and soil associated herbivores. All these components are very important for determining soil biodiversity. Although these components are relatively independent to each other but affects soil biodiversity.

Food Webs of Decomposer

A number of soil microbes attack the dead remains of plants and cause decomposition. In this process complex organic matters are converted into simple organic compounds. Compounds like sugars, starch and protein are decomposed first in the decompositions process and then cellulose, fatty substances and lastly lignin and woody substance are degraded. Protein when acted by microbes are converted into amino acids, ammonium salts, nitrates and nitrites. Humus, an intermediate product of decomposition process, is formed by microorganism in optimum physical conditions. In the decomposition process, a number of complex mineral compounds are also converted into simpler and soluble compounds. Organic acids and Carbon dioxide that are released by decomposition make insoluble phosphates and other unavailable compounds more easily available to plants.

Decomposition and mineralization of nutrients are controlled by the food web of decomposers. This includes bacteria and fungi (primary saprophytes) and the soil animals. Soil animals are nematodes, protozoa and predators (microfood-web), mites and spring tails (litter transformers or mesofauna) in which organic matter is consumed and transformed into organic structures (faecal pellets) and the earthworms (soil ecosystem engineer) which build organo-mineral structures that create habitats for smaller organisms.

Decomposition of dead organic matter primarily helps in the feeding and growth process of these microorganism and secondly, increases the nutrient contents of the soil. Bacteria and soil fungi (primary saprophytes) are main agents which bring about the process of decomposition in the soil.

Agricultural practices affect soil organism. Most affected agricultural practice is tillage. This is one of the principal disturbances associated with intensive agriculture. A study to evaluate the affect of tillage on soil microbes and fauna is conducted by Beare *et al.* (1992). They noticed that higher level consumers are more adversely affected by tillage than smaller ones. Larger organisms (Higher trophic levels) control lower organisms (lower trophic levels) by predation and other activities. Another study is also conducted by Hendrix *et al.* (1986) to compare the affect of conventional tillage (CT) and non-tillage (NT) on soil microflora. They showed that tillage favored bacteria.

These effects in term had important consequences for decomposition rates, nutrient mineralization and soil organic matter loss. In case of these microbial activities are more then NT system. These differences were supported by Beare *et al.* (1992). They also showed that fungal based food webs contributed to decomposition in NT system, while bacterial based food webs contributed in CT Systems.

In recent years there has been increased use of agricultural chemicals, notably pesticides and fertilizers. Thus agricultural chemicals have no doubt increased crop fields but they have contributed bad effect on soil micro-organism or food web of decomposes. No doubts, these are also important components of agricultural intensifications. There are instances in which application of pesticides have been shown to induce compositional shifts in components of the soil fauna, particularly those associated with plant litter (Hendrix and Parmelee, 1985). It is expected that the amount and quality of resource inputs resulting from selection of plant species, addition of mulches and removel of weeds have important effects on microbial biomass and activity. There are so many studies which indicate that biomass of soil food web are highly responsive to species composition of herbaceous plants.

Grazing may affect below ground food-web by determining the quantity and quality of organic matter returned to the soil. Thus grazing can alter the flow of carbon to the below ground system. It has been demonstrated that heavy grazing by sheep favors 'fast' cycles dominated by labile substrates and bacteria, while light grazing favours 'slow cycle's dominated by resistant substrate and fungi (Bardget, 1996).

Microorganism Associated with Nutrient Transformation

Most soil microbes govern essential steps in cycling of major nutrients such as nitrogen and sulpher. Major cations (Ca^{++}, Mg^{++}, K^{+}) and some other nutrients are cycling in nature directly in without involvement of biological transformation.

Altought the concentration of already available inorganic substances also effect the rate of decomposition of added matter on soil. In addition after decomposition from humus the elements, N,P,K, Na, Mg, Ca etc. are released in soil. Some amount is taken up by the growing micro- organisms and remainder is made available to plants.

Organic phosphorus represents an unusable form of the elements with respect to the plant. However, organic compounds are eventually decomposed and phosphorus is released in an organic form, which is readily taken up by the plant. Most of the phosphorus of soil is present in inorganic form mainly as the phosphate ions ($H_3 PO_4^-$) and (HPO_4^-).Calcium is major exchange cation of fertile soils. Most of the exchangeable calcium of the soil is adsorbed on to the surface of clay micelles. These micelles are disc shaped bodies with a surface enveloping layer of negative charges. The micelle as a whole, may be said to be negatively charged. The negative charges of the micelle attract cation such as H^+ and Ca^{++} rather strongly, these cation being readily adsorbed to the surface of the micelle. If the hydrogen ion concentration is raised, Ca ions will be released and the H ions will take them place. This phenomenon is known as cation exchange. Other cations such as Mg^{++} and K^+ may also become adsorbed to the surface of clay micelles.

The number of microbes involved in nutrient transformation process is uncertain and unsatisfactory. Both Bacterial and fungal species are involved in this process. Bacterial species in a gram of soil are several thousand. Out of these a small percentage can be cultivated. There are no reliable estimates of the comparable number of fungal species in soil. Even the measurements of relative contribution of bacteria and fungi to the microbial biomass is problematic.

Rhizobia are important bacteria in agriculture and fix atmospheric N_2 in symbiosis with legumes. They inhabit in root nodules of leguminous plants. *Azotabacter*, and *Clostridium pasteurianum* live free in soil and fix N_2 into nitrogenous compounds, such as nitrates and nitrites. Actionmycetes, fungi (*Aspergillus, Alternaria, Mucor, Penicillium* and *Rhizopus* etc.) and a number of blue green algae are also known to fix free atmospheric nitrogen and there by increase the fertility of the soil. *Anabaena, Nostoc* and *Microcystis* are important nitrogen fixing blue green algae. *Nitrosococcus, Nitrosolobus, Nitrosospira, Nitrosovibrio* are nitrifying bacteria.

Fungi and actinomycetes have not been found to be associated with denitrification. It is carried out only by certain bacteria such as *Pseudmonas, Bacillus, Paracoccus* etc. Denitrifying bacteria are abundant in Arabic fields and count for about a million per gram soil. Their population is higher in rhizosphere soil. It has been established that in one hectare of ordinary soil every year 25-50 kg. of nitrogen are fixed and in cultivated soil and in soil containing legume plants 35-60 kg. and 100-400 kg of nitrogen are fixed respectively.

Now it is clear that soil have a vast microbial diversity. It means that hundred of years of work will needed before the degree of saturation in sufficient to have confidence of reasonable coverage of the prokaryotes or fungi.

We have very little knowledge about the factor which regulate the microbial biodiversity in soil. Bacteria and fungi of soil related to the diversity of substrates, which in turn would be determined by the diversity of the above ground plant community.

Some heavy metals also affect the diversity of soil bacteria. Bacterial population was reduced with increasing heavy metal concentration (Reber, 1992). A study revealed that population of rhizobia reduced with the increase of metal concentration.

pH of the soil is also important in regulating the soil biodiversity. Diversity of rhizobia nodulating

Phaseolus vulgaris decreased in an acid soil as compared with a soil of near neutral pH. The majority of soil fungi are found in acidic soils. Actinomycetes prefer saline soils and soil bacteria grow fairly well in the neutral soils richly supplied with organic nutrients. Microorganism are found in the soil at variable depths.

It is estimated that in soil microflora bacteria form about 90 per cent of the total microbial population. Fungi and algae together represent only 1 per cent and actinomycetes cover 9 per cent.

Soil Associated Herbivores

A herbivore may be defined as a species with one or more life stages that reside in the soil for the primary purpose of nutrition by feeding on (Vascular) plant roots or other underground plant associated structures. As such a herbivorous species may be soil dwelling for its entire life cycle, but commonly in groups such as arthoropods one or more life stages occur above ground: Considering the richness, abundance and diversity of life-styles in above-ground systems, relatively few insect species have exploited the underground parts of plants as food resources. Only 10 of the 26 order of insects are well represented as below ground herbivores. Even in these large orders, below ground herbivory is only well developed in restricted families or subfamilies.

Except insects, there is a higher richness and abundance in nematodes below ground than above ground in terrestrial ecosystem because the free living stage of these organisms depend on water films for survival. Root herbivore is confined to only two of the 17 orders (Hooper, 1978).

Plants respond chemically and physiologically to herbivore attack and there is now ample evidence that feeding induced changes in the host plant reduce herbivore 'fitness'. (Green and Ryan, 1972; Edwarch *et al.*, 1992). It is known that a competitive effects exists between the herbivores occupying different parts of the plant. They compete by exploiting a common resource such as phloem sap.

Aphids feeding on roots compete with other aphid species galling the leaves without ever coming directly into contact with them (Mcran and Whitham, 1990). Likewise herbivores that are separated in time are potential competitors too, if they exploit a common resource or induce a persistent response in the plant (Strauss, 1991). Most interaction are asymmetrical, with one species gaining substantial benefits and the other being adversely affected. Mostly host plant's stress response is increased by root herbivores and the abundance of some foliar feeders may be affected (Masters *et al.*, 1990; Muller–Scharer, 1991). Conversely, defoliation by herbivers above ground can affect below ground herebivores by including shifts in C allocation that may be reflected in new equation in root-shoot ratio, root tissue turnover and root quality (C: N ratio, allelochemicals) Even more complex interactions my occur, such as when foliar herbivory-suppresses or stimulates infection of roots by mycorrihizal fungi (Masters *et al.*, 1990) that can confer resistance to nematodes or root feeding insects (Dehme, 1982). Clavicipitaceous fungi that commonly occurs as endophytic mutualists in grasses (Clay, 1988, 1996), may produce alkaloids and induce phytoalexins that provide, depending on the particular grass-plant association, defense against above-ground and/or below ground invertibrates and vertibrate herbivores. Inter specific competition is an important force structuring herbivore communities (Denno *et al.*, 1995) and these interspecific interactions are sensitive to environmental condition imposed by different agricultural management regimes.

The population dynamics of individual species are influenced by a complexity of species interactions, within and between trophic levels. It is to be expected that agricultural practices will impact on soil herbivore species and communities in an idiosyncratic manner because the outcome is highly sensitive to the faunal composition and the nature and strength of interactions among species.

Often the indirect effects of agricultural practice are larger than direct effects.

Importance of Microdiversity in Agriculture

Micro-diversity is very important for maintaining the key ecosystem–level functions (*e.g.* decomposition, nutrient mineralization). Decomposition and nutrients are needed for crop productivity. Decomposed matter from dead remain of plant and animals serve as total soil organic matter. If green manures and crop residues are added in soil, the size of microbial community gets increased.

The soil microbes have either beneficial or harmful effects on the development of plant. The microorganism are intimately associated with roots of plant, therefore, any toxic or beneficial substance produced by them has direct effect on plant. The soil microorganisms catalyse the reactions in vicinity of roots in soil and produce CO_2 and form organic acids that in term solubilize the inorganic nutrients of plants. Aerobic bacteria utilize O_2 and produce CO_2, therefore, lower O_2 and increase CO_2 tension that reduces root elongation and nutrient and water uptake. There are many soil microorganism which produce growth stimulating substances and release elements in organic forms through the process of mineralization, many soil organisms including soil fungi and bacteria produce growth stimulating substances such as Indole acetic acid, Gibberellin and Cytokinim in the soil. *Fusarium* species have been found to secrete Gibberellin and Gibberellic acid. In the absence of oxygen some soil microorganism secrete chemicals such as, aldehydes, organic acids, etc. which may show toxic effects on many plants. Toxin secreting microbe may be fungi, bacteria and algae. *Fusarium lini,* which causes wilt of flax secretes HCN, a deadly poisonous substance and *Fusarium udum* causing wilt of pigeon pea secretes fusaric acid in the roots of the host plants. The soil microorganism influence phosphorous availability to plant through the process of mineralization and immobilization. However when plant suffers from nutrient scarcity during summer in tropical areas the microorganisms release the immobilized nutrients. Therefore, they act as sink between soil and plant roots in nutrient poor system. Some soil microbes change the availability or toxicity of sulpher to plants. Many soil microbes improve aeration of soil. Burrowing worms are also helpful in improving the aeration and percolation. Bacteria, blue green algae and some other microorganisms secrete mucilaginous substances which bind the soil particles into soil aggregator. Now it is clear that soil micro diversity play a critical role in maintaining the productivity of agricultural fields.

Thus soil fertility is a sustainable power to produce good yields of high quality which is directly associated with soil micro-diversity.

References

Bargett R D (1996). Potential effect on the soil mycoflora of changes in the UK agricultural policy for upland grasslands. In: *Fungi and Environment Change,* (Eds) Frankland J C Magan N and Gadd G M. Cambridge University Press, Cambridge, pp. 163–183.

Beare M H, Parmelee RW, Hendrix PF, Cheng W, Coleman D C and Crossley DA (1992). Microbial and found interactions and effects on litter nitrogen and decomposition in agro-ecosystem. *Ecological Monographs,* 62: 569–591.

Clay K (1988). Clavicipitaceous fungal endophytes of grasses coevolution and the changes from parasitism to mutualism. In: *Coevolution of Fungi with Plants and Animals,* (Eds) Hawksworth D L and Pirozynski K. Academic Press, London, pp. 79–105.

Clay K (1996). Interactions among fungal endophytes, grasses and herbivores. *Researches on Population*

Ecology 38: 191–201.

Dehne H W (1982). Interaction between vesicular arbuscular mycorrhizal fungi and plant pathogen. *Phytopathology,* 72: 1115–1119.

Denno RF, McClure MS and Off JR (1995). Interspecific interactions in phytophagous insects: competition re-examined and resurrected. *Annual Review of Entomology,* 40: 297–331.

Edwards PJ, Wratten SD and Parker L (1992). The ecologial significance of rapid wound-induced changes in plants: Insects grazing and plant competition. *Oecopogia,* 91: 266–272.

Green TR and Ryan CA (1972). Wound-induced proteinase inhibitor in plant leaves: A possible defence mechanism against insects. *Science,* 175: 776–777.

Grime JP (1979). *Plant Strategies and Vegetation Processes.* John Wiley and Sons, Chichester.

Hendrix PF and ParmePee RW (1985). Decomposition, nutrient loss and microarthropad densities in herbicide-treated grass litter in a Georgia piedmont agroecosystem. *Soil Biology and Biochemistry,* 17: 421–428.

Hendix PF, Parmelee RW, CrossPey DA, Coleman DC, Odum EP and Groffman PM (1986). Detritus food webs in conventional and no tillage agro-ecosystems. *Bioscience,* 36: 374–380.

Hooper DJ (1978). Structure and classification of nematodes. In: *Plant Nematology,* (Ed) Southey JF. HMSO London for Ministry of Agriculture Fisheries and Food, pp. 3–45.

Huston MA (1994). *Biological Diversity: The Coexistence of Species on Changing Landscape.* Cambridge University Press, Cambridge.

Masters GJ, Brown VK and Gange AC (1990). Plant mediated interactions between above and below ground insect herbivores. *Oikos* 66: 148–151.

Mcran NA and Whitham TG (1990). Interspecific competition between root-feeding and leat-galling aphids mediated by host plant resistance. *Ecology,* 71: 1050–1058.

Muller Scharer H (1991). The impact of root herbivores as a function of plant density and competition, growth and fecundity of *Centaurea maculosa* in field plots. *Journal of Applied Ecology,* 28: 759–776.

Parton WJ, Schimel DS, Cole CV and Ojima DS (1987). Analysis of factors controlling soil organic matter level in Great Plains grasslands. *Soil Science Society of America Journal* 51: 1173–1179.

Reber HH (1992). Simultaneous estimates of the diversity and the degradative capability of heavy metal affected soil bacterial communities. *Biology and Fertility of Soils,* 13: 181–186.

Setala H and Huhta V (1991). Soil fauna increases *Betula pendula* growth: laboratory experiments with coniferous forest floor. *Ecology,* 72: 665–671.

Vavilov NI (1951). The origin variation immunity and breeding of cultivated plants. *Chionica Botanica,* 13: 139–248.

Wardle DA and Lavelle (1997). Linkages between soil biota, plant litter quality and decomposition. In: *Driven by Nature: Plant Litter Quality and Decomposition,* (Eds) Cadisch G and Giller KE. CAB International, Wallingford, pp. 107–124.

Soil Microflora, 2009 *Pages* **353–361**
Editor: **Rajan Kumar Gupta, Mukesh Kumar & Deepak Vyas**
Published by: **DAYA PUBLISHING HOUSE, NEW DELHI**

Chapter 29

Soil Microflora of Rohilkhand Division

Iqbal Habib and U.K. Chaturvedi
Department of Botany, Government Degree College, Budaun – 243 001, U.P.

ABSTRACT

Soil harbour different types of microorganisms which collectively form the "Soil Microflora". These microorganisms exert a profound effect on the fertility of soil. Although this flora comprises of Bacteria, Fungi, Protozoa and Algae etc. but here we are reporting some algal flora growing on the soils of six districts of Rohilkhand Division of Uttar Pradesh. These algae were found growing on the surface of moist soils and some of them are capable of fixing nitrogen in the paddy fields used for cultivation of rice crop. The largest number of them reported belong to Chlorophyceae, Cyanophycae and Bacillariophyceae while few of them belong to some other unimportant group. These algae bring about beneficial changes in the soil and thus help in the fertility of the soil. As such, by their photosynthetic activity they are helpful in increasing the organic matter in the soil besides the nitrogen fixation.

Keywords: *Soil microflora, Algae, Nitrogen fixation, Soil fertility.*

Introduction

There is a close relationship between man and the soil. All his daily needs are fulfilled either directly or indirectly through the soil. Soil is made up of living and non-living entities. The living entities constitute the 'Soil microflora' while the non-living comprising of minerals, air and water. It is precisely here that microorganisms exert a profound influence upon all the living forms that exist on the earth. The soil microflora includes Bacteria, Fungi, Protozoa, Algae and virusus. In addition to this a very large number of insects and nematodes are also present. The majority of these microbes inhabit in the top layer of the soil and their number decreases with the depth. A spoonful of soil contains billions of microorganisms.

The systematic identification of the form is based on standard works of Tilden (1910), Rao (1937) Venkataraman (1939), Waksman (1957), Desikachary (1959), Singh (1961), Stewart (1970, 1976) Philipose (1967) and Linton (1971), Venkataraman (1972), Alexander (1977), Swaminathan (1980), Roth (1982), Venkataraman and Becker (1985), Shukla *et al.* (1986), Dubey and Dwivedi (1988), Pawar and Daginawala (1991), Habib *et al.* (1988a, b, c, d, e,f 1992a, b) and Pandey (1989a, b, 1990), Habib (1990, 1993) and Chaturvedi *et al.* (1990).

In the present article, we have restricted ourselves to the "Algal soil microflora" growing in the six districts of Rohilkhand Division, in Uttar Pradesh, in India. On a conservative estimate, a total number of 154 Algae were reported by the authors and their account was published through a series of papers in the last twenty years. The authors were overwhelmed to report that a majority of these algae belong to the class chlorophyceae and Bacillariophyceae, though few of them belong to Euglenophyceae and Vaucheriaceae. It was seen that the rainy season was largely favourable for their extensive growth. It was also observed that their large numbers on soil had been beneficial in increasing the fertility of the soil. Their decayed masses increase the organic content in the soil while a large number of blue-green algae are able to fix atmospheric nitrogen in the soil and thus are helpful in increasing the fertility of soils, especially of the rice fields.

Algae are ubiquitous in their distribution. A very large number of them grow well in moist, well aerated and fertile soils. They also thrive in moist acid soils as well as in the alkaline and calcareous soils. Terrestrial species do not show very conspicuous growth, except in certain regions with pronounced rainy season when they form extensive coating on the soil and make the soil slippery by their presence. They may also grow beneath the surface of the soil.

Subterranean forms are generally confined to the upper 50 cm of the soil, although they have been found as deep as two meters below the surface. Certain algae may be found to be of heterotrophic growth in the dark and utilize carbohydrates present in soil. Perhaps, the precence of certain algal members in such deep layers of the soils may be due to the result of washing down the resting spores from the upper surface by rainy water.

In the tropical countries like India, rice fields afford a very hospitable habitat for the growth of blue-green algae. A number of them also grow in alkaline 'usar' soils. The cultivated loam soils are also very favourable for these algae. As these algae enjoy a wide distribution in all kind of habitats, they therefore are a highly successful group among the plant kingdom. In water logging condition, these cyanobacteria multiply, and fix atmospheric N_2 and release it into the surroundings in the form of nitrates and nitrites and these are converted into amino acids, proteins and other growth promoting substances (Stewart, 1970). The role of the blue-green algae, *e.g. Aulosira, Anabaena, Cylindrospermum, Nostoc, Plectonema, Tolypothrix* in the rice fields of India was realised much earlier (Singh, 1961). The research work which is being carried out in India mainly belong to the centre of Advanced Study in Botany, Banaras Hindu University, Varanasi; The central Rice Research Institute, Cuttack; The Indian Council of Agricultural Research, New Delhi. Venkataraman (1961), used the term 'algalization' for the process of application of blue-green algal culture in field as biofertilizer. He was the person who initiated algalization technology in India and demonstrated the way how this technology could be transferred to farmer level (Venkataraman, 1972). Department of Biotechnology centre of U.P. (Lucknow) has reported the increase in yield of paddy (about 12.5 q/ha) to be due to cyanobacterial biofertilizer.

The cyanobacterial biofertilizer not only increase the productivity and quality of paddy, but also reduce the harmful effects of the chemical fertilizers.

'Chlorophyceae' members also constitute the 'Soil microflora'. They exhibit a great variety of form, shape and thallus organisation. These green algal members are common in sub-aerial conditions, such as moist soils, moist walls, tree trunks and moist brick works etc. Some of the green algae grow as symbionts in association with fungi forming lichens.

The algae belonging to the class 'Bacillariophyceae' are commonly called 'Diatoms'. They are the jewels of the plant kingdom. These algal members are distributed wherever nature provides damp surface. They can be seen as a yellow scum on the surface of mud in ditches or ponds. They are also abundant on damp soil, rocky walls, dry cliffs, on the bark of trees etc. Their fossil sedimentary deposits are known as "Diatomaceous earth".

Members of the class 'Xanthophyceae anthophyceae', such as *Vaucheria, Botrydium* etc, though constitute a small group of 'soil microflora', yet their role in the making of soil microflora cannot be overlooked.

Following classes and the algal members belonging to them, have been reported here. These taxa, which have been reported here, were collected from different districts of Rohilkhand Division in Uttar Pradesh in India and constitute a very important 'Soil Microflora'.

Systematic Enumeration of Taxa

Class: Cyanophyceae

Aphanotheae bullosa (Menegh) Rabenh

Stichosiphon sansibaricum (Hiem.) Dinnet and Daily

Oscillatoria terebriformis Agardh ex Gomout

O. latevirens (Crouan) Gomont

O. chilkensis Biswas

O.limnetica Lemm

Phormidium incrustatum (Naeg.) Gomont

P. luridum (Kuetz.) Gomont

P. corium (Ag.) Gomont

P. faveolarum (Mont.) Gomont

Lyngbya martensiana Menegh ex Gomont

L. contorta Lemm

L. arboricola Bruhl et Biswas

L. contorta Lemm. f. *major* forma nov.

Aulosira fertillissima Ghose

Cylindrospermum indentatum West

C. doryphorum Bruhl et Biswas

C. majus kuetz ex Born et Flah

Nostoc humifusum Carmichael ex Born. et Flah

N. linckia (Roth.) Born et Flah

Anabaena ambigua Rao

A. iyengarii Bharadwaja. var. *unipora* Singh

Scytonema behneri Schmidle

S. rofmanni Ag. ex Born. et Flah

Calothrix membranacea Schmidle

C. thermalis (Schwabe) Hansg

Rivularia aquatica De Wilde

R. globiceps West

Gloeotrichia ghosei Singh

G. natans Rabenh ex Born. et Flah

G. pisum Thuret ex Born. et Flah

G. raciborskii Woloszynska var *koshiense* Rao

Class: Chlorophyceae

Characium acuminatum Braun et Kuetz

Ulothrix flaccida Kuetz

Ulothrix fimbriata Bold

Uronema terrestris Mitra

Hormidium subtile (Kuetz) Heering

H. flaccida (Kuetz) Braun

Schizomerix leibleinii Kuetzing

Cylindrocapsa conferta West

Cladophora profunda Brand var. *nordstedtiana* Brand

Gongrosira circinata (Borzi) Schmidle

Oedogonium westii (Tiffany et Brown) Tiffany

O. angustum (Hirn) Tiffany

O. globrum (Hall) Hirn

O. acrosporum (De Bary) Hirn. var *floridense* Wofle

O. crassiusculum (Witt.) Hirn. var *indica* Venkat

Spirogyra oudhensis Randhawa

S. singularis Nordst

Zygnema incauspicum Czurda

Stigeoclonium fractum Berthold

S. tenue (Agardh) Kuetz

Chaetophora elegans (Roth.) Agardh

Coleochaete soluta pringsheim

Sirogonium megasporum (Jao) Trans

Desmids

Closterium cynthia De Not var. *robustum* (West) Krieger

Cl. ehrenbergii Menegh var. *malin verianum* (De Not) Rab

Cl. pleurodermaticum West and West

Cl. recurvum Prescott

Pleurotaenium indicum (Grun) Lund

Euastrum spinulossum Delp var. *africanum* Nordst.

Cosmarium angulatum (Perty) Rab f. *majus* Grun

C. auriculatum Reinsch

C. bioculatum Breb

C. connatum Breb var. *depressum* I. Marrie

C. granatum Breb

C. granatum Breb var. *occellatum* West and West

C. impressulum Elfv.

C. javanicum Nordst

C. lundelli Delp

C. margaritatum (Lund) Roy et Biss

C. moniliforme (Turp) Ralfs

C. nudum Turner

C. obtusatum Schmidle

C. phaseolus Breb

C. pseudoconnatum Nordst

C. pseudopyramidatum Lund

C. quadrum Lund

C. scabrum Turner

Staurastrum pinnatum var. *subpinnatum f. robustum* Krieger

S. sebaldi Reinsch

Chlorococcales

Pediastrum simplex var. *duodenarium* (Bailey) Rabenh

P. ovatum (Ehr.) A. Br.

P. duplex Meyen

P. tetras (Ehr.) Ralfs

P. tetras (Ehr.) Ralfs var *excisum* (Reinsch) Hansg

T. regulare Keutz

Oocystis gigas Arch

O. irregularis (Petkof.) Printz

Nephrocytium lunatum West

N. agardhianusm West

*Botryococcus braunii.*Kuetz

Ankistrodesmus falcatus (Corda) Ralfs

Selenastrum gracile Reinsch

Coelastrum microporum Naeg

Crucigenia quadrata Morren

Scenedesmus obliquus (Turp.) Kuetz

S. dimorphus (Turp.) Kuetz

S. bijugatus (Turp.) Kuetz

S. prismaticus Bruchl et Biswas

S. armatus (Chodat) Smith var. *asymmetricus* Philipose

S. denticulatus Lagerheim var. *linearis* Hansg.

S. longus Meyen var. *naegelii* (Breb.) Smith

S. quadricauda (Turp.) Breb. var. *bicaudatus* Hansg

S. quactricauda (Turp.) Breb. var. *westii* Smith

Chlorococcum infusionum (Schrank) Menegh

Trebouxia humicola (Treboux) West et furitsch

Korshikoviella gracilips (Lemm.) Silva

Dicanthos belenophorus Korshikov

Golenkinia radiata Chodat

Class: Bacillariophyceae (Diatoms)

Melosira granulata (Ehr.) Ralfs

Cyclotella kuetzingiana Thwaites

C. meneghiniana Kuetz.

Synedra ulna (Nitz.) Ehr.

S. ulna (Nitz.) Ehr. var. *danica* (Kuetz.) Grun

Cocconeis pediculus Ehr.

Gyrosigma acuminatum (Kuetz.) Rabh.

G. scalproides (Rabh) Cleve

G. spencerii (Smith) Cleve

Stauroneis phoenicentron Ehr.

Navicula anglica Ralfs

N. decussis Ostrup

N. dicephala (Ehr.) Smith

N. exigua (Greg) Mueller

N. tuscula (Ehr.) Grun

Gomphonema parvulum. (Kuetz.) Grun

G. gracile Ehr.

G. sphaerophorum Ehr.

Cymballa cymbiformis Ag.

C. tumida (Breb.) var Heurek

C. ventricosa Kuetz.

Nitzschia angustata var. genuina Meister

N. clausii Hantz

Epithemia argus (Ehr.) Kuetz

Surirella linearis Smith

S. pseudolinearis Krasake

S. robusta Ehr.

Class: Euglenophyceae

Euglena acus Var. *hyalina* klebs

E. acus Ehr. var. *vanayei* Delfs

E. acutissima Lemm.

E. acutissima Lemm. var. *purva* Playf

E. alata Thomp.

E. allorgei Delf.

E. angusta Bern.

E. anabaena Manix

E. angusta Bern

E. clavata skuja

E. elenkinii polj.

E. fransundulata John

E. gracillis klebs var *urospora* cho. et pron

E. quentheri Goj.

E. inflata Massart.

E. lata Swirenko

References

Alexander M (1977). *Introduction to soil Microbiology*. John Wiley and Sons Inc, New York.

Chaturvedi UK, Habib I and Pandey UC (1990). Algae of Rohilkhand Division–XI UP, India. *Biojournal*, 2: 247–249.

Desikachary TV (1959). *Cyanophyta*. ICAR, New Delhi, pp. 686.

Dubey RC and Dwivedi RS (1988). *J Indian Bot Soc*, (67): 154–162.

Gupta RK, Kumar M and Paliwal GS (Eds). *Glimpses of Cyanobacteria*. Daya Publishing House, New Delhi, p. 78–86.

Habib I, Pandey UC and Shukla HM (1988a). Blue green algae from paddy fields of Bareilly district, UP, India. *Geobios New Reports*, 7: 157–159.

Habib I, Pandey UC and Shukla HM (1988b). Diatoms from paddy fields of Bareilly district. *Geobios New Reports*, 7: 160–161.

Habib I, Pandey UC and Shukla HM (1988d). Some chlorococcales from paddy fields of Bareilly district, UP India. *Geobios New Reports*, 15(2–3): 69–75.

Habib I and Pandey UC (1989a). Chlorococcales of Budaun–I. *Ad. Plant Sci.*, 2(2): 272–277.

Habib I and Pandey UC (1989b). On some desmids from paddy fields of Bareilly district, Uttar Pradesh, India. *J Phytol Res*, 2(2): 155–160.

Habib I and Pandey UC (1990). The Euglenophyceae of Bareilly, India. *Ad. Plant Sci*, 3(2): 245–250.

Habib I, Pandey UC and Chaturvedi UK (1992). An enumeration of Chlorococcales from Rampur District, UP, India. *Ad Plant Sci*, 5(1): 200–202.

Habib I and Pandey UC (1989). On some desmids from paddy fields of Bareilly district, UP, India. *J Phytol Res*, 2(2): 155–160.

Habib I and Pandey UC (1989e). An enumeration of chlorococcales from Mala forest, Pilibhit, UP, India. *Ad Plant Sci*, 2(2): 184–190.

Habib I and Pandey UC (1989f). Desmids of Shahjahanpur. *IBC*, 8: 83–87.

Habib I (1990). Diatoms from Moradabad district (UP), India. *Res J Pl Environ*, 6(2): 17–19.

Habib I (1993). Some fresh water diatoms from Rampur district, UP, India. *Bioved*, 4(1): 97–98.

Habib I, Shukla HM and Pandey UC (1992). A preliminary survey of Cyanophyceae of Mala forest, Pilibhit, UP. *J Econ Tax Bot*, 16(2): 367–371.

Nandi SK and Palni LMS (1992). In: *Microbial Activity in the Himalaya*, (Ed) RD Khulbe. Shree Almora Book Depot, Almora, p. 419–428.

Philipose MT (1967). *Chlorococcales*, ICAR, New Delhi, pp. 365.

Powar CB and Daginawala HF (1991). *General Microbiology*, Vol. II. Himalaya Publishing House, Bombay, p. 1–7.

Rao CB (1937). Myxophyceae of United Provinces–III. *Proc Indian Acad Sci*, 7: 339–375.

Roth FX (1982). In: *Advances in Agricultural Microbiology*, (Ed) Subba Rao NS. Oxford and IBH Publ Co, New Delhi, p. 663–676.

Singh RN (1961). *The Role of Blue-green Algae in Nitrogen Economy of Indian Agriculture.* ICAR, New Delhi, p. 125–137.

Stewart WDP (1976) (Ed). *Nitrogen Fixation of Free-living Organisms.* Cambridge, New York.

Stewart WDP (1970). *Nature (London)*, 214: 603.

Swaminathan MS (1980). *Second Annual Day Lecture*, NBRI, Lucknow.

Tilden JE (1910). *Myxophyceae in Minnesota Algae–I.*

Venkataraman GS (1939). A systematic account of some Indian diatoms. In: *Proc Indian Acad Sci*, 10: 193–268.

Venkataraman GS (1972). *Algal Biofertilizers and Rice Cultivation*, Today and Tomorrow Printers and Publ, New Delhi.

Venkataraman LV and Beeker EW (1985). *Biotechnology and Utilization of Algae: The Indian Experiance.* CFTRI, Mysore.

Soil Microflora, 2009 *Pages* ***362–369***
Editor: **Rajan Kumar Gupta, Mukesh Kumar & Deepak Vyas**
Published by: **DAYA PUBLISHING HOUSE, NEW DELHI**

Chapter 30

Role of Cyanobacteria in Amelioration of Soil

Pranita Jaiswal*
Department of Botany,
University of Delhi, Delhi – 110 007

ABSTRACT

Cyanobacteria are heterogeneous assemblage of photosynthetic organism, which successfully colonize diverse ecological habitats owing to their unique physiological characters and high adaptive capability. Many diazotrophic Cyanobacteria constituting dominant flora in rice field, not only contribute 25-30 kg N/hectare but also significantly improve physical chemical and biological property of soil, with residual effect on succeeding crops. Despite of their global recognition as a source of N in rice field, their potentiality as biological agent for remediation and amelioration of soil has not been completely exploited. The information available on the immense potential of these ubiquitous environmentally safe microorganisms in contributing towards sustainable productivity has been presented.

Keywords: *Amelioration; Biofertilizer; Cyanobacteria; Reclamation.*

Introduction

Cyanobacteria or blue green algae (BGA) are an ancient group of photosynthetic prokaryotes with oxygen evolving photosynthesis. They resemble bacteria on one hand in prokaryotic cellular organization and exhibit metabolic activity like higher plants on the other hand (Stanier and Cohen Bazire, 1977). They have been utilized as model system for understanding various physiological

* Corresponding Address:156, Rail Vihar, Sector-33, Noida – 201303;
E-Mail: pranitajaiswal@gmail.com; pranitajaiswal1@hotmail.com

processes (Jaiswal and Kashyap, 2002). They have ability to invade inhospitable habitats *e.g.* hot spring, volcanic soils, bare rocks; wide range of aquatic and damp environments *e.g.* arctic, freshwater, hyper-saline, deep sea, coastal soda-lakes, soils. Besides they are symbiotically associated with different types of plants ranging from bacteria to angiosperm. Cyanobacteria exhibit considerable morphological diversity ranging from unicell, cocoid and palmelloid to filamentous and branched forms. Under specific environmental stimuli many cyanobacteria initiate formation of two distinct types of cells-akinetes and heterocysts. Akinete represent the resting phase of cyanobacteria to combat adverse environmental conditions, having capacity to germinate on the onset of suitable conditions. Heterocysts are the main site of nitrogen fixation. Heterocystous cyanobacteria constitute the dominant component of microbial flora of rice field. The morphological difference is associated with the compartmentalization of autotrophy and diazotrophy *i.e.* vegetative cell is the site of oxygenic photosynthesis, while heterocyst is the site of nitrogen fixation (Wolk, 2000). This dual ability to photosynthesize and fix nitrogen simultaneously makes them an efficient biological system. The heterocystous forms commonly associated with N_2-fixation in rice field are *Anabaena, Nostoc, Cylindrospermum, Gloeotrichia, Scytonema.* Diazotrophy has also been reported in few unicellular, non-heterocystous *Cyanobacteria e.g. Aphanothece, Gloeocapsa (Gloeothece), Plectonema, Trichodesmium* etc. (Singh, 1973; Galon *et al.*, 1975;). The fixed nitrogen is made available to plants during the life cycle (under certain conditions) or after the death by decomposition of cells. The various beneficial effects of these microorganisms besides N-fixation include reduction in methane emission, transformation of P, Fe, Mn, Zn, Cu, pesticide degradation, improving soil stability, volatilization of ammonia, suppressing weeds etc. This compilation focuses on the potential of these unique microorganisms in amelioration of soil and their agricultural and environmental applications.

Nitrogen Fertilization

The fast growing human population has led to an increased use of chemical fertilizer, in an effort to enhance the production of agriculture and horticulture crops, to fulfill the growing demand. This has led to a gradual decline in soil productivity and nutrient deficiency. Cyanobacteria offer an easily manageable, eco-friendly and economically attractive alternative to chemical fertilizer

Rice is a staple food for more than 40 per cent of world population and is grown in about 43 m ha area in India (Pabbi, 1981). Application of BGA in field not only results in addition of N but also increase the productivity of rice upto 10-15 per cent. The compositions of algal biomass in rice field have been studied in many parts of the world including India (Singh, 1961; Bunt, 1961; Reynaud and Roger, 1978; Wilson and Alexender, 1979; Venkataraman 1981; Singh 1985; Halperin *et al.*, 1992). The agronomic potential of BGA in the field of agriculture was first recognized by De (1939). A study carried out at International Rice Research Institute (IRRI) Manila (Watanabe *et al.*, 1977) indicated that biological nitrogen fixation could sustain the fertility of flooded rice field during 12 years for 3 successive crops. In Japan algal inoculation resulted in a progressive increase in rice yield and was fond to be equivalent to application of 60 kg ammonium sulphate per hectare. Successful establishment of promising strains of algae has been reported to enhance the crop yield upto 30 percent (Singh, 1961; Venkataraman, 1972; Goyal, 1993). They provide 15-25 kg biologically fixed nitrogen, beside many other beneficial effects on soil quality (Goyal, 2000). Even in the presence of chemical fertilizer, the algal biofertilizer has been reported to show supplementary effect and reduce the fertilizer dose by 25-30 kg.

The basic method of BGA biofertilizer production involves, growing a mixture of N_2-fixing BGA along with carrier in an trough/tank/pit/field, then drying the mixed culture in sun. The dried flakes are then collected, packed and used to inoculate the rice field (Singh, 1981). The open air production system was easily prone to contamination and also the production was greatly affected by fluctuating environmental conditions (Pabbi *et al.*, 2000). During the last decade the technology has undergone various changes, which include

1. Multiplication of algal biomass in semi controlled condition
2. Minimal growth media for faster growth of organism
3. Development of new carrier material

The production technology has been substantially improved with introduction of new and cheap carrier material, that support higher microbial load and longer shelf life. (Shanmugasundaram, 1996; Pabbi and Kaushik, 1997; Pabbi *et al.*, 2000).

Phosphate Immobilization

Another chemical fertilizer, whose rising cost is becoming major problem to farmers, is phosphate. This motivated scientists to look for an alternative, naturally occurring, dependable, biodegradable phosphatic fertilizer. Many cyanobacterial strains can utilize extracellular insoluble phosphate under both CN^+ and CN^- conditions A number of other cyanobacteria *e.g. Anabaena, Nostoc, Tolypothrix, Aulosira and Anacystis* have been reported to solubilize extracellular insoluble phosphates (Bose *et al.*, 1971). They are reported to show increased levels of intracellular and cell surface alkaline phosphatase activity under phosphate starved conditions to solubilize polyphosphates (Healey, 1982; Nateshan and Shanmugasundaram, 1989). However, in *Anacystis nidulans* presence of both alkaline and acid phosphatases activity has been reported at lower phosphate levels in the medium (Gupta, 1983). The alkaline phosphatases are quite stable enzyme and require specific temperature and pH for their optimal activity (Banerjee and Sharma, 2005; Singh *et al.*, 2006) Singh *et al.*, (2006) reported differential response of NaCl stress on cellular and extracellular phosphomonoeasterase (PMEase) activity. They observed an increase in cellular PMEase activity at NaCl concentrations of 20mM, while concentrations higher than 20mM favoured release of extracellular PMEase in *Anabaena* oryzae. They further reported that the cyanobacterium required Ca^{2+} and Mg^{2+} for the activity of APase and metal Pb^{2+}, Cr^{6+} and Ni^{2+} severely inhibited enzyme activity.

Liberation of Plant Growth Promoting Substances

Cyanobacteria are known to liberate wide array of extracellular substances *e.g.* plant growth regulators, vitamins, amino acids, sugars etc., which have direct or indirect impact on plant growth. A number of cyanobacterial strains have been reported to produce extracellular amino acids (Fogg, 1952). Venkataraman and Saxena (1963) reported presence of aspartic acid, glutamic acid and alanine in *Anabaena azollae* and *Nostoc* sp. Singh and Trehan (1973) also reported predominance of aspartic acid in extracellular filtrates of *Aulosira fertilissima* and *Anacystis nidulans* along with proline, valine and glycine at various stages of growth of the culture. Kartikeyan *et al.* (2008) recently reported presence of an array of amino acids in culture filtrates of cyanobacterial strains isolated from wheat rhizosphere.

Among the growth regulators gibberllin, auxin, ethylene, cytokinin, absicic acid and jasmonic acid have been detected in cyanobacteria (Gupta and Agarawal, 1973; Ordog and Pulz, 1996; Stirk *et al.*, 1996). There is an increasing evidence that cyanobacteria produce plant hormones or biomolecules with plant hormone like activity. Cyanobacterial extract have been observed to promote somatic embryogenesis (Wake 1992; Bapat, 1996). Manickavelu *et al.* 2006) demonstrated role of cyanobacterial extracellular product in induction of organogenesis in rice callus. They compared the effect of 2,4-D and extracellular product and found that cyanobacterial extracellular metabolite not only brought about better root induction but also better proliferation of roots. Kartikeyan *et al.*, (2007) also reported growth-promoting effect of cyanobacterial strains isolated from wheat rhizosphere on plant. The production of phytohormone IAA (indole acetic acid) by several free-living and symbiotically competent cyanobacterial strains was documented by Sergeeva *et al.* (2002). They confirmed the accumulation and release of IAA by these strains immunologically and chemically by gas chromatography. Recently Prasanna *et al.* (2008) reported release of IAA by several *Anabaena* strains.

Soil Aggregation

Cyanobacterial inoculation is also known to improve the stability of soil due to excretion of polysaccharides, lipids etc. (Oikarinen, 1996). An improvement in both chemical and biological properties has been reported following cyanobacterial inoculation (Roger and Burns, 1994; Issa *et al.*, 2001; Thomas and Dougill, 2006). Roger and Burns (2001) reported 50-63 per cent increase in total C content, 111-120 per cent increase in total N content and overall 18 per cent increase in aggregate stability, following inoculation of soil with *Nostoc muscorum*. They further reported that the effect of cyanobacterial inoculation was more pronounced if the inoculated soil was left undisturbed prior to planting, while, homogenization of soil and irrigation significantly reduced the seedling emergence, still it was higher than uninoculated soil. Zulpa de Caire *et al.* (1997) reported in a greenhouse experiment that exopolysacharides from the cyanobacterium *Nostoc muscorum* increased the soluble C by 100 per cent, microbial activity by 366 per cent of sodic saline soil. In another study conducted with organically poor semi arid soil, a remarkable increase in the structural stability, nutrient status and productivity was recorded following application of indigenous cyanobacterial strains (Nisha *et al.*, 2007).

Reclamation of "Usar" Soil

'Usar' soil are saline (solonchak) and alkaline (solonetz) unproductive soils, found extensively throughout the India. They are characterized by higher salt concentration and/or high pH. Though salts are found in small quantity in all the soils, their concentrations in 'usar' soil reaches to an extent that it becomes unfavorable for growth of vegetation. The reclamation of these soils basically requires removal of exchangeable sodium and its replacement with calcium, which is done mainly by addition of chemical corrective *e.g.* gypsum, pyrite–a rather expensive method or using mechanical process e.g. flooding with ample amount of water and addition of organic matter. The abundant growth of N_2 fixing cyanobacteria in waterlogged soils fulfills these requirements.

The concept of using cyanobacteria for reclamation of 'usar' soil had been introduced as early as 1950 by R.N. Singh. Later on in eighties Kaushik and his coworkers (Kaushik and Krishna Murti, 1981; Kaushik and Subhashini, 1985, Kaushik, 1985, 1989, 1991, 1994); had done extensive study on various aspects of reclamative potential of cyanobacteria and reported that addition of autochthomas cyanobacterial strains to such a system improved soil quality by making it aerable, decreasing the pH, exchangeable Na/Ca and increase in N, P, organic matter and water holding capacity of soil over a period of time. Thomas and Apte (1984) reported a 12-35 per cent reduction in soil salinity of "Kharland" soil from coastal Maharashtra due to introduction of salt tolerant *Anabaena torulosa*. Aziz and Hashem (2003) in an study conducted in Bangladesh observed that cyanobacterial inoculation effectively improve the fertility of saline soil. Recently a mutant strain of an 'usar' land isolate–*Nostoc calcicola* has been reported to lower the pH of 'usar' land soil extract (Jaiswal *et al.*, 2008).

The mechanism employed by cyanobacteria for adaptation in such inhospitable habitats is still under debate. The various explanations given by scientists are based on the ability of cyanobacteria to avoid the salt stress either by restricted entry of Na^+ or Na^+ efflux. Accumulation of osmoregulatory compounds by alteration/modification of metabolic activity has also been suggested as one of the strategy to combat salt stress (Goel and Kaushik, 2002).

There exists ample information/evidences available on the role of cyanobacteria in reclamation of salt affected soils. Further there is a need to identify native strains of salt affected soils and study their physiological, biochemical, molecular basis and more importantly their suitability to be used as soil inoculants and develop a viable cost effective technology for field application by the farmers.

Conclusion

Cyanobacteria offer an effective, cheap and easily manageable, self-regenerating and eco-friendly alternative to the ever-increasing demand of chemical fertilizer. Their application provide not only fixed nitrogen but also improves soil quality by increasing P, water holding capacity, soil aggregation, buffer the soil against rapid change of pH, check weed proliferation, releases bio-molecules with plant growth promoting property etc. Besides, gradual build- up of their population results in soil fertility with residual effect on succeeding crops also, hence imparting long-term sustainability. Their reclamative potential can be utilized to bring million acres of land back into agricultural use. Concerted effort is needed to isolate and screen promising indigenous strains, identify their potential and develop them as region specific inoculants. The gap between researchers and farmers need to be bridged, by training them and making them aware of the 'wonder' these small microorganisms can do in the field.

Acknowledgements

The author is thankful to Council of Scientific and Industrial Research (CSIR), for providing financial assistance to carry out research activities. The facilities required provided by the Centre for Conservation and Utilization of Blue Green Algae, IARI, New Delhi is also gratefully acknowledged.

References

Aziz MA and Hashen MA (2003). Role of Cyanobacteria in improving fertility of alina soil. *Pakistan J Biol Sci,* 6: 1751–1752.

Banerjee M and Sharma D (2005). Comparative studies on growth and phospohatase activity of endolithic cyanobacterial isolates of chroococcidiopsis from hot and cold deserts. 15: 125–130.

Bapat VA, Iyer RK and Rao PS (1996). Effect of cyanobacterial extract on somatic embryogenesis in tissue cultures of sandalwood. *J Medicinal and Aromatic Plant Sci,* 18: 10–14.

Bose P, Nagpal US, Venkatraman GS and Goyal SK (1971). *Curr Sci,* 40: 165.

Bunt JS (1961). Nitrogen-fixing blue green algae in Australian rice soils. *Nature,* 192: 401, 479–480.

De PK (1939). The role of blue-green algae in nitrogen fixation in rice fields. *Proc Royal Soc London,* 127B: 121–139.

Fogg GE (1952). The production of extracellular nitrogenous substances by a blue green alga. *Proc Royal Soc London,* 139B: 372–397.

Gallon JR, Kurz WGW and LaRue TA (1975). The physiology of nitrogen fixation by a *Gloeocapsa* sp. In: *Nitrogen Fixation by Free Living Microorganisms,* (Ed) Stewart WDP. Cambridge University Press, Cambridge, p. 159–173.

Goel S and Kaushik BD (2002). Synthesis of cellular metabolites in response to salt stress by halotolerant and halosensitive *Nistoc muscorum*. *Indian J Microbiol,* 42: 101–106.

Goyal SK (1993). Algal biofertilizer for vital soil and free N. *Proc Indian Natl Sci Acad,* B59: 290.

Goyal SK (2000). Preparation of soil based algal inoculum. In: *Biofertilizers Blue Green Algae and Azolla* (Eds) Singh PK, Dhar, DW, Pabbi S, Prasanna R and Arora A. NCCUBGA, IARI New Delhi, p. 100–106.

Gupta AB and Agarwal PR (1973). Extraction, isolation and bioassay of a gibberelin like substances from *Phormidium foveolarum*. *Ann Bot,* 37: 737–741.

Gupta SL (1983). Acid and alkaline phosphatase activity in cyanobacterium *Anacystis nidulans* under copper stress. *Folia microbial,* 28: 458–462.

Halperin DR, De MS, De cano MSZ, De Mule and De Caire GZ (19932). Diazotrophic cyanobacteria from Argentine paddy fields. *Phycos,* 53: 135–142.

Healy FP (1982). In: *Biology of Cyanobacteria,* (Eds) Carr NG and Whitton BA. Oxford, Blackwell, p. 105.

Issa OM, Bissonnais Le, Defarge C and Trichet J (2001). Role of cyanobacterial cover on structural stability of sandy soils in the Sahelian part of western Niger. *Geoderma,* 101: 15–30.

Jaiswal P and Kashyap AK (2002). Isolation and characterization of two diazotrophic cyanobacteria tolerant to high concentrations of inorganic carbon. *Microbiol Res,* 157: 83–91.

Jaiswal P, Kashyap AK, Prasanna R and Singh PK (2008). Evaluating the potential of *N calcicola* and its bicarbonate resistant mutant in reclamation of usar soil. *Indian J Microbiol,* 48.

Kartikeyan N, Prasanna R, Lata N and Kaushik BD (2007). Evaluating the potential of plant growth promoting Cyanobacteria as inoculants for wheat. *Eu J Soil Biol,* 43: 23–30.

KartikeyanN, Prasanna R, Sood A, Jaiswal P, Nayak S and Kaushik BD (2008). Physiological characterization of electron microscopic investigations of cyanobacteria associated with wheat rhizosphere. *Folia Microbiol,* 48.

Kaushik BD (1985). Effect of native algal flora on nutritional and physico*chemical properties of sodic soils. *Acta Bot Indica,* 13: 143–147.

Kaushik BD (1989). Reclamative potential of cyanobacteria in salt-affected soils. *Phykos,* 28:101–109.

Kaushik BD (1991). Cyanobacterial response of crops in saline irrigated with saline ground water. In: *Current Trends in Limnology,* (Ed) Shastree NK. Narendra Pub House, 1: 201.

Kaushik BD (1994). Algalization of rice in salt-affected soils. *Ann Agril Res,* 14:105–106.

Kaushik BD and Krishna Murti GSR (1981). Effect of blue-green algae and gypsum application on physico-chemical properties of alkali soils. *Phykos,* 20: 91–94.

Kaushik BD and Subhashini D (1985). Amelioration of salt-affected soils with blue-green algae. II Improvement in soil properties. *Proc Indian Natl Sci Acad,* Part B 51: 386–389.

Manickavelu A, Nadarajan N, Ganesh SK, Ramalingam R, Raghuraman S and Gnanamalar RP (2006). Organogenesis induction in rice callus by cyanobacterial extracellular product. *African J Biotechnol,* 5: 437–439.

Natesan R and Shanmugasundaram SS (1989). Extracellular phosphate solubilization by the cyanobacterium *Anabaena* ARM310. *J Biosci,* 14: 203–208.

Nisha R, Kaushik A and Kaushik CP (2007). Effect of indigenous cyanobacterial application on structural stability and productivity of an organically poor semi-arid soil. *Geoderma,* 138: 49–56.

Olkarinen M (1996). Biological soil amelioration as the basis of sustainable agriculture and forestry. *Biol Fertil Soils,* 22: 342–344.

Ordog V and Pulz O (1996). Diurnal changes in cytokinin like activity in a strain of *Arthronema africanum*, determined by bioassays. *Algol Stud,* 82: 57–67.

Pabbi S (1981). Quality control parameters in BGA biofertilizers. In: *Mass Production of Blue Green Algal Biofertilizer,* (Eds) Singh PK and Pabbi S. NCCUBGA, IARI, New Delhi, p. 22–27.

Pabbi S and Kaushik BD (1997). Algal biofertilizer technology New concept and opportunities. *The Botanica,* 47: 43–51.

Pabbi S, Prasanna R, Dhar DW and Singh PK (2000). Algal biofertilizer to rice: Potential and constraints. In: *Biofertilizers Blue Green Algae and Azolla,* (Eds) Singh PK, Dhar, DW, Pabbi S, Prasanna R and Arora A. NCCUBGA, IARI New Delhi, p. 83–99.

Prasanna R, Lata N, Tripathi R, Gupta V, Chaudhary V, Middha S, Joshi M, Ancha R and Kaushik BD (2008). Evaluation of fungicidal activity of extracellular filtrates of Cyanobacteria-possible role of hydrolytic enzymes. *J Basic Microbiol,* 48: 186–194.

Renaud PA and Roger PA (1978). N_2-fixing algal biomass in Senegal rice fields. *Ecol bull Stockolm,* 26: 148–157.

Roger SL and Burns RG (1993). Changes in aggregate stability, nutrient status, indigenous microbial populations and seedling emergence, following inoculation of soil with *Nostoc muscorum. Biol Fert Soil,* 18: 209–215.

Sergeeva E, Liaimer A and Bergman B (2002). Evidence for production of the phytohoemone indole-3-acetic acid by Cyanobacteria. *Planta,* 215: 229–238.

Shanmugasundaram SS (1996). *Consolidated Report, Mission Mode Project on Technology Development and Demonstration of Algal Biofertilizer,* MKU, Madurai, India.

Singh PK (1973). Nitrogen fixation by the unicellular blue green alga *Aphanothece. Arch Microbiol,* 103: 297–302.

Singh PK (1981). Algal biofertilizer production: Current status. In: *Mass Production of Blue Green Algal Biofertilizer,* (Eds) Singh PK and Pabbi S. NCCUBGA, IARI, New Delhi, p. 1–6.

Singh PK (1985). Nitrogen fixation by blue green algae in paddy fields. In: *Rice research in India,* ICAR, New Delhi, p. 344–352.

Singh RN (1950). Reclamation of usar lands in India through bluegreen algae. *Nature,* 765: 325–326.

Singh RN (1961). *Role of Blue Green Algae in Nitrogen Economy of Indian Agriculture.* IARI, New Delhi.

Singh SK, Singh SS, Pandey VD and Mishra AK (2006). Factors modulating alkaline phosphatase activity in the diazotrophic rice field cyanobacterium *Anabaena oryzae*, 22: 927–935.

Singh VP and Trehan T (1973). Effect of extracellular product of *Aulisira fertilissima* on the growth of rice seedlings.*Plant Soil,* 38: 457–464.

Stanier RY and Cohen-Bazire G (1977). Phototrophic prokaryotes: the cyanobacteria. *Ann Rev Microbiol,* 31: 225–274.

Stirk WA, Ordog V and Staden J (1996). Identification for the cytokinin isopentenyladenine in a strain of *Arthronema africanum*. *J Physiol*, 35: 89–92.

Thomas AD and Dougill AJ (2006). Distribution and characteristics of cyanobacterial soil crusts in the Molopo Basin, South Africa. *J Arid Environ*, 64: 270–283.

Thomas J and Apte SK (1984). Sodium requirement and metabolism in nitrogen fixing Cyanobacteria. *J Biosci*, 6: 771–794.

Venkataraman GS (1972). *Algal Biofertilizer and Rice C*. Today and Tomorrow's Printers and publishers, New Delhi, p. 75.

Venkataraman GS (1981). *Blue Green Algae for Rice Production*. FAO Soils Bulletin No. 46.

Venkataraman GS and Saxena HK (1963). Studies on nitrogen fixation by blue green algae. IV Liberation of free amino acids in the medium. *Indian J Agri Sci*, 33: 22–24.

Wake H, Akasaka A, Umestsur H, Ozeki YRC, Shirmomura K and Matsunaga T (1992). Promotion of plantlet formation from somatic embryos of carrot treated with a high molecular weight extract from a marine cyanobacterium. *Pant Cell Rep*, 11: 62–65.

WatanabeI, Lee KK, Alimango BV, Sato M, Rosario DC and De Guzman MR (1977). Biological N_2-fixation in paddy field studied by the *in situ* acetylene reduction assays. *IRRI Res Pap Ser*, 3: 1–16.

Wilson JT and Alexander M (1979). Effect of soil nutrient status and pH on nitrogen fixing algae in flooded soil. *Soil Sci Amm J*, 43: 936–939.

Wolk CP (2000). Heterocyst formation in *Anabaena*. In: *Prokaryotic Development*, (Eds) Brun YV and Shimkets LJ. Am Soc Microbiol, Washington, USA.

Zulpa de Caire G, Storni de Cana M, Zaccaro de Mule MC, Palma RM and Colombo K (1997). Exopolysacharide of *Nostoc muscorum* (cyanobacteria) in the aggregation of soil particles. *J Appl Phycology*, 9: 249–253.

Subject Index

D

E

F

G

H

I

L

M

N

O

P